GEOLOGICAL SOCIETY SPECIAL PUBLICATION NO. 184

Continental Reactivation and Reworking

EDITED BY

J. A. MILLER
University of Cape Town, South Africa

R. E. HOLDSWORTH
University of Durham, UK

I. S. BUICK
Adelaide University, Australia

M. HAND
Adelaide University, Australia

2001

Published by

The Geological Society

London

THE GEOLOGICAL SOCIETY

The Geological Society of London was founded in 1807 and is the oldest geological society in the world. It received its Royal Charter in 1825 for the purpose of 'investigating the mineral structure of the Earth' and is now Britain's national society for geology.

Both a learned society and a professional body, the Geological Society is recognized by the Department of Trade and Industry (DTI) as the chartering authority for geoscience, able to award Chartered Geologist status upon appropriately qualified Fellows. The Society has a membership of 9099, of whom about 1500 live outside the UK.

Fellowship of the Society is open to persons holding a recognized honours degree in geology or a cognate subject, or not less than six years' relevant experience in geology or a cognate subject. A Fellow with a minimum of five years' relevant postgraduate experience in the practice of geology may apply for chartered status. Successful applicants are entitled to use the designatory postnominal CGeol (Chartered Geologist). Fellows of the Society may use the letters FGS. Other grades of membership are available to members not yet qualifying for Fellowship.

The Society has its own publishing house based in Bath, UK. It produces the Society's international journals, books and maps, and is the European distributor for publications of the American Association of Petroleum Geologists, (AAPG), the Society for Sedimentary Geology (SEPM) and the Geological Society of America (GSA). Members of the Society can buy books at considerable discounts. The Publishing House has an online bookshop (*http://bookshop.geolsoc.org.uk*).

Further information on Society membership may be obtained from the Membership Services Manager, The Geological Society, Burlington House, Piccadilly, London W1V 0JU (E-mail: *enquiries@geolsoc.org.uk*; tel: +44 (0)207 434 9944).

The Society's Web Site can be found at *http://www.geolsoc.org.uk/*. The Society is a Registered Charity, number 210161.

Published by The Geological Society from:
The Geological Society Publishing House
Unit 7, Brassmill Enterprise Centre
Brassmill Lane
Bath BA1 3JN, UK

(*Orders*: Tel. +44 (0)1225 445046
       Fax +44 (0)1225 442836)
Online bookshop: *http://bookshop.geolsoc.org.uk*

British Library Cataloguing in Publication Data
A catalogue record for this book is available from the British Library.

ISBN 1-86239-080-0 ✓
ISSN 0305-8719

Typeset by Aarontype Ltd, Bristol, UK

Printed by Alden Press, Oxford, UK

Distributors
USA
  AAPG Bookstore
  PO Box 979
  Tulsa
  OK 74101-0979
  USA
  Orders: Tel. +1 918 584-2555
       Fax +1 918 560-2652
       E-mail: *bookstore@aapg.org*

Australia
  Australian Mineral Foundation Bookshop
  63 Conyngham Street
  Glenside
  South Australia 5065
  Australia
  Orders: Tel. +61 88 379-0444
       Fax +61 88 379-4634
       E-mail: *bookshop@amf.com.au*

India
  Affiliated East-West Press PVT Ltd
  G-1/16 Ansari Road, Daryaganj,
  New Delhi 110 002
  India
  Orders: Tel. +91 11 327-9113
       Fax +91 11 326-0538
       E-mail: *affiliat@nda.vsnl.net.in*

Japan
  Kanda Book Trading Co.
  Cityhouse Tama 204
  Tsurumaki 1-3-10
  Tama-shi
  Tokyo 206-0034
  Japan
  Orders: Tel. +81 (0)423 57-7650
       Fax +81 (0)423 57-7651

Contents

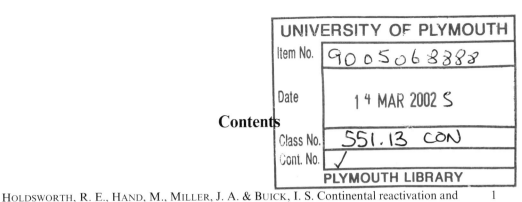

Acknowledgements

This volume is derived from presentations to the Oxbridge ... conference held in Alice Springs, Central Australia in July 1999 under the auspices of the Specialist Group in Geochemistry, Mineralogy and Petrology of the Geological Society of Australia with the support of the Tectonic Studies Group (TSG). Many of the ideas, concepts and examples discussed in this volume were first presented at this conference and its associated field trips, where they generated a great deal of discussion and debate. ... We hope that this volume will ...

Continental reactivation and reworking: an introduction

R. E. HOLDSWORTH[1], M. HAND[2], J. A. MILLER[3] & I. S. BUICK[4]

[1] Reactivation Research Group, Department of Geological Sciences, Durham University, Durham DH1 3LE, UK
[2] Department of Geology and Geophysics, Adelaide University, Adelaide, SA 5005, Australia
[3] Department of Geological Sciences, The University of Cape Town, Rondebosch 7700, Republic of South Africa
[4] Department of Earth Sciences, La Trobe University, Bundoora, VIC 3083, Australia

In contrast to oceanic lithosphere, the continents are manifestly composed of the products of tectonic processes whose cumulative duration spans much of the Earths history. Most continents contain Archaean nuclei that are enclosed by Proterozoic and Phanerozoic tectonic domains. The evolution of post-Archaean continental volumes has included additions of new continental material, but it has also involved repeated modification of parts of the existing continental lithosphere during periods of *tectonic rejuvenation*. This generally involves processes such as the formation of new structural fabrics, the overprinting of metamorphic assemblages and the generation and emplacement of magmas. Such behaviour can occur repeatedly throughout the geological record because the quartzofeldspathic continental crust cannot be subducted due to its relative buoyancy and weakness compared with its oceanic counterpart and the underlying lithospheric mantle. Thus, the character of the continents is significantly influenced by the way in which the existing lithosphere responds to new tectonothermal events that follow geologically significant cessations of activity for millions to hundreds of millions of years (Sutton & Watson 1986).

Existing continental lithosphere may be modified during its incorporation into new collisional systems, for example the involvement of the Hercynian 'basement' in the Alpine collision. However, the most dramatic manifestations of continental tectonic rejuvenation occur during intraplate orogeny, where a coherent pre-existing lithospheric volume undergoes large-scale failure. Notable modern examples of intraplate orogeny are the Cenozoic Tien Shan and the Mongolian Alti in north Asia, which are forming in response to the Himalayan collision (e.g. Hendrix et al. 1992; Dickson Cunningham et al. 1996), and also the Atlas Mountains of Morocco, which

are linked to the on-going Alpine collision (e.g. Ramandi 1998). In the ancient geological record, two of the best examples are the mid-Palaeozoic Alice Springs Orogeny and the Neoproterozoic to Palaeozoic Petermann Orogeny in central Australia (e.g. Sandiford & Hand 1998; Hand & Sandiford 1999). In recognition of the importance of intraplate orogeny as an expression of continental rejuvenation, a large amount of work has focused on the mechanisms leading to large-scale intraplate failure (e.g. Vilotte et al. 1982; England & Houseman 1985; Kuzsnir & Park 1987; England & Jackson 1989; Platt & England 1994; Tommasi et al. 1995; Ziegler et al. 1995, 1998; Avouac & Burov 1996; Neil & Houseman 1997; Sandiford & Hand 1998; Hand & Sandiford 1999; Pysklywec et al. 2000). A number of factors are likely to control the locus of tectonic activity, but there appear to be two first order controls: (1) temporal and spatial variations in the thermal state of the lithosphere (e.g. Sonder & England 1986; England 1987; Neil & Houseman 1997); and (2) the presence of pre-existing mechanical defects such as faults, shear zones or major compositional boundaries (e.g. Ziegler et al. 1995; Butler et al. 1997; Holdsworth et al. 1997).

The rejuvenation of pre-existing crust and lithosphere occurs largely via two related processes. *Reactivation* is normally considered to involve the rejuvenation of discrete structures (e.g. Holdsworth et al. 1997), whilst *reworking* involves the repeated focusing of metamorphism, deformation and magmatism into the same crustal- or lithospheric-scale volume. These are considered to be useful end-member definitions that describe the way in which continental lithosphere is modified. However, there is some ambiguity firstly because reworking and reactivation may represent broadly the same process operating at different scales and/or depths (see below),

From: MILLER, J. A., HOLDSWORTH, R. E., BUICK, I. S. & HAND, M. (eds) *Continental Reactivation and Reworking*. Geological Society, London, Special Publications, **184**, 1–12. 1-86239-080-0/01/$15.00 © The Geological Society of London 2001.

and secondly because there is some confusion regarding the use of these terms in the geological literature. Several studies refer to the 'reactivation' of mobile belts, but as these represent crustal volumes rather than discrete structures, they would be more accurately described as being reworked. Other terms used in the literature include 'renewal' and 'remobilisation', both of which refer to the superposition of younger geological events onto older geological systems.

The published geological literature suggests that continental reworking and reactivation can be expressed in a large number of ways, and are likely to arise for a range of complex reasons. Thus, attempts to characterize the style and distribution of reactivation and reworking in different continental settings should provide key datasets with which to evaluate the nature, distribution and dynamic controls of tectonic rejuvenation in the continental crust and lithosphere. In editing this book, two end-member approaches are recognized. The first takes a lithospheric-scale perspective and essentially considers how the continental lithosphere may respond to evolving first order variables, such as the density structure of the mantle and changes in convergence style and rate. The second approach is case-study oriented. The two approaches are complementary and wherever possible should be integrated. Case studies provide crucial temporal and spatial information concerning the thermo-mechanical evolution of the crust and mantle in a variety of settings. Geochronological studies are particularly important in this context as they can provide clear evidence for reworking or reactivation, and an absolute timescale of events. The conceptual approach can and should provide the stimulus for the collection of targeted datasets that seek to evaluate the interactions of potential behaviours and causative mechanisms.

Continental reworking

Continental reworking encompasses structural, metamorphic and magmatic processes that modify existing continental lithosphere at an orogenic scale. An important difference between reworking and reactivation is that older structures do not necessarily control the style, orientation and scale of later structures. This is evidenced by orogenic belts such as the Cambrian Prydz-Leeuwin orogenic system which linked Africa, Antarctica and Australia, incorporating Archaean and Proterozoic crustal domains into a single mountain belt (Veevers 2000). However, there may be important long-term linkages or feedback between apparently unrelated orogenic episodes. For example,

the distribution and intensity of crustal heat production is strongly controlled by the pattern and extent of denudation associated with orogenesis, which in turn may influence subsequent spatial and temporal variations in the lithospheric thermal regime (e.g. Sandiford & Hand 1998). Thus the consequences of one stage of continental evolution can potentially shape the pattern, style and distribution of subsequent events.

Since continental reworking involves diffuse lithospheric-scale deformation, the behaviour of the mantle lithosphere is likely to be central in determining the style and duration of the tectonic activity. The potential role of the mantle lithosphere in continental reworking is reviewed and investigated by **Houseman & Molnar** in the first paper of this volume. They suggest that localized lithospheric thickening associated with plate convergence can potentially affect the gravitational stability of a layered system in which a dense non-Newtonian lithosphere overlies a less dense fluid asthenosphere. Where instability arises, a relatively strong crust will lead to localized downwelling beneath the centre of the convergent zone, whilst a relatively weak crust will lead to downwelling on the margins of the convergent zone as the buoyant crustal layer resists thickening. The initial instability may then trigger rapid extension of the lithospheric mantle beneath the orogen which is driven by asymmetric cold downwellings that move away from the centre of the convergent zone. Using examples of modern orogens from Southern California, South Island New Zealand, the Mediterranean, and Central Asia, Houseman & Molnar show that there is strong evidence in each case that the mantle lithosphere has developed some form of instability that has led to at least part of it being replaced by hot asthenosphere. Interestingly, teleseismic tomographic images from these areas suggest that mantle lithosphere has been locally renewed following gravitational instability triggered by orogenic convergence. In this context, the time scales of reworking should be linked to the rates at which downwelling of gravitationally unstable lithosphere occurs. Although these time scales are not well known, Pysklywec *et al.* (2000) have suggested that crustal deformation driven by this process should have a duration of *c.* 25–40 Ma. **Rey** also considers the interplay between the convective and lithospheric mantle in terms of its potential role in controlling the style of crustal deformation. In contrast to Houseman & Molnar, he investigates the consequences of mantle behaviour for crustal extension, and evaluates the boundary conditions that may limit, or oppose spontaneous extension.

Although the dynamic behaviour of the mantle is likely to be of primary importance during continental reworking, the evolution of the crustal component of the system also potentially exerts long-term controls on the stability of continental lithosphere. For example, crustal heat production in the Australian Proterozoic is roughly twice the global Proterozoic average (Taylor & McLennan 1985), implying that lithospheric thermal regimes, and by inference mechanical strength, are strongly controlled by the thermal character of the crust (e.g. Sandiford & Hand 1998). The potential importance of the deep crustal evolution is highlighted by **Ryan**, who points out that as collisions thicken the crust by a factor of two or more, a large volume of continental material at the base of the orogen will experience eclogite-facies conditions. This dense crustal material may be partially subducted and lost to the system, or may sit isostatically below the Moho until partially exhumed during orogenic collapse. Using finite element models, he shows that the rate at which eclogite phase transformations take place can have profound buffering effects on the amount and duration of orogenic contraction. Such remnant orogenic roots may exist as seismically reflective mantle and provide a locus for subsequent rifting.

In some respects, continental reworking at convergent margins is a matter of chance that is determined by the presence and positions of the colliding continental blocks, arcs, etc. However, the spatial patterns of intraplate failure have attracted considerable attention (e.g. Ziegler *et al.* 1995; Hand & Sandiford 1999) from the perspective of trying to determine whether reworking is controlled by exclusively intraplate processes (e.g. Neil & Houseman 1999) or if it is driven by stresses originating at plate boundaries that are focused into a weak zone within the continental interior. One intriguing aspect of the spatial patterns of continental reworking relates to the reasons why the Archaean 'shields' have survived relatively intact, apparently exhibiting long-term strength. **Krabbendam** suggests that the susceptibility of continental lithosphere to reworking is largely determined by the geothermal gradient and rheological properties. Relatively strong orogenic lithosphere will occur if the crust has low rates of radiogenic heat production, or if the underlying sub-continental lithosphere remains thick. Low heat production rates may be a consequence of the average level of denudation associated with earlier orogenic events (e.g. Sandiford & Hand 1998), and therefore arise as a long-term consequence of the terrain history. Processes such as dehydration metamorphism and erosional thinning of the orogenic crust can

also strengthen the lithosphere, whilst proximity to 'strong' Archaean cratons and high mantle heat flow appear to markedly enhance susceptibility to reworking.

Within the context of a single period of orogenic evolution, many systems appear to record pulses of activity during which the rates and intensity of deformation, metamorphism and magmatism increase. In the context of the present volume, these pulses of tectonic activity could be thought of as short-term or short-spaced cycles of reworking if there has been a local cessation of tectonic activity. The existence of apparently discrete events is not surprising given the complexity of large orogenic systems and the non-uniformity of their boundary conditions. More controversial, however, is the suggestion that orogenic systems may display episodic (i.e. quasi-regular) pulses of behaviour (e.g. Stüwe *et al.* 1993). For example, Bell and co-workers have suggested that in a number of orogenic belts, foliations appear to have formed during repeated episodes of convergence and extension that can be correlated over large distances (e.g. Bell & Mares 1999). In some instances, these episodes appear to have occurred within the time scale of a single metamorphic cycle. Whether this apparent behaviour is truly episodic is yet to be determined, but if it is, the periodicity of the cycles holds importance clues regarding the dynamic behaviour of orogenic systems. **Lister *et al.*** highlight the episodicity of deformation and metamorphic mineral growth in orogenic belts and show that these events appear to take place at the same time over large length scales, even in apparently unrelated segments of the same orogenic belt. They relate this periodicity to switches between compressional orogenic surges and periods of lithospheric extension following accretion. They propose that the effect of these switches is greatest in back-arc environments where rollback of the subducting lithospheric slab ensures a large amount of lithospheric extension after each contractional accretion. This could explain why the majority of well-preserved examples of exhumed high-pressure metamorphic terranes and ophiolite sheets found in the geological record were formed in back-arc settings.

The mechanisms that could drive relatively short-term episodic behaviour within long-lived orogens may differ importantly from the processes that lead to the formation of new orogenic systems. In the former situation, the lithosphere is presumably in a non-equilibrium state. In the latter case, there may have been no tectonic activity for hundreds of millions of years, and the lithosphere may have been in a quasi-equilibrium condition.

Reactivation

In general, reactivation involves the structural modification of an existing feature without significantly changing its volume or orientation. Perhaps the most obvious example is the reactivation of a shear zone or fault system, in which a younger episode of deformation is localized within or along the boundaries of the existing structure. **Holdsworth *et al.*** review the nature, distribution and underlying controls of deformation processes and products in natural faults and shear zones. Reactivated faults and shear zones exposed in the deeply exhumed parts of ancient orogenic belts present particularly useful opportunities to study processes that influence the mechanical properties of long-lived fault zones at different palaeo-depths. A case study from Scotland is used to emphasize the important roles played by cataclasis and fluid flow in the deeper part of the frictional crust, processes that help to promote retrograde metamorphism and changes in deformation regime (see also Imber *et al.* 1997, 2001; Stewart *et al.* 2000). This in turn leads to changes in the depth and character of the frictional–viscous (brittle–ductile) transition along such faults, and to profound weakening that may account for the long-lived character of many crustal-scale structures, including the much studied San Andreas fault zone. In addition to the traditionally recognized lithological and environmental factors that influence fault rock rheology (e.g. composition, grain-size, temperature, pressure, etc), geometric factors such as fault size, orientation and interconnectivity are likely to be of equal importance in determining the repeated localization of displacements (e.g. Walsh *et al.* 2001). The longevity of continental lithosphere means that once major faults form, they are likely to persist unless there is pervasive re-structuring of the lithosphere. Thus, many continental fault zones have very long-lived movement histories (e.g. see Butler *et al.* 1997; Rutter *et al.* 2001 and associated papers).

In ancient settings, reactivation should refer to deformation events that are separated by more than 1 Ma (Holdsworth *et al.* 1997) since geochronological techniques generally cannot resolve events that are separated by shorter periods of time. In more recent settings, where the timing of events may be more precisely defined, it is useful to view reactivation as being distinct from recurrence. It is well established that many faults exhibit stick–slip behaviour, with short-lived periods of inactivity (*c.* 0.01 Ma) within what is essentially a continuous slip history (e.g. Rutter *et al.* 2001). In this case, reactivation can be viewed as being the accommodation of displacement along a structure formed during an earlier tectonic regime (Holdsworth *et al.* 1997).

Continental reactivation and reworking: a central Australian perspective

Central Australia arguably contains one of the best records of continental reactivation and reworking. This record is expressed by the formation of two spectacular orogenic systems: the late Neoproterozoic to early Palaeozoic Petermann Orogen (e.g. Sandiford & Hand 1998) and the mid Palaeozoic Alice Springs Orogen (e.g. Shaw *et al.* 1992). These intraplate orogenic events overprint complex Palaeo- and Mesoproterozoic orogenic belts, and thus central Australia can be viewed as a type example in which a series of superimposed episodes have occurred during continental rejuvenation. Several studies have focused on the factors that may have controlled the shifting patterns of continental reworking in central Australia and have emphasized the primary importance of temporal and spatial variations in the thermal regime (e.g. Sandiford & Hand 1998; Hand & Sandiford 1999; Fig. 1), and/or the role of the lithospheric mantle in driving deformation (Braun & Shaw 1998; Neil & Houseman 1999).

In the continued effort to understand the processes that lead to reworking of the central Australian lithosphere, **Roberts & Houseman** focus on the mid-Palaeozoic Alice Springs Orogeny, which is the last major event to have affected the region. They employ a thin viscous sheet approximation of continental lithosphere to investigate possible causative mechanisms for N–S shortening during the Alice Springs Orogeny, and the contemporaneous development of N–S extension in the Canning Basin in northwestern Australia. Assuming an orogenic time span of 100 Ma (e.g. Shaw *et al.* 1992), they impose a clockwise rotation of the northern boundary of a lithospheric plate in which there is an E–W trending internal weak zone corresponding to the location of the Alice Springs Orogen. They show that this can produce a realistic amount of crustal thickening in the region representing central Australia and thinning in the region representing the Canning Basin. The rotation may be induced either by eastward shear traction or by a clockwise bending moment. The degree of crustal thickening is primarily controlled by the relative dimension of the intracratonic weak zone, which is itself defined by Moho temperatures that are estimated to be

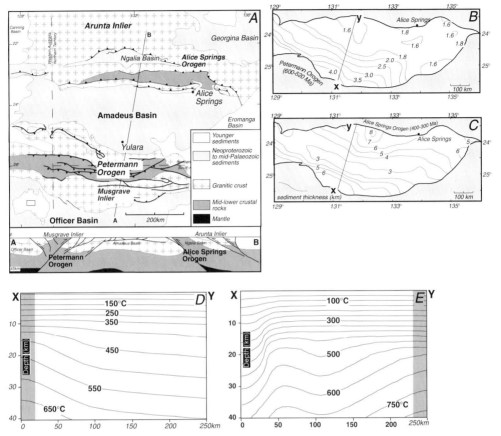

Fig. 1. (a) Generalized map of central Australia showing the location of major crustal blocks and Neoproterozoic to mid-Palaeozoic sedimentary basins that belonged to the formerly continuous Centralian Superbasin. Large-scale intraplate deformation during the late Neoproterozoic to Palaeozoic (580–530 Ma) Petermann Orogeny and the 400–300 Ma Alice Springs Orogeny resulted in exhumation of the Musgrave and Arunta Blocks from beneath the Centralian Superbasin. The Petermann Orogeny produced regional exhumation from depths in excess of 45 km, while maximum exhumation arising from the Alice Springs Orogeny is in the order of 25 km. **(b)** and **(c)** Generalized isopach maps within the Amadeus Basin prior to the onset of the Petermann Orogeny **(b)** and the Alice Springs Orogeny **(c)** (from Sandiford & Hand 1998). The isopach data indicate that prior to the Petermann Orogeny, the Musgrave Block was buried beneath a relatively thick sedimentary blanket that accumulated over an interval of c. 200 Ma. Following the Petermann Orogeny, the locus of sedimentation shifted such that the Arunta Block became the location of a long-lived depocentre where in excess of 8 km of sediment accumulated. The line of section used to model the thermal response to variations in sedimentary thickness shown in Figures 1d and 1e is indicated by X–Y. **(d)** and **(e)** Simplified 2D thermal structure across the Amadeus Basin along the line X–Y **(d)** prior to the Petermann Orogeny and **(e)** the Alice Springs Orogeny. The grey region in **(d)** indicates the location of the Petermann Orogen along the southern margin of the Amadeus Basin. Similarly, the grey region in **(e)** shows the location of the Alice Springs Orogeny. Both the Petermann and Alice Springs Orogeny appear to have been localized in regions with relatively elevated Moho temperatures, suggesting that the change in locus of intracratonic deformation between c. 530 Ma and 400 Ma may have reflected spatial and temporal changes in the lithospheric thermal regime. Thermal parameters: thermal conductivity; basin $= 2.25\,\mathrm{Wm^{-1}\,K^{-1}}$; basement $= 3\,\mathrm{Wm^{-1}\,K^{-1}}$; heat flow; crust $= 40\,\mathrm{mW\,m^{-2}}$, mantle $= 20\,\mathrm{mW\,m^{-2}}$. After Sandiford & Hand (1998) and Hand & Sandiford (1999).

c. 10% higher than adjacent stronger regions of lithosphere. **Braun & Shaw** present the results of a thin-plate model of the Australian continental lithosphere in which deformation is driven by velocities specified along the plate boundaries during a 200 Ma time period starting in the Ordovician (470 Ma). The results suggest that intracratonic deformation occurs at the interface between regions of contrasting litho-spheric strength and/or in zones of pre-existing weakness caused by previous tectonic activity. Strain localization can arise due to repeated deformation episodes in a volume of lithosphere and, in some cases, may result from the constructive interaction between plate tectonic forces that originate on separate plate margins.

Sandiford *et al.* propose a link between the locus of intraplate orogeny in central Australia and the presence of thick sedimentary successions deposited in widespread pre-orogenic intra-cratonic basins. This 'tectonic feedback' reflects the long-term thermal and mechanical consequences of changes in the distribution of heat producing elements induced by earlier tectonism. During intraplate orogenesis, deep crustal rocks were exposed by erosion of the heat-producing upper crust, leading to long-term cooling of the lithosphere. This causes a progressive strengthening of the lithosphere that eventually can be sufficient to terminate intraplate orogenesis in central Australia by providing a 'thermal lock'. The presence of cool, anomalously strong lithosphere may account for the extraordinary gravity anomalies that are associated with the central Australian intraplate orogens.

The role of crustal heat sources in controlling the long-term mechanical evolution of continental lithosphere is further explored by **McLaren & Sandiford**, who focus on the observed 300 Ma of episodic Palaeoproterozoic tectonic activity that occurred prior to effective Mesoproterozoic cratonisation in the Mt Isa domain of northern Australia. This tectonic activity resulted in dramatic changes in the heat production distribution in the crust. Felsic magmatism transferred heat producing elements from lower to mid-upper crustal levels, leading to lower crustal cooling, and a highly stratified heat production distribution. The long-term strengthening caused by the progressive concentration of heat-producing elements in the upper crust eventually led to effective cratonization due to the combined processes of erosional stripping of heat production and exhumation of the heat production concentrations. At various times prior to cratonization, strengthening was countered by episodes of basin formation that led to burial of existing crustal heat production and increases in total heat production due to the accumulation of sediments. The resulting thermal weakening effect may have helped to localize subsequent contractional deformation events expressed by the Mesoproterozoic Isan Orogeny.

Conceptual studies have resulted in a greater understanding of the mechanics of continental reworking and reactivation, but they rely crucially on datasets from orogenic belts that provide constraints on the duration of deformation, its kinematic character and the prevailing thermal regimes. **Hand & Buick** and **Scrimgeour & Raith** present case studies from central Australia that highlight different stages of the poly-tectonic evolution of central Australia. Hand & Buick present a multidisciplinary case study from the Reynolds–Anmatjira Range region of the Arunta Inlier, which has undergone four tecto-nothermal cycles over an interval of *c.* 1450 Ma. Up to three Proterozoic events produced a single regional foliation, axial planar to simple large-scale folds. Associated metamorphic assemblages increase in grade smoothly from northwest to southeast, producing a pattern of isograds that is remarkably similar to those formed during the later mid-Palaeozoic Alice Springs Orogeny. Stratigraphical and geochronological data show that the apparently simple metamorphic pattern in fact arises due to the superimposed metamorphic effects of four unrelated tecto-nothermal cycles. This illustrates that the apparent structural and metamorphic character of a terrain does not always offer a reliable guide to the degree of reworking, a feature that may in part arise due to the coaxial superimposition of regional strains during separate orogenic events (for a good example of this in a different setting, see Tavarnelli & Holdsworth 1999). Scrimgeour & Raith discuss two terrains in the eastern Arunta Inlier with distinct metamorphic histories that are separated by a crustal-scale shear system. They show that the juxtaposition of crustal units, with differing ages along a shear zone system, is broadly coincident with the inferred onset of south-vergent compressional deformation at the start of the Alice Springs Orogeny. Although the zone of Palaeozoic reworking is relatively narrow (<5 km wide), it appears to form the northern margin of Palaeozoic high-grade intraplate deformation in central Australia, and therefore represents a major tectonic boundary.

Geochronological and metamorphic perspectives on continental reworking

One of the major challenges faced when working in orogenic belts that have undergone reworking or reactivation is distinguishing between the products of different events. Failure to do this may result in erroneous deductions regarding the evolution of a terrain, and by inference incorrect assumptions regarding the processes that controlled the evolution. A good example of this comes from the late 1970s to late 1980s when there was a huge increase in the use of meta-morphic reaction textures to infer tectonother-mal evolutions. In many cases, metamorphic assemblages produced during unrelated meta-morphic events were linked together, producing apparent tectonic histories that had little rele-vance to the real evolution (e.g. Warren 1983; Hensen & Zhao 1995). In some cases, these apparent tectonic histories led to the postulation of causative mechanisms for tectonism that sub-sequent work has shown are unlikely to have applied (e.g. Williams et al. 1995).

The identification of features such as sim-ply metamorphosed cover sequences overlying polymetamorphic basement complexes provides compelling evidence for tectonic rejuvenation. In general, however, geochronology is relied upon heavily to identify the products of indivi-dual tectonic histories. In a review paper, **Parrish** illustrates how mineral chronometers, especially accessory minerals using the U/Pb isotopic sys-tem, can yield important information regarding the environmental conditions and duration of metamorphic-deformation events during the re-working of older rocks. A number of examples are presented from a wide variety of geological environments to illustrate the response of U/Pb isotope systematics within accessory minerals to superimposed deformation, metamorphism and/or mineral growth. **Manhica et al.** discuss the metamorphic and structural history of the Archaean to Palaeoproterozoic Kalahari Craton in central western Mozambique and the adjacent Mesoproterozoic rocks of the Mozambique Belt to the east. The new geochronological data pre-sented by these authors suggests that there are many similarities in the timing of events with those in western Dronning Maud Land, Ant-arctica which was adjacent to the study area prior to Gondwana breakup. **Zhao et al.** show that Archaean mafic granulites from the Palaeo-proterozoic Trans-North China Orogen and adjoining Archaean areas preserve textural evidence for two granulite facies events invol-ving contrasting P–T paths, one clockwise, the other anticlockwise. These events are correlated with recognized regional episodes in the adjacent Archaean rocks and are used to suggest that the polymetamorphic granulites were derived from the reworking of the 2.5 Ga metamor-phosed granulites during the final amalgama-tion of the North China Craton c. 1.8 Ga. **Paech** discusses the geological history of central Dron-ning Maud Land, Antarctica which consists of Grenvillian basement rocks intruded by char-nockitic granitoids and anorthosites during two Pan-African igneous episodes. The earlier epi-

sode (c. 600 Ma) was followed by Pan-African high-grade metamorphism, ductile deformation and later medium-grade retrogression (c. 570 to 520 Ma). This was followed by a second Pan-African igneous episode (c. 500 Ma) during which predominantly post-kinematic granitoids were intruded.

Reworking and reactivation of continental crust often involves important episodes of fluid flow that are in many cases, of central impor-tance in controlling the reactivation of shear zone and fault systems. **Cartwright et al.** dis-cuss the effects of fluid flow during the poly-metamorphic evolution of the Reynolds Range in the Arunta Inlier in central Australia. During contact metamorphism at around 1.78 Ga, igne-ous and surface-derived fluids interacted with the country rocks adjacent to granite plutons. During cooling following regional metamorph-ism at around 1.59 Ga, fluids were derived pre-dominantly from crystallization of partial melts reflecting internal fluid recycling. At around 340 Ma, during the Alice Springs Orogeny, surface-derived fluids infiltrated the middle crust. Much of the fluid flow was channelled into shear zones due to increases in intrinsic permeability caused by deformation or reaction enhancement. **Johnstone & Harris** use oxygen and carbon stable isotope data to examine fluid sources and fluid–rock interaction during early Cambrian thermal reworking of a c. 1200 to 900 Ma orogenic belt in western Dronning Maud Land, Antarctica.

Reworking and reactivation: different expressions of the same process?

An important issue that is not addressed specifically by the papers in this volume is the question of scale in determining whether an interval of continental rejuvenation should be regarded in terms of reactivation or reworking. This problem is well illustrated by the mid-Palaeozoic intracratonic Alice Springs Orogeny in central Australia. From a continental-scale perspective, the Alice Springs Orogen has the appearance of a region of diffuse continental deformation with dimensions of at least 1000 km by 250 km. However, at the map scale the vast bulk of deformation, and the associated meta-morphic expression of the tectonic rejuvenation, is localized within shear zones that can be shown in important cases to pre-date the Alice Springs Orogeny (e.g. Shaw & Black 1991). Between these shear zones, Palaeo- to Mesoproterozoic rocks are largely unaffected by later events. Therefore,

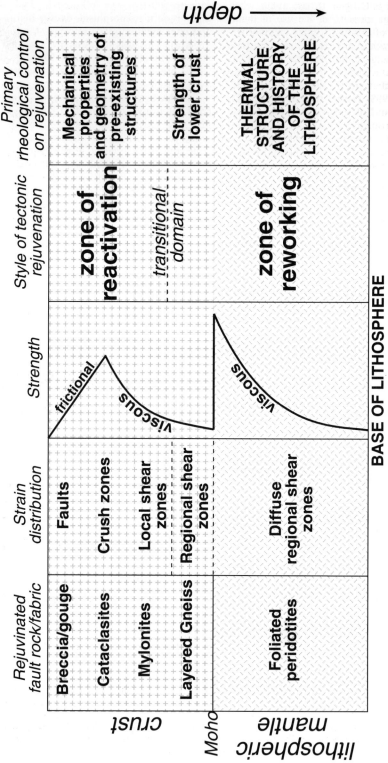

Fig. 2. Schematic diagram depicting the fault rocks/ fabrics, strain distribution, tectonic style and primary rheological controls during tectonic rejuvenation at different depths, and how they are related to the distribution of strength in the continental lithosphere. The boundary between the zones of reactivation and reworking is gradational. Its' nature and location may be largely determined by the strength of the lower crust. Note that the strength profile is for 'average' continental lithosphere: large scale tectonic processes, such as orogenesis, delamination and rifting, that perturb the geothermal gradient will lead to significant changes in the strength of the mantle section.

is the Alice Springs Orogeny an example of continental reworking, or is it reactivation involving a large number of existing shear zones? In the former case it would be relevant to ask why regional weakening of the lithosphere occurred in central Australia leading to the localization of deformation in that crustal volume (e.g. Sandiford & Hand 1998; Hand & Sandiford 1999). On the other hand, if the Alice Springs Orogeny is viewed from a reactivation perspective, strain localization mechanisms may have been controlled by shear zone geometry and rheology of the pre-existing fault rocks (e.g. Korsch *et al.* 1998).

Orogenic belts typically preserve superimposed tectonic, metamorphic and magmatic sequences that formed at different crustal depths during either burial or exhumation. In this context, reworking and reactivation could be viewed as different expressions of the same process (Fig. 2). The classical crustal fault zone model of Sibson (1977), and the results of numerous case studies (e.g. Scrimgeour & Close 1999; Flöttmann *et al.* 2001) suggest that deformation in the upper and middle crust is localized along frictional faults

or narrow viscous shear zones (the reactivation zone) that broaden downwards into a region of diffuse viscous flow (the reworking zone) (Fig. 2). Thus, reactivations of discrete shear zones and faults observed in surface exposures preserving shallower crustal deformation regimes should pass downwards into a zone of simultaneous reworking in the underlying lower crust. Similarly, reworking events in lower crustal rocks now exposed at the surface may have once been expressed as reactivations in the now eroded shallower parts of the ancient crust. Another example of where reworking and reactivation may simply be a different expression of the same process occurs during syn-tectonic magmatism. Melting of existing continental lithosphere is a dramatic thermal expression of reworking. However, the emplacement of magmatic rocks commonly exploits pre-existing structures and may promote transient fault weakening during emplacement (e.g. Handy *et al.* 2001). Thus the vastly different processes of magma generation (reworking) and emplacement (reactivation) are both essentially expressions of the thermal rejuvenation of existing crust.

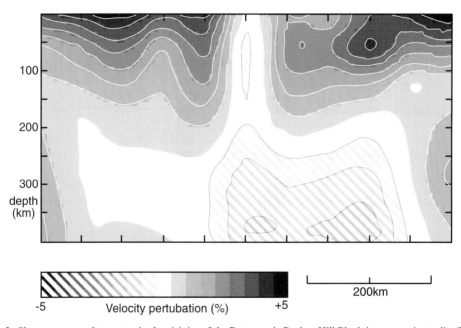

Fig. 3. Shear wave-speed structure in the vicinity of the Proterozoic Broken Hill Block in eastern Australia (from van der Hilst *et al.* 1998). The N–S cross-section is located at 143°E and shows large lateral variations in wave speed, and a low velocity zone extending from at least 400 km depth to the near surface separating two high velocity Proterozoic blocks. This low velocity zone may correspond to a region of elevated temperature or highly foliated mantle that is weak relative to the surrounding rocks. The low velocity structure occurs below the surface expression of the long-lived Tasman Line, which is the late Neoproterzoic rifted margin of the Australian continent.

More fundamentally, it is relevant to consider the distribution of strength in the continental lithosphere based upon extrapolation of experimental deformation studies (Fig. 2, central column; see Kohlstedt *et al.* 1995 for a review). These profiles are, at best, approximations of reality, but they do suggest that the majority of the lithospheric strength resides in the upper mantle for typical continental lithosphere (Molnar 1992). The existence of large-scale anisotropies in the upper mantle (e.g. Fig. 3; Houseman & Molnar 2001) suggests that many examples of continental reworking may simply be the diffuse expression of movements within long-lived zones or volumes of relative weakness located below the Moho. The 3-D distribution of strength in the lithospheric mantle is profoundly influenced by its' thermal structure and history, but it is likely that continental deformation will in general be driven by the development of broad shear zones in this region. These displacements will be transferred upwards into broad regions of reworking in the lower crust and reactivation at still shallower crustal depths (cf. Teyssier & Tikoff 1998, figs 8 & 9). If pre-existing weaknesses are present, particularly in the secondary load-bearing region of the mid-crust (Fig. 2), these will preferentially accommodate displacements, but their presence may not be essential to the process of tectonic rejuvenation.

In summary, the structural, metamorphic and compositional character of the continents is shaped by the way in which continental lithosphere undergoes tectonic rejuvenation. Rejuvenation is achieved by two end-member processes: reactivation and reworking. In reality, these may simply be different manifestations of the same process with the appropriateness of using these two terms determined by: (a) the observational scale applied to a specific problem; and/or (b) the depth at which the exposed deformation system formed in the lithosphere. As a result, reworking and reactivation convey important information regarding the styles and controls of tectonic rejuvenation, and are therefore fundamental in improving our understanding of continental tectonic processes. In many ways, these processes exemplify the fundamental differences that exist between plate tectonics, which is largely based on the dynamics of oceanic lithosphere, and the large-scale behaviour of continental lithosphere (Molnar 1988).

We would like to thank all the contributors to the present volume and those who organized and attended the Alice Springs conference and associated field excursions. Ken McCaffrey and Andy Morton are thanked for their comments on earlier versions of this manuscript.

References

AVOUAC, J. P. & BUROV, E. B. 1996. Erosion as a driving mechanism of intracontinental mountain growth. *Journal of Geophysical Research*, **101**, 17 474–17 769.

BELL, T. H. & MARES, V. M. 1999. Correlating deformation and metamorphism around orogenic arcs. *American Mineralogist*, **84**, 1727–1740.

BRAUN, J. & SHAW, R. D. 1998. Contrasting styles of lithospheric deformation along the northern margins of the Amadeus Basin, central Australia. *In*: BRAUN, J., DOOLEY, J., GOLEBY, B., VAN DER HILST, R. & KLOOTWIJK, C. (eds) *Structure and Evolution of the Australian Continent*. American Geophysical Union, Geodynamics Series, **26**, 139–156.

BUTLER, R. W. H., HOLDSWORTH, R. E. & LLOYD, G. E. 1997. The role of basement reactivation during continental deformation. *Journal of the Geological Society, London*, **154**, 69–71.

DICKSON CUNNINGHAM, W., WINDLEY, B. F., DORJNAMJAA, D., BADAMGAROV, J. & SAANDAR, M. 1996. Late Cenozoic transpression in southwestern Mongolia and the Gobi-Altai-Tien Shan connection. *Earth and Planetary Science Letters*, **140**, 67–81.

ENGLAND, P. C. 1987. Diffuse continental deformation; length scales, rates and metamorphic evolution. *Philosophical Transactions of the Royal Society of London*, **321**, 2–22.

ENGLAND, P. C. & HOUSEMAN, G. A. 1985. The role of lithospheric strength heterogeneities in the tectonics of Tibet and neighbouring regions. *Nature*, **315**, 297–301.

ENGLAND, P. C. & JACKSON, J. 1989. Active deformation of the continents. *Annual Reviews of Earth and Planetary Sciences*, **17**, 197–226.

FLÖTTMANN, T., HAND, M., CLOSE, D., EDGOOSE, C. & SCRIMGEOUR, I. 2001. Thrust tectonic styles of the intracratonic Alice Springs and Petermann orogenies, central Australia. *In*: McCLAY, K. R. (ed.) *Thrust Tectonics and Petroleum Systems*. American Association of Petroleum Geologists, Tulsa.

HAND, M. & SANDIFORD, M. 1999. Intraplate deformation in central Australia, the link between subsidence and fault reactivation. *Tectonophysics*, **305**, 121–140.

HANDY, M. R, MULCH, R., ROSENAU, M. & ROSENBERG, C. R. 2001. The role of fault zones and melts as agents of weakening, hardening and differentiation of the continental crust – a synthesis. *In*: HOLDSWORTH, R. E., STRACHAN, R. A., MAGLOUGHLIN, J. F. & KNIPE, R. J. (eds) *The Nature and Tectonic Significance of Fault Zone Weakening*. Geological Society, London, Special Publications, **186**, 303–330.

HENDRIX, M. S., GRAHAM, S. A., CARROLL, A. R., SOBEL, E. R., McKNIGHT, C. L., SCHULEIN, B. J. & WANG, Z. 1992. Sedimentary record and climatic implications of recurrent deformation in the Tian Shan; evidence from Mesozoic strata of the north Tarim, south Junggar, and Turpan basins, Northwest China. *Geological Society of America Bulletin*, **104**, 53–79.

HENSEN, B. J. & ZHOU, B. 1995. Retention of isotopic memory in garnets partially broken down during an overprinting granulite facies metamorphism: Implications for the Sm–Nd closure temperature. *Geology*, **23**, 225–228.

HOLDSWORTH, R. E., BUTLER, C. A. & ROBERTS, A. M. 1997. The recognition of reactivation during continental deformation. *Journal of the Geological Society, London*, **154**, 73–78.

HOUSEMAN, G. A. & MOLNAR, P. 2001. Mechanisms of lithospheric renewal associated with continental orogeny. *In*: MILLER, J. A., HOLDSWORTH, R. E., BUICK, I. S. & HAND, M. (eds) *Continental Reactivation and Reworking*. Geological Society, London, Special Publications, **184**, 13–37.

IMBER, J., HOLDSWORTH, R. E., BUTLER, C. A. & LLOYD, G. E. 1997. Fault-zone weakening processes along the reactivated Outer Hebrides Fault Zone, Scotland. *Journal of the Geological Society, London*, **154**, 105–110.

IMBER, J., HOLDSWORTH, R. E., BUTLER, C. A. & STRACHAN, R. A. 2001. A reappraisal of the Sibson–Scholz model: the nature of the frictional to viscous (brittle–ductile) transition along a long-lived, crustal-scale fault, Outer Hebrides, Scotland. *Tectonics*, **20**, in press.

KOHLSTEDT, D. L., EVANS, B. & MACKWELL, S. J. 1995. Strength of the lithosphere: Constraints imposed by laboratory experiments. *Journal of Geophysical Research*, **100**, 17 587–17 602.

KORSCH, R. J., GOLEBY, B. R., LEVEN, J. H. & DRUMMOND, B. J. 1998. Crustal architecture of central Australia based on deep seismic reflection profiling. *Tectonophysics*, **288**, 57–69.

KUZSNIR, N. J. & PARK, R. G. 1987. The extensional strength of the continental lithosphere: its dependence on geothermal gradient, and crustal composition and thickness. *In*: COWARD, M. P., DEWEY, J. F. & HANCOCK, P. L. (eds) *Continental Extensional Tectonics*. Geological Society, London, Special Publications, **28**, 35–52.

MOLNAR, P. 1988. Continental tectonics in the aftermath of plate tectonics. *Nature*, **335**, 131–137.

MOLNAR, P. 1992. Brace-Goetze strength profiles, the partitioning of strike-slip and thrust faulting at zones of oblique convergence and the stress-heat flow paradox of the San Andreas Fault. *In*: EVANS, B. & WONG, T.-F. (eds) *Fault Mechanics and Transport Properties in Rocks*. Academic Press Ltd, London, 359–435.

NEIL, E. A. & HOUSEMAN, G. A. 1997. Geodynamics of the Tarim Basin and the Tien Shan in central Asia. *Tectonics*, **16**, 571–584.

NEIL, E. A. & HOUSEMAN, G. A. 1999. Rayleigh–Taylor instability of the upper mantle and its role in intraplate orogeny. *Geophysical Journal International*, **138**, 89–107.

PLATT, J. P. & ENGLAND, P. C. 1994. Convective removal of the lithosphere beneath mountain belts; thermal and mechanical consquences. *American Journal of Science*, **294**, 307–336.

PYSKLYWEC, R. N., BEAUMONT, C. & FULLSACK, P. 2000. Modelling the behaviour of the continental mantle lithosphere during plate convergence. *Geology*, **28**, 655–658.

RAMANDI, F. 1998. Geodynamic implications of the intermediate depth earthquakes and volcanism in the intraplate Atlas Mountains (Morocco). *Physics of the Earth and Planetary Interiors*, **108**, 245–260.

RUTTER, E. H., HOLDSWORTH, R. E. & KNIPE, R. J. 2001. The nature and tectonic significance of fault zone weakening: an introduction. *In*: HOLDSWORTH, R. E., STRACHAN, R. A., MAGLOUGHLIN, J. F. & KNIPE, R. J. (eds) *The Nature and Tectonic Significance of Fault Zone Weakening*. Geological Society, London, Special Publications, **186**, 1–11.

SANDIFORD, M. & HAND, M. 1998. Controls on the locus of intraplate deformation in central Australia. *Earth and Planetary Science Letters*, **162**, 97–110.

SCRIMGEOUR, I. & CLOSE, D. 1999. Regional high-pressure metamorphism during intracratonic deformation: the Petermann Orogeny, central Australia. *Journal of Metamorphic Geology*, **17**, 557–572.

SHAW, R. D. & BLACK, L. P. 1991. The history and tectonic implications of the Redbank Thrust Zone, central Australia, based on structural, metamorphic and Rb-Sr isotopic evidence. *Australian Journal of Earth Sciences*, **38**, 307–332.

SHAW, R. D., ZEITLER, P. K., MCDOUGALL, I. & TINGATE, P. R. 1992. The Palaeozoic history of an unusual intracratonic thrust belt in central Australia based on ^{40}Ar–^{39}Ar, K–Ar and fission track dating. *Journal of the Geological Society, London*, **149**, 937–954.

SIBSON, R. H. 1977. Fault rocks and fault mechanisms. *Journal of Geological Society, London*, **133**, 191–213.

SONDER, L. & ENGLAND, P. C. 1986. Vertical averages of rheology of the continental lithospheric; relation to thin sheet parameters. *Earth and Planetary Science Letters*, **77**, 81–90.

STEWART, M., HOLDSWORTH, R. E. & STRACHAN, R. A. 2000. Deformation processes and weakening mechanisms within the frictional-viscous transition zone of major crustal-scale faults: insights from the Great Glen Fault Zone, Scotland. *Journal of Structural Geology*, **22**, 543–560.

STÜWE, K., SANDIFORD, M. & POWELL, R. 1993. Episodic metamorphism and deformation in low-pressure high temperature terranes. *Geology*, **21**, 829–832.

SUTTON, J. & WATSON, J. V. 1986. Architecture of the continental lithosphere. *Philosophical Transactions of the Royal Society, London*, **A317**, 5–12.

TAVARNELLI, E. & HOLDSWORTH, R. E. 1999. How long do structures take to form in transpression zones? A cautionary tale from California. *Geology*, **27**, 1063–1066.

TAYLOR, S. R. & MCLENNAN, S. M. 1985. *The continental crust: its composition and evolution*. Blackwell Scientific Publications, Oxford.

TEYSSIER, C. & TIKOFF, B. 1998. Strike–slip partitioned transpression of the San Andreas Fault system: a lithospheric scale approach. *In*: HOLDSWORTH, R. E., STRACHAN, R. A. & DEWEY, J. F. (eds) *Continental Transpressional and Transtensional Tectonics*. Geological Society, London, Special Publications, **135**, 143–158.

TOMMASI, A., VAUCHEZ, A. & DAUDRÈ, B. 1995. Initiation and propogation of shear zones in a heterogeneous continental lithosphere. *Journal of Geophysical Research*, **100**, 22 083–22 101.

VAN DER HILST, R. D., KENNET, B. L. N. & SHIBUTANI, T. 1998. Upper mantle structures beneath Australia from portable array deployments. *In*: BRAUN, J., DOOLEY, J., GOLEBY, B., VAN DER HILST, R. & KLOOTWIJK, C. (eds) *Structure and Evolution of the Australian Continent*. American Geophysical Union, Geodynamic Series, **26**, 39–58.

VEEVERS, J. J. 2000. *Billion year Earth History of Australia and Neighbours in Gondwanaland*. GEMOC Press, Macquarie University, Australia.

VILOTTE, J. P., DAIGNIERES, M., MADARIAGA, R. & ZIENKIEWICZ, O. 1982. The role of a heterogeneous inclusion in continental collision. *Physics of the Earth and Planetary Interiors*, **36**, 236–259.

WALSH, J. J., CHILDS, C., MEYER, V., MANZOCCHI, T., IMBER, J., NICOL, A., TUCKWELL, G., BAILEY, W. R., BONSON, C. G., WATTERSON, J., NELL, P. A. & STRAND, J. 2001. Geometric controls on the evolution of normal fault systems. *In*: HOLDSWORTH, R. E., STRACHAN, R. A., MAGLOUGHLIN, J. F. & KNIPE, R. J. (eds) *The Nature and Tectonic Significance of Fault Zone Weakening*. Geological Society, London, Special Publications, **186**, 157–170.

WARREN, R. G. 1983. Metamorphic and tectonic evolution of granulites, Arunta Block, central Australia. *Nature, London*, **305**, 300–303.

WILLIAMS, I. S., BUICK, I. S. & CARTWRIGHT, I. 1995. An extended episode of early Mesoproterozoic metamorphic fluid flow in the Reynolds Range, central Australia. *Journal of Metamorphic Geology*, **14**, 29–47.

ZIELGER, P. A., CLOETINGH, S., & VAN WEES, J. D. 1995. Dynamics of intra-plate compressional deformation: the Alpine foreland and other examples. *Tectonophysics*, **252**, 7–59.

ZIELGER, P. A., VAN WEES, J. D. & CLOETINGH, S. 1998. Mechanical controls on collision related compressional intraplate deformation. *Tectonophysics*, **300**, 103–129.

Mechanisms of lithospheric rejuvenation associated with continental orogeny

GREGORY HOUSEMAN[1,2] & PETER MOLNAR[3,4]

[1] *Department of Earth Sciences, P.O. Box 28E, Monash University, Clayton, VIC 3180, Australia*
[2] *Present address: School of Earth Sciences, University of Leeds, Leeds LS2 9JT, UK (e-mail: greg@earth.leeds.ac.uk)*
[3] *Department of Earth, Atmospheric and Planetary Sciences, Massachusetts Institute of Technology, Cambridge, MA 02139, USA*
[4] *Present address: Department of Geological Sciences, Cooperative Institute for Research in Environmental Science, Campus Box 399, University of Colorado, Boulder, Colorado 80309, USA*

Abstract: Gravitational instability of the continental lithospheric mantle is often associated with orogenic activity. Recent theoretical and experimental developments in the understanding of the convective instability of a dense layer, with non-Newtonian viscosity (representing lithosphere) above a less dense fluid layer (representing asthenosphere) are reviewed. These developments offer an explanation for why the continental lithospheric mantle might be generally mechanically stable in spite of a thermally induced density stratification, which one might expect to be unstable. Gravitational stability of this system depends on the initial amplitude of a disturbance to the stratified system, a disturbance that is most likely to be provided by localized lithospheric thickening associated with plate convergence. If the constitutive law that describes the deformation of dry olivine is applicable to the subcontinental mantle, the perturbation required to produce instability could be created by localized horizontal shortening of the order of 10%. If the wet olivine flow law is applicable, the required amount of shortening may be on the order of only 1%, in each case provided that it occurs in a time short compared with the thermal diffusion timescale of the lithosphere. The long-term stabilization of continental lithosphere may thus be associated with dehydration. Under circumstances of localized lithospheric convergence, the buoyancy of the continental crust plays an important role in determining the form of downwelling. If the crust is strong compared to mantle lithosphere, instability generally takes the form of localized downwelling beneath the centre of the convergent zone. If the crust is weak compared to mantle lithosphere, downwelling commences on the margins of the convergent zone as the buoyant crustal layer resists thickening. The initial instability may then trigger rapid extension of the lithospheric mantle beneath the convergent orogen. The extension is driven by asymmetric cold downwellings that move away from the centre of the convergent zone in a way that bears some resemblance to a delaminating slab or a retreating subduction zone. With these results in mind, some of the geological and geophysical evidence for lithospheric instability in modern orogens of Southern California, the South Island of New Zealand, the Mediterranean, and Central Asia are reviewed. Seismological evidence from Southern California and New Zealand suggest that these young orogens provide examples of lithospheric instability, in which downwelling occurs beneath the centre of the convergent zone where the crustal thickening is maximum. In contrast, the Alboran Sea and Tyrrhenian Sea basins show that extension has followed convergence as downwelling has retreated away from the convergent zone, and lithospheric mantle beneath the centre of convergence apparently has been replaced by asthenosphere. The Tien Shan and Tibetan Plateau provide large modern-day examples of continental convergence. In each case there is strong evidence that the mantle lithosphere has undergone some form of instability that has led to at least part of it being replaced by hot asthenosphere. Images provided by teleseismic tomography of variations of seismic wave speeds beneath these orogens suggest that mantle lithosphere has been locally renewed following gravitational instability triggered by orogenic convergence.

From: MILLER, J. A., HOLDSWORTH, R. E., BUICK, I. S. & HAND, M. (eds) *Continental Reactivation and Reworking*. Geological Society, London, Special Publications, **184**, 13–38. 1-86239-080-0/01/$15.00 © The Geological Society of London 2001.

Oceanic lithosphere clearly participates in the mantle convection system, and is renewed on a timescale of order 100 Ma. The behaviour and role of the mantle part of continental lithosphere in large-scale geodynamics, on the other hand, are more enigmatic. Although it is covered by low-density crust that ranges in age from recent to Archaean, the longevity of the crust does not necessarily imply a comparable longevity of the lithospheric mantle. Xenoliths from South Africa suggest that lithospheric mantle there is 3 Ga old (Richardson et al. 1984), and there are many other examples of mantle being at least Proterozoic in age. In contrast, the Basin and Range Province of the western US, for example, demonstrates extensive magmatism, high heat flow, and low seismic wave speeds, which suggest that the lithosphere has been recently heated throughout, if not replaced by asthenosphere. It is deduced that continental lithosphere commonly appears to be relatively isolated from the underlying mantle convection system, but occasionally interacts catastrophically with it. This interaction may be referred to as lithospheric rejuvenation because it involves at least partial replacement of older colder lithospheric mantle with hot asthenosphere. The aims of this paper are to review some concepts that have been proposed to explain lithospheric rejuvenation in the continental environment, to examine the implications of Rayleigh–Taylor instability as a physical model that explains a range of rejuvenation mechanisms, and to discuss the geological development of some regions in the light of this model.

Among mechanisms for heating the continental lithosphere, lithospheric extension (e.g. Sleep 1971; McKenzie 1978) introduces hot asthenosphere beneath a continental rift zone, the lithosphere subsequently cools and incorporates the additional material. Extension can occur in response to an increase in gravitational potential energy brought about by lithospheric heating or by uplift associated with convective upwelling in the mantle (e.g. England & Houseman 1989; Molnar et al. 1993; Platt & England 1994). Slab detachment may also be an important process in driving lithospheric extension in basin systems on an overriding continental plate (Wortel & Spakman 1992). Reheating might be due partly to conductive heating and partly to the replacement of lithosphere by asthenosphere (e.g. Stüwe & Sandiford 1995). In general, however, conductive heating of thickened lithosphere cannot be an important process unless the plate remains stationary over an enhanced heat source for periods that are comparable to the diffusive time constant of the lithosphere

(L^2/κ), where L is the thickness of the plate and κ is its thermal diffusivity.

Perhaps the most important mechanism of lithospheric rejuvenation in continental regions is associated with convergence and the subsequent gravitational instability of the continental lithosphere. The gravitational stability of an unstably stratified fluid depends essentially on two physical factors: (1) the density contrast at the boundary between the two layers; and (2) the viscosities of the two layers. If the mantle lithosphere were of the same composition as the underlying asthenosphere, it would be potentially unstable simply because its temperature is lower than that of asthenosphere at the same depth, and the corresponding thermal contraction would make it more dense than asthenosphere at the same pressure. The observation that continental lithospheric mantle is generally stable therefore has usually been attributed to compositional buoyancy (e.g. Jordan 1975; Griffin et al. 1998). The positive compositional buoyancy is assumed to offset the negative thermal buoyancy. We argue, however, that another property of lithosphere, non-Newtonian viscosity, might account for its stability. Under a non-Newtonian mechanism, strain-rate varies as some power ($n > 1$) of the stress, so that the effective viscosity (the local ratio of stress to strain-rate) varies as a negative power of the strain-rate. For small driving stresses, as in marginal stability, the strain-rate approaches zero much faster than the stress. Under these circumstances an unstable density stratification might persist until disrupted by some large perturbation to the stratification imposed by external forces.

The obvious mechanism that might provide this large destabilizing perturbation is continental collision or, on a smaller scale, intraplate convergence. Plate collision (involving continents or arcs) is the major process by which mountains are built. When an ocean closes, continental shortening may continue near the suture. Intraplate convergence may also be activated by plate-boundary orogeny, even though the site of activation is far removed from plate boundaries, as for example in the Tien Shan (e.g. Molnar & Tapponnier 1975; Hendrix et al. 1992) and Central Australia (Hand & Sandiford 1999). In a convergent environment, both crust and mantle are thickened by faulting near the surface and presumably by ductile flow at depth. As it thickens, the cold and negatively buoyant mantle lithospheric root creates a large destabilizing stress. This thickening of mantle lithosphere provides a perturbation large enough to drive the instability over the threshold required to initiate a rejuvenation event. Orogeny may endure for periods of order 1 to 100 Ma but, on timescales that exceed 100 Ma it is likely that thermal diffusion will dissipate the destabilizing lithospheric root as it forms.

The existence of a threshold for stability (small perturbations decay, whereas large perturbations grow rapidly) concurs with the deformation of the lithosphere being governed by a non-Newtonian viscosity. For non-Newtonian viscosity especially, the growth of the mechanical instability is likely to be defeated by thermal diffusion until localized convergence induces a significant departure from horizontal stratification. Once a sufficiently large thickness perturbation has been induced, however, the growth follows a strongly non-linear development, which ultimately concludes with rapid removal of a large fraction of the thickened mantle lithosphere beneath the convergent zone (e.g. Houseman & Molnar 1997; Conrad 2000). Dislocation creep in olivine may therefore determine the mechanical stability of the continental mantle lithosphere and thus explain the localization in space, and episodicity in time, of those regions affected by lithospheric instability.

Several mechanisms that remove continental lithospheric mantle have been described (Fig. 1),

including mantle delamination (Bird 1978, 1979), convective thinning of the lithosphere (Houseman et al. 1981), and subduction of the mantle lithosphere (e.g. Beaumont et al. 1994; Willett & Beaumont 1994). In each case it is assumed that buoyancy differences drive the deformation. In mantle delamination (Fig. 1a), mantle lithosphere progressively peels away from crust and is replaced by asthenosphere. The delamination point migrates across the thinned region (Bird 1978, 1979). Convective thinning (Fig. 1b) occurs as the classic Rayleigh–Taylor instability, with the dense layer descending in viscous drips, at least where perturbations to the base of the layer are not so small that thermal diffusion can erase them. Growth is driven by negative buoyancy but perhaps triggered by imposed convergence. Subduction of mantle lithosphere (Fig. 1c) is viewed primarily as a response to convergence driven by plate boundary stress. In this concept the mantle lithosphere detaches from the crust at a point of discontinuity, and convergence in the crust continues by means of conjugate shear zones (Beaumont et al. 1994; Willett & Beaumont 1994), while the lithospheric mantle undergoes a kind of subduction. Mantle subduction in this context is generally assumed to develop from an earlier phase of subduction of an oceanic plate, though detachment of the subducted oceanic slab (e.g. Wortel & Spakman 1992; Yoshioka & Wortel 1995) is perhaps more probable. In common with delamination or convective thinning, the results of slab detachment can include both an uplift of the surface in response to the load removed from the remaining lithosphere and a heating by the replacement of the slab by asthenosphere at shallow depths (see also von Blanckenburg & Davies 1995; van der Meulen et al. 1998).

The application of Rayleigh–Taylor instability to the convective thinning concept rests on a firm dynamical basis (Conrad & Molnar 1997, 1999; Houseman & Molnar 1997; Houseman et al. 1981; Molnar et al. 1998; Neil & Houseman 1999; Conrad 2000), and can be used to provide quantitative predictions, whereas the application of the other concepts has depended essentially on an imposed kinematic framework. The processes of delamination and convective thinning both lead to local lithospheric renewal, but some of their other consequences are different. For convective thinning, the mantle lithosphere from adjoining regions flows into the downwelling as the instability grows. Eventually mantle lithosphere is replaced as it is thinned and stretched, even though crustal convergence may continue. In contrast, delamination describes a process that is not necessarily

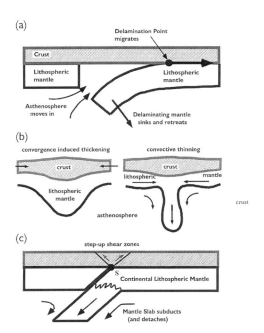

Fig. 1. Proposed mechanisms by which continental lithospheric mantle is renewed: (**a**) delamination (after Bird 1978, 1979); (**b**) convective thinning (after Houseman et al. 1981); and (**c**) mantle subduction (after Beaumont et al. 1994).

preceded by crustal thickening, but in which mantle lithosphere peels away from the base of the crust as a coherent sheet, without necessarily undergoing major internal deformation. Hot asthenosphere moves into the gap between mantle and crustal layers as it opens. The locus of deformation migrates away from its starting point and the mantle lithosphere is progressively replaced as the delamination proceeds (Bird 1978, 1979). Slab detachment may have similar consequences in terms of lithospheric heating, because of both the removal of a long heavy slab and its replacement by hot asthenosphere at shallow depths beneath an orogenic belt, but it is not necessarily associated with the removal of continental lithospheric mantle. These mechanisms are not necessarily mutually exclusive; each may occur in different tectonic settings, and combinations of them can occur in some situations (e.g. Pysklywec et al. 2000).

In the following, we review and summarize three key aspects of the gravitational stability of the mantle lithosphere system, illustrating how this information can help us to understand the observed geological history and deep structure of convergent mountain belts. Considered first is the instability of a perturbation to the thickness of a dense layer (representing mantle lithosphere) over a lighter layer (representing asthenosphere), when the dense layer satisfies a non-Newtonian constitutive relation. Because the density distribution that drives the instability arises from temperature differences, its growth may require a large initial perturbation. We then consider the effects of externally imposed shortening and thickening on the development of the instability, and in particular how this process influences the growth of the instability over time. Numerical experiments show that the buoyancy of the crust has an important influence on the location and shape of mantle downwelling beneath a zone of localized convergence. In the final section, we review evidence from several regions suggesting that the lower part of the mantle lithosphere has been removed by gravitational instability triggered by convergence.

Convective thinning as a finite amplitude instability

To follow is a summary of the theory of convective instability of the continental lithospheric mantle that has been developed in a sequence of recent papers by Conrad & Molnar (1997), Houseman & Molnar (1997), Molnar et al.

(1998), Neil & Houseman (1999), Conrad & Molnar (1999), Houseman et al. (2000) and Conrad (2000). We aim here to focus on the essential physical reasoning and some of the key results, so inevitably some aspects have been oversimplified while others have been ignored. The reader will find a more complete description in the papers above, but beware that in some cases these papers have followed different conventions for scaling dimensionless quantities (such as the growth rate factor $C(kh)$ described below). This theory develops the concept of a critical Rayleigh number for a convective boundary layer introduced by Howard (1966). The concept has previously been applied to the stability of oceanic lithosphere (Parsons & McKenzie 1978), and Houseman et al. (1981) used the idea of a boundary layer Rayleigh number to predict convective thinning rates for thickened continental mantle lithosphere. The recent developments have allowed the theory to be extended to quantify the stability of layers with non-Newtonian viscosity and vertically stratified density and viscosity variation.

On geological timescales the continental lithospheric mantle may be viewed as an unstably stratified system of layers of viscous fluid. To simplify, we may represent the continental lithosphere as a 3-layered system. The middle layer in this model, representing the lithospheric mantle, is potentially unstable because its density is greater than that of the underlying asthenosphere. Lithospheric mantle is denser than asthenosphere because it is cooler by up to 800°C. The density difference caused by this temperature difference varies from $\Delta\rho = \Delta T\alpha\rho = 800\,\text{K} \times 3 \times 10^{-5}\,\text{K}^{-1} \times 3300\,\text{kg m}^{-3} = 80\,\text{kg m}^{-3}$ to zero as the temperature increases with depth through the lithosphere. Overlying all is a relatively low-density layer, representing the continental crust, which has a stabilizing influence.

Neglecting, for the time being, the effect of the low-density crustal layer and the effect of thermal diffusion, we examine the gravitational stability of this system. The mathematics describing the gravitational instability of a two-layer unstably stratified fluid system, Rayleigh-Taylor instability, are well known (e.g. Chandrasekhar 1961). The system is unconditionally unstable because any small deflection of the interface that separates the two layers causes a flow system to develop, which in turn causes the deflection to increase in magnitude. The dense upper layer is thickened at some locations and thinned elsewhere and, while the deflection of the interface is small, the rate of flow is proportional to the horizontal gradients of thickness. Assuming incompressible flow, both deflection Z and deflection rate $w = dZ/dt$ increase in proportion to their own magnitudes:

$$w = \frac{dZ}{dt} = q_Z Z \quad \text{and} \quad \frac{dw}{dt} = q_w w \quad (1)$$

where, for Newtonian viscosity

$$q = q_Z = q_w = \frac{gh\Delta\rho}{2\eta} C_1(kh) \quad (2)$$

h is the thickness of the upper layer; g is acceleration due to gravity; η is viscosity; $C_1(kh)$, is a dimensionless function of order unity that describes the dependence of growth rate on the horizontal wavenumber $k = 2\pi/\lambda$, and λ is the wavelength of the disturbance (Fig. 2). The deflection Z therefore grows exponentially, and the coefficient q is referred to as the growth rate. Equation (2) states that growth is faster for a heavier upper layer (whether it be more dense or thicker), for larger g and for a less viscous layer. Growth is fastest for the wavenumber at which $C_1(kh)$ is a maximum. At shorter or longer wavelengths the viscous dissipation is relatively large compared to the accompanying decrease in gravitational potential energy, and the growth rate is less (Fig. 2). Because growth is exponential, perturbations with shorter or longer wavelengths are rapidly overwhelmed by perturbations with maximum growth rates. If the upper layer has a much greater viscosity than the lower layer, and no slip

is allowed at the top boundary, the growth rate is maximum ($C_1(kh) \approx 0.32$) for a harmonic disturbance whose wavelength is about 3 times the thickness of the upper layer (Conrad & Molnar 1997). If the upper layer has a density that varies linearly from zero at the base of the layer to $\Delta\rho$ at the top, (1) and (2) remain valid, but the function $C_1(kh)$ is modified. The maximum growth rate ($C_1(kh) \approx 0.108$) then occurs for a wavelength about 2.4 times the thickness, and $C_1(kh)$ decreases with increasing wavenumber at a greater rate than for constant density. Similarly, changes to the viscosity stratification within the layer can be accounted for by re-evaluation of the function $C_1(kh)$ (Conrad & Molnar 1997).

Suppose that a linear density variation in the upper layer is caused by temperature stratification. Thermal diffusion then acts on any harmonic deflection of the isothermal surfaces, causing the deflection to decay at a rate that depends on horizontal wavenumber k and thermal diffusivity κ, approximately expressed by (e.g. Conrad & Molnar 1997):

$$\frac{1}{Z}\frac{dZ}{dt} = -\frac{\kappa}{h^2}\left(\frac{\pi^2}{4} + k^2 h^2\right) \quad (3)$$

(The $\pi^2/4$ term arises for fixed temperature upper boundary condition and insulating lower boundary; it would be zero for purely insulating boundaries or π^2 for a temperature fixed at both boundaries.) The instability grows if the mechanical growth rate q in (2) exceeds the thermal decay rate in (3):

$$\frac{g\Delta\rho h}{2\eta} C_1(kh) - \frac{\kappa}{4h^2}(\pi^2 + 4k^2 h^2) > 0 \quad (4)$$

or,

$$Ra_1 = \frac{g\Delta\rho h^3}{2\eta\kappa} > \frac{\pi^2 + 4k^2 h^2}{4C_1(kh)} = Ra_{1c} \quad (5)$$

Allowing for the difference in boundary conditions, this inequality is just a restatement of the well-known criterion due to Rayleigh for the convective stability of a fluid layer. The non-dimensional ratio on the left of the inequality is recognized as a Rayleigh number for the layer and the term on the right defines the critical Rayleigh number Ra_{1c}. The most unstable mode occurs with the dimensionless wavenumber kh for which the critical Rayleigh number is minimum. The minimum value of the function Ra_{1c} defined in (5) is of order 50, for $kh \approx 1.5$ ($\lambda \approx 4h$), obtained using the function $C_1(kh)$ for constant viscosity and linear density. Conrad & Molnar (1999) obtained a similar value for

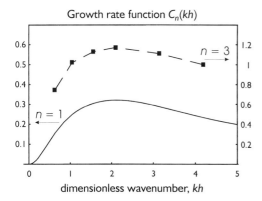

Growth rate function $C_n(kh)$

Fig. 2. Variation of the functions $C_n(kh)$ (equations 2 and 8) that describe the dependence on wavenumber kh of the growth rate of the Rayleigh–Taylor instability of a dense layer of constant density and viscosity over a low viscosity half-space, for Newtonian viscosity ($n = 1$; left hand scale; based on linear theory described by Conrad & Molnar (1997)) and non-Newtonian viscosity ($n = 3$; right hand scale; based on numerical experiments of Houseman & Molnar (1997).

Ra_{1c} from numerical experiments of convective instability. The effect of thermal diffusion stabilizes the layer at shorter wavelengths and therefore the wavelength of the least stable mode is greater than that of Rayleigh–Taylor instability. This analysis shows that the stability of a stratified layer of Newtonian viscosity depends only on whether its critical Rayleigh number (5) is exceeded. If so, a disturbance that can remove half or more of the layer can grow from a noise signal of any amplitude.

If the lithosphere has a non-Newtonian viscosity, however, a fundamentally different conclusion is reached: the growth rate, and therefore the stability of the layer, depend on the magnitude of the initial disturbance. Assuming that viscosity depends on strain-rate (or equivalently on deviatoric stress):

$$\eta = \frac{B}{2}\,\dot{E}^{(1-n)/n} = \frac{B^n}{2}\,\Theta^{(1-n)} \qquad (6)$$

where, $\dot{E} = \sum_{i,j}\dot{\varepsilon}_{ij}\dot{\varepsilon}_{ij}$ and $\Theta = \sum_{i,j}\tau_{ij}\tau_{ij}$ are second invariants of strain-rate and deviatoric stress respectively, and B and n are material constants. In what follows, values of $n \approx 3$ are taken as representative of olivine, the principal constituent of the mantle. An exact solution to the growth equations (1) can no longer be obtained because $\dot{E}$ has a complex spatial dependence. Suppose, however, the assumption is made that the magnitude of the strain-rate is approximated by the vertical speed at the base of the layer, divided by the layer thickness

$$\dot{E} \approx \frac{w}{h} \qquad (7)$$

As for Newtonian viscosity, the rate of change of the deflection (1) is proportional to the deflection, and the form of the growth rate in (2) for w is assumed. Substituting (7) into (6), thence into (2), thence into (1), the following equation is obtained, which describes the time-dependence of the rate of deflection w for Rayleigh–Taylor instability of a non-Newtonian layer (Houseman & Molnar 1997):

$$\frac{dw}{dt} = \frac{g\Delta\rho}{B}\,h^{1/n}C_n(kh)w^{(2n-1)/n} \qquad (8)$$

On integration the solution is:

$$w = \left[\left(\frac{n-1}{n}\right)\frac{g\Delta\rho}{B}\,h^{1/n}C_n(kh)(t_b - t)\right]^{n/(1-n)} \qquad (9)$$

where the integration constant t_b represents the time when the deflection rate approaches infinity. Equation (9) may be integrated once more to obtain the deflection function Z;

$$\frac{Z}{h} = \left[(n-1)\left\{\frac{hg\Delta\rho}{nB}\,C_n(kh)\right\}^n(t_b - t)\right]^{1/(1-n)} \qquad (10)$$

and the integration constant t_b may then be evaluated in terms of the deflection Z_0 at time zero:

$$t_b = \frac{1}{(n-1)}\left\{\frac{nB}{hg\Delta\rho C_n(kh)}\right\}^n\left(\frac{Z_0}{h}\right)^{(1-n)} \qquad (11)$$

Thus, power-law growth occurs from the outset for Rayleigh–Taylor instability in a layer of non-Newtonian viscosity, with slow deformation of the layer during most of the development of the instability but smooth acceleration into a brief period of catastrophic growth towards the end (Fig. 3a, b). Numerical experiments, both for constant rheological parameters (Houseman & Molnar 1997) and for depth-dependent rheological parameters (Molnar et al. 1998), demonstrate the validity of equations (9)–(11) for Rayleigh–Taylor instability. For $n = 3$, Z^{-2} decreases approximately linearly with time towards the instability time t_b (Fig. 3c), as does $w^{-2/3}$ (Fig. 3d). The wavenumber-dependent term for non-Newtonian instability $C_n(kh)$ includes a scaling factor of order 1 introduced with the approximation (7), which must be evaluated numerically. The maximum value of $C_3(kh)$ is approximately 1.1 at $kh \approx 2$ for constant density (Houseman & Molnar 1997). If density varies linearly from 0 to $\Delta\rho$ across the layer, the above approximate theory appears to work equally well, though $C_3(kh)$ is modified with its maximum value reduced to $c.0.32$. Empirically it is found, for a layer of constant density and viscosity, that the function $C_n(kh)$, normalized by its maximum value and raised to the power $(n + 1)/2$ is approximately independent of n.

The growth of the non-Newtonian instability may be described as super-exponential because the graph of $\log Z$ v. time is concave upward; the growth rate increases monotonically with time. From (9) and (10), the respective growth rates for deflection and deflection rate increase in inverse proportion to the time to instability:

$$q_z = \frac{1}{Z}\frac{dZ}{dt} = [(n-1)(t_b - t)]^{-1} \qquad (12)$$

and

$$q_w = \frac{1}{w}\frac{dw}{dt} = \left[\left(\frac{n-1}{n}\right)(t_b - t)\right]^{-1}$$

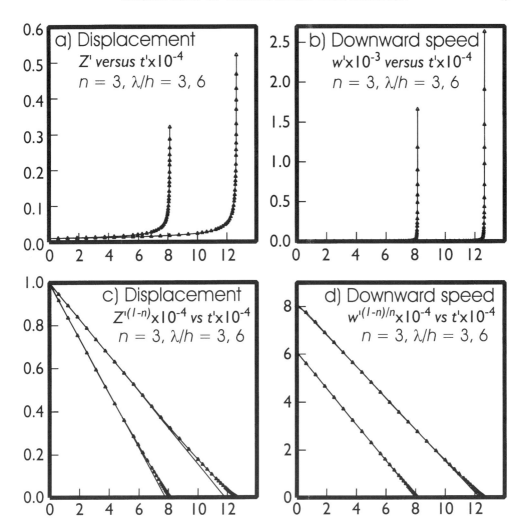

Fig. 3. Variation with time of maximum downward displacement (**a**) and maximum downward speed (**b**) for a layer with non-Newtonian viscosity ($n = 3$) (Houseman & Molnar 1997). Displacement raised to the power $(1 - n)$ decreases approximately linearly with time (**c**), as does speed raised to the power of $(1 - n)/n$. Data are shown from 2 numerical experiments: $\lambda = 3h$ ($kh \approx 2$; the fast developing experiment) and $\lambda = 6h$ ($kh \approx 1$). Data points represent parameter values at sequential time levels in the finite element calculation. The thin straight lines in (**c**) and (**d**) represent best-fits to the early parts of the data.

The expression in (12) for q_Z gives the growth rate for deflection of isotherms, in the absence of thermal diffusion. If the layer is to be unstable, however, the perturbations must grow at a sufficiently fast rate to overcome the stabilizing effects of thermal diffusion (Conrad & Molnar 1997, 1999). As for Newtonian viscosity the criterion for convective instability is that the mechanical growth rate of Z in (12) exceeds its thermal decay rate (3):

$$\frac{1}{(n-1)(t_b - t)} > \frac{\kappa}{4h^2}(\pi^2 + 4h^2k^2) \quad (13)$$

If this criterion is met at $t = 0$ then clearly it will be met for all $t > 0$. The layer is therefore unstable if

$$t_b < \frac{4h^2}{\kappa(\pi^2 + 4h^2k^2)(n-1)} \quad (14)$$

Substituting (11) into (14), a criterion for convective instability is obtained that defines a critical Rayleigh number for a layer with non-Newtonian viscosity (Conrad & Molnar 1999):

$$Ra_n = \left(\frac{g\Delta\rho h}{nB}\right)^n \left(\frac{h^2}{\kappa}\right)\left(\frac{Z_0}{h}\right)^{n-1}$$

$$> \frac{\pi^2 + 4k^2h^2}{4[C_n(kh)]^n} = Ra_{nc} \qquad (15)$$

where Ra_n is a non-Newtonian Rayleigh number, and Ra_{nc} is the critical value for which convection occurs. Minimizing the Ra_{3c} function with respect to wavenumber kh, using an estimate of $C_3(kh)$ based on numerical experiments yields a value for the critical Rayleigh number of order 200, as found by Conrad & Molnar (1999). If $n = 1$, we recover from (15) the stability criterion for Newtonian viscosity, (5). The conceptual difference between (15) and (5) lies in the term $(Z_0/h)^{n-1}$, by which the criterion for convective instability of a non-Newtonian layer depends on the amplitude of the initial deflection of the density contours. If Z_0 is defined as the deflection of the base of the lithosphere, then the ratio (Z_0/h) may be interpreted as a measure of lithospheric thickening. For the simple case of an initial linear density gradient undergoing rapid uniform shortening, the layer will be unstable only if this thickening measure exceeds a critical value given by:

$$\left(\frac{Z_0}{h}\right)^{n-1} > \left(\frac{\kappa}{h^2}\right)\left(\frac{nB}{g\Delta\rho h}\right)^n Ra_{nc} \qquad (16)$$

Thus, the combination of non-Newtonian viscosity and temperature-dependent density provides the requirements for a finite amplitude gravitational instability: a possibly large destabilizing perturbation to the system is needed in order that the growth rates of the gravitational instability can overcome the stabilizing effects of thermal diffusion. In the absence of that large deflection, however, the lithospheric mantle will be gravitationally stable in spite of its negative buoyancy.

The preceding analysis applies to a lithospheric layer of constant viscosity coefficient (B for $n > 1$ or η for $n = 1$). The consequences for convective instability of a strength stratification associated with the temperature stratification were also investigated by Conrad & Molnar (1999). They showed, using numerical experiments, that the simplified theory of convective instability described above may be adapted to describe the case of a viscosity coefficient that decays exponen-

tially with depth through the lithosphere. To do so, they introduced a dimensionless quantity referred to as available buoyancy:

$$F_n = \frac{1}{h}\left(\frac{B_m}{\Delta\rho}\right)^n \int_0^h [\rho(z)/B(z)]^n \, dz, \qquad (17)$$

in which the scaling constant B_m represents the viscosity coefficient at the base of the dense layer. The negative buoyancy at each depth is thus weighted by the inverse of the relative vicosity coefficient. Conrad & Molnar (1999) showed that the theory from which (15) and (16) are derived applies equally well to a layer in which viscosity varies exponentially, if the Rayleigh number in (15) is multiplied by $(n+1)/F_n$. The factor $(n+1)$ is required for consistency with the constant B case, for which $F_n = 1/(n+1)$. The stability criterion (16) may then be rewritten:

$$Ra_n = (n+1)F_n\left(\frac{g\Delta\rho h}{nB_m}\right)^n \left(\frac{h^2}{\kappa}\right)\left(\frac{Z_0}{h}\right)^{n-1}$$

$$> Ra_{nc} \qquad (18)$$

or, in terms of the magnitude of perturbation required to promote instability:

$$\left(\frac{Z_0}{h}\right)^{n-1} > \left(\frac{\kappa}{h^2}\right)\left(\frac{nB_m}{g\Delta\rho h}\right)^n \frac{Ra_{nc}}{(n+1)F_n} \qquad (19)$$

To illustrate the magnitude of the perturbation required for gravitational instability to occur, consider an example calculation based on the stability criterion (19) and values of F_n estimated by Conrad & Molnar (1999). For temperature increasing linearly with depth through the mantle lithosphere, $\Delta\rho \approx 80 \, \text{kg m}^{-3}$, and using a temperature-dependent viscosity coefficient based on the olivine rheology of Karato et al. (1986), Conrad & Molnar estimated $F_n = 5.5 \times 10^{-5}$ for wet olivine ($n = 3$), $F_n = 8.8 \times 10^{-6}$ ($n = 3.5$) for dry olivine, and $B_m \approx 1.9 \times 10^9 \, \text{Pa s}^{1/n}$ in either case. The parameter F_n depends on n, the activation energy, and the total temperature contrast across the layer, and is sensitive to the shape of the geotherm, but is not specifically dependent on layer thickness h. Using the above parameter values, together with $\kappa = 10^{-6} \, \text{m}^2 \, \text{s}^{-1}$, $g = 10 \, \text{m s}^{-2}$, and $Ra_{nc} = 200$, the dependence of Z_0 on h described by (18) for both dry olivine (Fig. 4a) and wet olivine (Fig. 4b) may be plotted. For example, the perturbation required to drive instability of a dry olivine layer initially 80 km thick is $c. 12$ km, but for wet olivine the necessary perturbation is only about 1 km. For a lithospheric thickness of 250 km, the perturbation required for destabilization is reduced to only 3 km for dry olivine, and 150 m for

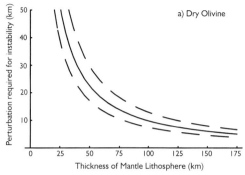

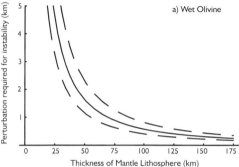

Fig. 4. Perturbation Z_0 required for instability to occur, as a function of the thickness of the mantle lithosphere, based on rheological parameters summarised by Karato *et al.* (1986) and using available buoyancy estimates of Conrad & Molnar (1999) for (a) dry olivine ($n = 3.5$, $F_n = 8.8 \times 10^{-6}$), and (b) wet olivine ($n = 3.0$, $F_n = 5.5 \times 10^{-5}$). In each case $\Delta\rho = 80 \, \text{kg m}^{-3}$, the solid line is for the nominal value of F_n and the two dashed lines represent the dependence when F_n is halved (upper curve) or doubled (lower curve).

wet olivine. In deriving the above relations, the damping effect of the viscous asthenospheric layer has been neglected, so the magnitude of the perturbation required for instability may have been underestimated. For example, if the growth-rate function C_n in (15) is reduced by a factor of 2 Ra_{nc} is increased by 2^n and the perturbation Z_0/h in (16) is increased by the factor $2^{n/(n-1)}$. There is also some uncertainty in the applicable rheological parameters, which may be represented here as an uncertainty in the value of F_n. Figure 4a and b also shows therefore the dependence of Z_0/h on h when F_n is increased or decreased by a factor of 2.

The time required for the instability to develop depends on the degree of supercriticality of the boundary layer Rayleigh number Ra_n (18). If the magnitude of the disturbance Z_0 is only slightly greater than that required for marginal

stability, the instability time will approach the thermal time constant of the lithosphere (on the order of 100 Ma). If the perturbation is much larger however, diffusion may be neglected. The growth time may then be estimated from the time required for Rayleigh–Taylor instability, modified from (11) by Conrad & Molnar (1999) to incorporate the concept of available buoyancy (17).

$$t_b = \frac{1}{(n-1)F_n} \left\{ \frac{nB_m}{hg\Delta\rho C_n} \right\}^n \left(\frac{Z_0}{h} \right)^{(1-n)} \quad (20)$$

Conrad & Molnar showed that the wavenumber function C_n in (20) is approximately constant (≈ 0.45) for a range of depth-dependent viscosity stratifications.

The theory outlined above for the convective instability of a stratified non-Newtonian layer is consistent with the apparent stability of continental lithosphere. The lithosphere is evidently more easily destabilised as its thickness increases. For a wet-olivine lithosphere relatively minor disturbances may trigger lithospheric renewal, but if the dry-olivine constitutive law is applicable a major disturbance of the lithosphere is required for gravitational instability. In either case, externally forced convergence in a zone of restricted horizontal extent, presumably associated with plate-boundary convergence is the most probable cause of such an event.

Effect of an external stress field on the temporal development of convective thinning

The previous discussion has considered the conditions necessary for growth of thickness perturbations of a potentially unstable layer on which no significant levels of externally imposed deviatoric stress act. In the Earth, such conditions explain the general stability of continental lithosphere, but of more interest here are those circumstances in which the unstable layer is subjected to an externally imposed strain field induced, for example, by plate boundary forces. Externally imposed convergence, in addition to creating and enhancing thickness variations that may destabilize the layer, decreases the effective viscosity of a non-Newtonian layer. Pervasive strike-slip deformation may have a similar influence on the effective viscosity, since the viscosity (6) depends only on the magnitude of strain-rate, not on the style of deformation.

For the unstressed layer described in the preceding section, the effective viscosity (6) in the layer decreases linearly in time as the perturbation grows and rates of deformation increase. For a large imposed background strain rate, however, as occurs with horizontal shortening or with strike-slip deformation, the effective viscosity of the layer initially is constant in time, with its magnitude determined by that strain rate, $\dot{E}_x$. The initial stage of growth of a thickness perturbation is exponential, with a growth rate described by (4) (e.g. Conrad & Molnar 1997). The criterion for stability may be derived from (5) and (6) and expressed:

$$\frac{g\Delta\rho h^3}{\kappa B} \dot{E}_x^{(n-1)/n} > Ra_{1c} \qquad (21)$$

At this stage stability depends on the magnitude of $\dot{E}_x$ but is independent of the magnitude of the perturbation. No matter how small the disturbance is, it will grow exponentially if the imposed strain rate is large enough.

Numerical experiments carried out by Molnar et al. (1998) on an unstable layer with a small thickness perturbation and undergoing horizontal shortening at a constant rate not only confirmed the predicted exponential growth, but also yielded growth rates consistent with those predicted from linear stability analysis (Conrad & Molnar 1997). Obviously, more rapid horizontal shortening leads to more rapid growth, for the effective viscosity is smaller with faster shortening. When strain rates caused by the gravitationally driven flow become comparable to those imposed by the background shortening of the layer, growth becomes super-exponential, as given by (9) and (10). Numerical experiments confirm the sequence of both phases of growth and show that the amplitude of the perturbation at which growth switches from exponential to super-exponential depends upon the material properties and rate of shortening (Molnar et al. 1998). For values typical of the earth, the perturbation grows exponentially until it corresponds approximately to a doubling of the thickness of the layer. Conrad (2000) examined the effect of horizontal shortening on a convectively unstable layer and determined the conditions under which growth occurred or did not. He found that under some conditions exponential growth did not play a significant role but was rapidly by-passed by super-exponential growth.

In a later section, geological examples where strike-slip deformation dominates that associated with convergence are discussed. In such cases the initial stages of growth of the convective thinning instability may be represented using a Newtonian viscosity model for the lithosphere are represented.

Effect of crustal buoyancy on the mode of convective thinning

Imposed convergence in a continental environment is accompanied by other processes, the most important of which is crustal thickening. To understand the role of the buoyant crust, it is easiest first to consider its effect on Rayleigh–Taylor instability (assuming density to be determined by temperature, but omitting thermal diffusion). If there is no imposed horizontal convergence, the effect of a buoyant crustal layer on the development of a gravitational instability in the mantle lithosphere is relatively minor (Neil & Houseman 1999). Gravitational instability of the lithospheric mantle may, however, drive significant crustal thickening. For typical lithospheric parameters, the deflection of the Moho δm is on the order of 6% of the deflection of the interface δh between lithosphere and asthenosphere during the early development of a gravitational instability (Neil & Houseman 1999). Crustal and lithospheric thickening factors are defined as:

$$f_c = \frac{m + \delta m}{m}; \qquad f_L = \frac{h + \delta h}{h} \qquad (22)$$

with m and h, the original thickness of crust and lithosphere respectively, it follows that

$$\frac{(f_c - 1)}{(f_L - 1)} = \frac{h}{m}\frac{\delta m}{\delta h} \approx 0.25, \qquad (23)$$

for typical parameters $m/h \approx 0.25$ and $\delta m/\delta h = 0.06$. If the crustal layer is much stronger than the mantle, $\delta m/\delta h$ can be much less than 0.06, because the strong crust behaves like a rigid upper boundary to the unstable mantle layer. Neil & Houseman (1999) found that if the crust is no stronger than the mantle, however, $\delta m/\delta h$ does not exceed about 0.06 for plausible density distributions. From (23) it is clear that the crustal thickening that would naturally accompany lithospheric instability is much less (relative to the same degree of lithospheric thickening) than is usually assumed for uniform shortening of crust and mantle. Thus, it is not surprising that, if gravitational instability of the mantle lithosphere is a dynamically important process, the relative buoyancy of the crustal layer plays a major role.

In order to investigate the interaction of crustal thickening and lithospheric instability, Houseman et al. (2000) carried out numerical experiments on Rayleigh–Taylor instability under conditions of forced localized convergence driven by far-field plate interactions. In those experiments, the model lithosphere is represented in a 2-D vertical section by a low density crustal layer

Initial density and velocity fields

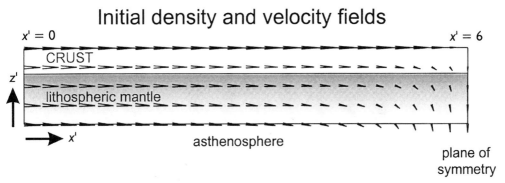

Fig. 5. Initial density and velocity fields for 2D numerical experiments shown in Figures 6 and 7. At $t = 0$ density is stratified with a low density layer ($\rho_c' = (\rho_c - \rho_a)/\Delta\rho$) representing crust, and a dense layer representing mantle lithosphere (densities are expressed relative to an asthenospheric column). Density increases linearly from $\rho' = 0$ at the base of the dense layer to $\rho' = (\rho - \rho_a)/\Delta\rho = 2$ at the Moho. Convergence is forced by means of a plate like velocity condition on the left hand end, and on the top boundary ($u' = U_0'$ from $x' = x/h = 0$ to $x' = 5$) decreasing smoothly to $u' = 0$ at $x' = 6$. The plane of symmetry at $x' = 6$ means that the solution is reflected in that boundary. The lower boundary, representing the base of the lithosphere, has a hydrostatic stress condition, representing the mechanical effect of a low-viscosity, constant-density asthenosphere.

overlying a mantle lithospheric layer, whose density is greater than that of the asthenosphere (Fig. 5). Convergence is forced by means of plate-like velocity conditions on the side boundaries and on the upper boundary, producing localized shortening in a central convergent zone. Houseman *et al.* (2000) experimented with a model consisting of crust and lithospheric mantle layers with constant density and constant viscosity coefficient. For the present study (Figs 6 and 7), some of their experiments were rerun, with density decreasing linearly through the lithospheric mantle layer, though still with constant viscosity coefficients in both crust and mantle layers. The linear decrease of density through the lithospheric mantle layer (from $\rho = 2\Delta\rho$ at the Moho to $\rho = 0$ at the base of the lithosphere) is a better representation of the distribution of negative buoyancy caused by temperature variation in the lithosphere.

In their experiments, with forced localized convergence, Houseman *et al.* (2000) found that the buoyancy of the crustal layer can produce a critical transition in the form of downwelling. If the crust is strong relative to the lithospheric mantle, or if the convergence rate is fast relative to the natural timescale for development of the gravitational instability, the instability develops as expected, with downwelling beneath the centre of the convergent zone where the light, top layer thickens most (as in Fig. 6). On the other hand, if the crust is weak, or if the convergence rate is sufficiently low, then the single central downwelling is replaced by a pair of downwellings on the margins of the convergent zone (as in

Fig. 7). After a brief initial period, the buoyancy of the crust in the centre of the convergent zone resists further thickening, and the downwelling is pushed outwards to the margins of the convergent zone, even when the boundary conditions continue to favour centralized downwelling. Most of the experiments of Houseman *et al.* (2000) used Newtonian viscosity, but a few with non-Newtonian viscosity demonstrate qualitatively similar behaviour. This similarity should not be surprising given the results of the previous section that a layer of non-Newtonian material under an imposed external stress field behaves as if viscosity were Newtonian during initial stages of growth. The change from constant density to depth-dependent density in the mantle lithosphere introduces some differences to their description, but changes are relatively minor. The main difference is that the core of the unstable mass is centered at a somewhat higher level in the lithosphere, consistent with the initial buoyancy distribution. Though it is slower to develop, the drip that eventually forms draws a somewhat greater fraction of the mantle lithosphere into the gravitational instability.

With either density distribution, the change in form of downwelling from a single centralized downwelling to multiple lateral downwellings depends most clearly on the ratio of the imposed convergence rate to the rate at which the buoyancy-driven instability would develop:

$$U_0' \approx \left(\frac{U_0}{h}\right)\left(\frac{B}{hg\Delta\rho}\right)^n \qquad (24)$$

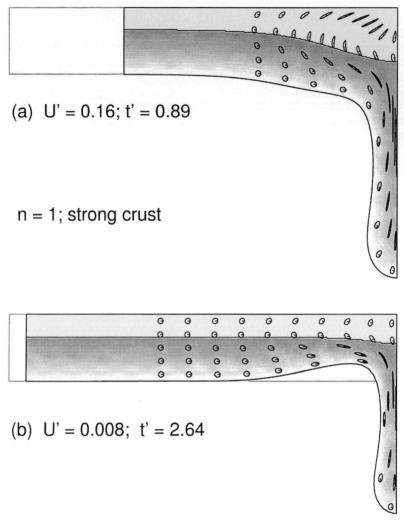

Fig. 6. Finite deformation and the growth of the gravitational instability for experiments in which $n = 1$ and both layers have constant viscosity, but the upper layer is 5 times stronger than the lower layer. Density in the lower layer is indicated by the grey shading, and is initially linear with depth as described in Figure 5. When the imposed boundary velocity is fast relative to the growth of the instability (**a**) there is no significant lithospheric thinning but considerable crustal thickening. When it is slow (**b**), crustal thickening is minor and significant lithospheric thinning occurs adjacent to the downwelling. The background rectangle represents the outline of the undeformed layer, and strain markers are inscribed as circles at $t' = 0$.

When U'_0 is large, plate boundary forces, more than the negative buoyancy of the lithosphere, drive the centralized downwelling in the mantle, and both crust and mantle layers are thickened to a similar degree in the central convergent zone. Thus, f_c/f_L in the convergent zone is of order one, much greater than is predicted by (23). If the convergence is driven quickly (or equivalently the strength of the entire lithosphere is relatively high), the gravitational instability develops rela-

tively slowly and in a central downwelling. On the other hand, if convergence is driven slowly (or strength of crust and mantle is low), the initially large values of f_c/f_L in the convergent zone imply an excess crustal buoyancy that resists flow into the central downwelling. Consequently, separate downwellings then form on either side of the initial zone of thickened crust. Subsequent crustal thickening above the down-wellings is then found to be consistent with the

n = 1; weak crust

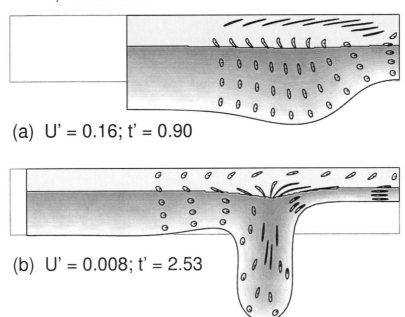

(a) U' = 0.16; t' = 0.90

(b) U' = 0.008; t' = 2.53

Fig. 7. Finite deformation and the growth of the gravitational instability for experiments in which $n=1$ and both layers have constant viscosity, but the upper layer is 5 times weaker than the lower layer. For the weak crust, downwelling first occurs on the sides of the convergent zone and migrates away from the convergent zone as it develops . When convergence is fast (**a**) thickening of the mantle lithosphere occurs over a broad region. When convergence is slow (**b**) the instability may cause extreme lithospheric thinning beneath the centre of the convergent zone. Density, boundary conditions, and finite strain indicators are as described in Figures 5 and 6. Note that here, as in Figure 6, only the left half of the symmetric solution is shown.

expression (23). The value of U'_0, at which this transition takes place, is of the order 0.02 when crust and mantle layers have equal Newtonian viscosities and constant densities (Houseman *et al.* 2000).

The second, perhaps more important, dimensionless quantity determining the form that downwelling takes beneath a zone of forced convergence is the ratio of viscosity coefficients of the lighter (crustal) to heavier (mantle lithosphere) layer $\eta'_c = \eta_c/\eta_m$ (or for $n > 1$, $B'_c = B_c/B_m$). If the crustal layer is relatively strong ($\eta'_c > 1$) the timescale for any buoyancy-driven flow within the crustal layer is correspondingly increased. If boundary-driven deformation continues, stresses within the crust are increased, but the effect of crustal buoyancy in resisting central downwelling in the underlying layer is diminished. Significant crustal thickening above the central mantle downwelling (in excess of that given by (23)) may then be supported by the strength of the crust while the mantle instability grows (e.g. Fig. 6a).

If the crustal layer is relatively weak ($\eta'_c < 1$) however, the tendency of downwellings to occur on the margins of the convergent zone is strongly enhanced. A factor of 5 reduction in η'_c, or 2 in B'_c with $n=3$ causes a dramatic change in the style of the mantle convective thinning instability (e.g. Fig. 7b). Having a weak crustal layer above the mantle layer has a physical effect similar to setting a free-slip condition on the surface of the mantle layer. The effect of a free-slip condition on the Rayleigh–Taylor instability is to enhance the rate of growth and to increase the horizontal wavelength of the unstable flow field (Neil & Houseman 1999).

A third parameter affecting this transition in the style of gravitational instability is the ratio of (positive) crustal buoyancy to the negative buoyancy of lithospheric mantle:

$$\rho'_c = \frac{\rho_c - \rho_a}{\Delta \rho} < 0 \qquad (25)$$

where, ρ_c and ρ_a are the densities of crust and asthenosphere, respectively. As discussed above, the buoyancy of the crust, expressed by the numerator of (25) resists thickening, and hence downwelling of the heavier underlying layer. When the magnitude of ρ_c' is greater, the transition to multiple downwellings occurs at a larger value of U_0' (for fixed η_c') or at a larger value of η_c' (for fixed U_0').

In some experiments when multiple downwellings develop, the lower layer (representing mantle lithosphere) beneath the centre of the convergent zone undergoes rapid extension driven by the downwelling sheets of dense material on either side. Extension of this lower layer in the central region is enhanced as the two downwellings continue to migrate outwards (e.g. Fig. 7b). If this process occurred in the Earth, the thinned lithospheric layer would be underplated by hot asthenosphere that moves up from beneath. As in these experiments, the rapid extension of the lithospheric mantle beneath the convergent zone might occur even while external boundary conditions continue to force convergence. The buoyancy of the weak crustal layer, combined with its inability to support large deviatoric stresses, means that consequent crustal thickening (or thinning) is minor, relative to what occurs in the mantle lithosphere. Crustal thickening associated with continued convergence would occur on the edges of the belt.

Discussion and geological examples

The numerical experiments of Rayleigh–Taylor instability described above and the theory underlying this process bear on at least three basic aspects of continental lithosphere: the inherent stability of ancient cratons; the magnitude of lithospheric thinning that occurs in mountain belts that have undergone considerable crustal thickening; and the distribution of that thinning. First, non-Newtonian viscosity, which characterizes most rock-forming minerals and olivine in particular, makes the rate of growth of gravitational instability dependent on the magnitude of a perturbation to lithospheric thickness. Unlike the case of Newtonian viscosity, small perturbations will grow so slowly that lateral diffusion of heat can erase them before they grow. Thus, the stability of ancient cratons may result more from the non-Newtonian viscosity of the lithosphere than from either the temperature dependence of viscosity or from chemical differences between continental lithosphere and asthenosphere.

The brief description in the preceding section has also highlighted two contrasting styles of lithospheric instability triggered by convergence in the continental environment, which may mani-

fest themselves differently in the amount and location of thinning of mantle lithosphere. The form of lithospheric downwelling is determined by the relative importance of stresses caused by buoyancy and those caused by plate convergence. If viscous stress, rather than buoyancy of crust, dominates the resistance to convergence, a narrow central downwelling develops. Thinning of mantle lithosphere is relatively modest, occurs on either side of the downwelling, and becomes significant only after the instability has grown to the stage where downwelling rates exceed imposed convergence. On the other hand, when buoyancy dominates, the central downwelling is suppressed in favour of downwellings beneath the margins of the convergent zone. If the crust is relatively weak, dramatic lithospheric thinning occurs beneath the centre of the convergent zone. This latter phenomenon exhibits some aspects in common with the process of delamination, as Bird (1978) conceived it. A large fraction of the dense, lower layer (mantle lithosphere) may be removed by the convective downwellings, and the affected region progressively increases in area outward from the initial convergent zone. In the central region the upper layer (crust) would first thicken, and then warm and rise as the high-temperature low-strength substrate (asthenosphere) replaces the unstable lower layer (mantle lithosphere). The convergence rate, which varies greatly in different tectonic environments, is important in determining which style of instability develops, but crustal and lithospheric strength are probably even more important because those parameters may vary by orders of magnitude due to variation of lithospheric thermal structure (Molnar & Tapponnier 1981; Sonder & England 1986). We may therefore expect to see the range of instability styles described in these experiments reflected in the geological histories of different regions.

Southern California

The Transverse Ranges of Southern California represent an apparently clear example of the first type of lithospheric instability. This region, which straddles the big bend of the San Andreas Fault system, is dominated by strike-slip deformation, but the obliquity of the fault zone to the relative plate motion vector implies also a large component of convergence (10 to $20 \, \text{mm a}^{-1}$) perpendicular to the local orientation of the fault zone. This convergence has continued for around 5 Ma and has produced relatively modest local thickening of the crust (Kohler & Davis 1997). Teleseismic tomography of the lithosphere and

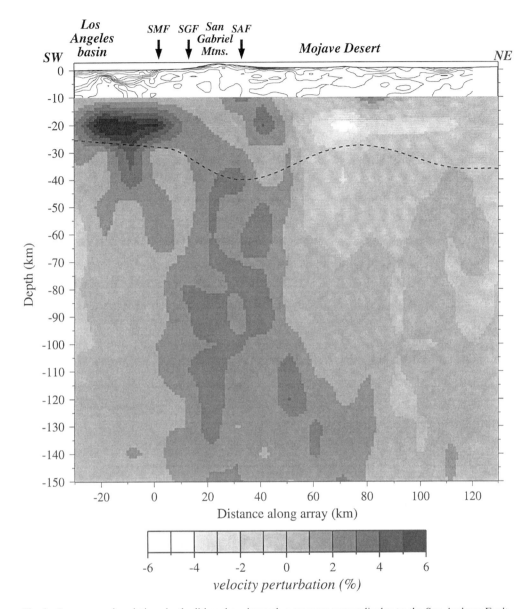

Fig. 8. *P*-wave speed variations in the lithosphere beneath a traverse perpendicular to the San Andreas Fault and the Transverse Ranges of Southern California. This image, from Kohler (1999), shows a slice through a 3-D tomographic solution derived from teleseismic arrival time residuals. Regions of positive wave speed anomaly are associated with lower temperatures and negative buoyancy. The dashed line is the Moho, from Kohler & Davis (1997).

upper mantle beneath the Transverse Ranges (Fig. 8) reveals a steeply dipping high-speed structure (Raikes 1980), which appears to extend hundreds of kilometers into the asthenosphere (Humphreys *et al.* 1984; Humphreys & Clayton 1990; Kohler 1999). This tongue of seismically fast material has been interpreted to represent a sheet of high-speed lithospheric material down-welling beneath the centre of the convergent zone. (Humphreys *et al.* 1984; Humphreys & Hager 1990; Houseman *et al.* 2000). Convective downwelling beneath the Transverse Ranges thus could be interpreted as an example of the type of gravitational instability illustrated in Figure 6.

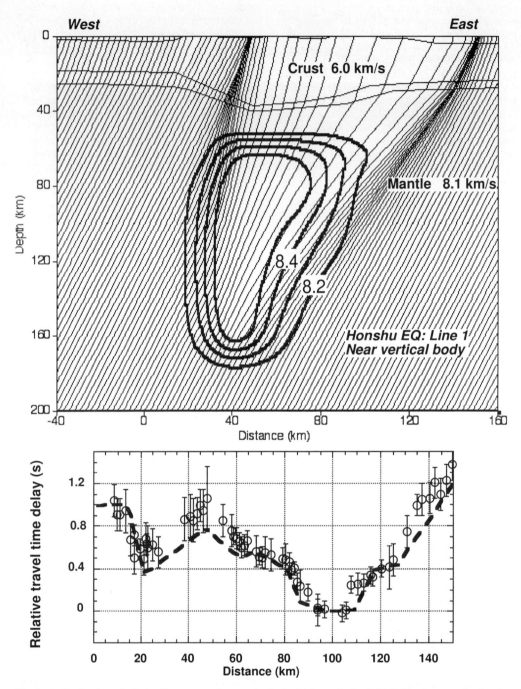

Fig. 9. A distribution of relative *P*-wave speed anomaly (in km/s) beneath the SIGHT line 1 across the South Island of New Zealand that is consistent with the observed relative arrival time delays for an event in Honshu, as computed by Stern *et al.* (2000). The dashed line in the lower diagram shows computed relative arrival time delays for the structure and ray paths shown in the upper diagram. The relatively steep dip of the anomalously fast material is required to match the observed residuals for this event.

Comparison of the geometry of the downwelling region with the results of numerical experiments implies a value of U_0' from which the effective lithospheric viscosity in the region of the San Andreas Fault system was estimated to be less than about 10^{21} Pa s (Houseman et al. 2000). This estimate of lithospheric viscosity was found to be consistent with that determined by Flesch et al. (2000), who matched the geodetically determined deformation field of the southwestern US with a model that included plate boundary stresses and internal buoyancy forces associated with crustal thickness variations. The relatively low effective viscosity of the lithosphere in this convergent zone may well be attributable to the interaction of a large component of distributed strike-slip shear associated with the San Andreas Fault system and a non-Newtonian lithospheric viscosity (Houseman et al. 2000).

Southern Alps, New Zealand

Although deformation of the South Island of New Zealand has been largely strike-slip (e.g. Sutherland 1999), convergence of c. 100 km since 6–7 Ma between the Pacific and Australian plates in this region has built the Southern Alps (Walcott 1998). The pattern of P-wave arrival times at stations in New Zealand demonstrates that a high-speed zone underlies the Southern Alps (Stern et al. 2000). Two dense linear arrays that span the South Island installed to measure the crustal structure of the Southern Alps, using explosive and air-gun sources (Stern et al. 1997; Holbrook et al. 1998; Kleffmann et al. 1998; Scherwath et al. 1998), also recorded teleseismic P-waves from three earthquakes to the northwest. These P-wave arrival times from events with a limited range of distances and azimuths cannot resolve deep structure in detail but can be used to test hypotheses. Stern et al. (2000) showed that P-wave speed distributions based on a simple model of asymmetric subduction misfit the arrival times. An almost symmetrical high-speed zone, however, c. 100 km wide and extending to a depth of almost 200 km beneath the eastern part of the Southern Alps, beneath the thickest crust, is consistent with observed arrival times (Fig. 9). Moreover, the thickness of the crustal root determined by travel times for controlled-source crustal refractions and reflections exceeds that calculated assuming Airy isostasy. Thus a dense mass within the uppermost mantle must compensate for the excess crustal thickness. A simple calculation shows that the required density anomaly is consistent both with P-wave travel times and with the

amount of lithospheric shortening and thickening calculated from the relative plate movement (Stern et al. 2000). Thus, it appears that a single downwelling zone underlies the Southern Alps, though its dimensions are not yet well constrained by observation.

The existence of central downwellings beneath two regions where large components of strike-slip deformation have occurred raises the question of whether that strike-slip shear dictates the location of downwelling. The strike-slip deformation field is orthogonal to the imposed convergence and thus the two strain-fields may be considered independent. Stern et al. (2000) and Houseman et al. (2000) both suggested that the strike-slip shear may have weakened the underlying mantle lithosphere by imposing a relatively large background strain-rate on the non-linear viscosity. Shear heating may also have been a factor (Stern et al. 2000). Although shear-induced weakening of the mantle lithosphere may to some degree control the location of downwelling, our numerical experiments show that central downwelling may occur even in the absence of localized lithospheric weakening, and indeed is likely if the crust is strong relative to the lithospheric mantle.

Alboran Sea

The Alboran Sea illustrates perhaps better than anywhere a setting where mantle lithosphere almost surely was thickened and subsequently has been drastically thinned. Late Cretaceous and early Cenozoic convergence between Europe and Africa apparently built a mountain belt whose remnants today comprise the Betic Cordillera in southern Spain and the Rif in northern Morocco (Fig. 10) (e.g. Doublas & Oyarzun 1989; Platt & Vissers 1989; Platt et al. 1998). Subsequent stretching, normal faulting, and thinning of the crust, caused the Earth's surface between the two belts to subside below sea level forming the Alboran Sea basin (Vissers et al. 1995; Platt et al. 1996; Argles et al. 1999). Seismic refraction profiles show a thin crust (15–20 km) (Hatzfeld & Ben Sari 1977; Working Group 1978) between the Betic Cordillera and Rif, where thicknesses reach nearly 40 km (Working Group 1977; Banda & Ansorge 1980; Banda 1988; Banda et al. 1993). Such a difference implies an extensional strain $\beta \approx 2$–3 (Watts et al. 1993).

Analyses of metamorphic rock obtained by drilling the floor of the Alboran Sea indicate that it had lain at depths of 30–40 km (pressures of 0.8–1.05 GPa) with temperatures of 500–600°C; subsequently, during exhumation these rocks

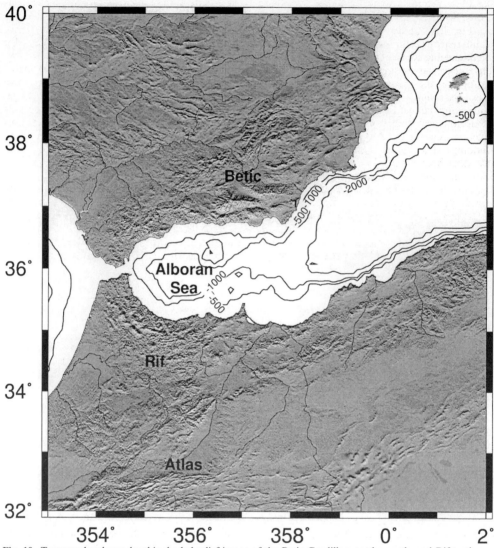

Fig. 10. Topography shown by this shaded relief image of the Betic Cordillera to the north, and Rif to the south, of the Alboran Sea Basin, indicates structural trends in which approximately north-south convergence around the edges of the basin accompanied east-west extension within the basin. Sea-floor depths greater than 1000 m in the basin indicate thin (15–20 km) extended crust.

warmed to 700°C as they reached depths of only 10–15 km (Platt *et al.* 1996; Soto & Platt 1998). Complete exhumation apparently occurred in only a few million years near 18–20 Ma. Simultaneously, thrust faulting on the margins of the belt continued (e.g. Lonergan 1993; Vissers *et al.* 1995). Turner *et al.* (1999) showed from analyses of rare earth elements and isotopic ratios ($^{87}Sr/^{86}Sr$ and $^{143}Nd/^{144}Nd$) that some magmas, particularly those emplaced before lithospheric thinning occurred, seem to be derived from asthenosphere, but those emplaced *c.* 10 Ma after the

thinning seem to have formed by melting of lithosphere. From the amount of heating implied by both the metamorphic assemblages and the igneous rock that formed in this region, Platt *et al.* (1998) concluded that virtually all of the mantle lithosphere must have been removed. The basic analyses of both Rayleigh–Taylor (Houseman & Molnar 1997; Molnar *et al.* 1998) and convective (Conrad & Molnar 1999; Conrad 2000) instability, which omit the crustal layer, cannot account for the degree of thinning of the mantle lithosphere inferred by Platt *et al.* (1998).

The allowance for a weak buoyant layer above (crust), however, not only permits such thinning (e.g. Fig. 7b), but also is qualitatively similar to the inference of thinning near the axis of the belt and thickening adjacent to it. Externally imposed convergence has continued throughout the development of the Alboran Sea, producing the thrust belts which, reflected in the present-day topography, almost encircle the basin (Fig. 10).

Seismological studies of the uppermost mantle structure beneath the Alboran Sea and its surroundings also suggest that mantle lithosphere has either been removed or greatly thinned beneath the Alboran Sea, and that cold material descends beneath the adjacent regions. First, high-frequency S_n phases are attenuated (Seber et al. 1996a), as commonly occurs where mantle lithosphere has been thinned or removed (e.g. Molnar & Oliver 1969). Moreover, unusually low P_n speeds (7.5–7.0 km/s) below the Alboran Sea (Hatzfeld & Ben Sari 1977; Working Group 1978; Banda 1988), suggest high temperatures there. Relatively early P-wave arrivals in the western part of the Rif in Morocco from teleseisms to the east-northeast imply a high-speed body, and tomographic inversion of traveltime residuals yields such a body between 35 and 150 km beneath the Rif and plunging to the north (Seber et al. 1996b; Calvert et al. 2000). Similarly, travel times recorded in southern Spain also call for a high-speed zone in the upper mantle beneath this area (Blanco & Spakman 1993; Plomerová et al. 1993). Calvert et al. (2000) and Seber et al. (1996a, b) inferred from the low-speed zone at the base of the crust and the deeper high speed body that mantle lithosphere had delaminated, in the sense of Bird (1978, 1979).

The upper mantle structure beneath the Alboran Sea is remarkable in that the lithospheric mantle immediately beneath the crust appears to have been displaced downwards by hot asthenosphere, but the distribution of seismicity below 50 km (Hatzfeld & Frogneux 1981; Seber et al. 1996a) and anomalously high P-wave speed (Seber et al. 1996b) imply that cold mantle lithosphere remains at relatively shallow depths in the upper mantle. At depths below 200 km the anomalously fast region is poorly defined but it appears to be broad and is probably centrally positioned beneath the Alboran Sea Basin (e.g. Calvert et al. 2000; Wortel & Spakman 1992). In terms of the simplified numerical experiments described above, the evolving marginal downwellings that result from Rayleigh–Taylor instability occurring beneath a weak crustal layer provide some insight into how extension and subsidence may have begun in

the middle of the orogen and migrated outwards to the present-day compressional margins (Docherty & Banda 1995). Although the pair of downwellings does not match the present-day central position of the deep velocity anomaly, we should bear in mind that the simplified model assumed a passive upper mantle. An active mantle circulation associated with the large-scale plate movements could well draw separate small-scale downwellings into a central downflow at greater depths.

Tyrrhenian Sea

Extension in the Tyrrhenian Sea has also been associated with post-orogenic collapse in a zone flanked by compressional belts and makes an interesting comparison with the Alboran Sea, though retreating subduction also plays a role here (Faccenna et al. 1996). The evolution of the Tyrrhenian Sea and its present upper mantle structure have some elements in common with the Alboran Sea, though the low-speed zone beneath the thinned lithosphere appears more extensive in depth and horizontal range (Wortel & Spakman 1992), while the high-speed zone associated with the subducted slab is continuous and associated with seismicity to depths of around 500 km. Compared with the Alboran Sea, it is perhaps easier to interpret the evolution of the Tyrrhenian in terms of marginal downwelling following a convective thinning event that began as a Rayleigh–Taylor instability beneath the eastern margin of the Apeninnic orogenic belt (e.g. van der Meulen et al. 1998).

Tien Shan

Nowhere does intracontinental mountain building occur more rapidly or on a larger scale than in the Tien Shan. Some 1800 km long and up to 300 km wide at its western end, it has absorbed a fraction of India's convergence with Eurasia. More importantly, the rate of shortening at the western end of $c. 20$ mm a^{-1} (Abdrakhmatov et al. 1996) accommodates nearly half of the current India–Eurasia convergence rate ($c. 44$ mm a^{-1}) at that longitude (DeMets et al. 1994). Both earthquakes (e.g. Tapponnier & Molnar 1979; Ghose et al. 1998) and active faulting (e.g. Makarov 1977; Sadybakasov 1990) also indicate approximately N–S crustal shortening throughout the belt, not just on its margins. Thus, deformation in the Tien Shan differs markedly from narrower belts like the Transverse Ranges or the Southern Alps of New Zealand, and the style of mantle deformation beneath may also differ.

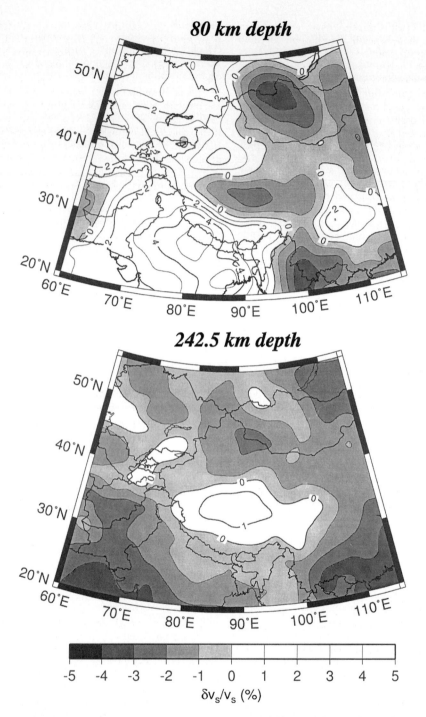

Fig. 11. Variations in *S*-wave speed beneath central Asia at depths of (**a**) 80 km and (**b**) 242.5 km. This image shows two horizontal slices through the upper mantle obtained from the inversion of fundamental mode Love and Rayleigh waves, as described by Villaseñor *et al.* (2000). As with *P*-wave speed anomalies, an increase in wave speed is associated with a decrease in temperature, other factors being constant.

Although crust is thicker beneath the western Tien Shan than beneath the neighboring lowlands, it appears that the region is underlain by a relatively hot upper mantle. Gravity anomalies over the western Tien Shan confirm the existence of isostatic equilibrium (e.g. Burov et al. 1990), and receiver functions require relatively thick crust (e.g. Kosarev et al. 1993; Bump & Sheehan 1998). Roecker et al. (1993), however, showed that P-wave speeds in the underlying upper mantle are quite low, $<8.0\,km/s$, and therefore suggest that the mantle is unusually hot. They recognized two possibilities: that the area was hot, and therefore presumably weak, before the present mountain belt formed, or that convective instability had removed mantle lithosphere. In support of the latter, Chen et al. (1997) showed that the 410 km discontinuity beneath the Tien Shan is elevated and hence partly compensates for body-wave travel time delays due to the low-speed uppermost mantle. They interpreted the elevated boundary as an indication that detached mantle lithosphere has sunk to a depth of about 400 km. In support of the former, Sobel & Arnaud (2000) dated isolated outcrops of basalt and compiled other data from the western Tien Shan that show basaltic volcanism to have occurred for many tens of millions of years before the current mountain belt formed. They inferred that the area had been weakened by such volcanism. At present the seismic wave-speed structure of the upper mantle beneath the belt is too poorly resolved to indicate the shape or distribution of possible downwellings, but the absence of a lithospheric root after significant large-scale convergence suggests that mantle lithosphere has been removed by the convective thinning process.

Tibetan Plateau

Finally, the Tibetan Plateau is probably the extreme example (at least in scale) of a convergent orogen in which large-scale convective replacement of the upper mantle is thought to have occurred (Houseman et al. 1981). Seismological measurements (summarized, for instance, by Molnar 1988) indicate a thick low-speed zone in the upper mantle beneath northern Tibet in particular (e.g. Fig. 11a). The low-speed zone is associated with high attenuation, recent volcanism, and high elevation, all of which indicate a hot upper mantle. England & Houseman (1989) suggested that the onset of E–W extension in Tibet in the Miocene was caused by the convective thinning of the lithosphere beneath Tibet. There is broad concurrence of evidence from the

present-day geophysical and geological structure, and from the climatic and stratigraphic record, that this event occurred at around 8 Ma (e.g. Harrison et al. 1992; Molnar et al. 1993). The horizontal length scale of the region affected by this deformation is so much greater than the least stable wavelength that multiple downwellings are likely to have occurred. Seismic images of the upper mantle beneath Tibet are probably not yet sufficiently well resolved to provide clear constraints on the lithospheric instability mechanism. The upper mantle structure implied by P_n travel-time tomography (McNamara et al. 1997) and by S-wave speed anomalies determined from the inversion of fundamental mode Love and Rayleigh waves (Villaseñor et al. 2000; Fig. 10a) suggests, however, that extreme thinning of the mantle lithosphere has probably occurred beneath Northern Tibet. The surface wave inversion solutions at greater depth (Fig. 11b) show relatively high S-wave speed beneath the plateau, as if the cold material has been stripped from the lithosphere but remains yet at relatively shallow depths.

Basaltic volcanism erupted since c. 10 Ma, especially in northern Tibet, supports the image of a thinner lithosphere in northern Tibet than in the southern part (e.g. Arnaud et al. 1992; Turner et al. 1993, 1996). This basalt is characterized by isotopic ratios of $^{87}Sr/^{86}Sr$ that suggest long durations of radioactive decay. High percentages of incompatible rare-earth elements, including potassium, also suggest slow source enrichment by small melt fractions. The implied longevity of the source of this basalt suggests melting of lithosphere rather than asthenosphere, and hence that complete removal of mantle lithosphere has not occurred.

Conclusions

This review has attempted to emphasize firstly the understanding of the mechanical process involved when convective thinning occurs in an unstably stratified lithospheric layer whose deformation is governed by a non-Newtonian constitutive law. Such a layer may be unstable, but the instability requires a finite amplitude perturbation to get it started, most likely to occur in the form of significant localized horizontal shortening driven by plate convergence. If density is controlled by temperature, small deflections of the isodensity surfaces decay by thermal diffusion faster than they grow by gravity driven flow unless the deflections are externally forced. If a dry olivine constitutive law is appropriate for the continental lithosphere, it is likely that the externally forced horizontal shortening must be on the order of 10% or greater and must occur quickly relative to the thermal time constant of the lithosphere (say within 10 Ma). Hence, the apparent long-term stability of continental lithospheric mantle (Jordan 1975) is quite consistent with its being due to non-Newtonian viscosity, even supposing that there is no intrinsic compositional difference between lithospheric mantle and asthenosphere. Our calculations suggest that a wet olivine lithosphere could be more easily destabilized by orogenic shortening. Thus, the stabilization of the sub-continental lithospheric mantle may ultimately depend on loss of volatiles.

Second, the role of crustal buoyancy in governing the form of gravitational instability in the mantle has been reviewed. The influence of crustal buoyancy is relative to the magnitude of stress caused by an imposed convergence field. If the crustal buoyancy is relatively small (or imposed convergence is fast, or the crust is strong) downwelling develops beneath the centre of the convergent zone where significant crustal thickening takes place. On the other hand, if the crust is relatively buoyant (or imposed convergence is slow, or the crust is weak compared to the mantle), it resists thickening, to the point that downwelling commences on the margins of the convergent zone. The lithospheric mantle beneath the orogenic belt may thin by a large factor as the marginal downwellings develop. Rapid extensional thinning of the mantle lithosphere beneath the orogen is associated with asymmetric off-axis downwelling, which migrates outward from the orogenic center in a way that is similar to a delaminating slab or a retreating subduction zone. An important conclusion from this section is that the mechanism of gravitational instability may take different forms in different environments because of regional differences in physical or thermal parameters that govern the rheological strength and the density of the lithosphere.

Third, several geological examples where lithospheric shortening in the continental environment may have triggered instability have been briefly summarized. These brief summaries illustrate the range of convective thinning styles that appear to be manifest in the modern continental environment. Mantle downwellings beneath the Transverse Ranges of California and the South Island of New Zealand appear to illustrate the central downwelling instability that develops beneath a relatively narrow orogen when the crust is strong compared to the mantle. On a somewhat larger scale, the Alboran and Tyrrhenian Seas appear to illustrate a quite different

style where convergent orogeny has been succeeded by dramatic lithospheric thinning and heating, even while convergence has continued on the margins of the orogenic zone. The Mediterranean examples bring to mind the asymmetric migrating downwellings and extreme lithospheric thinning observed in the Rayleigh–Taylor experiments with a weak buoyant crust, though clearly the simple models so far discussed do not explain all the important physical aspects of the observed upper mantle structure in these regions. The evidence from the western Tien Shan and from the Tibetan Plateau also implies that large-scale lithospheric instability has occurred in those regions, and possibly has occurred via multiple downwellings on the margins of the orogeny.

The above list of examples is hardly exhaustive, but included are some of the better studied examples of recent orogenic events, that probably involve gravitational instability of the lithosphere. Increasingly, new data provided by teleseismic P-wave tomography are proving important in the interpretation of processes and structures in the upper mantle. Certainly the tomographic technique provides only an image of the current structure, too often poorly resolved. Yet some structures revealed by these images can be explained only in terms of a mantle lithosphere that interacts tectonically with the underlying mantle. Such images will be very persuasive if they prove to be reliable.

Much of the work reviewed here was carried out either by or in collaboration with C. P Conrad, E. A. Roberts née Neil, and M. D. Kohler. We thank M. Ritzwoller and A. Villaseñor for providing Fig. 11, T. Stern for Fig. 9 and M. Kohler for Figure 8. We also thank L. Evans for continuing support and development of the *basil/sybil* finite deformation package and C. Beaumont, R. Pysklywec, and K. Stüwe for thorough, constructive reviews that improved the manuscript. The GMT package (Wessel & Smith 1995) was used to prepare some figures. This work was supported in part by the National Science Foundation under grants EAR-9406026 and EAR-9725648.

References

ABDRAKHMATOV, K. YE., ALDAZHANOV, S. A., HAGER, B. H. *ET AL.* 1999. Relatively recent construction of the Tien Shan inferred from GPS measurements of present-day crustal deformation rates. *Nature*, **384**, 450–453.

ARGLES, T. W., PLATT, J. P. & WATERS, D. J. 1999. Attenuation and excision of a crustal section during extensional exhumation: the Carratraca Massif, Betic Cordillera, southern Spain.

Journal of the Geological Society, London, **156**, 149–162.

ARNAUD, N. O., VIDAL, PH., TAPPONNIER, P., MATTE, PH. & DENG, W. M. 1992. The high K₂O volcanism of northwestern Tibet: geochemistry and tectonic implications. *Earth and Planetary Science Letters*, **111**, 351–367.

BANDA, E. 1988. Crustal parameters in the Iberian peninsula. *Physics of the Earth and Planetary Interiors*, **51**, 222–225.

BANDA, E. & ANSORGE, J. 1980. Crustal structure under the central and eastern part of the Betic Cordillera. *Geophysical Journal of the Royal Astronomical Society*, **63**, 515–532.

BANDA, E., GALLART, J., GARCÍA-DUEÑAS, V., DAÑOBEITIA, J. J. & MAKRIS, J. 1993. Lateral variation of the crust in the Iberian peninsula: new evidence from the Betic Cordillera. *Tectonophysics*, **221**, 53–66.

BEAUMONT, C., FULLSACK, P. & HAMILTON, J. 1994. Styles of crustal deformation caused by subduction of the underlying lithosphere. *Tectonophysics*, **232**, 119–132.

BIRD, P. 1978. Initiation of intracontinental subduction in the Himalaya. *Journal of Geophysical Research*, **83**, 4975–4987.

BIRD, P. 1979. Continental delamination and the Colorado Plateau. *Journal of Geophysical Research*, **84**, 7561–7571.

BLANCO, M. J. & SPAKMAN, W. 1993. The P-wave velocity structure of the mantle below the Iberian peninsula: evidence for subducted lithosphere below southern Spain. *Tectonophysics*, **221**, 13–34.

BUMP, H. A. & SHEEHAN, A. F. 1998. Crustal thickness variations across the northern Tien Shan from teleseismic receiver functions. *Geophysical Research Letters*, **25**, 1055–1058.

BUROV, E. V., KOGAN, M. G., LYON-CAEN, H. & MOLNAR, P. 1990. Gravity anomalies, the deep structure, and dynamic processes beneath the Tien Shan. *Earth and Planetary Sciences Letters*, **96**, 367–383.

CALVERT, A., SANDVOL, E., SEBER, D., BARAZANGI, M., ROECKER, S., MOURABIT, T., VIDAL, F., ALGUACIL, G. & JABOUR, N. 2000. Geodynamic evolution of the lithosphere and upper mantle beneath the Alboran region of the western Mediterranean: Constraints from travel time tomography. *Journal of Geophysical Research*, **105**, 10871–10898.

CHANDRASEKHAR, S. 1961. *Hydrodynamic and Hydromagnetic Stability*. Oxford University Press, Oxford.

CHEN, Y. H., ROECKER, S. W. & KOSAREV, G. L. 1997. Elevation of the 410 km discontinuity beneath the central Tien Shan: Evidence for a detached lithospheric root. *Geophysical Research Letters*, **24**, 1531–1534.

CONRAD, C. P. 2000. Convective instability of thickening mantle lithosphere. *Geophysical Journal International*, **143**, 52–70.

CONRAD, C. P. & MOLNAR, P. 1997. The growth of Rayleigh–Taylor-type instabilities in the lithosphere for various rheological and density structures. *Geophysical Journal International*, **129**, 95–112.

CONRAD, C. P. & MOLNAR, P. 1999. Convective instability of a boundary layer with temperature- and strain-rate-dependent viscosity in terms of 'available buoyancy.' *Geophysical Journal International*, **139**, 51–68.

DOCHERTY, C. & BANDA, E. 1995. Evidence for the eastward migration of the Alboran Sea based on regional subsidence analysis: a case for basin formation by delamination of the subcrustal lithosphere? *Tectonics*, **14**, 804–818.

DEMETS, C., GORDON, R. G., ARGUS, D. F. & STEIN, S. 1994. Effects of a recent revision to geomagnetic reversal timescale on estimates of current plate motions. *Geophysical Research Letters*, **21**, 2191–2194.

DOUBLAS, M. & OYARZUN, R. 1989. Neogene extensional collapse in the western Mediterranean (Betic-Rif Alpine orogenic belt): Implications for the genesis of the Gibraltar arc and magmatic activity. *Geology*, **17**, 430–433.

ENGLAND, P. C. & HOUSEMAN, G. A. 1989. Extension during continental convergence with application to the Tibetan Plateau. *Journal of Geophysical Research*, **94**, 17561–17579.

FACCENNA, C., DAVY, P., BRUN, J.-P., FUNICIELLO, R., GIARDINI, D., MATTEI, M. & NALPAS, T. 1996. The dynamics of back-arc extension: an experimental approach to the opening of the Tyrrhenin Sea. *Geophysical Journal International*, **126**, 781–795.

FLESCH, L. M., HOLT, W. E., HAINES, A. J. & SHEN-TU, B. 2000. Dynamics of the Pacific-North American Plate boundary in the western United States. *Science*, **287**, 834–836.

GHOSE, S., HAMBURGER, M. W. & AMMON, C. J. 1998. Source parameters of moderate-size earthquakes in the Tien Shan, central Asia, from regional moment tensor inversion. *Geophysical Research Letters*, **25**, 3181–3184.

GRIFFIN, W. L., O'REILLY, S. Y., RYAN, C. G., GAUL, O. & IONOV, D. A. 1998. Secular variation in the composition of subcontinental lithospheric mantle: geophysical and geodynamic implications. *In*: BRAUN, J., DOOLEY, J. J., GOLEBY, B., VAN DER HILST, R. & KLOOTWIJK, C. (eds) *Structure and Evolution of the Australian Continent*. Geodynamics Series, **26**, 1–26.

HAND, M. & SANDIFORD, M. 1999. Intraplate deformation in central Australia, the link between subsidence and fault reactivation. *Tectonophysics*, **305**, 121–140.

HARRISON, T. M., COPELAND, P., KIDD, W. S. F. & YIN, A. Raising Tibet. *Science*, **255**, 1663–1670.

HATZFELD, D. & BEN SARI, D. 1977. Grands profils sismiques dans la region de l'arc de Gibraltar. *Bulletin de la Societe Géologique de France*, 7, 749–756.

HATZFELD, D. & FROGNEUX, M. 1981. Evidence of intermediate depth earthquakes around the Gibraltar area. *Nature*, **292**, 443–445.

HENDRIX, M. S., GRAHAM, S. A., CARROLL, A. R., SOBEL, E. R., MCKNIGHT, C. L., SCHULEIN, B. J.

& WANG, Z. X. 1992. Sedimentary record and climatic implications of recurrent deformation in the Tian Shan: Evidence from Mesozoic strata of the north Tarim, south Junggar and Turpan basins, northwest China. *Geological Society America Bulletin*, **104**, 53–79.

HOLBROOK, W. S., OKAYA, D., STERN, T., DAVEY, F., HENRYS, S. & VAN AVENDONK, H. 1998. Deep seismic profiles across the Pacific–Australian plate boundary, South Island, New Zealand. *EOS Transactions of the American Geophysical Union*, **79**(45) (suppl.), F901.

HOUSEMAN, G., MCKENZIE, D. P. & MOLNAR, P. 1981. Convective instability of a thickened boundary layer and its relevance for the thermal evolution of continental convergent belts, *Journal of Geophysical Research*, **86**, 6115–6132.

HOUSEMAN, G. & MOLNAR, P. 1997. Gravitational (Rayleigh–Taylor) instability of a layer with non-linear viscosity and convective thinning of continental lithosphere. *Geophysical Journal International*, **128**, 125–150.

HOUSEMAN, G., NEIL, E. A. & KOHLER, M. D. 2000. Lithospheric instability beneath the Transverse Ranges of California. *Journal of Geophysical Research*, **105**, 16237–16250.

HOWARD, L. N. 1966. Convection at high Rayleigh numbers. *In*: GÖRTLER, H. (ed.) *Proceedings of the 11th International Congress of Applied Mathematics*. Springer, New York, 1109–1115.

HUMPHREYS, E. D. & CLAYTON, R. W. 1990. Tomographic image of the southern California mantle. *Journal of Geophysical Research*, **95**, 19725–19746.

HUMPHREYS, E., CLAYTON, R. W. & HAGER, B. H. 1984. A tomographic image of mantle structure beneath southern California. *Geophysical Research Letters*, **11**, 625–627.

HUMPHREYS, E. D. & HAGER, B. H. 1990. A kinematic model for the late Cenozoic development of southern California crust and upper mantle. *Journal of Geophysical Research*, **95**, 19747–19762.

JORDAN, T. 1975. The continental tectosphere. *Reviews of Geophysics and Space Physics*, **13**, 1–12.

KARATO, S.-I., PATERSON, M. S. & FITZGERALD, J. D. 1986. Rheology of synthetic olivine aggregates: influence of grain size and water. *Journal of Geophysical Research*, **91**, 8151–8176.

KLEFFMANN, S., DAVEY, F., MELHUISH, A., OKAYA, D. & STERN, T. 1998. Crustal structure in the central South Island from the Lake Pukaki seismic experiment. *New Zealand Journal of Geology and Geophysics*, **41**, 39–50.

KOHLER, M. D. 1999. Coupled crust-mantle deformation beneath the San Gabriel Mountains in the Southern California Transverse Ranges. *Journal of Geophysical Research*, **104**, 15025–15041.

KOHLER, M. D. & DAVIS, P. M. 1997. Crustal thickness variations in southern California from Los Angeles Region Seismic Experiment passive phase teleseismic travel times. *Bulletin of the Seismological Society of America*, **87**, 1330–1344.

KOSAREV, G. L., PETERSEN, N. V., VINNIK, L. P. & ROECKER, S. W. 1993. Receiver functions for the Tien Shan analog broadband network: Contrasts

in the evolution of the structures across the Talasso–Ferghana fault. *Journal of Geophysical Research*, **98**, 4437–4448.

LONERGAN, L. 1993. Timing and kinematics of deformation in the Malaguide complex, internal zone of the Betic Cordillera, southeast Spain. *Tectonics*, **12**, 460–476.

MAKAROV, V. I. 1977. New tectonic structures of the Central Tien Shan (in Russian). *In*: *Transactions Order of the Red Banner Geology Institute, Volume 307*. Akademia Nauk, Moscow.

MCNAMARA, D. E., WALTER, W. R., OWENS, T. J. & AMMON, C. J. 1997. Upper mantle velocity structure beneath the Tibetan Plateau from Pn travel-time tomography. *Journal of Geophysical Research*, **102**, 493–505.

MCKENZIE, D. P. 1978. Some remarks on the development of sedimentary basins, *Earth and Planetary Science Letters*, **40**, 25–32.

MOLNAR, P. 1988. A review of geophysical constraints on the deep structure of the Tibetan Plateau, the Himalaya and the Karakoram, and their tectonic implications. *Philosophical Transactions of the Royal Society of London*, **A326**, 33–88.

MOLNAR, P., ENGLAND, P. C. & MARTINOD, J. 1993. Mantle dynamics, the uplift of the Tibetan Plateau, and the Indian monsoon. *Reviews of Geophysics*, **31**, 357–396.

MOLNAR, P., HOUSEMAN, G. & CONRAD, C. 1998. Rayleigh–Taylor instability and convective thinning of mechanically thickened lithosphere: effects of non-linear viscosity decreasing exponentially with depth and of horizontal shortening of the layer. *Geophysical Journal International*, **133**, 568–584.

MOLNAR, P. & OLIVER, J. 1969. Lateral variations of attenuation in the upper mantle and discontinuities in the lithosphere. *Journal of Geophysical Research*, **74**, 2648–2682.

MOLNAR, P. & TAPPONNIER, P. 1975. Cenozoic tectonics of Asia: Effects of a continental collision, *Science*, **189**, 419–426.

MOLNAR, P. & TAPPONNIER, P. 1981. A possible dependence of tectonic strength on the age of the crust in Asia. *Earth and Planetary Science Letters*, **52**, 107–114.

NEIL, E. A. & HOUSEMAN, G. A. 1999. Rayleigh–Taylor instability of the upper mantle and its role in intraplate orogeny. *Geophysical Journal International*, **138**, 89–107.

PARSONS, B. & MCKENZIE, D. 1978. Mantle convection and the thermal structure of the plates. *Journal of Geophysical Research*, **83**, 4485–4496.

PLATT, J. P. & ENGLAND, P. C. 1994. Convective removal of lithosphere beneath mountain belts: Thermal and mechanical consequences. *American Journal of Science*, **294**, 307–336.

PLATT, J. P. & VISSERS, R. L. M. 1989. Extensional collapse of thickened continental lithosphere: A working hypothesis for the Alboran Sea and Gibraltar Arc. *Geology*, **17**, 540–543.

PLATT, J. P., SOTO, J.-I., COMAS, M. C. & LEG 161 SHIPBOARD SCIENTISTS. 1996. Decompression and high-temperature–low-pressure metamorphism in the exhumed floor of an extensional basin, Alboran Sea, western Mediterranean. *Geology*, **24**, 447–450.

PLATT, J. P., SOTO, J.-I., WHITEHOUSE, M. J., HURFORD, A. J. & KELLY, S. P. 1998. Thermal evolution, rate of exhumation and tectonic significance of metamorphic rocks from the floor of the Alboran extensional basin, western Mediterranean. *Tectonics*, **17**, 671–689.

PLOMEROVÁ, J., PAYO, G. & BABUSHKA, V. 1993. Teleseismic P-residual study in the Iberian peninsula. *Tectonophysics*, **221**, 1–12.

PYSKLYWEC, R. N., BEAUMONT, C. & FULLSACK, P. 2000. Modeling the behaviour of the continental mantle lithosphere during plate convergence. *Geology*, **28**, 655–658.

RAIKES, S. A. 1980. Regional variations in upper mantle structure beneath southern California. *Geophysical Journal of the Royal Astronomical Society*, **63**, 187–216.

RICHARDSON, S. H., GURNEY, J. J., ERLANK, A. J. & HARRIS, J. W. 1984. Origin of diamonds in old enriched mantle. *Nature*, **310**, 198–202.

ROECKER, S. W., SABITOVA, T. M., VINNIK, L. P., BURMAKOV, Y. A., GOLVANOV, M. I., MAMATKANOVA, R. & MUNIROVA, L. 1993. Three-dimensional elastic wave velocity structure of the western and central Tien Shan. *Journal of Geophysical Research*, **98**, 15779–15795.

SADYBAKASOV, I. 1990. *Neotectonics of High Asia* (in Russian). Nauka, Moscow.

SCHERWATH, M., STERN, T., OKAYA, D., DAVIES, R., KLEFFMANN, S., DAVEY, F. & SIGHT TEAM. 1998. Crustal structure from seismic data in the vicinity of the Alpine fault, New Zealand, Results from SIGHT Line II. *EOS Transactions of the American Geophysical Union*, **79**(45) (suppl.), F901.

SEBER, D., BARAZANGI, M., IBENBRAHIM, A. & DEMNATI, A. 1996a. Geophysical evidence for lithospheric delamination beneath the Alboran Sea and Rif-Betic mountains. *Nature*, **379**, 785–790.

SEBER, D., BARAZANGI, M., TADILI, B. A., RAMDANI, M., IBENBRAHIM, A. & BEN SARI, D. 1996b. Three-dimensional upper mantle structure beneath the intraplate Atlas and interplate Rif mountains of Morocco. *Journal of Geophysical Research*, **101**, 3125–3138.

SLEEP, N. H. 1971. Thermal effects of the formation of Atlantic continental margin by continental break up. *Geophysical Journal of the Royal Astronomical Society*, **24**, 325–350.

SOBEL, E. R. & ARNAUD, N. 2000. Cretaceous–Paleogene basaltic rocks of the Tuyon basin, NW China and the Kyrgyz Tian Shan: the trace of a small plume. *Lithos*, **50**, 191–215.

SONDER, L. J. & ENGLAND, P. 1986. Vertical averages of rheology of the continental lithosphere: relation to thin sheet parameters. *Earth and Planetary Science Letters*, **77**, 81–90.

SOTO, J.-L. & PLATT, J. P. 1998. Petrological and structural evolution of high-grade metamorphic rocks from the floor of the Alboran sea Basin, western Mediterranean. *Journal of Petrology*, **40**, 21–60.

STERN, T. A., WANNAMAKER, P. E., EBERHART-PHILLIPS, D., OKAYA, D., DAVEY, F. J. & THE SIGHT WORKING GROUP. 1997. Mountain building and active deformation studies in New Zealand. *EOS Transactions of the American Geophysical Union*, **78**(32), **329**, 335–336.

STERN, T., MOLNAR, P., OKAYA, D. & EBERHART-PHILIPS, D. 2000. Teleseismic P-wave delays and modes of shortening the mantle lithosphere beneath South Island, New Zealand. *Journal of Geophysical Research*, **105**, 21615–21631.

STÜWE, K. & SANDIFORD, M. 1995. Mantle-lithospheric deformation and crustal metamorphism with some speculations on the thermal and mechanical significance of the Tauern event, Eastern Alps. *Tectonophysics*, **242**, 115–132.

SUTHERLAND, R. 1999. Cenozoic bending of New Zealand basement terranes and Alpine Fault displacement: a brief review. *New Zealand Journal of Geology & Geophysics*, **42**, 295–302.

TAPPONNIER, P. & MOLNAR, P. 1979. Active faulting and Cenozoic tectonics of the Tien Shan, Mongolia and Baykal Regions. *Journal of Geophysical Research*, **84**, 3425–3459.

TURNER, S., HAWKESWORTH, C., LIU, J., ROGERS, N., KELLEY, S. & VAN CALSTEREN, P. 1993. Timing of Tibetan uplift constrained by analysis of volcanic rocks. *Nature*, **364**, 50–54.

TURNER, S. P., ARNAUD, N., LIU, N., ROGERS, N., HAWKESWORTH, C., HARRIS, N., KELLEY, S. P., VAN CALSTEREN, P. & DENG, W. 1996. Post-collision, shoshonitic volcanism on the Tibetan Plateau: Implications for convective thinning of the lithosphere and the source of ocean island basalts. *Journal of Petrology*, **37**, 45–71.

TURNER, S. P., PLATT, J. P., GEORGE, R. M. M., KELLEY, S. P., PEARSON, D. G. & NOWELL, G. M. 1999. Magmatism associated with orogenic collapse of the Betic-Alboran domain, SE Spain. *Journal of Petrology*, **40**, 1011–1036.

VAN DER MEULEN, M. J., MEULENKAMP, J. E. & WORTEL, M. J. R. 1998. Lateral shift of Apenninic foredeep depocentres reflecting detachment of subducted lithosphere. *Earth and Planetary Science Letters*, **154**, 203–219.

VILLASEÑOR, A., RITZWOLLER, M. H., LEVSHIN, A. L., BARMIN, M. P., SPAKMAN, W., TRAMPERT, J. & ENGDAHL, E. R. 2001. Shear velocity structure of central Eurasia from inversion of surface wave velocities. *Physics of Earth and Planetary Interiors*, **123**, 169–184.

VISSERS, R. L. M., PLATT, J. P. & VAN DER WAL, D. 1995. Late orogenic extension of the Betic Cordillera and the Alboran domain: A lithospheric view. *Tectonics*, **14**, 786–803.

VON BLANCKENBURG, F. & DAVIES, J. H. 1995. Slab breakoff: a model for syncollisional magmatism and tectonics in the Alps. *Tectonics*, **14**, 120–131.

WALCOTT, R. I. 1998. Modes of oblique compression: Late Cenozoic tectonics of the South Island of New Zealand. *Reviews of Geophysics*, **36**, 1–26.

WATTS, A. B., PLATT, J. P. & BUHL, P. 1993. Tectonic evolution of the Alboran Sea basin. *Basin Research*, **5**, 153–177.

WESSEL, P. & SMITH, W. H. F. 1995. New version of the Generic Mapping Tools released. *EOS Transactions of the American Geophysical Union*, **76**, 329.

WILLETT, S. D. & BEAUMONT, C. 1994. Subduction of Asian lithospheric mantle beneath Tibet inferred from models of continental collision. *Nature*, **369**, 642–645.

WORKING GROUP FOR DEEP SEISMIC SOUNDING IN SPAIN (1974–1975). 1977. Deep seismic soundings in southern Spain. *Pure and Applied Geophysics*, **115**, 721–735.

WORKING GROUP FOR DEEP SEISMIC SOUNDING IN THE ALBORAN SEA 1974. 1978. Crustal seismic profiles in the Alboran Sea – Preliminary results. *Pure Applied Geophysics*, **116**, 167–180.

WORTEL, M. J. R. & SPAKMAN, W. 1992. Structure and dynamics of subducted lithosphere in the Mediterranean region. *Proceedings of Koninklijke Nederlandse Akademie van Wetenschappen*, **95**, 325–347.

YOSHIOKA, S. & WORTEL, M. J. R. 1995. Three-dimensional numerical modeling of detachment of subducted lithosphere. *Journal of Geophysical Research*, **100**, 20 223–20 244.

The role of deep basement during continent–continent collision: a review

PAUL D. RYAN

Department of Geology, National University of Ireland, Galway, Ireland
(e-mail: ryan@alisanos.nuigalway.ie)

Abstract: Structural, geophysical and metamorphic studies show that collisional orogeny thickens the crust by a factor of two or more. A large volume of continental material at the base of the orogen is, therefore, subject to eclogite facies conditions. Phase equilibration results in a loss of buoyancy and thermodynamic heating of this crustal root. This dense crustal material may be partially subducted, as in the Alps or the Himalayas, and lost to the system. Alternatively, it may rest isostatically below the Moho until it is partially exhumed during orogenic collapse, as in the Scandinavian Caledonides or the Tonbai-Dabie Mountains. Remnant orogenic roots may exist as seismically reflective mantle and provide a locus for subsequent Wilson Cycle rifting. The rate at which these phase transformations take place may have a profound buffering effect on the amount and duration of orogenic contraction. Isostatically compensated transient 2-dimensional finite element thermal models are presented, which seek to place some limits on these processes. It is interesting to speculate whether more is learnt about the process of orogeny from a single exhumed eclogitic boudin or from mapping nappe complexes.

The topography of mountain belts developed during continent–continent collision is supported by a deep crustal root. Deep seismic reflection and refraction profiles across Cenozoic (e.g. Hirn *et al.* 1984*a,b*; Valasek *et al.* 1991; Zhao *et al.* 1993) and ancient orogens (e.g. Mathews & Hirn 1984; Hynes & Snyder 1995) have demonstrated that throughout the Phanerozoic these roots have extended to depths of 70 km or more. Exhumed ultra-high pressure (UHP) metamorphic terranes developed within such orogens (Smith 1984; Chopin 1987; Wang *et al.* 1989) contain coesite, implying that continental crust has been buried to even greater depths, exceeding 90 km. This contribution will argue that the phase assemblages of these roots, particularly the portion below 40 km, has a profound effect upon orogenic evolution. The transformation of lithologies within this root to the eclogite facies during convergence reduces buoyancy and buffers topography. Such phase changes may also exert controls on the subduction and delamination of continental crust and affect the P–T–t path of the orogeny. Crustal roots of mantle density occurring beneath the seismic Moho may play a role in cyclic re-opening of sutured oceans. Retrogression of such large volumes of rock result in a gain in buoyancy that contributes to the collapse process. Dewey *et al.* (1993) calculated the magnitude of the buoyancy effects during both burial and

exhumation using a one dimensional airy isostatic model. Ryan & Dewey (1997) have modelled the long-term thermal weakening of the lithosphere caused by eclogitized crustal material beneath the seismic Moho. This study integrates these two approaches to model the topographical and thermal effects of eclogite facies transformations during continental collision. This approach allows the robustness of the earlier topographical models to be tested.

Coherent eclogite facies terranes

Some collapsed orogens contain regions of tens of thousands of square kilometres of continental lithologies, which were subjected to eclogite facies conditions during continental collision (Fig. 1). The eclogite facies assemblages in these terranes are typically preserved in mafic boudins within a lower grade, usually amphibolite facies, matrix showing structural evidence for horizontal extension. This is probably because the basaltic rocks tend to form early boudins which resist subsequent fluid penetration and alteration. However, radiometric and petrological studies show that the rocks of the matrix were also subjected to eclogite facies conditions during peak metamorphism (see, for example, Krogh 1977; Dewey *et al.* 1993; Ames *et al.* 1996); the

From: MILLER, J. A., HOLDSWORTH, R. E., BUICK, I. S. & HAND, M. (eds) *Continental Reactivation and Reworking.*
Geological Society, London, Special Publications, **184**, 39–55. 1-86239-080-0/01/$15.00 © The Geological Society of London 2001.

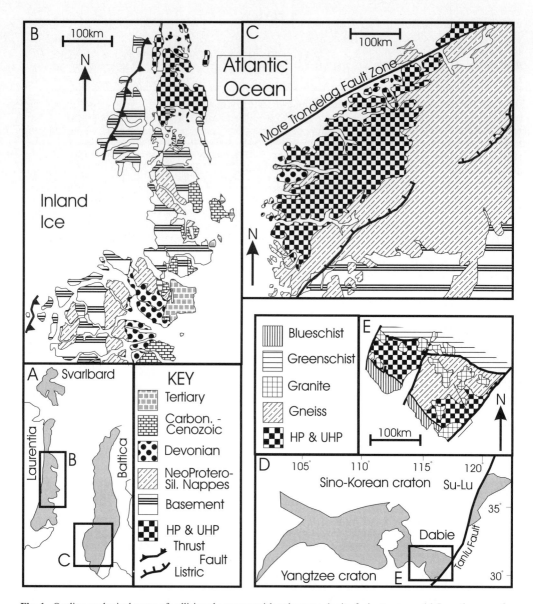

Fig. 1. Outline geological maps of collisional orogens with coherent eclogite facies terranes. (**a**) Location map for Greenland and Baltic plotted using pre-Atlantic opening reconstruction. The Caledonides are shaded. (**b**) Geological map of the Western Gneiss Region, Norway. (**c**) Geological map of the north-east Greenland Caledonides. (**d**) Location map showing the Triassic Dabie–Qinling orogeny. (**e**) Outline geological map of the Dabie Shan region. All geological maps are drawn to the same scale.

current fabric of the terrane being mainly controlled by the process of exhumation which took place during the collapse of the orogen (Andersen & Jamtveit 1990; Chauvet *et al.* 1992; Ames *et al.* 1993; Steltenphol *et al.* 1993; Andersen 1998). A relict high pressure gradient may occur across such terranes which indicates that they have been buried and exhumed wholesale (Krogh 1977) or without apparent substantial disruption of their burial geometry (Cuthbert *et al.* 2000).

These terranes, in which all crustal rocks have undergone HP or UHP metamorphism during prograde metamorphism are here termed coherent eclogite facies terranes (CEFT). They should be distinguished from eclogite bearing terranes formed within subduction zone complexes, which pre-date continent–continent collision (e.g. Eoalpine eclogites of the Sesia zone within the European Alps), are tectonically imbricated with lower grade metamorphic, usually blue-schist facies rocks (e.g. the Franciscan mélange, Cloos 1982), and do not affect large regions of older continental basement. Such terranes undergo a much lower bulk density change ($\pm 5\%$ for transformations to and from the blueschist facies) than CEFTs (see below) and will always remain positively buoyant with respect to the asthenosphere.

CEFTs were developed during: the Caledonian (Silurian) Baltica–Laurentia collision in the Western Gneiss region (WGR) of Norway (Krogh 1977; Smith 1984) (Fig. 1c) at 420–400 Ma (Griffin & Brueckner 1985) and NE Greenland (Gilotti 1993) at 405 ± 24 Ma (Brueckner et al. 1998) (Fig. 1b); the collision of Gondwana with Laurussia during the Variscan orogeny of Europe (Carswell 1990; Le Pichon et al. 1997); and in the Dabie-Qinling-Su Lu orogen developed during the Triassic collision of the Sino-Korean and Yangtze cratons (Wang et al. 1989; Hacker et al. 1998) (Fig. 1e). Such terranes may have developed beneath the Himalayas and Tibet (Sapin & Hirn 1997), and the Alps (Butler 1986). Preservation of Pan-African and Grenvillian continental eclogites suggests that this process may have been active since the Neo-Proterozoic (Sanders et al. 1987; Sanders 1989; Bernard-Griffiths et al. 1991; Castaing et al. 1993; Indares 1993; Möller 1998, 1999).

However, the conditions necessary for UHP metamorphism may not have occurred until the Proterozoic–Phanerozoic boundary (Maruyama & Liou 1998). Higher geotherms of perhaps 15–$25°\mathrm{C\,km^{-1}}$ (Maruyama & Liou 1998) and consequent lower lithospheric strengths would make it difficult to form very deep roots during the Archaean or perhaps the Palaeoproterozoic. The positive slope of the granulite to eclogite facies reactions in P–T space mean that a greater thickness of root would be required before such transformations could occur, and lithosphere with a lower strength would be less likely to support the topography associated with such a root.

CEFTs and buoyancy within the orogen

Transformation to eclogite facies assemblages is associated with an increase in density of c. 6.7% to 21.1% for a range of crustal lithologies (Table 1). If an orogen develops an eclogitized crustal root then the continental lithosphere will loose buoyancy (Richardson & England 1979; Dewey et al. 1993). This effect is illustrated in Figure 2, which shows the increase in mean lithospheric density associated with the development of an eclogitized root. The lithospheric density (ρ_l) is calculated using the following equation (Dewey et al. 1993):

$$\rho_l = \frac{\int_0^{C_z} \rho_c(z)\delta z + \int_{C_z}^{L_z} \rho_m(z)\delta Z}{l_z}$$

Table 1. *Model and measured densities for eclogite facies rocks and their precursors*

Rock type	Initial density $\mathrm{kg\,m^{-3}}$	Granulite facies density $\mathrm{kg\,m^{-3}}$	Eclogite facies density $\mathrm{kg\,m^{-3}}$	Percentage density increase	Reference
granodiorite	2760		3100	12.3	Bosquet et al. (1997)
		2870		8.0	
andesite	2910		3460	18.9	Bosquet et al. (1997)
		3040		13.8	
basalt	2940		3560	21.1	Bosquet et al. (1997)
		3270		8.9	
anorthositic gabbro	3055		3260	6.7	Austrheim & Mörk (1988)
anorthosite	2785		3060	9.9	Austrheim & Mörk (1988)
gabbro	3166		3480	9.9	Austrheim & Mörk (1988)
gabbro anorthosite	2870		3190	11.1	Austrheim & Mörk (1988)
mafic mangerite	2980		3210	7.7	Austrheim & Mörk (1988)
pelitic restite	2900		3400	17.2	Hynes & Snyder (1995)

The two values for the percentage change in model density for each rock type calculated by Bosquet et al. (1997) represent eclogitization of upper crustal rocks (upper value) and their garnet granulite facies equivalents (lower value). Austrheim & Mörk (1988) measured the change in density in the field at the granulite–eclogite transition. The value for a pelitic restite (Hynes & Snyder 1995) is a model value calculated from a kyanite–garnet containing pelite after 57% melting at 0.8 GPa.

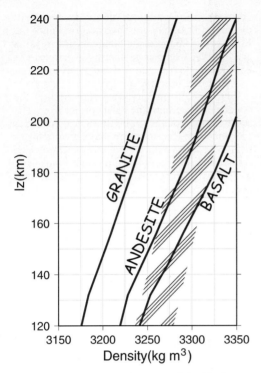

Fig. 2. Plots showing the increase in lithospheric density for a lithosphere of initially 120 km with a continental crust of 40 km. The model assumes that the crust between 30 and 40 km is converted to granulite and that below 40 km to eclogite facies assemblages using the density values of Bosquet *et al.* (1997). The lithosphere is thickened by vertical pure shear by a factor of 2. There is no erosion. The crust is assumed to have a mean temperature of 300°C and the mantle of 950°C. The likely range of asthenospheric densities is patterned.

where ρ_c, ρ_l and ρ_m are the densities of the crust, lithosphere and mantle, respectively; C_z and l_z are the thicknesses of the crust and lithosphere; and z is the depth.

The simple model presented in Figure 2 considers a 40 km thick crust of granodioritic, andesitic or basaltic composition completely transformed to granulite below 30 km and eclogite below 40 km. Homogenous pure shear is assumed, the lithosphere is initially 120 km in thickness and densities are taken from Bosquet *et al.* (1997). The crustal rocks are assumed to have a compressibilities of 10^{-11} Pa^{-1} and $8.0 * 10^{-12}$ Pa^{-1}, and a thermal expansivities of $2.5 * 10^{-5}$ $°K^{-1}$ and $3.0 * 10^{-5}$ $°K^{-1}$ for the crust and mantle, respectively. The likely range of densities of the asthenosphere (3250–3290 kg m^{-3} at 120 km, equivalent to 3284–3324 kg m^{-3} at normal temperature and pressure) is hachured. The

mean density of lithosphere with basaltic crust can exceed the likely range for the density of the asthenosphere. A continental lithosphere with andesitic crust might become neutrally buoyant with respect to an asthenosphere of relatively low density when the lithosphere is stretched to a thickness of 190 km (Fig. 2) and consequently the crust to a thickness of 63 km. This means that further thickening will not lead to continued uplift and the orogen would subside below sea level. A lithosphere with a granodioritic crust will always remain positively buoyant with respect to the asthenosphere. These curves assume wholesale instantaneous metamorphic conversion, which for kinetic reasons is unlikely (Austrheim *et al.* 1997), and must represent an extreme end member. They do illustrate, however, that increasing thickening is associated with increasing lithospheric density which will tend to buffer orogenic uplift.

Dewey *et al.* (1993) have modelled the topographical consequences of transforming a significant volume of the root below 40 km to eclogite facies assemblages during convergence, and retrogressing them to amphibolite facies assemblages during exhumation. The prograde reactions allow shortening to continue after the formation of topography of c. 3 km because a steady-state is reached where the increase in density of the orogenic root compensates for any further thickening without producing more uplift. If such phase changes do not take place then the orogenic root is unlikely to exceed 60–80 km in thickness (Dewey 1988; Molnar *et al.* 1993) as the resultant gravitational forces, produced by the excess topography, will exceed those driving the convergence.

During exhumation the dense eclogitic root acts as a 'gravitational battery'. The retrogression associated with orogenic collapse and exhumation decreases the density of the root, increasing buoyancy and driving uplift. It is probably this effect which is primarily responsible for the uplift of coherent eclogite terranes (Dewey *et al.* 1993). Austrheim (1990, 1991) makes similar arguments and points out that the loss of feldspar during eclogitization produces a marked loss in strength in the root, which may contribute to the buffering process by promoting lower crustal flow. Le Pichon *et al.* (1997) argue that thermal relaxation beneath the Tibetan plateau has retrogressed eclogite facies rocks formed earlier during the collision of India with Asia. They calculate that the heating over 40 Ma caused by the thickening of the crust and detachment of the mantle root in an orogen without substantial exhumation will transform eclogite into granulite facies assemblages within the orogenic root and the resultant

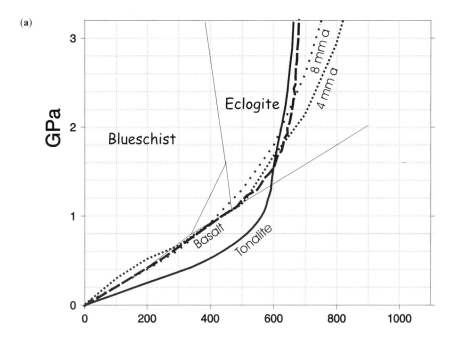

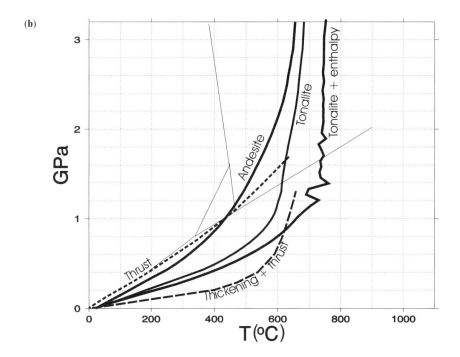

Fig. 3. (a) Shows model geotherms for the Alps assuming continental subduction at 4 mm a^{-1} and 8 mm a^{-1} (Bosquet *et al.* 1997) and for a vertical stretch orogen with a crust of granodioritic composition ('granite') and basaltic ('basalt') compositions (this work) after 32 Ma of convergence with phase change to eclogite facies below 40 km. (b) Geotherms for different orogenic models. 'Thrust', model where the crust thickens by thrusting and the lithospheric mantle by pure shear (Midgley & Blundell 1997). 'Thickening and thrust', an orogen with vertical thickening and thrusting (Jamieson *et al* 1997). 'Andesite' and 'Tonalite' represent curves for a vertical stretch orogen after 32 Ma of convergence with crusts of andesitic and tonalitic composition respectively with phase change to eclogite facies below 40 km. 'Tonalite + enthalpy', a vertical stretch model after 32 Ma of convergence for a tonalitic crust with phase change that allows for the enthalpy of the reaction *anorthite* ⇌ *garnet + kyanite + quartz*.

buoyancy gain will account for 2.5 km of the topography of Tibet. Le Pichon *et al.* (1997) also propose that the conversion of Variscan eclogite facies rocks in the Massif Centrale of France to granulite facies during collapse indicates that an equivalent process took place beneath a Tibetan style plateau in central Europe during the late Carboniferous.

The formation of coherent eclogitic terranes

The crust can be thickened during collisional orogeny by a variety of mechanisms. Modelled geothermal gradients for continental subduction (Bosquet *et al.* 1997), a thickening orogenic wedge (Le Pichon *et al.* 1997), underplating during subduction (Huerta *et al.* 1996), vertical thickening and thrusting (Jamieson *et al.* 1997),

crustal overthrusting (Midgley & Blundell 1997) or vertical stretch (this work, see below) are consistent with 'medium temperature' eclogite facies conditions, that is temperatures in the order of 550°C at a pressure of 1 GPa (Carswell 1990), forming in the orogenic root (Fig. 3).

The occurrence of coesite, which forms at depths of >90 km and temperatures around 610–700°C (Fig. 4), giving mean geothermal gradients of *c.* 5°C to 7°C per km, leads most authors to assume that continental crust is subducted. This conclusion is supported by the lack of seismic evidence for modern orogenic roots extending to such depths. However, if rocks at the base of the crustal root are all in the eclogite facies then they will have seismic velocities nearer that of the mantle than the crust and it may not be possible to image them (see Austrheim 1990; Hynes & Snyder 1995; Sapin & Hirn 1997, for

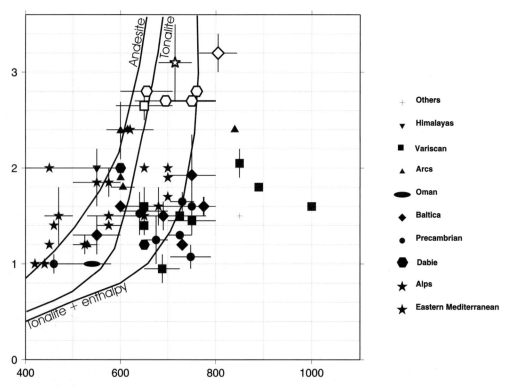

Fig. 4. Plots of pressure temperature estimates for eclogites in collisional orogens. Points for the Alps, Baltica and Dabie (ornaments have hollow centers) at pressures of 2.8 GPa or above represent coesite bearing assemblages. The data plotted is taken from Andersen & Jamtveit (1990), Bernard-Griffiths *et al.* (1993); Boufette & Caron (1991); Camacho *et al.* (1997); Castelli (1991); Chauvet *et al.* (1992); Chopin (1987); Clarke *et al.* (1997); Cliff *et al.* (1998); Cotkin *et al.* (1988); Di Vincenzo *et al.* (1997); El-Din *et al.* (1990); Erdmer *et al.* (1998); Fry & Barnicoat (1987); Gardien (1993); Gomez-Pugnaire & Fernandez-Soler (1987); Guillot *et al.* (1997); Indares (1993); Jamtveit (1987); Kienast *et al.* (1991); Liati & Seidel (1996); Liu & Liou (1995); Mercier *et al.* (1991); Nicollet (1989); O'Brien (1993); Page (1992); Perchuk & Phillippot (1997); Pognante & Spencer (1991); Sanders (1989); Schliestedt (1986); Schmadicke (1991); Wang & Liou (1991, 1993); Wang *et al.* (1989); Waters (1989).

discussion of this problem). If large volumes of continental crust were carried down a subduction zone to a depth of 100 km, buoyancy considerations (see below) require that it had undergone phase changes which made it negatively buoyant. It would also be overlain by the lithospheric mantle of the hanging wall and would be difficult to exhume (see Andersen *et al.* 1991). Most mechanisms appeal to subducted

eclogitized crust being accreted onto the hanging wall and rising under its own buoyancy. This suggests that the upper layers of the subducting slab are most likely to be exhumed as they are the most buoyant and structurally highest. Also, such a mechanism is inherently asymmetric and eclogites during collision should be restricted to footwall lithologies. Pre-collision eclogite facies terranes are probably exhumed in

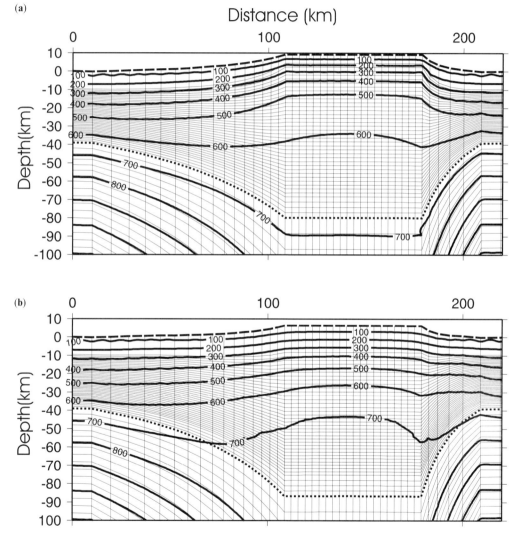

Fig. 5. (**a**) Finite element mesh for a vertical stretch orogen after 32 Ma of convergence at a strain rate of $10^{-15} \, \text{sec}^{-1}$. The crust is of a tonalitic composition. No phase transformations take place and a topography of almost 10 km is developed. Thermal contours are drawn using a transient 2-dimensional finite element thermal model. Assumptions are given in Table 2. (**b**) Finite element mesh and transient 2-dimensional finite element thermal model for a lithosphere with the same initial structure as Figure 5a, but which allows for both the volume changes and the enthalpy associated with the *anorthite* ⇌ *garnet + kyanite + quartz* reaction. Mass is conserved in each quadrilateral cell. Assumptions are given in Table 2.

this manner. However, in western Norway and NE Greenland (Fig. 1), syn-collisional Silurian eclogites form in coherent terranes in continental basement beneath, or within, nappe piles with opposite senses of transport during convergence: to the east in Scandinavia and to the west in Greenland. The increase in pressure in the WGR is systematic towards the west, the supposed root zone of the orogen (Krogh 1977; Cuthbert et al. 2000). These geometries are consistent with eclogite facies metamorphism during homogenous thickening. However, mantle peridotites emplaced along basement cover contacts of the UHP terrane of the WGR are perhaps best explained by continental subduction (Cuthbert et al. 2000). Also the CEFTs of Greenland and Norway are offset by some 600 km on a post-collision reconstruction (Fig. 1A), a figure that has been reduced by post-collision sinistral strike-slip. This geometry could be attributed to two continental subduction systems of opposite polarity, but would require the existence of an unreported transform separating these terranes. The geology of CEFTs is complex and greatly modified by the exhumation process, however, at least in the case of the Silurian Greenland–Baltica collision there is evidence that supports the contention that CEFTs can develop in orogenic roots, should a suitable geothermal gradient exist.

The likelihood of achieving a sufficiently low geothermal gradient during collision within an orogenic root formed by vertical stretching is investigated below. Figure 5a presents a transient finite element model for an isostatically compensated, asymmetric orogeny, with crustal thickening by vertical stretch and no eclogitic phase transformations within the orogenic root. The advection–conduction heat equation was solved following the method outlined in Ryan & Dewey (1997) and erosion and topography were analysed using the method of Dewey et al. (1993). Assumptions made in construction of this figure are given in Table 2. The resultant geothermal gradient at the centre of the orogeny for crusts composed of basalt, andesite and tonalite are plotted on Figures 3a and b along with those obtained by Bosquet et al. (1997), assuming continental subduction at $8 \, \mathrm{mm \, a^{-1}}$ or $4 \, \mathrm{mm \, a^{-1}}$. The geotherms obtained by Jamieson et al. (1997) for thickening and thrusting, and Midgley & Blundell (1997) for an overthrust orogen are plotted on Figure 3b. All geotherms pass through the continental eclogite field, whilst those for continental subduction and vertical stretch geotherms pass through the coesite stability field (Figs 3a, b and 4). The subduction related geotherms (4 and $8 \, \mathrm{mm \, a^{-1}}$) first pass through the blueschist facies and eventually yield higher temperatures at 3 GPa (Fig. 3a).

Table 2. *Assumptions made in the finite element models presented in Figures 5 and 8*

Heat productivity for upper crust	$1.4 * 10^{-6} \, \mathrm{W \, m^{-3}}$
Heat productivity for lower crust	$0.7 * 10^{-6} \, \mathrm{W \, m^{-3}}$
Heat productivity of mantle	$2.3 * 10^{-8} \, \mathrm{W \, m^{-3}}$
Strain rate	$10^{-15} \, \mathrm{sec^{-1}}$
Crustal thickness	35–40 km
Thermal diffusivity of upper crust	$7.78 * 10^{-7} \, \mathrm{m^2 \, sec^{-1}}$
Thermal diffusivity of lithospheric mantle	$7.45 * 10^{-7} \, \mathrm{m^2 \, sec^{-1}}$
Thermal diffusivity of asthenosphere	$8.10\text{--}11.30 * 10^{-7} \, \mathrm{m^2 \, sec^{-1}}$
Time steps	0.5 Ma
Density of crust	2775, 2802 and $2850 \, \mathrm{kg \, m^{-3}}$
Density of eclogitized crust	3080, 3370 and $3560 \, \mathrm{kg \, m^{-3}}$
Density of mantle	$3330 \, \mathrm{kg \, m^{-3}}$
Enthalpy of $an \rightleftharpoons gt + ky + qz$	$40 \, \mathrm{KJ \, mole^{-1}}$
Half-life of conversion to eclogite	5 Ma
Fixed temperature at surface	$20°C$
Fixed temperature at 240 km	$1440°C$
Erosion constant	1.0
Heat capacity of crust	$2.5 * 10^6 \, \mathrm{J \, m^{-3}}$
Heat capacity of mantle	$3.0 * 10^6 \, \mathrm{J \, m^{-3}}$
Thermal expansion coefficient of the crust	$2.5 * 10^{-5} \, \mathrm{K^{-1}}$
Thermal expansion coefficient of the	$3.0 * 10^{-5} \, \mathrm{K^{-1}}$
Compressibility of the crust	$10^{-11} \, \mathrm{Pa^{-1}}$
Compressibility of the mantle	$8.0 * 10^{-12} \, \mathrm{Pa^{-1}}$
Percentage melting at threshold temperatures:	
675, 700, 740, 780, 820, $\geq 860°C$	15%, 35%, 50%, 60%, 65%, 70%

Their initial lower temperature portion is due to the upper radiogenic crust of the upper crustal wedge not being significantly thickened. The lower temperatures of the vertical stretch model (basalt and tonalite) at 3.0 GPa may be because this geotherm was calculated using a transient, not a steady state solution. Both analyses yield results probably 100–250°C higher at 3.0 GPa than that encountered in subduction zones, such temperatures are typical of 'medium tempera-ture' eclogites formed during continental colli-sion (Carswell 1990). The higher geothermal gradient of the continental (Bosquet *et al.* 1997), as opposed to oceanic subduction, model (e.g. Peacock 1995) is due to the low rates of subduc-tion, the presence of higher concentrations of heat producing elements, and the thickening of the upper crustal material in the hanging wall. The vertical stretch model also produces higher gradients because of progressive thickening of the radiogenic upper crust, plus the relatively low rate of the penetration of the mantle root into the asthenosphere (1.3 mm a^{-1} to 3.2 mm a^{-1} as the model evolved). It is, therefore, possible that coherent eclogite facies terranes can be formed in the hanging wall and the foot wall lithologies of orogens with significant components of verti-cal thickening.

Role in metamorphism

Recent thermo-mechanical models of orogens (Stüwe 1998; Jamieson *et al.* 1997) show that heating during continent–continent collision due to the thickening of the radiogenic crust is often inadequate to account for the observed metamorphic parageneses. Other sources of heat suggested are frictional heating (Stüwe 1998), ad-vection by igneous intrusions (Dewey & Mange 1999), or accretion of material with relatively high heat productivity to the base of the meta-morphic pile (Huerta *et al.* 1996; Jamieson *et al.* 1997). Such models principally investigate the effects of radiogenic heating and cooling by subduction and erosion, and discount the effects of the enthalpy of metamorphic reactions. Eclo-gite facies phase transformations during conver-gence affect the crust in the root of an orogen beneath *c.* 40 km, a volume of rock that might exceed that of the non-eclogitized material above. These reactions have negative enthalpy as they tend to reduce the volume for a given mass and, hence, tend to heat the root. The affect of such enthalpy changes on metamorphic heat-ing is modelled (Fig. 5b) by assuming that the root is comprised of anorthite and undergoes the *anorthite ⇌ garnet + kyanite + quartz* reac-

tion with a half life of 5 Ma when it exceeds depths of 40 km. The reaction is assumed to have an enthalpy of −40 KJ $mole^{-1}$ at NTP, pressure and temperature dependence is taken from Hol-loway & Wood (1988). The rate of thermo-dynamic heating is governed by the following equation:

$$\Delta H_{t_{(i+1)}} = \Delta H \cdot (e^{-\lambda \cdot t_i} - e^{-\lambda \cdot t_{(i+1)}})$$

where,

$$\lambda = \ln(2)/t_{1/2}$$

(ΔH), is the total enthalpy; ($\Delta H_{t_{(i+1)}}$), is the aliquot of the enthalpy attributed to time step ($t_{(i+1)}$); and $t_{1/2}$, is the half-life. Other assump-tions are as for Figures 3a and 5. Mass is con-served in the finite element grid and the volume of the cells altered appropriately.

The resultant geotherm (tonalite + enthalpy, Figs 3b and 4) passes through the P-T field of most coherent eclogite facies terranes. This phase transformation increases the temperature by some 50°C at 1 GPa, and up to 100°C at 2 GPa, over that of models which do not account for the enthalpy of HP and UHP transforma-tions. The oscillations in the curve at 1.0–1.5 GPa are caused by the instabilities created in the mathematical model used to solve the conduc-tion–advection equation, when the first aliquots of heat of reaction are added to the system. They provide an estimate of the errors inher-ent in such calculations, which are in the order of ±30°C. Another interesting feature of this model is the near vertical geotherm at pressures between 2 and 3 GPa. This is attributed to the thickening of the cool lithospheric mantle under-lying the crustal root, which offsets conductive heating from below making the rocks at 3 GPa colder than they would be in a steady state model. Enthalpic heating in the layers above 2 GPa makes these rocks hotter than the steady state model. The result of these two effects is that there is little temperature change between 2 and 3 GPa. This analysis is not intended to be exhaustive, merely to demonstrate that the enthalpy of meta-morphic transformations within a deep crustal root can affect P–T–t gradients and should be incorporated into thermo-mechanical models of collisional orogenies.

The role of fluids

Fluids are essential to many of the prograde and retrograde metamorphic transformations (Austrheim 1990; Pennacchioni 1996; Austrheim *et al.* 1997) involving in the creation and destruc-tion of eclogitic terranes. Evidence from the

Bergen Arcs region of Norway shows that transformations upon the penetration of fluid are fast enough to be associated with seismic activity (Austrheim *et al.* 1996). This suggests that there may be considerable metamorphic overshoot and the availability of fluids, especially in dry basement terranes, may control the formation of eclogite facies assemblages. However, if transformations in the root are limited by lack of fluids, mean crustal density will not increase and excessive topography will develop. Thickening sufficient to develop coesite will produce mean regional elevations of 7 km, rising to 10 km during the collapse phase (figure 7a of Dewey *et al.* 1993). Such topography, which is higher than values recorded today, is unlikely to form as it would be limited by the potential energy gradients it would create (Dewey 1988). Thus the amount of overshoot must either be limited, or the presence of a dry root must inhibit vertical stretch during collision to *c.* 1.7, the value required for a topography of 3 km (Dewey *et al.* 1993), unless the crust is extremely mafic or the asthenosphere of relatively low density. The absence of fluids in rocks at eclogite facies conditions can, therefore, lead to a relatively early cessation of convergence, whilst access to abundant fluids allows continued convergence without developing excessive topography.

Subducting continental crust

Material balance calculations (e.g. Butler 1986; Laubscher 1988; Pfiffner 1992; Marchant & Stampfli 1997) suggests that there is a shortage of deep basement within Cenozoic collisional orogens. This has been attributed to the subduction of continental crust immediately prior to final collision. Although an alternative view is offered by Ménard *et al.* (1991), who suggest that, within the limits of error, convergence was accommodated by thickening of an extended crust and the development of an 50 km orogenic root. Substantial volumes of continental crust can only be subducted if its buoyancy is reduced (McKenzie 1969). Ignoring slab pull and mantle drag forces, it is assumed that the necessary condition for subduction of continental lithosphere is that it should have negative buoyancy with respect to the asthenosphere. Dewey (1988) argues that continental lithosphere becomes neutrally buoyant when the crust (C_z) to overall lithospheric (l_z) thickness ratio is *c.* $\leq$0.16, depending upon the density assumed for the asthenosphere. It may, therefore, be possible to subduct thinned crust, such as a continental margin or lower crust, which has been tectonically

stripped of upper crust, as is suggested for the European upper crust in the Alps (Laubscher 1988; Pfiffner 1992).

The mean density of the lithosphere (ρ_l) is given by:

$$\rho_l = \frac{(\rho_c * C_z + \rho_m * (C_z - l_z))}{l_z}$$

where, ρ_c and ρ_m are the mean densities of the crust and mantle, respectively. If ρ_c is increased by eclogite facies metamorphism, then greater thicknesses of crust can be subducted. Figure 6 shows that a mean crustal density of 3000 to 3170 kg m^{-3} is needed for 2/3 of a normal thickness continental crust ($C_z/l_z \simeq 0.2$) to subduct, a value that may not be unreasonable for the Alps (see Laubscher 1988; Pfiffner 1992). Assuming that the crust has a tonalitic composition implies that much of the lower and middle crust of the subducting slab is in eclogite facies. The decrease in strength at the granulite–eclogite transition (Boundy *et al.* 1992) may provide a

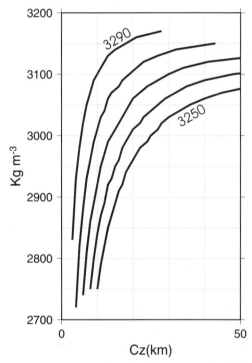

Fig. 6. Curves showing the thickness of crust (Cz) of a given density attached to 80 km of lithospheric mantle that can be subducted. Curves are drawn assuming neutral buoyancy with asthenospheres of density 3250, 3260, 3270, 3280 and 3290 kg m^{-3} equivalent to densities of 3284, 3294, 3304, 3314 and 3324 kg m^{-3} at NTP. The crust is assumed to have a mean temperature of 300°C and the mantle of 950°C.

suitable surface for delamination to take place (Wijbrans *et al.* 1993), allowing the upper crust to rise buoyantly, whilst the eclogitized middle and lower crust are subducted. If the crust is dry and metamorphic fluids cannot penetrate then metamorphic transformation to eclogite facies is unlikely and only the lowermost proportion of the continental crust can be subducted, unless pulled down by an attached oceanic slab (Molnar & Gray 1979). However, this will put the subducting plate into tension and probably result in its failure and the consequent cessation of continental subduction. The thickness of continental crust that can be subducted during the early phases of collision is, at least, partly controlled by basement composition and metamorphic history.

Delamination

Several authors (see Meissner & Mooney 1998; Gao *et al.* 1999 for refs) have pointed out that the average continental crust is more mafic than that of island arcs. The lowermost seismic crustal layer with P wave velocities of greater than $7.2 \, \mathrm{km \, sec^{-1}}$ is often missing under orogens. Late tectonic granite composition, including negative Eu anomalies, suggests that the melt was in equilibrium with a dense eclogitic restite phase, which is now not imaged seismically. They, therefore, suggest that much of the basic eclogites or eclogitic restites in the lower crust have delaminated. An appropriate time for this to occur is during orogenesis when the lower crust may be deeply buried, eclogitized and, if basaltic in composition, negatively buoyant. Gao *et al.* (1999) calculated using geochemical constraints that the Dabie–Qinling orogeny has lost between 38 and 74 km of its lower crust beneath Dabie Shan. This is supported by petrophysical and seismic studies (Kern *et al.* 1999) which indicate that large volumes of mafic eclogite do not now exist beneath Dabie Shan, but these results do not exclude the possibility that much of the crust underlying Dabie is an amphibolitized coherent crustal eclogite terrane, such as the WGR.

However, the thermal and isostatic consequences of such delamination must be taken into account. If the lower crust is mechanically attached to the upper plate then its delamination (unless by lateral flow; Meisnner & Mooney 1998) must also remove the lithospheric mantle. A simple isostatic calculation balanced out to the mid-oceanic ridges,

$$h = \frac{(\rho_l - \rho_a)}{(\rho_a)} * l_z - 2.64$$

shows that for a crust of 30–50 km thickness, total instantaneous delamination would produce a topography (h) of some 2–5 km above sea level (Fig. 7). Airy isostasy need only be assumed in this case as the lithosphere would be too weak to have any effective elastic thickness. Also, the exposure of the Moho to asthenospheric temperatures would produce significant volumes of melt. Figure 8 was produced using a two-dimensional transient finite element solution to the conduction–advection equation (see Ryan & Dewey 1997, for the methodology). A 40 km continental crust overlying 90 km of lithospheric mantle was initially allowed to equilibrate for 10 Ma and then all the mantle nodes from 160–320 km horizontally and 40–130 km vertically were to set to 1330°C instantaneously (Fig. 8a). The volumes of melt were then calculated using muscovite dehydration melting models following

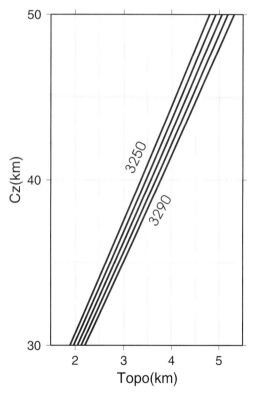

Fig. 7. Model curves showing the topography that would be developed by a crust of tonalitic composition ($2810 \, \mathrm{kg \, m^{-3}}$) if the lithospheric mantle were delaminated. Curves are drawn for asthenospheres of density 3250, 3260, 3270, 3280 and $3290 \, \mathrm{kg \, m^{-3}}$ equivalent to densities of 3284, 3294, 3304, 3314 and $3324 \, \mathrm{kg \, m^{-3}}$ at NTP. The crust is assumed to have a mean temperature of 400°C and the mantle of 1050°C.

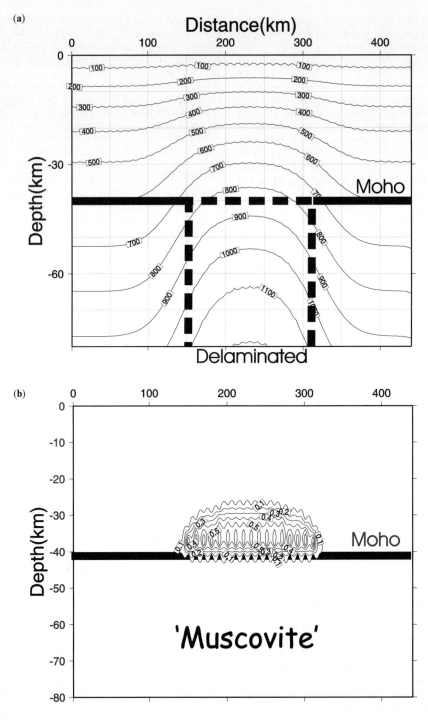

Fig. 8. (a) Transient 2-dimensional finite element thermal model 15 Ma after delamination of the lithospheric mantle between 160 and 320 km. The crust is 40 km in thickness. (b) Contour plot for the proportion of total melt produced 35 Ma after delamination of the lithospheric mantle between 160 and 320 km. The model assumes muscovite dehydration melting after Zen (1995). Melt is produced up to 35 Ma after delamination. Assumptions are given in Table 2.

the method of Zen (1995). A detailed discussion of this approach is given elsewhere (Ryan & Soper in press). The resultant melt volumes are contoured (Fig. 8b). This analysis suggests that between 20 and 50% of the lower 10–13 km of the lower crust will melt, producing extensive volcanism and plutonism for up to 35 Ma after delamination. This may well have been the case in parts of the Variscides of Europe where some 50% of the crust is either granite or migmatite. This conclusion is supported by the presence of very high temperatures and relatively low pressure eclogites (1000°C, 1.6 GPa, Fig. 4) in the Gfohl Nappe (Medaris *et al.* 1998). However, xenolith evidence suggests that not all of the eclogitized lower crust has delaminated (Mengel 1992). The transition from eclogite to granulite facies in the Massif Centrale suggests thermal relaxation of a thickened crust rather than delamination (Le Pichon *et al.* 1997). Similarly, the formation of early Cretaceous gneiss domes and granitic plutons in Dabie Shan (Wang *et al.* 1989; Faure *et al.* 1998) may be related to delamination triggered by subduction along the margin of the Yangtze Block. The absence of extensive migmatites and granites suggest that this process did not take place in the WGR or NE Greenland.

The nature of the continental crust may control this process. In the Variscides, there is evidence of wide spread mafic vulcanism associated with the break-up of the Gondwana margin in the early Palaeozoic (Floyd *et al.* 2000). Volcanic margins with extensive underplating would have become sites where delamination could take place as eclogitization of significant volumes of basalt would produce an orogenic root that was negatively buoyant with respect to the lithosphere (Fig. 2).

Role in orogenic cyclicity

Ryan & Dewey (1997) suggest that if a substantial volume of eclogite facies rocks of crustal composition remain in isostatic equilibrium with the mantle and are not exhumed during orogenic collapse, they may provide a mechanism for the cyclic re-opening of closed oceans. Such material is likely to have a bulk density of *c.* 3330 kg m^{-3} if it comprises eclogite facies rocks of 50% basalt and 50% granodioritic composition (Table 1), and consequently a P wave velocity in the order of 8 km sec^{-1}. Such material would be imaged seismically as resting beneath the Moho (Griffin & O'Reilly 1987; Hynes & Snyder 1995) and may account for upper mantle reflectivity (Austrheim 1990). These crustal lithologies have a different rheology and a higher heat productivity from that of the mantle they replace. Initially, their lower strength would account for much of the relative weakness of the collapsed orogen compared to its foreland. However, after 300 Ma the radiogenic heat produced by these crustal rocks residing in the mantle amplifies this relative weakness by a factor of 2 to 3 times (Ryan & Dewey 1997) making the old orogen a target for future rifting.

Conclusions

Metamorphic transformations in an orogenic root below 40 km can alter the density of the crust such that it can approach, or even exceed, that of the mantle. This greatly reduces buoyancy during convergence and can, if the correct metamorphic conditions exist, allow vertical stretches to exceed 2.0 without producing excessive topography. The formation of coherent eclogite facies terranes, in which all rocks have undergone HP or UHP metamorphism during convergence may be caused by crustal subduction or by burial within an orogenic root. The broadly coeval CEFTs in the Caledonides of Norway and East Greenland, which lie in nappe stacks of opposite vergence, were probably formed in an orogenic root. Numerical models show that conditions for HP and UHP metamorphism can exist in such a root beneath 40 km and that the associated enthalpy will assist heating within the orogenic pile. Formation of eclogite facies assemblages may assist continental subduction and loss of significant amounts of footwall basement during the early stages of collision. Lithospheres of similar composition and structure may respond differently during convergence depending on the availability of suitable metamorphic fluids.

Eclogitized crustal roots may influence late orogenic history. Whole scale eclogitization of a mafic lower crust, perhaps formed by underplating during earlier rifting, may lead to delamination. Orogenic collapse will then be associated with rapid uplift and high degrees or partial melting and fragmentation of the CEFT. A more gradual retrogression of a less mafic root will tend to buffer subsidence during collapse and assist in exhuming CEFTs. Some continental eclogites, with densities nearer that of the mantle than the crust, may remain beneath the Moho after collapse and cooling of the orogen. They replace mantle rheologies and heat productivities with those of the crust. This leads to a long term weakening of ancient orogenic lithosphere with respect to that of the foreland and may provide an explanation for the Wilson Cycle reopening of sutured oceans.

Research into the controls that crustal roots exert upon the evolution of an orogen is at an early stage. However, existing models suggest that phase transformations within such roots have a significant effect and should be taken into account when modelling collisional orogens. This contribution argues that the existence of a low geothermal gradient within an HP–UHP terrane is not sufficient to prove the mechanism of crustal subduction. Rather, field relationships must be used to establish whether such terranes develop by vertical stretch or crustal subduction, or both. One of the problems in testing such models in ancient orogens is that the process of exhumation destroys much of the collisional fabric within the deep basement to the orogeny. HP and UHP assemblages often only comprise a small percentage of supposed orogenic roots. The coexistence of protolith, HP and UHP assemblages and extensive retrograde assemblages indicate that metamorphic equilibrium was never established. The model presented here requires that at least a significant volume of deep crust underwent transformation to HP and UHP assemblages during collision. Whether or not this happened may depend as much upon the availability of fluids as on P and T conditions. It may be possible to test this model in modern orogens where remotely sensed data exists for that portion of the root still in place. Careful study of exhumed HP and UHP boudins may tell us a great deal about the early history of an orogeny.

References

AMES, L., TILTON, G. R., ZHOU, G. 1993. Timing of collision of the Sino-Korean and Yangtse cratons: U–Pb zircon dating of coesite-bearing eclogites. *Geology*, **21**, 339–342.

AMES, L., ZHOU, G. & XIONG, B. 1996. Geochronology and isotopic character of ultrahigh-pressure metamorphism with implications for collision of the Sino-Korean and Yangtze cratons, central China. *Tectonics*, **15**, 472–489.

ANDERSEN, T. B. 1998. Extensional tectonics in the Caledonides of southern Norway, an overview. *Tectonophysics*, **285**, 333–351.

ANDERSEN, T. B. & JAMTVEIT, B. 1990. Uplift of deep crust during orogenic extensional collapse: a model based on field studies in the Sogn-Sunnfjord region of western Norway. *Tectonics*, **9**, 1097–1111.

ANDERSEN, T. B., JAMTVEIT, B., DEWEY, J. F. & SWENSSON, E. 1991. Subduction and education of continental crust: major mechanisms during continent–continent collision and orogenic extensional collapse, a model based upon the south Norwegian Caledonides. *Terra Nova*, **3**, 303–310.

AUSTRHEIM, H. 1990. The granulite–eclogite facies transition: a comparison of experimental work and a natural occurrence in the Bergen Arcs, western Norway. *Lithos*, **25**, 163–169.

AUSTRHEIM, H. 1991. Eclogite formation and dynamics of crustal roots under continental collision zones. *Terra Nova*, **3**, 492–499.

AUSTRHEIM, H. & MÖRK, M. B. E. 1988. The lower continental crust of the Caledonian mountain chain: evidence from former deep crustal sections in Norway. *Norges Geologisk Undersökelse Special Publication*, **3**, 102–133.

AUSTRHEIM, H., ERAMBERT, M. & BOUNDY, T. M. 1996. Garnets recording deep crustal earthquakes. *Earth and Planetary Science Letters*, **39**, 223–238.

AUSTRHEIM, H., ERAMBERT, M., & ENGVIK, A. K. 1997. Processing of crust in the root of the Caledonian continental collision zone: the role of eclogitization. *Tectonophysics*, **273**, 129–153.

BERNARD-GRIFFITHS, J., PEUCAT, J.-J. & MENOT, R.-P. 1991. Isotopic (Rb–Sr, U–Pb and Sm–Nd) and trace element geochemistry of eclogites from the pan-African Belt: a case study of REE fractionation during high-grade metamorphism. *Lithos*, **27**, 43–57.

BERNARD-GRIFFITHS, J., PEUCAT, J. J. & OHTA, Y. 1993. Age and nature of protoliths in the Caledonian blueschist- eclogite complex of western Spitsbergen: a combined approach using U–Pb, Sm–Nd and REE whole-rock systems. *Lithos*, **30**, 81–90.

BOSQUET, R., GOFFÉ, B., HENRY, P., LE PICHON, X. & CHOPIN, C. 1997. Kinematic, thermal and petrological model of the central Alps: Lepontine metamorphism in the upper crust and eclogitization of the lower crust. *Tectonophysics*, **273**, 105–127.

BOUFFETTE, J. & CARON, J.-M. 1991. Trajets metamorphiques prograde et retrograde dans des eclogites piemontaises (Ouest du Massif du Rocciavre, Alpes cottiennes) (Metamorphic prograde and retrograde P–T paths in Piemontese ophiolites, western Rocciavre Massif, Cottic Alps) *Comptes Rendus – Academie des Sciences, Serie II*, **312**, 1459–1465.

BOUNDY, T. M., FOUNTAIN, D. M. & AUSTRHEIM, H. 1992. Structural development and petrofabrics of eclogite facies shear zones, Bergen Arcs, western Norway: implications for deep crustal deformational processes. *Journal of Metamorphic Geology*, **10**, 127–146.

BRUECKNER, H. K., GILOTTI, J. A. & NUTMAN, A. P. 1998. Caledonian eclogite-facies metamorphism of early Proterozoic protoliths from the North-East Greenland Eclogite Province. *Contributions to Mineralogy and Petrology*, **130**, 103–120.

BUTLER, R. W. H. 1986. Thrust tectonics, deep structure and crustal subduction in the Alps and Himalayas. *Journal of the Geological Society, London*, **143**, 857–873.

CAMACHO, A., COMPSTON, W., McCULLOCH, M. & McDOUGALL, I. 1997. Timing and exhumation of eclogite facies shear zones, Musgrave Block, central Australia. *Journal of Metamorphic Geology*, **15**, 735–751.

CARSWELL, D. 1990. Eclogites and eclogite facies: definitions and classification. *In*: CARSWELL, D. A. (ed.) *Eclogite Facies Rocks*. Blackie, Glasgow, 1–13.

CASTAING, C., TRIBOULET, C., FEYBESSE, J. L. & CHEVREMONT, P. 1993. Tectonometamorphic evolution of Ghana, Togo and Benin in the light of the Pan-African/Brasiliano orogeny. *Tectonophysics*, **218**, 323–342.

CASTELLI, D. 1991. Eclogitic metamorphism in carbonate rocks: the example of impure marbles from the Sesia–Lanzo zone, Italian western Alps. *Journal of Metamorphic Geology*, **9**, 61–77.

CHAUVET, A., KIENAST, J. R., PINARDON, J. L. & BRUNEL, M. 1992. Petrological constraints and PT path of Devonian collapse tectonics within the Scandian mountain belt (Western Gneiss Region, Norway). *Journal of the Geological Society, London*, **149**, 383–400.

CHOPIN, C. 1987. Very-high-pressure metamorphism in the western Alps: implications for subduction of continental crust. *Philosophical Transactions of the Royal Society of London*, **321**, 183–197.

CLARKE, G. L., AITCHISON, J. C. & CLUZEL, D. 1997. Eclogites and blueschists of the Pam Peninsula, NE New Caledonia: a reappraisal. *Journal of Petrology*, **38**, 843–876.

CLIFF, R. A., BARNICOAT, A. C. & INGER, S. 1998. Early Tertiary eclogite facies metamorphism in the Monviso ophiolite. *Journal of Metamorphic Geology*, **16**, 447–455.

CLOOS, M. 1982. Flow mélanges: numerical modeling and geologic constraints on their origin in the Franciscan subduction complex, California. *Geological Society of America Bulletin*, **93**, 330–345.

COTKIN, S. J., VALLEY, J. W. & ESSENE, E. J. 1988. Petrology of a margarite-bearing meta- anorthosite from Seljeneset, Nordfjord, western Norway: implications for the P–T history of the Western Gneiss Region during Caledonian uplift. *Lithos*, **21**, 117–128.

CUTHBERT, S. J., CARSWELL, D. A., KROGH-RAVNA, E. J. & WAIN, A. 2000. Eclogites and eclogites in the Western Gneiss Region, Norwegian Caledonides. *Lithos*, **52**, 165–195.

DEWEY, J. F. 1988. The extensional collapse of orogens. *Tectonics*, **7**, 1123–1139.

DEWEY, J. F. & MANGE, M. 1999. Petrography of Ordovician and Silurian sediments in the western Irish Caledonides: tracers of a short-lived Ordovician continent-arc collision orogeny and the evolution of the Laurentian Appalachian–Caledonian margin. *In*: MAC NIOCAILL, C. & RYAN, P. D. (eds) *Continental Tectonics*. Geological Society, London, Special Publications, **164**, 55–107.

DEWEY, J. F., RYAN, P. D. & ANDERSEN, T. B. 1993. Orogenic uplift and collapse, crustal thickness, fabrics and metamorphic phase changes: the role of eclogites. *In*: PRICHARD, H. M., ALABASTER, T., HARRIS, N. B. W. & NEARY, C. R. (eds) *Magmatic Processes and Plate Tectonics*. Geological Society of London Special Publications, **76**, 325–343.

DI VINCENZO, G., RICCI, C. A., PALMERI, R., TALARICO, F. & ANDRIESSEN, P. A. M. 1997. Petrology and geochronology of eclogites from the Lanterman Range, Antarctica. *Journal of Petrology*, **38**, 1391–1417.

EL-DIN, E. A., COLEMAN, R. G. & LIOU, J. G. 1990. Eclogites and blueschists from northeastern Oman: petrology and P–T evolution. *Journal of Petrology*, **31**, 629–666.

ERDMER, P., GHENT, E. D., ARCHIBALD, D. A. & STOUT, M. Z. 1998. Paleozoic and Mesozoic high-pressure metamorphism at the margin of ancestral North America in central Yukon. *Bulletin of the Geological Society of America*, **110**, 615–629.

FAURE, M., LAIN, W. & SUN, Y. 1998. Doming in the southern foreland of the Dabieshan (Yangtze block, China). *Terra Nova*, **10**, 307–311.

FLOYD, P. A., WINCHESTER, J. A., SESTON, R., KRYZA, R. & CROWLEY, Q. G. 2000. Review of geochemical variation in Lower Palaeozoic metabasites from the NE Bohemian Massif: intracratonic rifting and plume-ridge interaction. *In*: FRANCKE, W., HAAK, V., ONCKEN, O. & TANNER, D. (eds) *Orogenic Processes: Quantification and Modelling in the Variscan Belt*. Geological Society, London, Special Publications, **157**, 155–174.

FRY, N. & BARNICOAT, A. C. 1987. The tectonic implications of high-pressure metamorphism in the western Alps. *Journal of the Geological Society, London*, **144**, 653–659.

GAO, S., ZHANG, B., JIN, Z. & KERN, H. 1999. Lower crustal delamination in the Qinling–Dabie orogenic belt. *Science in China, Series D: Earth Sciences*, **42**, 423–433.

GARDIEN, V. 1993. Les reliques petrologiques de haute a moyenne pression des series du Vivarais oriental (Est du Massif Central francais) (High to medium pressure relics in the eastern Vivarais series, eastern part of the French Massif Central). *Comptes Rendus – Academie des Sciences, Serie II*, **316**, 1247–1254.

GILOTTI, J. A. 1993. Discovery of a medium-temperature eclogite province in the Caledonides of northeast Greenland. *Geology*, **21**, 523–526.

GOMEZ-PUGNAIRE, M. T. & FERNANDEZ-SOLER, J. M. 1987. High-pressure metamorphism in metabasites from the Betic Cordilleras (SE Spain) and its evolution during the Alpine orogeny. *Contributions to Mineralogy and Petrology*, **95**, 231–244.

GRIFFIN, W. L. & BRUECKNER, H. K. 1985. Rb–Sr and Sm–Nd studies of Norwegian eclogites. *Chemical Geology*, **52**, 249–271.

GRIFFIN, W. L. & O'REILLY, S. Y. 1987. Is the continental Moho the crust-mantle boundary? *Geology*, **15**, 241–244.

GUILLOT, S., DE SIGOYER, J., LARDEAUX, J. M. & MASCLE, G. 1997. Eclogitic metasediments from the Tso Morari area (Ladakh, Himalaya): evidence for continental subduction during India–Asia convergence. *Contributions to Mineralogy and Petrology*, **128**, 197–212.

HACKER, B. R., IRELAND, T., WALKER, D., SHUWEN, D., RATSCHBACHER, L. & WEBB, L. 1998. U/Pb zircon ages constrain the architecture of the ultrahigh-pressure Qinling-Dabie Orogen, China. *Earth and Planetary Science Letters*, **161**, 215–230.

HIRN, A., LÉPINE, J. C., JOBERT, G. *ET AL.* 1984*a*. Crustal structure and variability of the Himalayan border of Tibet. *Nature*, **307**, 23–25.

HIRN, A, NERCESSIAN, A., SAPIN, M. *ET AL.* 1984*b*. Lhasa block and bordering sutures: a continuation of a 500 km Moho traverse through Tibet. *Nature*, **307**, 25–28.

HOLLOWAY, J. R. & WOOD, B. J. 1988. *Simulating the Earth: experimental geochemistry*. HarperCollins, London.

HUERTA, A. D., ROYDEN, L. H. & HODGES, K. V. 1996. The interdependence of deformational and thermal processes in mountain belts. *Science*, **273**, 637–639.

HYNES, A. & SNYDER, D. B. 1995. Deep-crustal mineral assemblages and potential for crustal rocks below the Moho in the Scottish Caledonides. *Geophysical Journal International*, **123**, 1–17.

INDARES, A. 1993. Eclogitized gabbros from the eastern Grenville Province: textures, metamorphic context, and implications. *Canadian Journal of Earth Sciences*, **30**, 159–173.

JAMIESON, R. A., BEAUMONT, C., FULLSACK, P. & LEE, B. 1997. Barrovian regional metamorphism: where's the heat? *In*: TRELOAR, P. J. & O'BRIEN, P. J. (eds) *What Drives Metamorphism and Metamorphic Reactions?* Geological Society, London, Special Publications, **138**, 23–51.

JAMTVEIT, B. 1987. Metamorphic evolution of the Eiksunddal eclogite complex, western Norway, and some tectonic implications. *Contributions to Mineralogy and Petrology*, **95**, 82–99.

KERN, H., POPP, T., JIN, S., GAO, S. & JIN, Z. 1999. Petrophysical studies on rocks from the Dabie ultrahigh-pressure (UHP) metamorphic belt, Central China: Implications for the composition and delamination of the lower crust. *Tectonophysics*, **301**, 191–215.

KIENAST, J. R., LOMBARDO, B., BIINO, G. & PINARDON, J. L. 1991. Petrology of very-high-pressure eclogitic rocks from the Brossasco-Isasca Complex, Dora-Maira Massif, Italian western Alps. *Journal of Metamorphic Geology*, **9**, 19–34.

KROGH, E. J. 1977. Evidence for a Precambrian continental collision in western Norway. *Nature*, **267**, 17–19.

LAUBSCHER, H. 1988. Material balance in Alpine orogeny. *Geological Society of America Bulletin*, **100**, 1313–1328.

LE PICHON, X., HENRY, P. & GOFFÉ, B. 1997. Uplift of Tibet: from eclogites to granulites – implications for the Andean Plateau and the Variscan Belt. *Tectonophysics*, **273**, 57–76.

LIATI, A. & SEIDEL, E. 1996. Metamorphic evolution and geochemistry of kyanite eclogites in central Rhodope, northern Greece. *Contributions to Mineralogy and Petrology*, **123**, 293–307.

LIU, J. & LIOU, J. G. 1995. Kyanite-anthophyllite schist and southwest extension of the Dabie Mountains ultrahigh- to high-pressure belt. *Island Arc*, **4**, 334–346.

MARCHANT, R. H. & STAMPFLI, G. M. 1997. Subduction of the continental crust in the Western Alps. *Tectonophysics*, **269**, 217–235.

MARUYAMA, S. & LIOU, J. G. 1998. Initiation of ultrahigh pressure metamorphism and its significance on the Proterozoic–Phanerozoic boundary. *Island Arc*, **7**, 6–35.

MATHEWS, D. & HIRN, A. 1984. Crustal thickening in the Himalayas and the Caledonides. *Nature*, **308**, 497–498.

MCKENZIE, D. P. 1969. Speculations on the consequences and causes of plate motions. *Geophysical Journal of the Royal Astronomical Society*, **18**, 1–32.

MEDARIS, L. G., MISAR, Z., FOURNELLE, J. H., GHENT, E. D. & JELINEK, E. 1998. Prograde eclogite in the Gfohl Nappe, Czech Republic: new evidence on Variscan high-pressure metamorphism. *Journal of Metamorphic Geology*, **16**, 563–576.

MEISSNER, R. & MOONEY, W. 1998. Weakness of the lower continental crust: a condition for delamination, uplift and escape. *Tectonophysics*, **296**, 47–60.

MENARD, G., MOLNAR, P. & PLATT, J. P. 1991. Budget of crustal shortening and subduction of continental crust in the Alps. *Tectonics*, **10**, 231–244.

MENGEL, K. 1992. Evidence from xenoliths for the composition of the lithosphere. *In*: BLUNDELL, D., FREEMAN, R. & MUELLER, S. (eds) *The European Geotraverse*. Cambridge University Press, Cambridge, 91–102.

MERCIER, L., LARDEAUX, J.-M. & DAVY, P. 1991. On the tectonic significance of retrograde P-T-t paths in eclogites of the French Massif Central. *Tectonics*, **10**, 131–140.

MIDGLEY, J. P. & BLUNDELL, D. J. 1997. Deep seismic structure and thermo-mechanical modelling of continental collision zones. *Tectonophysics*, **273**, 155–167.

MÖLLER, C. 1999. Sapphirine in SW Sweden: A record of Sveconorwegian (-Grenvillian) late-orogenic tectonic exhumation. *Journal of Metamorphic Geology*, **17**, 127–141.

MÖLLER, C. 1998. Decompressed eclogites in the Sveconorwegian (-Grenvillian) orogen of SW Sweden: petrology and tectonic implications. *Journal of Metamorphic Geology*, **16**, 641–656.

MOLNAR, P., ENGLAND, P. & MARTINOD, J. 1993. Mantle dynamics, uplift of the Tibetan plateau, and the Indian monsoon. *Review of Geophysics*, **31**, 357–396.

MOLNAR, P. & GRAY, D. 1979. Subduction of continental lithosphere: some constraints and uncertainties. *Geology*, **7**, 58–62.

NICOLLET, C. 1989. L'eclogite de Faratsiho (Madagascar): un cas exceptionnel de metamorphisme de haute-P–basse-T au Proterozoique Superieur (The Faratsiho eclogite, Madagascar: a record of late Proterozoic high-pressure–low-temperature metamorphism). *Precambrian Research*, **45**, 343–352.

O'BRIEN, P. J. 1993. Partially retrograded eclogites of the Munchberg Massif, Germany: records of a multi-stage Variscan uplift history in the Bohemian Massif. *Journal of Metamorphic Geology*, **11**, 241–260.

PAGE, L. M. 1992. Pressure-temperature-time constraints for the Seve Nappe of the Singis-Tjuoltajaure area central Norrbotten Caledonides, Sweden: implications for early Caledonian marginal Baltica. *Geodinamica Acta*, **5**, 3–16.

PEACOCK, S. M. 1995. Ultrahigh-pressure metamorphic rocks and the thermal evolution of continent collision belts. *Island Arc*, **4**, 376–383.

PENNACCHIONI, G. 1996. Progressive eclogitization under fluid-present conditions of pre-Alpine mafic granulites in the Austroalpine Mt Emilius Klippe (Italian Western Alps). *Journal of Structural Geology*, **18**, 549–561.

PERCHUK, A. & PHILIPPOT, P. 1997. Rapid cooling and exhumation of eclogitic rocks from the Great Caucasus, Russia. *Journal of Metamorphic Geology*, **15**, 299–310.

PFIFFNER, A. 1992. The Alpine Orogeny. *In*: BLUNDELL, D., FREEMAN, R. & MUELLER, S. (eds) *The European Geotraverse*. Cambridge University Press, Cambridge, 180–190.

POGNANTE, U. & SPENCER, D. A. 1991. First report of eclogites from the Himalayan belt, Kaghan Valley (northern Pakistan). *European Journal of Mineralogy*, **3**, 613–618.

RICHARDSON, S. W. & ENGLAND, P. C. 1979. Metamorphic consequences of crustal eclogite production in overthrust zones. *Earth and Planetary Science Letters*, **42**, 183–190.

RYAN, P. D. & DEWEY, J. F. 1997. Continental eclogites and the Wilson Cycle. *Journal of the Geological Society, London*, **154**, 437–442.

RYAN, P. D. & SOPER, N. J. 2001. Modeling anatexis in intra-cratonic rift basins: an example from the Neoproterozoic of the Scottish Highlands. *Geological Magazine*, in press.

SANDERS, I. S. 1989. Phase relations and P-T conditions for eclogite-facies rocks at Glenelg, northwest Scotland. *In*: DALY, J. S., CLIFF, R. A. & YARDLEY, B. W. D. (eds) *Evolution of Metamorphic Belts*. Geological Society, London, Special Publication, **43**, 513–517.

SANDERS, I. S., DALY, J. S., DAVIES, G. R. 1987. Late Proterozoic high-pressure granulite facies metamorphism in the north-east Ox inlier, northwest Ireland. *Journal of Metamorphic Geology*, **5**, 69–85.

SAPIN, M. & HIRN, A. 1997. Seismic structure and evidence for eclogitization during the Himalayan convergence. *Tectonophysics*, **273**, 1–16.

SCHLIESTEDT, M. 1986. Eclogite–blueschist relationships as evidenced by mineral equilibria in the high-pressure metabasic rocks of Sifnos (Cycladic Islands), Greece. *Journal of Petrology*, **27**, 1437–1459.

SCHMADICKE, E. 1991. Quartz pseudomorphs after coesite in eclogites from the Saxonian Erzgebirge. *European Journal of Mineralogy*, **3**, 231–238.

SMITH, D. C. 1984. Coesite in clinopyroxene in the Caledonides and its implications for geodynamics. *Nature*, **310**, 641–644.

STELTENPOHL, M. G., CYMERMAN, Z., KROGH, E. J.; Kunk, M. J. 1993. Exhumation of eclogitized continental basement during Variscan lithospheric delamination and gravitational collapse, Sudety Mountains, Poland. *Geology*, **21**, 1111–1114.

STÜWE, K. 1998. Heat sources of Cretaceous metamorphism in the eastern Alps – a discussion. *Tectonophysics*, **287**, 251–269.

VALASEK, P., MUELLER, S., FREI, W. & HOLLINGER, K. 1991. Results of NFP20 seismic reflection profiling along the Alpine sector of the European Geotraverse (EGT). *Geophysics Journal International*, **105**, 85–102.

WANG, X. & LIOU, J. G. 1991. Regional ultrahigh-pressure coesite-bearing eclogitic terrane in central China: evidence from country rocks, gneiss, marble, and metapelite. *Geology*, **19**, 933–936.

WANG, X. & LIOU, J. G. 1993. Ultra-high-pressure metamorphism of carbonate rocks in the Dabie Mountains, central China. *Journal of Metamorphic Geology*, **11**, 575–588.

WANG, X., LIOU, J. G. & MAO, H. K. 1989. Coesite-bearing eclogite from the Dabie Mountains in central China. *Geology*, **17**, 1085–1088.

WATERS, C. N. 1989. The metamorphic evolution of the Schistes lustres ophiolite, Cap Corse, Corsica. *In*: DALY, J. S., CLIFF, R. A. & YARDLEY, B. W. D. (eds) *Evolution of Metamorphic Belts*. Geological Society, London, Special Publications, **43**, 557–562.

WIJBRANS, J. R., VAN WEES, J. D., STEPHENSON, R. A. & CLOETINGH, S. A. P. L. 1993. Pressure–temperature–time evolution of the high-pressure metamorphic complex of Sifnos, Greece. *Geology*, **21**, 443–446.

ZEN, E. A. 1995. Crustal magma generation and low-pressure high-temperature regional metamorphism in an extensional environment: possible application to the Lachlan Belt, Australia. *American Journal of Science*, **295**, 851–874.

ZHAO, W. J., NELSON, D. K. & THE INDEPTH TEAM. 1993. Deep seismic reflection evidence for continental underthrusting beneath southern Tibet. *Nature*, **366**, 557–559.

When the Wilson Cycle breaks down: how orogens can produce strong lithosphere and inhibit their future reworking

MAARTEN KRABBENDAM

Australian Crustal Research Centre, Department of Earth Sciences, Monash University, Melbourne VIC 3800, Australia
Present address: British Geological Survey, Murchison House, West Mains Road, Edinburgh EH9 3LA, UK (e-mail: mkrab@bgs.ac.uk)

Abstract: Although poly-cyclicity is common, many orogens show a remarkable lack of reworking. In this paper, a review of some factors that may either enhance or inhibit reworking of orogens is presented. As a general rule, orogens are unlikely to rift and rework if their lithospheric strength is higher than adjacent lithosphere. The strength of the lithosphere is strongly dependent on the geothermal gradient and the rheology of the rocks; both these factors can depend on the preceding orogenic evolution, even several hundred Ma after orogenesis. Strong orogenic lithosphere is expected if the crust is composed of material with a low radiogenic heat production capacity, such as island arcs, or if the underlying sub-continental lithosphere is still thickened, as in the Urals. Extensive dehydration metamorphism, a concentration of radiogenic heat production in the upper crust and erosional thinning of the orogenic crust can also strengthen the lithosphere and inhibit reworking. However, proximity of Archean cratons and anomalously high mantle heat flow appear to strongly enhance susceptibility to reworking.

To understand orogenic poly-cyclicity and re-working, it is important not only to study factors that favour reworking, but also to study factors that *inhibit* reworking of orogens. Many orogenic belts show evidence of extensive large-scale reworking (for instance the Pan-African reworking of Kibaran orogens and the Alpine reworking of Variscan orogens). Other orogens, however, show a remarkable lack of reworking. The Limpopo Belt, for example, has not been reworked since 2200 Ma and the Urals did not rift during the break-up of Laurasia. This raises the question: Why are some orogens reworked and others not?

Reworking can occur in different ways, on different scales and for different reasons. In this contribution only reworking on a large, orogen scale is dealt with. Reworking can involve horizontal crustal extension or shortening, or both. In intra-continental settings, large-scale reworking can affect a particular domain or zone of a lithospheric plate if this domain is weaker than other domains, and plate boundary stresses are transferred through the plate to reach the weak domain. In the absence of substantial plate boundary stresses, a particular domain within a plate may also be reworked if it has a different gravitational potential energy (and may also be weaker) than adjacent domains within the same plate (e.g. Neil & Houseman 1999). Large-scale reworking can also, and probably most commonly, involve rifting and continental break-up

along an older orogen (Wilson 1966). Subsequent ocean closure and collision results in thorough reworking of the first orogen, typically several 100 Ma after the first orogenic cycle. This paper focuses on such 'Wilson Cycle-type' reworking. The critical stage in the Wilson Cycle is the continental break-up event, because all oceans will eventually close. Continental break-up occurs along zones that are either weaker or have a higher gravitational potential (for instance because of the arrival of a mantle plume) than adjacent lithosphere, or both.

Many papers on reworking of orogens have focussed on the reactivation of pre-existing anisotropies, such as faults and fabrics during rifting (e.g. Ring 1994; Vauchez et al. 1997). This phenomenon is undoubtedly important, particularly so on the small scale and at shallow levels in the brittle field. It needs to be appreciated however that virtually all of the Earth's crust is orogenic, therefore, most of the Earth's lithosphere is mechanically anisotropic in some form or another. Krabbendam & Barr (2000) showed that during the break-up of Gondwana only half of the total rift length was sub-parallel to pre-existiing orogenic anisotropies. Half of Gondwana rifts cross-cut orogenic trends at high angles or cut trough cratonic nucleii. Less than half of the total length of Mesoproterozoic or younger orogens within Gondwana was reworked during supercontinent break-up (Krabbendam & Barr 2000). This suggests that the role of pre-existing

From: MILLER, J. A., HOLDSWORTH, R. E., BUICK, I. S. & HAND, M. (eds) *Continental Reactivation and Reworking*. Geological Society, London, Special Publications, **184**, 57–75. 1-86239-080-0/01/$15.00 © The Geological Society of London 2001.

anisotropies during rifting may be somewhat overrated and that other factors, discussed here, are equally important.

This paper focuses on additional reasons why some orogens experienced subsequent rifting and break-up whilst others did not. The answer to this problem is in part related to the (long-term) lithospheric strength of the orogen with respect to adjacent lithosphere. Thus, it is useful to compare the lithospheric strength of that particular orogen with the strength of the adjacent lithosphere (Fig. 1). Reviewed here are some factors that may either inhibit or enhance reworking of orogens, by comparing the strength of several different lithospheric models. These factors can be divided into two groups: factors that are unrelated to orogenesis (proximity to Archean cratons and anomalously high Moho heat flow), and factors that are related to orogenesis, for example distribution of radioactive elements, dehydration metamorphism and the thickness of subcontinental lithospheric mantle. Comparing the North Atlantic Caledonides and the Urals will test the applicability of this approach. First, however, it must be determined how reworked and non-reworked orogens can be objectively compared.

Comparing reworked and non-reworked orogens

A thoroughly reworked terrane provides very limited geological information about the previous orogenic cycle(s) that such a terrane has undergone. Structures are transposed, metamorphic assemblages are overprinted, rocks may be (re)melted and isotopic resetting occurs. It is therefore difficult to objectively compare the first cycle of a reworked orogen with an orogen that has not been reworked. For example, much more information can be retrieved from the monocyclic Palaeoproterozoic Halls Creek Orogen in Northern Australia than from the Palaeoproterozoic SW Scandinavian Event in West Norway, which has been intensely reworked during the Caledonian Orogeny.

One solution to this problem is to look at orogens that have rifted and are *going to be further reworked in the future*. These orogens are currently positioned along passive margins and have rifted during the break-up of Pangea (see Fig. 2). These belts have experienced a first orogenic cycle and the first stages of reworking (the rifting and continental break-up stages) of the second Wilson Cycle, but not yet the subsequent stages of subduction, ocean closure and collision. In contrast, several orogenic belts did not rift during the break-up of Pangea (see Fig. 2) and, so far, many of these orogens have not been reworked. Some of these orogens may have been in an unfavourable orientation with respect to Pangea rift extension directions. Other orogens, however, may simply have a lithospheric strength higher than that of other orogenic belts. Future rifting and subsequent collision of these orogens is possible, but reworking is less likely than in the orogens currently positioned along passive margins.

Lithospheric strength under horizontal extension

In this paper, the strength under horizontal extension of particular domains of lithosphere relative

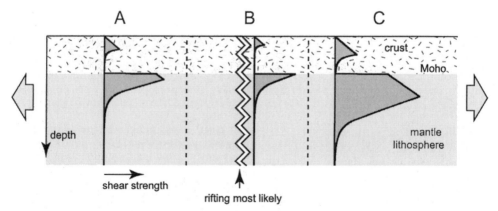

Fig. 1. A portion of continental lithosphere is subdivided into three domains, each with a different strength under horizontal extension. The profiles show schematically the variation of strength with depth; the area enclosed by each profile (the lithospheric strength integrated over depth) is a measure of the total strength of the lithosphere. Domain B is the weakest domain, is most likely to rift under horizontal extension and most susceptible to reworking. Possible causes of the different lithospheric strengths are discussed in the text.

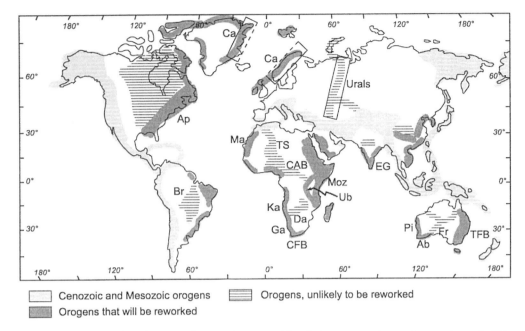

Fig. 2. Global distribution of orogens. Orogens that are positioned along present-day passive margins and will be reworked in the future include the Albany Belt (Ab), Appalachians (Ap), Caledonides (Ca), Cape Fold Belt (CFB), Eastern Ghats (EG), Gariep Belt, (Ga), Kaoko Belt (Ka), Mauretanide Belt (Ma), Mozambique Belt (Moz), Pinjarra Belt (Pi) and the Tasman Fold Belt (TSB). Orogens that are positioned in present-day plate interiors have not rifted during Pangea break-up and are less likely to be reworked in the future. These orogens include the Brasiliano Belt (Br), Central African Belt (CAB), Damara Belt (Da), Fraser Belt (Fr), Trans-Saharan Belt (TS) and the Urals and most of the Canadian Shield. See text for discussion concerning the Ubende Belt (Ub).

to others is discussed (Fig. 1). The strength of the lithosphere is the maximum differential stress $(\sigma_1 - \sigma_3)$ that can be supported by the lithosphere subjected to horizontal extension at a constant strain rate (in this paper assumed to be $10^{-15}\,\text{s}^{-1}$). The strength of the lithosphere depends mainly on the geothermal gradient, the rheology of the rocks involved, and the thickness of the crust and the lithospheric mantle (e.g. Brace & Kohlstedt 1980; Kusznir & Park 1987). An orogenic cycle can affect the values of these parameters in a variety of ways, even several 100 Ma after cessation of orogenesis. For instance, the geothermal gradient depends on the amount and distribution of radioactive elements in the crust, which may change due to orogenesis (e.g. Sandiford & Hand 1998). The crust also may remain overthickened for several 100 Ma after cessation of deformation, as in the Urals (see below). The effects on the strength of the lithosphere of variations in lithospheric and crustal thickness, distribution of radioactive elements in the crust and changing rheologies in the crust are assessed.

Calculation of lithospheric strength

The strength of the upper crust and, at low temperatures, also the upper mantle is constrained by the brittle failure envelope and mainly depends on the effective lithostatic pressure. The strength of the lower crust and lower lithospheric mantle is assumed to be constrained by the crystal plastic behaviour of quartz and olivine respectively, and can be calculated from power-law creep laws that strongly depend on temperature. For simplicity, the lithosphere is assumed to deform by depth invariant pure shear.

The brittle failure strength of the upper crust and mantle, derived from the Coulomb failure criterion (e.g. Brace & Kohlstedt 1980), is given as follows:

$$(\sigma_1 - \sigma_3) = 0.577(\rho g z - P_f) \qquad (1)$$

where, g, is the gravitational acceleration; ρ, the density; z, the depth; and P_f, the pore fluid pressure (assumed to be 0). Different densities were used for the crust, mantle and, in some cases, the lower crust (Table 1, 3). Ductile

Table 1. *Constants used for all thermal and rheological models*

Constants	Symbol	Value
Thermal conductivity	k	$2.25\,\mathrm{Wm^{-1}\,K^{-1}}$
Gravitational acceleration	g	$9.8\,\mathrm{m\,s^{-2}}$
Gas constant	R	$8.314\,\mathrm{J\,mol^{-1}\,K^{-1}}$
Average strain rate	$\dot{\varepsilon}$	$10^{-15}\,\mathrm{s^{-1}}$
Density of the mantle	ρ_m	$3300\,\mathrm{kg\,m^{-3}}$
Radioactive heat production, mantle	H_m	$0\,\mu\mathrm{Wm^{-3}}$
Power law pre-exponent, mantle	A_m	$2860\,\mathrm{MPa^n\,s^{-1}}$
Power law sensitivity, mantle	n_m	3.6
Power law activation, mantle	Q_m	$535\,\mathrm{kJ\,mol^{-1}}$
Mantle Dorn Law stress threshold	σ_d	$8500\,\mathrm{MPa}$
Mantle Dorn Law strain rate	$\dot{\varepsilon}_d$	$5.70\,10^{11}\,\mathrm{s^{-1}}$
Mantle Dorn Law activation	Q_d	$540\,\mathrm{kJ\,mol^{-1}}$

deformation of the crust and mantle (except in the upper mantle, see below) is modelled as power law creep:

$$(\sigma_1 - \sigma_3) = (\dot{\varepsilon}A)^{1/n}\exp(Q/nRT(z)) \qquad (2)$$

where, $\dot{\varepsilon}$, is the strain rate; A, the power law coefficient; n, the power law exponent; Q, the power law activation energy; R, the gas constant; and $T(z)$, the temperature at depth z. The values of A_c, Q_c and n_c (Tables 1 to 3) represent the rheology of quartz (Brace & Kohlstedt 1980; Ranalli 1992), except for dry granulitic lower crust where values where taken from Wilks & Carter (1990). The values of A_m, Q_m and n_m for the mantle were determined from laboratory experiments on olivine (Brace & Kohlstedt 1980).

In those parts of the upper mantle where $(\sigma_1 - \sigma_3) > 200\,\mathrm{MPa}$, ductile deformation is modelled as Dorn law creep:

$$(\sigma_1 - \sigma_3) = \sigma_d(1 - ((RT(z)/Q_d\ln(\dot{\varepsilon}/\dot{\varepsilon}_d))^{0.5} \qquad (3)$$

where, σ_d, is the mantle Dorn plasticity law stress threshold; Q_d, the mantle Dorn plasticity law activation energy; and $\dot{\varepsilon}_d$, the mantle Dorn plasticity law strain rate (Ranalli 1992). The strength at a given depth is either the brittle or the ductile strength, whichever is lower. The total strength of the lithosphere is calculated by integrating the strength over depth (e.g. Kusznir & Park 1987).

Table 2. *Parameters and calculated results used for single layer models*

Geothermal – Parameters	Symbol	Unit	Reference model	Archean model	High heat flow prov. model	Felsic granulite model	Ural 50 km model	Ural 35 km model
Surface heat flow	Q_0	$\mathrm{mW\,m^{-2}}$	55	40	68	41	35	37
Total crustal thickness	Z_{moho}	km	35	35	35	35	50	35
Density crust	ρ_c	$\mathrm{kg\,m^{-3}}$	2800	2800	2800	2800	2800	2800
Heat production crust	H_c	$\mu\mathrm{W\,m^{-3}}$	1.0	0.7	1.0	0.6	0.45	0.445
Crustal rheology								
Crustal power law pre-exponent	A_c	$\mathrm{MPa^n\,s^{-1}}$	6.7×10^{-6}	6.7×10^{-6}	6.7×10^{-6}	8.0×10^{-3}	6.7×10^{-6}	6.7×10^{-6}
Crustal power law sensitivity	n_c		2.4	2.4	2.4	3.1	2.4	2.4
Crustal power law activation energy	Q_c	$\mathrm{kJ\,mol^{-1}}$	156	156	156	243	156	156
Calculated results:								
Heat flow at Moho	Q_d	$\mathrm{mW\,m^{-2}}$	20	15.5	33	20	19	21
Moho temperature	T_{moho}	°C	582	431	784	474	527	454
Base lithosphere ($T=1300°\mathrm{C}$)	L_z	km	116	160	70		189	124
Integrated lithospheric strength	σ_{tot}	GNm	107	312	13	221	276	207
See Figure			3a	3b	3c	5c	9a	9b

Table 3. *Parameters and calculated results used for two-layer models*

Geothermal – Parameters	Symbol	Unit	Reference model	Layered reference model	Felsic lower crust model	Mafic lower crust model
Surface heat flow	Q_o	mW m^{-2}	55	55	55	40
Total crustal thickness	Z_{moho}	km	35	35	35	35
Crustal parameters						
Thickness upper crust	D_{uc}	km	n/a	20	20	20
Density upper crust	ρ_{uc}	kg m^{-3}	2800	2800	2800	2800
Heat production upper crust	H_{uc}	µW m^{-3}	1.0	1.3	1.3	0.7
Thickness lower crust	D_{lc}	km	n/a	15	15	15
Density lower crust	ρ_{lc}	kg m^{-3}	2800	2850	2850	2850
Heat production lower crust	H_{lc}	µW m^{-3}	1.0	0.6	0.6	0.4
Crustal rheology						
The rheology of the upper crust is modelled with a quartz rheology (see Table 2)						
Power law pre-exponent, lower crust	A_{lc}	MPa^n s^{-1}	6.7×10^{-6}	6.7×10^{-6}	8.0×10^{-3}	1.4×10^{4}
Power law sensitivity, lower crust	n_{lc}			2.4	3.1	4.2
Power law activation energy, lower crust	Q_{lc}	kJ mol^{-1}	156	156	243	445
Calculated results:						
Heat flow at Moho	Q_d	mW m^{-2}	20	20	20	20
Temperature of Moho	T_{moho}	°C	582	536	536	446
Base lithosphere ($T = 1300$°C)	L_z	km	116	121	121	131
Integrated lithospheric strength	σ_{tot}	GNm	107	156	161	316
See Figure			3a	3d	5a	5c

Calculation of geothermal gradients

Because of the temperature dependency of power-law creep (see equation [2]), the geothermal gradient exerts an important control on the lithospheric strength. The geothermal gradient depends on the amount and distribution of radiogenic heat production in the crust, the heat flow across the Moho (hereafter called 'Moho heat flow') and the heat conductivity. Rifting, where part of a second Wilson cycle, commonly takes place between 200 and 500 Ma after orogenesis. In these cases, it is justified to assume an equilibrium geothermal gradient just before rifting. I solve the steady-state heat flow equation, assuming constant temperature at the surface ($=0$°C) and constant surface heat flow. The geothermal gradient in crust with a uniform distribution of radiogenic heat production is calculated as:

$$T(z) = -Hz^2/2k + Q_0z/k$$
$$\text{for } 0 \le z \le z_{moho} \quad (4)$$

where, $T(z)$, is the temperature at depth z; H, is the radiogenic heat production; Q_0, the surface heat flow; and k, the thermal conductivity

(e.g. Fowler 1990). The radiogenic heat production in the mantle is taken as zero, so that:

$$T(z) = (\delta t/\delta z)|_{z=z_{moho}}(z - z_{moho}) + T_{z_{moho}} \quad (5)$$

where $(\delta t/\delta z)|_{z=z_{moho}}$ and $T_{z_{moho}}$ are calculated from equation [4].

For some lithospheric models an upper crust that is enriched and a lower crust that is depleted in radioactive elements are invoked. For these two-layer models (e.g. 'felsic and mafic models', see below) the geothermal gradients is calculated by:

$$T(z) = -H_az^2/2k + Q_0z/k$$
$$\text{for } 0 \le z \le za \quad (6)$$

$$T(z) = [-H_a(z - z_a)^2]/2k$$
$$+ (\delta t/\delta z)|_{z=z_a}(z - z_a) + T_{z_a}$$
$$= [-H_a(z - z_a)^2]/2k$$
$$+ [(H_az_a + Q_0)/k](z - z_a)$$
$$- H_az^2/2k + Q_0z_a/k$$
$$\text{for } z_a < z \le z_{moho} \quad (7)$$

where, H_a and H_b, are the radiogenic heat production for the upper and lower crust, respectively; and z_a, is the depth of the bottom of the upper crust. Constants used for all thermal and rheological models are shown in Table 1; parameters and calculated results for single layer crustal models are shown in Table 2 and for two-layered crustal models in Table 3.

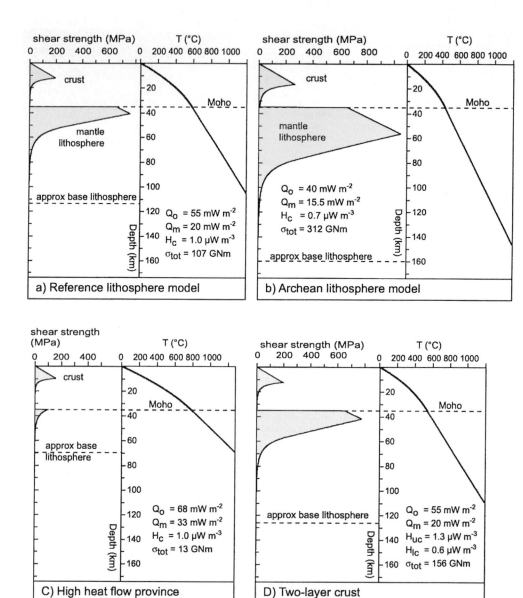

Fig. 3. Rheological profiles and geothermal gradients of four lithospheric models. See text for explanation and Tables 1 & 2 for parameters used. A quartz rheology has been applied to the crust in all four models. (**a**) The reference lithospheric model. (**b**) Archean lithospheric model, representing the strength and geothermal gradient of an Archean lithosphere at present-day conditions. (**c**) Lithospheric model of a high heat flow province. (**d**) Lithospheric model with a crust with an upper crust enriched and a lower crust depleted in radioactive elements. All parameters other than surface heat flow (Q_0), mantle heat flow (Q_m) and radioactive heat production (H_c) are the same as the reference model.

Strength of the reference lithosphere

The strength of a reference lithosphere was calculated based upon a normal geothermal gradient (with $Q_0 = 55\,mW\,m^{-2}$ and a uniformly distributed radiogenic heat production of $1.0\,\mu Wm^{-3}$), a crustal thickness of 35 km and a quartz rheology throughout the crust (see also Tables 1, 2 and Fig. 3a). The integrated strength of this reference model is 107 GNm, for a strain rate of $10^{-15}\,s^{-1}$. The absolute value of this figure may not be precise (mainly due to the uncertainties of the values of the constants Q, A and n) but it serves as a benchmark against which the other lithospheric models are compared.

Factors unrelated to orogenesis

In the following sections, some factors that affect the geothermal gradient and the strength of the lithosphere are evaluated, and hence may determine whether a particular domain of the lithosphere is susceptible to reworking. Factors that are unrelated to orogenic processes (proximity to Archean cratons and anomalously high Moho heat flow) are assessed below.

Proximity to Archean cratons

Archean lithosphere is colder and therefore thicker and stronger than younger lithosphere. Such lithosphere is gravitationally stable because Archean mantle lithosphere is depleted in iron and has a lower density than younger mantle lithosphere at the same temperature (e.g. Durrheim & Mooney 1991). Figure 3b shows a geothermal gradient and rheological profile of Archean lithosphere under present-day conditions. The surface heat flow is an average for Archean cratons (Sclater et al. 1980) and the radiogenic heat production is an average of measurements from the Superior Province, Canada (Ashwal et al. 1987). Archean lithosphere is stronger than any of the other lithospheric models presented in the rest of the paper and about three times stronger than the reference lithosphere (Fig. 3). This implies that most terranes *adjacent* to Archean lithosphere will be relatively weak and susceptible to reworking, regardless of their own evolution. Interestingly, the presence of weak orogenic belts may in turn actually increase the longevity and stability of thick Archean lithosphere, by buffering cratons from tectonic stresses (Lenardic et al. 2000).

An example of an orogenic belt situated between two Archean cratons is the Paleoproterozoic Ubende Belt (Fig. 2) in Tanzania. The

Ubende Belt comprises mafic amphibolite- and granulite-facies rocks (Theunissen et al. 1996) and is probably significantly stronger than the reference lithosphere (see below). The Ubende Belt was, however, reactivated during the Neoproterozoic and again during Cenozoic rifting of East Africa (Theunissen et al. 1996). This reworking was *primarily* guided by its position between the Tanzania Craton in the east and the Zambia (or Bangweulu) Block in the west. In other settings, such a strong belt may not have been reworked.

Anomalously high Moho heat flow

Many continental regions have high surface heat flow, for instance $80–120\,mW\,m^{-2}$ for large parts of Germany (Cermák 1995) and $80–110\,mW\,m^{-2}$ in northern Algeria (Lesquer et al. 1990). Some of these high heat flow provinces cannot be directly explained by previous orogenesis, by high radiogenic heat production of the crust or by fluid movements in the upper crust. Instead, they may be caused by anomalously high Moho heat flow ($40–60\,mW\,m^{-2}$). Mantle plumes may be responsible for such high heat flow provinces, but only if the lithospheric plate moved slowly with respect to mantle plumes. Whatever their cause, provinces with high surface *and* mantle heat flow are likely to have a low lithospheric strength.

The strength of lithosphere of such high heat flow provinces is calculated with conservative values of surface heat flow ($68\,mW\,m^{-2}$) and Moho heat flow ($33\,mW\,m^{-2}$, Table 2, Fig. 3c). Such lithosphere is about seven times weaker than the reference lithosphere (Fig. 3a), making it very susceptible to rifting and reworking. If caused by (slow moving) mantle plumes, the increase in gravitational potential energy may create horizontal deviatoric stresses and cause extension, rifting and, ultimately, continental break-up (e.g. Houseman & Hegarty 1987).

Factors related to orogenesis

In this section, some factors that are related to orogenesis are evaluated. If reworking would occur, these factors apply to the first orogenic cycle. Some effects are directly related to orogenesis such as the extent of dehydration metamorphism and partial melting. Some factors depend on how the thickened orogenic crust is thinned (by erosion or by late-orogenic extension). Other factors ultimately depend on the evolution of the orogenic system before collision, such as whether many island arcs occurred in the ocean that was closed.

Amount and distribution of radiogenic heat production

The amount of radiogenic heat producing elements in a particular crustal domain is of obvious importance to the geothermal gradient (see equation (4)) and hence to the strength of the lithosphere. A crustal domain with low radiogenic heat production will have a relatively shallow geothermal gradient and low surface heat flow (assuming the same or lower Moho heat flow), and therefore be relatively strong. Low radiogenic heat production is not only expected in Archean cratons (caused by long-term radioactive decay), but also in orogenic crust rich in mafic material that results from abundantly incorporated oceanic and island arc material (e.g. the Urals and the Arabian–Nubian Shield). Thus, the pre-collisional evolution of the first orogenic cycle may determine whether an orogen is susceptible to rifting and reworking. Orogeny resulting from closure of a wide ocean basin with numerous island arcs (e.g. Urals) will produce a strong orogen. Conversely, an orogen involving closure of a small ocean basin without island arcs and composed mainly of continental material with relatively high radiogenic heat production (e.g. Alps and Damara Belt) will produce a hot and weak lithosphere.

The vertical distribution of heat production is commonly assumed to decrease exponentially with depth (e.g. Haack 1983). Although partial melting and granulite-facies metamorphism of the lower crust provide plausible mechanisms for depletion of radioactive elements of the lower crust (Taylor & McLennan 1985), there is no *a priori* geochemical reason why the vertical distribution should follow such a neat mathematical formula. The vertical distribution of heat production can in fact have a distinct layering. It is also possible that an upper crustal layer has a *lower* heat production than a middle crustal layer, for instance because of (overthrust?) greenstone/island arc material (Ashwal et al. 1987) or because of deposition of sediments poor in radioactive elements (Sandiford & Hand 1998). An upper crustal layer with low heat production and low thermal conductivity can trap heat produced in the middle crust. Such blanketing results in heating of the crust and lithosphere and, hence, in *weak* lithosphere (Sandiford & Hand 1998).

Here, the effect of an uneven distribution of radiogenic heat production on the lithospheric strength is calculated from a two-layered crust with high heat production in the upper crust and low heat production in the lower crust (Fig. 3d). The total amount of radiogenic heating, surface heat flow and the Moho heat flow are the same as in the reference lithosphere, but the upper crust has a radiogenic heat production that is twice as high as the lower crust (Table 3). The lithosphere with the two-layered crust is almost 1.5 times *stronger* than the reference lithosphere (Fig. 3a), because the concentration of heat production in the upper crust does not contribute much to heating of the lithosphere. Therefore, enrichment of radioactive elements in the upper crust with concomitant depletion in the lower crust eventually *strengthens* the lithosphere, if no blanketing occurs.

Dehydration metamorphism

Metamorphism and partial melting, caused by orogenesis, can result in large-scale dehydration of the lower crust, particularly when temperatures increase above 600–800°C and granulite-facies conditions are reached. These anhydrous ('water under-saturated') metamorphic rocks contain virtually no free water (Yardley & Valley 1989). The absence of free water and the scarcity of hydrous minerals (which are generally weaker phases) in granulite-facies rocks are expected to result in a higher strength than lower grade rocks of comparable composition, once a normal geothermal gradient is re-established. Deformation experiments of felsic and mafic granulites by Wilks & Carter (1990) suggest that these rocks are stronger (at similar temperatures) than quartz and granite-based rheologies. In addition to dehydration metamorphism, partial melting caused by high-temperature metamorphism will preferentially remove radioactive elements from the lower crust and transport these elements together with escaping melt to higher crustal levels. This will further strengthen the lithosphere because of lower heat production (see previous section).

P–T paths of three imaginary orogenic cycles are shown in Figure 4. Terrane (a), characteristic of medium pressure and temperature evolution, reaches a maximum temperature of about 500°C and metamorphic assemblages will still contain about 1% water within its hydrous minerals (Bousquet et al. 1997). Terrane (b), characteristic of high pressure and medium temperature evolution, reaches a maximum temperature of about 650°C; wet granite melting will occur and considerable dehydration may occur. In such high pressure–medium temperature terranes, the exhumation mechanism and exhumation rates will determine how much heating will occur after peak-pressure conditions (e.g. Grasemann et al. 1998). Terrane (c) experiences high temperature metamorphism and reaches a maximum

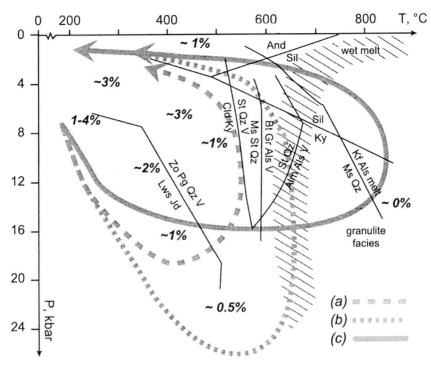

Fig. 4. P–T diagram with water content (wt.%) within stable mineral assemblages of granodioritic composition (after Bousquet *et al.* 1997). Three P–T paths are shown. P–T path (**a**) reaches T-max of about 500°C; assemblages will still contain about 1% water. P–T path (**b**) reaches T-max of about 650°C; wet granite melting will occur and considerable dehydration may occur. P–T path (**c**) reaches T-max of about 800°C well into the granulite-facies; the rocks will become virtually anhydrous. Some dehydration reactions are indicated (after Yardley 1989; Evans 1990).

temperature of about 800°C well into the granulite-facies. At these conditions, dehydration is severe and the terrane becomes virtually anhydrous. A terrane that underwent granulite-facies metamorphism in an 'anticlockwise' ('clockwise' in Fig. 4) manner, such as Namaqualand (Waters 1989), will also dehydrate; so the direction of the P–T path is not relevant here.

Two rheological profiles, with lower crustal rheologies of felsic and mafic granulite respectively, are shown in Figure 5a and b. The Moho heat flow is the same as in the reference lithosphere model and in the two-layered model with crustal quartz rheology (Figs 3 and 6). The lithosphere with felsic granulite in the lower crust (Fig. 5a) is only a few percent stronger than the two-layer lithosphere with a quartz rheology (Fig. 3d), whereas the lithosphere with lower crust of mafic granulite (Fig. 5b) is twice as strong.

An entire crustal section composed of granulite-facies material can result from an orogen that has undergone granulite-facies metamorphism in its root and is subsequently thinned by

erosion. After long-term cooling, the low radiogenic heat production throughout the crust will result in a relatively shallow geothermal gradient. Such a lithosphere will be stronger than an orogen that did not experience substantial high-temperature metamorphism. This mechanism would have a particularly strong effect when erosion rather than extensional collapse thinned the orogenic crust, because erosion predominantly removes the upper crust (see below).

Presence of thickened sub-continental lithospheric mantle

Many models of the Wilson Cycle argue that orogens are weak zones because the thickened crust supposedly displaces the strong sub-continental lithospheric mantle. This ignores the fact that the lithospheric mantle may also thicken during orogenesis (e.g. Stüwe & Sandiford 1995). For weakening to occur in these cases, the lithospheric mantle needs to be thinned. This can occur by slow conductive heating or by convective

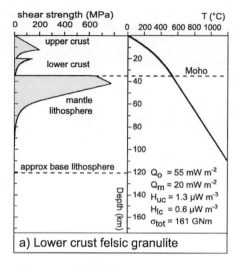

a) Lower crust felsic granulite

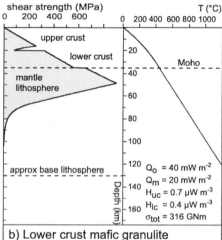

b) Lower crust mafic granulite

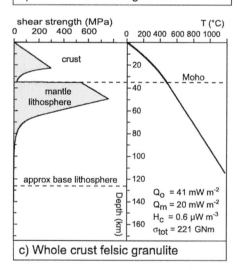

c) Whole crust felsic granulite

thinning of the lithospheric mantle (Houseman et al. 1981). Once initiated, convective thinning of the lithospheric mantle is a rapid process (Houseman & Molnar 1997). Orogens with a lithospheric thickening factor of 2 are likely to experience convective thinning of the lithospheric mantle (Houseman et al. 1981; Houseman & Molnar 1997) and become lithospheric weak zones. Such orogens also heat up relatively quickly due to their doubled heat production capacity. (This weakening effect maybe, however, short-lived, see below.)

An orogen that experienced only limited thickening may not experience convective thinning of the lithospheric mantle (e.g. Urals, see Berzin et al. 1996 and further below). In such an orogen, a thickened lithospheric mantle would lower the mantle Moho heat flow, resulting in strong lithosphere. Krabbendam & Barr (2000) specifically modelled the thermal evolution of such orogens and suggested that such orogens (with a thickening factor of up to about 1.3) may remain a lithospheric strong zone for up to 500 Ma after orogenesis.

Thinning of orogenic crust: Erosion v. late-orogenic extension

Thickened orogenic crust can be thinned to normal thickness (30–40 km) by two (end member) processes: erosion and late- to post-orogenic extension (e.g. Dewey 1988). The manner and timing of thinning of thickened orogenic crust can affect the long-term strength of the orogenic lithosphere, particularly if the crust has a distinct layered distribution of radiogenic heating. Four factors are important.

Firstly, erosion can only thin the crust by removing the top of the crust. If the upper crust is strongly enriched in radioactive elements, the more enriched portions of the crust are preferentially removed, so that an orogen becomes depleted in radiogenic heat producing elements. In contrast, the radiogenic upper and middle crust are more likely to be preserved by late-orogenic extension. An extended orogen may

---

Fig. 5. Rheological profiles and geothermal gradients of three lithospheric models with different crustal rheologies. The mantle heat flow (Q_m) is the same for all models. See text for explanation and Tables 1, 2 & 3 for parameters used. (**a**) Model of lithosphere with a lower crust of felsic granulite and an upper crust with a quartz rheology. (**b**) Model of lithosphere with a lower crust of mafic granulite and an upper crust with a quartz rheology granulite. (**c**) Model of lithosphere with the entire crust as felsic granulite. Rheological parameters of granulite after Wilks & Carter (1990).

thus have a heat producing capacity similar to that of the crust before orogenesis, and maybe in the long-term weaker than an eroded orogen. This effect is independent of the cause of late-orogenic extension.

A second factor is the longevity of the orogen in its thickened state. A long-lived orogen has more time to heat up by radiogenic heating than a short-lived one. This is expected to result in extensive high temperature metamorphism and associated dehydration and partial melting of the lower crust of the orogens concerned (P–T path c in Fig. 7). Paradoxically, after crustal thinning and thermal relaxation, this may contribute to a higher lithospheric strength due to dehydration and granulitization (see above). Orogens are expected to be long-lived if they are thinned solely by erosion or if there is a long time-lag between crustal thickening and crustal thinning, caused by convective removal of thickened mantle lithosphere (e.g. Houseman & Molnar 1997).

However, some orogens (such as the Scandinavian Caledonides, see below) are relatively short-lived (< 50 Ma) as they are thinned by late-orogenic extension that immediately followed crustal thickening. In such orogens, typically represented by large areas of high pressure-low temperature metamorphic rocks, radiogenic heating and hence dehydration and partial melting will have been limited (P–T path b in Fig. 4), which may result in a long-term crustal weakness.

Thirdly, late-orogenic extension or orogenic collapse (Dewey 1988) is commonly thought to result from convective thinning of the lithospheric mantle (Dewey 1988; England & Houseman 1988). Cold and strong lithospheric mantle is partially replaced by hot, weak asthenospheric mantle, which will quickly heat the remaining lithospheric mantle. This leads to weakening of the entire lithosphere and to a large increase in its gravitational potential energy, which may trigger orogenic collapse and thinning of the crust. After crustal thinning however, thermal re-equilibration may restore the sub-continental lithospheric mantle to its original (pre-orogenic) thickness and strength. The thermal weakening effect of convective thinning of the lithospheric mantle, therefore, may only be significant in the first 100 Ma or so after orogenesis (e.g. Ryan & Dewey 1997). Note that not all late-orogenic extension is thought to be caused by convective removal of mantle lithosphere (Rey et al. 1997; Fossen 2000).

Finally, erosion is commonly a slower process than extensional collapse. Orogenic crust thinned by erosion will have more time to heat up by radioactive decay. In the short term this results in a weaker lithosphere, but after a prolonged time (after crustal thinning and thermal relaxation) it

may lead to stronger lithosphere, because dehydration metamorphism will be more extensive and a large part of the crust would have a granulitic rheology.

The effects of erosional thinning of orogens on the long-term lithospheric strength is approximated here by calculating the strength of a lithosphere with a crust that has a low radiogenic heat production ($0.6 \mu Wm^{-3}$) and a rheology of felsic granulite (Fig. 5c). Such a crust could result from a doubling (by pure shear) of a crust with layered heat production (as in Fig. 3d) of which the upper half is removed by erosion, so that the original granulite-facies lower crust now encompasses the entire crustal section. This is a reasonable scenario, given that many Proterozoic mobile belts are now characterised by extensive outcrops of granulite-facies rocks. Such lithosphere (Fig. 5c) has a low surface heat flow and is almost two times stronger than the reference lithosphere (compare with Fig. 3a) and 1.3 times stronger than the original layered crust as shown Figure 3d.

Testing the hypotheses: North Atlantic Caledonides v. Urals

In this section, the evolution and architecture of the North Atlantic Caledonides (comprising the Scandinavian and East Greenland Caledonides) and the Urals (Fig. 2) are compared to examine some of the factors that has been assessed above. The Caledonides are presently situated along the passive margins of the North Atlantic. Opening of the North Atlantic took place in the Early Tertiary and followed the strike of the Caledonide Orogen. The Caledonides have thus experienced the first stages of reworking (the rifting and break-up phases) and are likely to be subjected to a future collision event, when the North Atlantic closes. In contrast, the Urals have not rifted. Is it possible to conclude that the Cenozoic break-up of Laurasia occurred along the Caledonides because it represented weak lithosphere, whereas the Urals remained intact because it was and is a relatively strong zone?

North Atlantic Caledonides

The Caledonides were formed by the collision of Baltica and Laurentia, following the closure of the Iapetus Ocean. Early orogenic events occurred at c. 460 Ma and c. 500 Ma (Boundy et al. 1997; Essex et al. 1997). At 440 Ma, new ocean crust was still being formed, now found within the Solund–Stavfjord Ophiolite Complex (Dunning & Pedersen 1988). The main 'Scandian' Phase of the orogeny occurred during the Silurian (c. 430–410 Ma) and was associated with closure of the Iapetus Ocean and final collision of Baltica and Laurentia.

The architecture of the Caledonides is summarized as follows (Fig. 6). On the East Greenland side, quartzo-feldspathic basement rocks (Archean to Mesoproterozoic) are overthrust by Neoproterozoic to Lower Paleozoic sedimentary rocks, mainly representing Laurentian passive margin sediments (Henriksen 1985; Andresen et al. 1998). No arc-related or ophiolitic rocks occur within the East Greenland Caledonides. On the Scandinavian side, quartzo-feldspathic continental basement of the Baltic continent (with 1800–1600 Ma and 1200–900 Ma ages, and reworked during the Caledonian; Kullerud et al. 1986) is overlain by a series of thrust sheets that have been subdivided into four super-units. These are, from bottom to top (Gee & Sturt 1985; Stephens & Gee 1989; Milnes et al. 1997):

- The Lower and Middle Allochthon, comprising rift and shelf-facies sequences of Baltic provenance and gneissic basement slivers;
- the Upper Allochthon, comprising island arcs, ophiolitic sequences and other exotic terranes;
- the Uppermost Allochthon, comprising exotic terranes of Laurentian affinity, as determined from fossil provenance studies (Pedersen et al. 1992).

Although these thrust sheets occupy a large area, they are relatively thin (< 5–8 km), so that in cross-section (Fig. 6b) the bulk of the Scandinavian Caledonides is composed of reworked quartzo-feldspathic continental basement rocks (Hossack et al. 1985; Milnes et al. 1997). Continental basement crops out not only on the foreland side, but also in several basement windows, the largest of which is the Western Gneiss Region. The Western Gneiss Region contains abundant eclogite boudins that locally indicate burial to depth exceeding 90 km (Wain 1997). This high pressure to ultrahigh pressure terrane was exhumed by transtensional deformation (Krabbendam & Dewey 1998) whereby thrusts were reactivated as extensional detachments (Fossen 1992). The late-orogenic transtension was associated with Early Devonian oblique plate divergence of Baltica and Laurentia (Rey et al. 1997; Krabbendam & Dewey 1998; Fossen 2000), possibly but not necessarily related to some form of convective thinning of the lithospheric mantle (Andersen et al. 1991). Cooling ages of basement rocks underneath the extensional detachments indicate that by 400–390 Ma (Chauvet & Dall-

meyer 1992; Berry et al. 1994; Andersen 1998) the crustal thickness had returned to nomal (c. 35 km) values. Thus, the Scandinavian Caledonide crust was thickened and thinned in a period no longer than 45 Ma and possibly as short as 30 Ma.

As a consequence of this short life span and the rapid thinning of the orogen, associated with (near) isothermal decompression, heating of the basement gneisses was restricted and the gneisses underwent only minor dehydration metamorphism or partial melting occurred. Much of the basement gneisses contain hydrous amphibolite-facies assemblages (Bryhni 1989; Krabbbendam & Wain 1997), rather than mechanically stronger, anhydrous granulite-facies rocks.

Ryan & Dewey (1997) postulated that an eclogite-facies root of continental crustal material still resides underneath continental crust of normal density in the Caledonides. The effect of such an eclogite-facies root is an increase in the total radiogenic heat producing capacity (because the effective radiogenic crustal thickness is larger than normal), resulting in long term lithospheric weakness. Such a root would also displace strong mantle lithosphere.

Heat flow measurements in the Norwegian Caledonides are between 40 and $50\,\text{mW}\,\text{m}^{-2}$ (Balling 1995). These values are uncorrected for paleoclimate so maybe somewhat higher. Close to the coast, some heat flow measurements are higher (50–$60\,\text{mW}\,\text{m}^{-2}$). It is uncertain to what extent the surface heat flow has been affected by the Tertiary rifting event and whether the measured surface heat flow is representative for the surface heat flow just prior to Tertiary rifting.

The features relevant for the lithospheric strength, after the Caledonide orogeny but before the Tertiary rifting, can be summarized as follows:

- Crustal thinning by late-orogenic extension, leaving much of the radioactive upper to middle crust intact;
- composition dominated by relatively high heat producing felsic continental crust, with only limited amount of mafic island arc material incorporated in the orogen;
- late-orogenic extension resulting in fast exhumation of buried rocks, limiting the effect of dehydration metamorphism and leaving much of the crust with hydrous assemblages;
- possible convective thinning of the lithospheric mantle;
- the possible presence of deep buried eclogitic crust, increasing the heat producing capacity.

This suggests that the Scandinavian Caledonide Orogen was weakened by the Scandian phase of

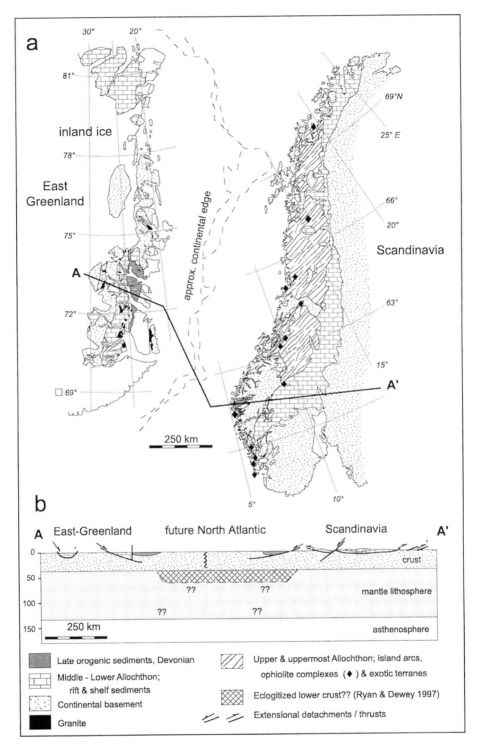

Fig. 6. (a) Simplified map of the East Greenland and Scandinavian Caledonides, in a pre-Atlantic fit. See Figure 2 for location. (b) Simplified cross-section through the North Atlantic Caledonides with the effects of the opening of the North Atlantic removed. Mainly after Haller (1985), Henriksen (1985), Roberts & Gee (1985), Pedersen *et al.* (1992), Fossen (1992), Andersen (1998) and Andresen *et al.* (1998). Eclogitized lower crust was postulated by Ryan & Dewey (1997). $V \approx 2H$.

orogeny, with a resultant strength that was the same or lower than the reference lithosphere.

The Urals

The Urals are a 2000 km long orogenic belt of Palaeozoic age, caused by the collision of the Eastern European platform in the west with one or more island arcs and micro-continents and finally the Kazakhstan continent in the east (Zonenshain *et al.* 1984; Puchkov 1997). The Urals comprises, from west to east, the following features (Figs 7 and 8, e.g. Ivanov *et al.* 1975, Puchkov 1997; Brown *et al.* 1998).

(a) The Uralian (or Pre-Uralian) Foredeep, containing Carboniferous shelf deposits, overlain by Permian flysch and molasse sequences were deposited as a response to mountain building further east. This basin overlies Precambrian continental basement of the East European continent.

(b) The West Uralian Zone comprises mainly folded and thrusted Palaeozoic shelf and rift sediments representing the former passive margin sediments of the East European Continent.

(c) The Central Uralian Zone contains folded metasediments of mainly Late-Proterozoic age, although some Palaeozoic and Archean rocks also occur. In the Maksyotov Complex high-pressure metamorphism ($P \leq 16$ kbar) occurred at $c.375$ Ma (e.g. Matte *et al.* 1993). This high-pressure complex appears to have been exhumed during continuous plate convergence by a combination of normal movement along the Main Uralian Fault and thrust movement on a large thrust further west (e.g. Matte & Chemenda 1996; Hetzel *et al.* 1998).

(d) The Main Uralian Fault marks the boundary of the Palaeozoic East European continent and accreted island arc, oceanic and micro-continental terranes to the west. It is a prominent suture that runs along the entire length of the orogen. Many ophiolite bodies occur along the fault, which itself is marked by low-grade serpentinite melanges. Ophiolitic remnants that occur east of the Main Uralian Fault are allochthonous klippen (Savelieva & Nesbitt 1996).

(e) The Tagil–Magnitogorsk zone comprises island arcs and ophiolite complexes. The Magnitogorsk Arc is of Ordovician to Late Devonian

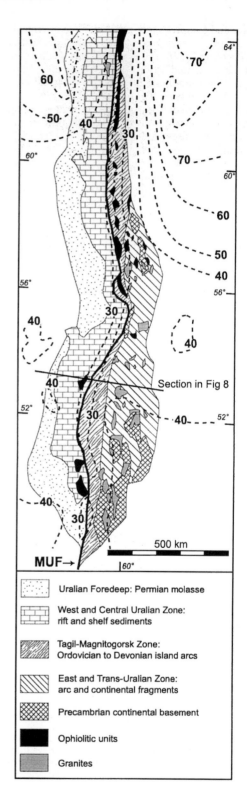

Fig. 7. Tectonic map of the Urals (modified after Ivanov *et al.* 1975; Savieleva & Nesbitt 1996), with heat flow contours in mW m^{-2} (after Kukkonen *et al.* 1997). Note the large amount of island arc and oceanic material present. See Figure 2 for location. MUF, Main Uralian Fault.

Legend:
- Uralian Foredeep: Permian molasse
- West and Central Uralian Zone: rift and shelf sediments
- Tagil-Magnitogorsk Zone: Ordovician to Devonian island arcs
- East and Trans-Uralian Zone: arc and continental fragments
- Precambrian continental basement
- Ophiolitic units
- Granites

age and comprises low-grade mafic to intermediate igneous rocks and abundant volcaniclastics (Puchkov 1997). The Tagil–Magnitogorsk zone is characterized by high density and low radiogenic heat production (Kukkonen *et al.* 1997).

(f) The eastern part of the Urals is composed of fragments of Precambrian continental crust, in part overlain by Palaeozoic sediments and volcanics. Fragments of island arc and ophiolite complexes also occur. Further east is the poorly studied Trans–Uralian Zone that consists of Palaeozoic volcanic and plutonic complexes with marine and terrigenous sediments, possibly representing an Andean Style margin to the Kazakhstan continent (Puchkov 1997).

The collision of the Magnitogorsk island arc against the East European Continent occurred in the Devonian. After a period of tectonic quiescence, further collision of terranes from the east and final closure of the 'Ural Ocean' occurred during the Permian (Brown *et al.* 1997, 1998). Seismic evidence suggests that the present-day crustal thickness is up to 55 km (Carbonell *et al.* 1996; Puchkov 1997; Steer *et al.* 1998). The Uralian crust has been in this thickened state for at least 250 Ma after cessation of orogenesis, since no substantial crustal deformation occurred since Permian collision (Puchkov 1997). Fission track analysis suggests that erosion rates since the

Triassic have been < 0.05 mm a^{-1} (Seward *et al.* 1997). In contrast to the Caledonides, no significant late- or post-orogenic extension appears to have occurred in the Urals (Hetzel *et al.* 1998; Steer *et al.* 1998).

The distribution of surface heat flow in the Urals is characterized by a narrow (50–100 km wide) zone of very low heat flow (< 30 mW m^{-2}) that roughly coincides with the outcrop of the Tagil–Magnitogorsk Zone (Fig. 7, Kukkonen *et al.* 1997). Most of the remainder of the Urals and adjacent areas have heat flow between 30 and 40 mW m^{-2}. These are very low values for a Palaeozoic orogen (cf. Sclater *et al.* 1980). This anomalously low heat flow can be partly explained by low radiogenic heat production in the central Urals (Kukkonen *et al.* 1997). However, low radiogenic heat production alone is not adequate to explain the very low surface heat flow: the Moho heat flow must also be low (Fig. 8). A thick and cold lithospheric mantle can explain a low Moho heat flow and can also explain the high crustal thickness (<50 km) in conjunction with the low topography (<1 km average, Knapp *et al.* 1996) and the small amplitude of the negative gravity anomaly across the Urals (Kruse & McNutt 1988). Very deep (55 s two-way travel time, *c.* 200 km) seismic mantle reflectors have been interpreted as the base of the thickened

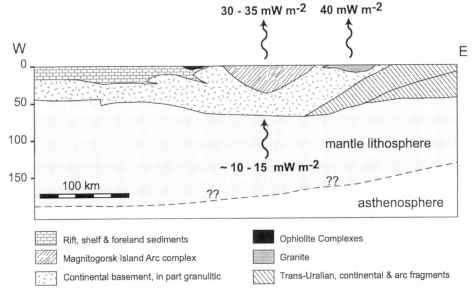

Fig. 8. Schematic cross-section through the Urals, based on surface geology, seismic profiling and density modelling of gravity anomalies (after Berzin *et al.* 1996), with heat flow indicated. The low surface heat flow also implies a low mantle heat flow, which is calculated with a uniform crustal radiogenic heat production of 0.45 μW m^{-3}. This provides evidence for the presence of thickened mantle lithosphere underneath the Urals. $V = 1.4H$.

lithospheric mantle (Knapp *et al.* 1996). It is thus likely that neither late- or post-orogenic extension took place nor convective thinning of the lithospheric mantle and that the lithospheric mantle is still thickened underneath the Urals (Berzin *et al.* 1996; Steer *et al.* 1998).

Summarizing, the features of the Urals relevant for the lithospheric strength are:

• Low radiogenic heat production, caused by the large proportion of island arc and oceanic material incorporated into the orogen;
• probable presence of thickened mantle lithosphere, resulting in a low Moho heat flow;
• post-orogenic thinning by very slow erosion; absence of late-orogenic extension.

These features suggest that the Urals have an anomalously high lithospheric strength.

Considering the low heat flow ($35\,\mathrm{mW\,m^{-2}}$) and surface heat production ($0.45\,\mu\mathrm{W\,m^{-3}}$, after Kukkonen *et al.* 1997) and the thickened crust, but assuming the other parameters the same as the reference lithosphere (Fig. 3a), the lithosphere strength of the Urals is 2.5 times stronger than the reference lithosphere (Fig. 9a). This estimate is conservative because neither the higher density (which would increase the strength in the brittle field, e.g. equation (1)), nor the probably stronger rheology of the more mafic crust have been taken into account. Some 200 to 300 Ma after collision the Urals represent a lithospheric strong zone.

Despite the low present-day topography, the Urals will eventually be restored to a thinner crust, most likely by slow erosion. Radioactive heating in the crust and conductive heating of the lithospheric mantle will, in time, create a somewhat steeper geothermal gradient. The geothermal gradient and rheology profile for a 'future' Urals (with a crustal thickness of 35 km) is shown in Figure 9b. In this model, the radioactive heating is somewhat lower than today and the mantle and surface heat flow are somewhat higher. The total lithospheric strength of a 'future' Urals is calculated to be twice as high as the reference lithosphere, because of the cooler geothermal gradient. Thus, an orogen like the Urals, rich in island arc material and thinned by erosion, is likely to be a lithospheric strong zone for several 100 Ma after orogenesis.

Discussion and conclusions

The location of rifting, and hence reworking, is determined by many factors. The direction of the external stress field with respect to preexisting anisotropies (Ring 1994; Vauchez *et al.* 1997) is of obvious importance. The arrival of a mantle plumes commonly leads to rifting; however, rifting may also occur without such plumes. In these cases, rifting and reworking are expected along orogens that are weaker than adjacent

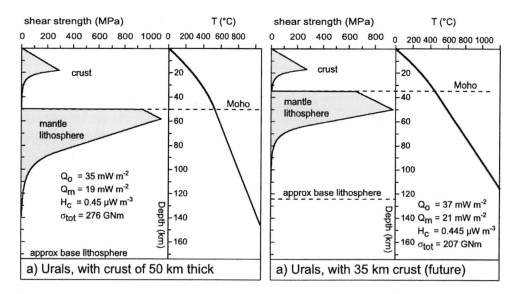

Fig. 9. (**a**) Rheological profile and geothermal gradient of the present-day Urals lithosphere model, with a crustal thickness of 50 km. (**b**) Rheological profile and geothermal gradient of the 'future' Urals lithospheric model with 35 km crustal thickness, after (several) 100 Ma of erosion. A quartz rheology has been applied to the crust. See text for explanation and Tables 1 & 2 for parameters used.

lithosphere. Many orogens will be weaker than adjacent lithosphere and, therefore, reworking of orogens is common.

However, some orogenic processes may render orogenic lithosphere *stronger* than adjacent lithosphere. Processes that can contribute to a relatively high lithospheric strength include enrichment of radioactive material in the upper crust with concomitant depletion of the lower crust; widespread dehydration metamorphism and orogenic thinning by erosion rather than by late-orogenic extension. Strong lithosphere may also result if the orogen contains abundant material with a low radiogenic heat production capacity, such as island arcs; or if its underlying lithospheric mantle is still thickened. This review shows that reworking of orogens is not inevitable and that not all orogens are lithospheric weak zones.

Terence Barr, Pete Betts, David Giles, Greg Houseman and Gordon Lister are thanked for critically reading a previous version of the manuscript. Ralf Hetzel and Basil Tikoff are thanked for constructive reviews. Jodie Miller is thanked for organizing a very stimulating conference in the back of beyond. Funding by the Australian Research Council is acknowledged.

References

ANDERSEN, T. B. 1998. Extensional tectonics in the Caledonides of southern Norway, an overview. *Tectonophysics*, **285**, 333–351.

ANDERSEN, T. B., JAMTVEIT, B., DEWEY, J. F. & SWENSSON, E. 1991. Subduction and eduction of continental crust: major mechanisms during continent-continent collision and orogenic extensional collapse, a model based on the south Norwegian Caledonides. *Terra Nova*, **3**, 303–310.

ANDRESEN, A., HARTZ, E. H. & VOLD, J. 1998. A late orogenic extensional origin for the infracrustal gneiss domes of the East Greenland Caledonides (72°–74°N). *Tectonophysics*, **285**, 353–369.

ASHWAL, L. D., MORGAN, P., KELLEY, S. A. & PERCIVAL, J. A. 1987. Heat production in an Archean crustal profile and implications for heat flow and mobilization of heat-producing elements. *Earth and Planetary Science Letters*, **85**, 439–450.

BALLING, N. 1995. Heat flow and thermal structure of the lithosphere across the Baltic Shield and northern Tornquist Zone. *Tectonophysics*, **244**, 13–50.

BERRY, H. N., LUX, D. R., ANDRESEN, A. & ANDERSEN, T. B. 1994. Argon 40–39 dating of rapidly uplifted high pressure rocks during late-orogenic extension in southwestern Norway. *Geological Society of America: Abstracts with Programs*, **25**(6) A–477.

BERZIN, R., ONCKEN, O., KNAPP, J. H., PÉREZ-ESTAÚN, A., HISMATULIN, T., YUNUSOV, N. & LIPILIN, A.

1996. Orogenic evolution of the Ural Mountains: Results from an integrated seismic experiment. *Science*, **274**, 220–221.

BOUNDY, T. M., MEZGER, K. & ESSENE, E. J. 1997. Temporal and tectonic evolution of the granulite-eclogite association from the Bergen Arcs, western Norway. *Lithos*, **39**, 159–178.

BOUSQUET, R., GOFFÉ, B., HENRY, P., LE PICHON, X. & CHOPIN, C. 1997. Kinematic, thermal and petrological model of the Central Alps: Lepontine metamorphism in the upper crust and eclogitisation of the lower crust. *Tectonophysics*, **231**, 105–127.

BRACE, W. F. & KOHLSTEDT, D. L. 1980. Limits on lithospheric stress imposed by laboratory experiments. *Journal of Geophysical Research*, **B 85**(11), 6248–6252.

BROWN, D., ALVAREZ MARRON, J., PÉREZ-ESTAÚN, A., GOROZHANINA, Y., BARYSHEV, V. & PUCHKOV, V. 1997. Geometric and kinematic evolution of the foreland thrust and fold belt in the Southern Urals. *Tectonics*, **16**, 551–562.

BROWN, D. L., JUHLIN, C., ALVAREZ MARRON, J., PÉREZ-ESTAÚN, A. & OSLIANSKI, A. 1998. Crustal-scale structure and evolution of an arc-continent collision zone in the Southern Urals, Russia. *Tectonics*, **17**, 158–171.

BRYHNI, I. 1989. Status of the supracrustal rocks in the Western Gneiss Region, S Norway. In: GAYER, R. A. (ed.) *The Caledonide Geology of Scandinavia*. Graham & Trotman, London, 221–228.

CARBONELL, R., PÉREZ-ESTAÚN, A., GALLART, J., DIAZ, J., KASHUBIN, S., MECHIE, J., STADTLANDER, R., SCHULZE, A., KNAPP, J. H. & MOROZOV, A. 1996. Crustal root beneath the Urals; wide-angle seismic evidence. *Science*, **274**, 222–224.

CERMÁK, V. 1995. A geothermal model for the Central Segment of the European Geotraverse. *Tectonophysics*, **244**, 51–55.

CHAUVET, A. & DALLMEYER, R. D. 1992. 40Ar/39Ar mineral dates related to Devonian extension in the southwestern Scandinavian Caledonides. *Tectonophysics*, **210**, 155–177.

DEWEY, J. F. 1988. Extensional collapse of orogens. *Tectonics*, **7**, 1123–1139.

DUNNING, G. R. & PEDERSEN, R. B. 1988. U/Pb dating of ophiolites and arc-related plutons of the Norwegian Caledonides: implications for the development of Iapetus. *Contributions to Mineralogy and Petrology*, **98**, 13–23.

DURRHEIM, R. J. & MOONEY, W. D. 1991. Archean and Proterozoic crustal evolution: Evidence from crustal seismology. *Geology*, **19**, 606–609.

ENGLAND, P. C. & HOUSEMAN, G. 1988. The mechanics of the Tibetan Plateau. *Philosophical Transactions of the Royal Society of London*, **A 326**, 301–320.

ESSEX, R. M. GROMET, L. P., ANDRÉASSON, P. G. & ALBRECHT, L. 1997. Early Ordovician U–Pb metamorphic ages of the eclogite-bearing Seve Nappes, northern Scandinavian Caledonides. *Journal of Metamorphic Geology*, **15**, 665–676.

EVANS, B. W. 1990. Phase relations of epidote-blueschists. *Lithos*, **25**, 3–23.

FOSSEN, H. 1992. The role of extensional tectonics in the Caledonides of south Norway. *Journal of Structural Geology*, **14**, 1033–1046.

FOSSEN, H. 2000. Extensional tectonics in the Caledonides; synorogenic or postorogenic? *Tectonics*, **19**, 213–224.

FOWLER, C. M. R. 1990. *The Solid Earth, an introduction to global geophysics.* Cambridge University Press, Cambridge.

GEE, D. G. & STURT, B. A. (eds) 1985. *The Caledonide Orogen – Scandinavia and related areas.* John Wiley & Sons, Chichester.

GRASEMANN, B., RATSCHBACHER, L. & HACKER, B. 1998. Exhumation of ultrahigh-pressure rocks: Thermal boundary conditions and cooling history. *In*: HACKER, B. R. & Liou, J. G. (eds) *When Continents Collide: Geodynamics and Geochemistry of Ultrahigh-Pressure Rocks.* Kluwer Academic Publishers, Dordrecht, 117–139.

HAACK, U. 1983. On the content and vertical distribution of K, Th and U in the continental crust. *Earth and Planetary Science Letters*, **62**, 360–364.

HALLER, J. 1985. The East Greenland Caledonides; reviewed. *In*: GEE, D. G. & STURT, B. A. (eds) *The Caledonide Orogen – Scandinavia and related areas.* John Wiley & Sons Chichester, 1031–1046.

HENRIKSEN, N. 1985. The Caledonides of central East Greenland 70°–76°N. *In*: GEE, D. G. & STURT, B. A. (eds) *The Caledonide Orogen – Scandinavia and related areas.* John Wiley & Sons, Chichester, 1095–1113.

HETZEL, R., ECHTLER, H. P., SEIFERT, W., SCHULTE, B. A. & IVANOV, K. S. 1998. Subduction- and exhumation-related fabrics in the Paleozoic high-pressure-low-temperature Maksyutov Complex, Antingan area, Southern Urals, Russia. *Geological Society of America Bulletin*, **110**, 916–930.

HOSSACK, J. R., GARTON, M. R. & NICKELSEN, R. P. 1985. The geological section from the foreland up to the Jotun thrust sheet in the Valdres area, South Norway. *In*: GEE, D. G. & STURT, B. A. (eds) *The Caledonide Orogen – Scandinavia and related areas.* John Wiley & Sons, Chichester, 443–456.

HOUSEMAN, G. A., MCKENZIE, D. P. & MOLNAR, P. 1981. Convective instability of a thickened boundary layer and its relevance for the thermal evolution of continental convergent belts. *Journal of Geophysical Research*, **86**(B7), 6115–6132.

HOUSEMAN, G. A. & HEGARTY, K. A. 1987. Did rifting on Australia's southern margin result from tectonic uplift? *Tectonics*, **6**, 515–527.

HOUSEMAN, G. A. & MOLNAR, P. 1997. Gravitational (Rayleigh–Taylor) instability of a layer with non-linear viscosity and convective thinning of continental lithosphere. *Geophysical Journal International*, **128**, 125–150.

IVANOV, S. N., PERFILYEV, A. S., YEFIMOV, A. A., SMIRNOV, G. A., NECHEUKHIN, V. M. & FÉRSHTATER, G. B. 1975. Fundamental features in the structure and evolution of the Urals. *American Journal of Science* **275-A**, 107–130.

KNAPP, J. H., STEER, D. N., BROWN, L. D. *ET AL.* 1996. Lithosphere-scale seismic image of the Southern Urals from explosion-source reflection profiling. *Science*, **274**, 226–228.

KRABBENDAM, M. & WAIN, A. 1997. Late-Caledonian structures, differential retrogression and structural position of (ultra)high-pressure rocks in the Nordfjord-Stadlandet area, Western Gneiss Region. *Norges geologiske undersøkelse Bulletin*, **432**, 127–139.

KRABBENDAM, M. & DEWEY, J. F. 1998. Exhumation of UHP rocks by transtension in the Western Gneiss Region, Scandinavian Caledonides. *In*: HOLDSWORTH, R. E., STRACHAN, R. A. & DEWEY, J. F. (eds) *Continental transpressional and transtensional tectonics.* Geological Society, London, Special Publications, **135**, 159–181.

KRABBENDAM, M. & BARR, T. 2000. Proterozoic orogens and the break-up of Gondwana: Why did some orogens not rift? *Journal of African Earth Sciences*, **31**, 35–49.

KRUSE, S. & MCNUTT, M. 1988. Compensation of Paleozoic orogens: a comparison of the Urals to the Appalachians. *Tectonophysics*, **154**, 1–17.

KUKKONEN, I. T., GOLUUANOVA, I. V., KHACHAY, Y. V., DRUZHININ, V. S., KOSAREV, A. M. & SCHAPOV, V. A. 1997. Low geothermal heat flow of the Urals fold belt; implication of low heat production, fluid circulation or palaeoclimate? *Tectonophysics*, **276**, 63–85.

KULLERUD, L., TØRUDBAKKEN, B. O. & ILEBEKK, S. 1986. A compilation of radiometric age determinations from the Western Gneiss Region, South Norway. *Norges geologiske undersøkelse Bulletin*, **406**, 17–42.

KUSZNIR, N. J. & PARK, R. G. 1987. The extensional strength of the continental lithosphere: its dependence on geothermal gradient and crustal composition and thickness. *In*: COWARD, M. P., DEWEY, J. F. & HANCOCK, P. L. (eds) *Continental extensional tectonics.* Geological Society, London, Special Publications, **28**, 35–52.

LENARDIC, A., MORESI, L. & MUEHLHAUS, H. 2000. The role of mobile belts for the longevity of deep cratonic lithosphere; the crumple zone model. *Geophysical Research Letters*, **27**, 1235–1238.

LESQUER, A., TAKHERIST, D., DAUTRIA, J. M. & HADIOUCHE, O. 1990. Geophysical and petrological evidence for the presence of an 'anomalous' upper mantle beneath the Sahara basin (Algeria). *Earth and Planetary Science Letters*, **96**, 407–418.

MATTE, P. & CHEMENDA, A. I. 1996. A mechanism for exhumation of high pressure metamorphic rocks during continental subduction in Southern Urals. *Comptes Rendu de l' Académie des Sciences de Paris*, **323 IIa**, 525–530.

MATTE, P., MALUSKI, H., CABY, R., NICOLAS, A., KEPEZHINSKAS, P. & SOBOLEV, S. 1993. Geodynamic model and 39 Ar/ 40 Ar dating for the generation and emplacement of the high pressure (HP) metamorphic rocks in SW Urals. *Comptes*

Rendu de l' Académie des Sciences de Paris, **317**
IIa, 1667–1674.

MILNES, A. G., WENNBERG, O. P., SKAR, Ø. &
KOESTLER, A. G. 1997. Contraction, extension
and timing in the South Norwegian Caledonides:
the Sognefjord transect. *In*: BURG, J. P. & FORD, M.
(eds) *Orogeny through time*. Geological Society,
London, Special Publications, **121**, 123–148.

NEIL, E. A. & HOUSEMAN, G. A. 1999. Rayleigh–
Taylor instability of the upper mantle and its role
in intraplate orogeny. *Geophysical Journal International*, **138**, 89–107.

PEDERSEN, R. B., BRUTON, D. L. & FURNES, H.
1992. Ordovician faunas, island arcs and ophiolite
in the Scandinavian Caledonides. *Terra Nova*, **4**,
217–222.

PUCHKOV, V. N. 1997. Structure and geodynamics of
the Uralian orogen. *In*: BURG, J. P. & FORD, M.
(eds) *Orogeny through time*. Geological Society,
London, Special Publications, **121**, 201–236.

RANALLI, G. 1992. Average lithospheric stresses
induced by thickness change: a linear approximation. *Physics of the Earth and Planetary Interiors*,
69, 263–269.

REY, P., BURG, J. P. & CASEY, M. 1997. The
Scandinavian Caledonides and their relationship
to the Variscan belt. *In*: BURG, J. P. & FORD, M.
(eds) *Orogeny through time*. Geological Society,
London, Special Publications, **121**, 179–200.

RING, U. 1994. The influence of preexisting structure
on the evolution of the Cenozoic Malawi rift
(East African rift system). *Tectonics*, **13**, 313–326.

ROBERTS, D. & GEE, D. G. 1985. An introduction to
the structure of the Scandinavian Caledonides. *In*:
Gee, D. G. & Sturt, B. A. (eds) *The Caledonide
Orogen – Scandinavia and related areas*. John
Wiley & Sons, Chichester, 55–68.

RYAN, P. D. & DEWEY, J. F. 1997. Continental eclogites and the Wilson Cycle. *Journal of the Geological Society, London*, **154**, 437–442.

SANDIFORD, M. & HAND, M. 1998. Australian Proterozoic high-temperature, low-pressure metamorphism in the conductive limit. *In*: TRELOAR, P. J.
& O'BRIEN, P. J. (eds) *What drives metamorphism
and metamorphic relations?* Geological Society,
London, Special Publications, **138**, 109–120.

SAVELIEVA, G. N. & NESBITT, R. W. 1996. A synthesis of the stratigraphic and tectonic setting of
the Uralian ophiolites. *Journal of the Geological
Society, London*, **153**, 525–537.

SCLATER, J. G., JAUPART, C. & GALSON, D. 1980. The
heat flow through oceanic and continental crust
and the heat loss of the earth. *Reviews of
Geophysics and Space Physics*, **18**, 269–311.

SEWARD, D., PÉREZ-ESTAÚN, A. & PUCHKOV, V. N.
1997. Preliminary fission-track results from the

Southern Urals; Sterlitamak to Magnitogorsk.
Tectonophysics, **276**, 281–290.

STEER, D. N., KNAPP, J. H., BROWN, D., ECHTLER,
H. P., BROWN, D. L. & BERZIN, R. 1998. Deep
structure of the continental lithosphere in an
unextended orogen; an explosive-source seismic
reflection profile in the Urals (Urals Seismic
Experiment and Integrated Studies (URSEIS
1995)). *Tectonics*, **17**, 143–157.

STEPHENS, M. B. & GEE, D. G. 1989. Terranes and
polyphase accretionary history in the Scandinavian Caledonides. *In*: DALLMEYER, R. D. (ed.)
Terranes in the Circum-Atlantic Paleozoic orogens.
Geological Society of America Special Paper, **230**,
17–30.

STÜWE, K. & SANDIFORD, M. 1995. Mantle-lithospheric
deformation and crustal metamorphism with some
speculations on the thermal and mechanical significance of the Tauern Event, Eastern Alps.
Tectonophysics, **242**, 115–132.

TAYLOR, S. R. & MCLENNAN, S. M. 1985. *The
continental crust: its composition and evolution*.
Blackwell, Oxford.

THEUNISSEN, K., KLERKX, J., MELNIKOV, A. &
MRUMA, A. 1996. Mechanisms of inheritrance
of rift faulting in hte western branch of the East
African Rift, Tanzania. *Tectonics*, **15**, 776–790.

VAUCHEZ, A., BARRUOL, G. & TOMASSI, A. 1997. Why
do continents break-up parallel to ancient orog-
neic belts? *Terra Nova*, **9**, 62–66.

WAIN, A. 1997. New coesite-eclogite occurrences in
western Norway: the nature of an ultrahigh pressure province in the Western Gneiss Region.
Geology, **25**, 927–930.

WATERS, D. J. 1989. Metamorphic evidence for the
heating and cooling path of Namaqualand granulites. *In*: DALY, J. S., CLIFF, R. A. & YARDLEY,
B. W. D. (eds) *Evolution of metamorphic belts*.
Geological Society, London, Special Publications,
43, 357–363.

WILKS, K. R. & CARTER, N. L. 1990. Rheology of
some continental lower crustal rocks. *Tectonophysics*, **182**, 57–77.

WILSON, J. T. 1966. Did the Atlantic Ocean close and
then reopen? *Nature*, **211**, 676–681.

YARDLEY, B. W. D. 1989. *An introduction to metamorphic geology*. Longman Scientific & Techini-
cal, Harlow.

YARDLEY, B. W. D. & VALLEY, J. W. 1997. The
petrologic case for a dry lower crust. *Journal of
Geophysical Research*, **102**(B6), 12 173–12 185.

ZONENSHAIN, L. P., KORINEVSKY, V. G., KAZMIN,
V. G., PECHERSKY, D. M., KHAIN, V. V. & MAT-
VEENKOV, V. V. 1984. Plate tectonic model of the
South Urals development. *Tectonophysics*, **109**,
95–135.

From lithospheric thickening and divergent collapse to active continental rifting

P. REY

School of Geosciences, The University of Sydney, NSW 2006, Australia

Abstract: The Aegean Sea, the Alboran Sea, and the Basin and Range Province suggest that continental lithosphere following gravitational collapse may end up being thinner than it was before convergence and thickening. In order to assess the condition leading to the development of finite lithosphere thinning following convergence and convective thinning, the strength of the continental lithosphere, the gravitational force, and the rate of gravity-driven flow (spreading rate) are calculated during and after continental collision. One-dimensional numerical experiments, presented here, assume that the deformation is homogeneous, that erosion is a function of strain rate and elevation, and that thermal relaxation involves no lateral conduction of heat. Results show that if 43% of the lower lithospheric mantle is dragged into the convective mantle (convective thinning), gravitational collapse may lead to a lithosphere thinner than the initial lithosphere (pre-thickening lithosphere), provided that gravitational collapse is accommodated by the passive displacement of the surrounding lithosphere (free boundary collapse). When a slightly larger volume of lithospheric mantle is removed, a phase of extension leading to a necking instability and the formation of an active rift follows collapse. The presence of fixed boundaries and/or horizontal compressive stresses strongly reduces the spreading rate and opposes finite lithosphere thinning and therefore active rifting. It is suggested that back-arc extension occurring in continental settings could exemplify post-collapse active rifting.

Thermal thinning of the lithospheric mantle (above a mantle plume for instance) has been proposed as the driving mechanism for active continental rifting (Le Pichon 1983; Turcotte & Emerman 1983; Le Pichon & Alvarez 1984). Although normal stresses imposed at the base of the lithosphere by the plume may be important (Houseman & England 1986), active continental rifting is usually related to gravity-related tensile stresses induced by thinning of the lithospheric mantle. Alternatively, thinning of the lithospheric mantle can be achieved through convective thinning following lithospheric thickening (Houseman *et al.* 1981). This process increases the gravitational potential energy of the thickened lithosphere producing tensile horizontal stresses, the gravity-driven flow that reduces lateral variation of gravitational potential energy, that may trigger divergent collapse. As collapse proceeds, the thickened crust tends to recover its normal thickness and, assuming homogeneous deformation, the lithospheric mantle becomes thinner than it was before convergence and thickening. Therefore, the rise of a mantle plume underneath a lithosphere of normal thickness and divergent collapse, following thickening and convective thinning, may result in similar vertical lithospheric geometry where a continental crust of normal thickness overlies a thin lithospheric mantle (Fig. 1). Tensile horizontal stresses that arise from this geometry may be strong enough to induce self-enhanced lithospheric thinning. Should the extensional strain rates be too slow, this thinning will be limited by thermal relaxation and cooling, leading to aborted rifts. Should convergence occur in the next couple of hundred Ma, this thinned lithosphere will, most likely, localize contractional deformation. In contrast, fast extensional strain rates could lead to active continental rifting and the formation of continental margins. In both cases orogenic collapse could explain why orogenic belts tend to be the sites of successive orogenesis.

Much has been written about gravitational collapse (e.g. England & McKenzie 1982, 1983; Molnar & Chen 1982, 1983; Coney & Harms 1984; Sonder *et al.* 1987; Dewey 1988; England & Houseman 1988, 1989; Molnar & Lyon-Caen 1988; Sun & Murrell 1989; Rey *et al.* 2001), and its thermal and mechanical consequences (e.g. Platt & England 1993; Rey 1993; Mareschal 1994; Costa & Rey 1995). One aspect that still needs to be addressed is how collapsed orogenic lithospheres such as in the Basin and Range, the Aegean Sea, the Alboran Sea, and the Carpathian

From: MILLER, J. A., HOLDSWORTH, R. E., BUICK, I. S. & HAND, M. (eds) *Continental Reactivation and Reworking.* Geological Society, London, Special Publications, **184**, 77–88. 1-86239-080-0/01/$15.00 © The Geological Society of London 2001.

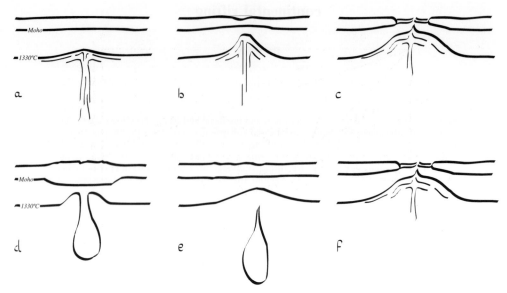

Fig. 1. Thinning of the lithospheric mantle related to a thermal plume (**a**, **b**), or following the convective thinning of the thermal boundary layer of the lithosphere (**d**, **e**), may result in active continental rifting (**c**, **f**).

Basin, end up being thinner than they were before thickening. Through simple 1-D numerical experiments, this paper explores the evolution of the spreading strain rate as the key parameter for mapping the evolution of both integrated lithospheric strength and gravitational force driving extension. Some of the experiments suggest that, should convective thinning occur, the crust and the lithosphere will end up being significantly thinner and weaker for at least a few hundreds of Ma, and therefore prone to tectonic reactivation. In some favourable cases, some experiments show that collapse could even be followed by self-enhanced extension leading to a necking instability and the formation of a continental rift.

Modelling approach, main assumptions and simplifications

Numerical experiments on simplified lithosphere systems provide quantitative insights into orogenic processes, and allow investigation of the consequences of some theoretical processes. The numerical design involves building a simplified lithosphere, whose vertical geometry changes under the action of four processes: (i) plane strain homogeneous deformation driven convergence; (ii) erosion/sedimentation; (iii) thermal relaxation; and (iv) gravitational spreading. In order to maintain isostatic equilibrium, these four processes are implemented successively over small increments of time, a Crank–Nicholson finite differences scheme being used for the

treatment of the thermal relaxation. In conjunction, a thin sheet approximation is used which considers stresses and strain rates in terms of their vertical average through the lithosphere (e.g. Sonder & England 1986; England & Houseman 1988, 1989; Ranalli 1992, 1995). Given knowledge of the rheology of the continental lithosphere, and knowing the gravitational force acting on the deformed lithosphere, one can calculate the strain rate of the gravity-related flow and that of the convergence-related flow. It is assumed for simplicity that both flows occur in two perpendicular directions.

The evolution of the geometry of the lithosphere can be portrayed as a path within the fc–fl plane (Fig. 2), where fc, is the thickness ratio of the thickened crust to that of the reference crust (z_c), and fl is the thickness ratio of the thickened lithosphere to that of the reference lithosphere (z_l) (Sandiford & Powell 1990; Sandiford et al. 1992). During deformation, the deforming lithosphere follows a path in the fc–fl plane which starts at $fc = fl = 1$ (i.e. an undeformed reference state). Convergent orogens attain higher fc and fl values, while thinned lithosphere evolves to fc and fl values < 1 (Fig. 2). The triangular field at high fc and low fl values portrays the condition where the lithosphere consists entirely of crust (i.e. where $fc \cdot zc \leq fl \cdot zl$). This paper employs a slight variation in the representation of the fc–fl plane to that originally defined by Sandiford & Powell (1990). Here the fc–fl plane is oriented so that the crustal component of the lithosphere is lying

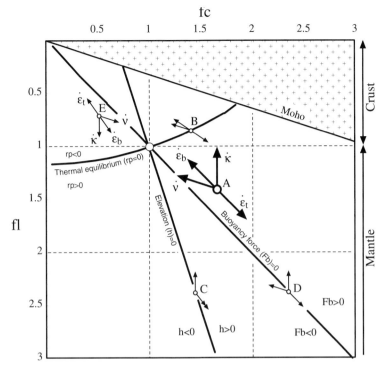

Fig. 2. Four elementary processes may affect the vertical geometry of the lithosphere (i.e. the thickness of the crust and that of the lithospheric mantle). These processes can be represented as vectors in the *fc–fl* plane. For example the lithophere A is subjected to the action of erosion (ν), tectonic thickening ($\dot{\varepsilon}_t$), gravity-driven flow ($\dot{\varepsilon}_b$), and thermal relaxation (κ). The path followed by the lithosphere A in the *fc–fl* plane depends on the magnitude of the four vectors and their respective orientation. The length of each vector represents its rate. Lithosphere E is also subjected to the action of the four processes but all the vectors have a direction opposite to lithosphere A. Indeed, the gravity-driven flow is divergent and tends to thin the lithosphere when the gravitational force is positive (the deformed lithosphere has an excess of gravitational potential energy). However, when the gravitational force is negative the gravity-driven flow has an opposite effect. When the elevation is positive, erosion tends to reduce the thickness of the crust. In contrast, when elevation is negative, sedimentation tends to increase the thickness of the crust. Thermal relaxation may induce the thickening or the thinning of the lithospheric mantle depending of the sign of the thermal potential (rp, Sandiford & Powell 1991). Lithosphere B C, and D represent cases where only three processes are active.

above the mantle component. In this orientation the *fc–fl* plane becomes more illustrative, with the column above any point of the *fc–fl* plane representing a lithospheric column with its mantle and crustal component.

What follows describes the main assumptions related to the four processes that affect the vertical geometry of the continental lithosphere.

Tectonic thickening

A reference lithosphere ($z_c = 35$ km, $z_l = 104$ km), in isostatic and mechanical equilibrium with a column beneath the mid-ocean ridge, is submitted to a tectonic force (*Fd*) to achieve a constant strain rate of 10^{-15} s^{-1}. As the litho-

sphere becomes thicker and stronger, *Fd* must increase to maintain a constant strain rate. It is assumed that the maximum magnitude for *Fd* is 30×10^{12} Nm^{-1} (e.g. Bott 1993). When this value is reached the strain rate decreases as thickening proceeds. When the strain rate is no longer of significance, one can expect thickening to be partitioned into adjacent portions of the orogen leading to the development of a plateau. In the *fc–fl* plane, homogenous deformation is represented by a vector parallel to the line that joins the origin of the *fc–fl* plane (*fc* = *fl* = 0, top left corner of the *fc–fl* plane) to the point representing the deforming lithosphere. This vector is directed away from the origin in the case of homogeneous thickening, and towards it during homogeneous thinning.

Erosion/sedimentation

Erosion is mainly dependent on the wavelength of the topography (sharp relief implying fast erosion rate), and its magnitude (Ruxton & McDougall 1967; Ahnert 1970). Most estimations of erosional rates fall in the range of 0.2 to 1 mm a^{-1} (Clark & Jaeger 1969; England & Richardson 1977; Pinet & Souriau 1988; Mercier et al. 1991). The rate of erosion is assumed to be proportional to: (i) the strain rate (fast strain rates produce sharp relief); and (ii) the difference in elevation between that of the reference lithosphere (+874 m) and that of the deformed lithosphere. This is consistent with the observation that orogenic plateaux are characterized by a relatively flat topography, a slow strain rate, and little erosion. If, at some stage, the surface of the lithosphere drops below sea level, it is assumed that sedimentation maintains a constant water–sediment ratio of 2:1 in the basin.

In the fc–fl plane, erosion/sedimentation is represented by a vector parallel to the boundary between the crust and the mantle (Moho line in Fig. 2), since these processes do not affect the thickness of the sub-continental mantle lithosphere. During erosion, the vector is directed towards the origin of the fc–fl plane (the crust becomes thinner and therefore fc decreases). During sedimentation it is directed away from the origin of the fc–fl plane.

Thermal relaxation

During lithospheric thickening, isotherms are displaced vertically producing a thermal anomaly. Assuming that a constant heat flow is maintained at the base of the lithosphere, defined by the isotherm 1330°C, the relaxation of this thermal anomaly tends to reduce the thickness of the lithospheric mantle. This thermal thinning increases the surface elevation and therefore the tensile gravitational force. During and following thickening, thermal relaxation also tends to increase the average temperature of the lithosphere and therefore tends to reduce its strength.

In the fc–fl plane, the vector representing thermal relaxation is parallel to the fl axis, since this process does not affect the thickness of the crust. The vector is oriented towards higher fl values when thermal relaxation increases the thickness of the lithosphere (following thinning), and in the opposite direction when it decreases the thickness of the lithospheric mantle (following thickening).

Gravity-driven flow

Having assumed that the lithosphere behaves as a viscous thin sheet, the gravitational flow affects the crust and the lithospheric mantle in the same way. The homogeneous gravity-driven flow is represented in the fc–fl plane by a vector parallel to the line that joins the origin of the fc–fl plane and the point representing the deforming lithosphere. This vector is directed away from the origin of the fc–fl plane in the case of convergent gravitational flow (thickening is enhanced by convergent flow), and towards it in the case of divergent flow.

Divergent gravitational spreading can be accommodated by either the thickening of the surrounding lithosphere (fixed boundary collapse) or by its passive displacement (free boundary collapse, Rey et al. 2001). Depending on the accommodation mechanism, it is assumed that the spreading rate is buffered by either the shortening rate of the surrounding lithosphere, or the extensional rate of the collapsing lithosphere.

Physical model

What follows describes the parameters that enter in the definition of the balance of forces, and how these parameters change during deformation. To facilitate comparisons with studies that have investigated force balance in convergent orogen, the numerical experiments presented in this paper use similar physical parameters, simplifications and approximations (Le Pichon & Alvarez 1984; Sandiford & Powell 1991; Ranalli 1992; Zhou & Sandiford 1992). Table 1 gives a list of symbols, values and parameters used in this paper.

Temperature profile

The density and strength profile, as well as the gravitational force are dependent on temperature. Steady state and transient geotherms are derived from the one-dimensional diffusion–advection equation:

$$\frac{\partial T}{\partial t} = \frac{\kappa \cdot \partial^2 T}{\partial z^2} + \frac{A}{\rho \cdot Cp} - \nu \frac{\partial T}{\partial z} \qquad (1)$$

where, κ, represents thermal diffusivity; A, the radiogenic heat production; ρ, the density; Cp, the heat capacity; and ν, the velocity. Boundary conditions involve a constant heat flow entering the base of the lithosphere, and a constant temperature at the surface. The velocity of the medium relative to its surface is a function of

Table 1. *List of parameter values*

Gravity g, m.s^{-2}	9.81
Crust thermal expansion coefficient α_c, 10^{-5} K^{-1}	3.5
Mantle thermal expansion coefficient $\alpha_m = a_0 + a_1 \cdot T + a_2 \cdot T$, K^{-1}	$a_0 = 2.697 \cdot 10^{-3}$
	$a_1 = 1.0192 \cdot 10^{-8}$
	$a_2 = -0.1282$
Mantle bulk modulus β_m, MPa	130
Lithosphere thermal diffusivity κ, μm^2 s^{-1}	0.97
Lithosphere thermal conductivity, Wm^{-1} K^{-1}	3.1
Crust volumetric heat production, μW m^{-3}	0.80
Mantle volumetric heat production, μW m^{-3}	0
Surface temperature T_o, K	273
Temperature at the base of the lithosphere Tl, K	1603
Heat Flow at the base of the lithosphere Qo, μW m^{-2}	34.9
Crustal thickness zc, km	35
Lithosphere thickness zl, km	104
Neutral buoyant level of asthenosphere za, km	-3.6
Lithospheric mantle density ($T = To$) ρ_{lmo}, kg m^{-3}	3370
Asthenospheric mantle density ($T = To$) ρ_{ao}, kg m^{-3}	3390
Crust mass density @ $z = h$, ρ_a, kgm^{-3}	2670
Crust mass density @ $z = zc$, ρ_b, kgm^{-3}	2950
Water mass density, ρ_w, kgm^{-3}	1030
Ratio pore pressure to overburden stress λ,	0.36
Universal gas constant R, J mol^{-1} K^{-1}	8.3144
Crust power law sensitivity n_c,	3
Crust power law activation enthalpy Q_c, kJmol^{-1}	190
Crust power law pre-exponent A_c, MPa^{-3} s^{-1}	$5 \cdot 10^{-6}$
Mantle power law sensitivity n_m,	3
Mantle power law activation enthalpy Q_m, kJmol^{-1}	520
Mantle power law pre-exponent A_m, MPa^{-3} s^{-1}	$7 \cdot 10^4$
Mantle Dorn plasticity law activation enthalpy Q_d, kJmol^{-1}	540
Mantle Dorn plasticity law stress treshold σ_d, MPa	8500
Mantle Dorn plasticity law strain rate $\dot{\varepsilon}_d$, s^{-1}	$3.05 \cdot 10^{11}$

deformation, erosion/sedimentation, and gravity-driven flow. It was implemented iteratively over small time intervals.

Density profile and surface elevation

The density profile enters in the calculation of the surface elevation (h) of the lithosphere, which in turn is necessary for calculation of the gravitational force. For the crust a linear density profile is used. It increases from the top of the crust (ρ_a) to the Moho (ρ_b), and is dependent on temperature (α_c the coefficient of thermal expansion).

$$\rho_{\text{crust}}(z) = \rho_a + \frac{\rho_b - \rho_a}{fc \cdot zc} \cdot (1 - \alpha_m \cdot T(z)) \quad (2)$$

The density of the mantle is in addition pressure dependent (β_m the coefficient of compressibility). The density profile for the lithospheric mantle is given by:

$$\rho_{lm}(z) = \rho_{lmo} \cdot (1 - \alpha_m \cdot T(z) + \beta_m \cdot P) \quad (3)$$

where, ρ_{lmo} is the density of lithospheric mantle at room temperature. For the density profile in

the asthenosphere, the density of the asthenophere ρ_{ao} at room temperature is substituted in equation (3).

To calculate the surface elevation, the lithospheric column is assumed to be in isostatic balance with a column beneath the mid-ocean ridge whose elevation (za) represents the hydrostatic level of the asthenosphere. Following Le Pichon & Alvarez (1984), the hydrostatic level of the asthenosphere is assumed to be close to 3600 m. Therefore, the reference mid-ocean ridge column is formed of $za = 3600$ m of water (density ρ_w) overlying ($flzl + h - za$) metres of asthenospheric mantle.

When the surface of the deforming lithosphere is below sea level, it is assumed that one third of the basin ($h/3$) is filled with sediments.

Gravitational force

For simplicity flexural stresses induced by the elastic behaviour of the lithosphere are not considered here. With depth (z) increasing downward from a sea level origin, the gravitational potential energy (*Pe*) per unit area of a litho-spheric column is given by the integral of the vertical stress (σ_{zz}) from the bottom of the column at depth *flzl* + *h* to its top which is at an elevation *h* (when the column stands above sea level) or 0 (when it is below sea level):

$$Pe = \int_{flzl+h}^{surface} \rho(z')gz'\,dz' \qquad (4)$$

where, $\rho(z')$, is the density profile; *g*, is the gravitational acceleration; z', is the integration variable.

Assuming local isostatic compensation, the gravitational force per unit length that the deformed lithosphere and the surrounding litho-sphere apply to one another is given by the con-trast in gravitational potential energy between the reference and deformed lithospheric column. It is the difference between the integrals of the vertical stress profile down to a compensation level (*L*) beneath the lithosphere.

$$Fg = \Delta Pe = \Delta\left[\int_L^S \left(\int_z^s \rho(z'')g\,dz''\right)dz\right] \quad (5)$$

S, represents either the elevation of the refer-ence lithosphere or that of the modified litho-sphere, whichever is higher. *L*, is the bottom of the reference lithosphere (*zl* + *h*) or that of the modified lithosphere (*flzl* + *h*), whichever is the deepest.

Strength profile

In the rheological profile, frictional sliding describes failure mechanisms at low temperature and high strain rate (in the upper crust and the upper mantle (Sibson 1974):

$$\sigma_1(z) - \sigma_3(z) = \beta \cdot \rho(z) \cdot g \cdot (z-h) \cdot (1-\lambda) \quad (6)$$

where, *g*, the gravitational acceleration; λ, the ratio of fluid pore pressure to the normal stress; and β, a parameter dependent on the type of faulting.

At high temperatures, the viscous deforma-tion of the crust is modelled as power law creep:

$$\sigma_1(z) - \sigma_3(z) = \left(\frac{\dot{\varepsilon}}{A_c}\right)^{1/n} \cdot \exp\left(\frac{Q_c}{n \cdot R \cdot T(z)}\right), \quad (7)$$

and in the mantle as power law creep and Dorn law creep depending on ($\sigma_1 - \sigma_3$):

$$\sigma_1(z) - \sigma_3(z)$$

$$= \begin{cases} \left(\dfrac{\dot{\varepsilon}}{A_m}\right)^{1/n} \cdot \exp\left(\dfrac{Q_m}{n \cdot R \cdot T(z)}\right) \\ \qquad\qquad (\sigma_1 - \sigma_3) < 200\,\text{MPa} \\[6pt] \sigma_d \cdot \left(1 - \sqrt{\dfrac{R \cdot T(z)}{Q_d}} \cdot \ln\left(\dfrac{\dot{\varepsilon}_d}{\varepsilon Y}\right)\right) \\ \qquad\qquad (\sigma_1 - \sigma_3) > 200\,\text{MPa} \end{cases} \quad (8)$$

Assuming a pure shear deformation and a constant strain rate with depth, the vertical inte-grated strength in extension (*Fedt*) and contrac-tion (*Fedc*) of the lithosphere are respectively given by:

$$Fedt = \int_{fl.zl}^{h_d} (\sigma_1(z) - \sigma_3(z))\,dz \quad \text{with } \beta = 0.75 \quad (9)$$

$$Fedc = \int_{fl.zl}^{h_d} (\sigma_1(z) - \sigma_3(z))\,dz \quad \text{with } \beta = 3,$$

using the minimum of (6) and (7) for the crust, and the minimum of (6) and (8) for the mantle.

Lithospheric geometry through time

The increase in crustal and whole lithospheric thicknesses over a small time interval (Δt) due to homogeneous pure shear are given by:

$$\dot{\varepsilon}_t \cdot f_c z_c \cdot \Delta t, \quad (10)$$

and

$$\dot{\varepsilon}_t \cdot f_l z_l \cdot \Delta t \quad (11)$$

respectively. The thickening rate ($\dot{\varepsilon}_t$) derives from balancing the tectonic force with the strength in contraction of the deforming litho-sphere. Because of gravity-driven flow, the lithosphere tends also to thin. This thinning is given by:

$$\dot{\varepsilon}_s \cdot f_l z_l \cdot \Delta t \quad (12)$$

When divergent gravitational collapse is accommodated by the passive displacement of the surrounding lithosphere, the spreading rate ($\dot{\varepsilon}_s$) derives from balancing the gravitational force with the strength in extension of the deform-ing lithosphere. Alternatively, when gravitation-al collapse is accommodated by the shortening of the lithosphere surrounding the orogenic do-main, the spreading rate is buffered by either the strength in extension of the collapsing litho-spheric column or the strength in contraction of the surrounding lithosphere.

Presentation of the results

In all of the experiments, the reference lithosphere is shortened at a strain rate of 10^{-15} s^{-1} over a period of 30 Ma before it is unloaded. This shortening produces a 75 km thick crust and a 222 km thick lithosphere. In the first experiment, the thickened lithosphere is allowed to evolve under the action of erosion, gravitational force, and thermal relaxation. In the following experiments the thickness of the lithospheric mantle is instantaneously reduced to simulate the convective removal of the lower part of the lithosphere immediately after the end of convergence. The tectonic histories that unfold strongly depend on the amount of lithospheric mantle removed.

Stable Thermal Boundary Layer during convergence

The evolution of the vertical geometry of the deforming lithosphere is shown in Figure 3. At the end of convergence the gravitational force is compressive and relatively small (1.70×10^{12} Nm^{-1}) compared to the strength in con-

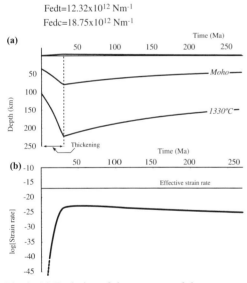

Fedt=12.32x10^{12} Nm^{-1}
Fedc=18.75x10^{12} Nm^{-1}

Fig. 3. (a) Evolution of the geometry of the lithosphere assuming a homogeneous tectonic thickening. Thickening last for 30 Ma, the thermal boundary layer is stable. Erosion and thermal relaxation slowly bring back the lithosphere towards its original geometry over a time scale of a few hundreds of million years. (b) Evolution of the spreading rate. At no time is the spreading rate higher than the effective strain rate (10^{-17} s^{-1}). Divergent collapse is therefore not significant.

traction of the thickened lithosphere ($c. 32.88\times 10^{12}$ Nm^{-1}, strengths are given for a nominal strain rate of 10^{-15} s^{-1}). Through time, the strength of the lithosphere decreases but remains higher than that of the reference lithosphere. In contrast, the gravitational force remains nearly constant. This suggests that over time the decay of the excess in gravitational potential energy stored in the crust is balanced by the decay through thermal relaxation of an equivalent deficit in gravitational potential energy stored in the lithospheric mantle. The lithosphere tends to slowly recover its initial geometry. 260 Ma after the end of convergence the continental crust and the lithosphere are still 45 km and 150 km thick respectively, and the lithosphere is stronger than before deformation (Fig. 3).

The tectonic evolution of the reference lithosphere can be summarized in two phases. In the first phase the lithosphere thickness increases during convergence. The second starts when the system is unloaded. During this phase the lithosphere slowly recovers its initial geometry over a few hundreds of million years. At no stage during its evolution is the orogenic domain weaker than the reference lithosphere, and at no stage is the gravitational force stronger that the strength of the deformed lithosphere. This evolution is therefore characterized by an absence of significant collapse and the formation of a strong lithosphere that resists tectonic reactivation.

Post-convergence removal of the Thermal Boundary Layer

In this experiment the lower part of the lithosphere is removed at the end of convergence ($t_o + 30$ Ma). Before convective thinning, the continental crust is then 75 km thick, the temperature at the Moho is $c. 640°C$, the elevation of the lithosphere is 4650 m, and its strength in extension has increased up to 24.5×10^{12} Nm^{-1}. The gravitational force is low and compressive (1.75×10^{12} Nm^{-1}). Contrasting lithosphere geometries (Fig. 4) develop depending on the magnitude of the convective thinning.

In the first example the entire lithosphere is thinned to 130 km, following the removal of the lower 92 km of the lithospheric mantle (Fig. 4 a$_{1-2}$). As a result, the gravitational force switches from being slightly compressive to strongly extensive (7.2×10^{12} Nm^{-1}), whereas the strength in extension of the lithosphere is strongly reduced (5×10^{12} Nm^{-1}). The elevation increases to about 6600 m, and the temperature at the Moho increases to 780°C. Assuming that the surrounding undeformed lithosphere

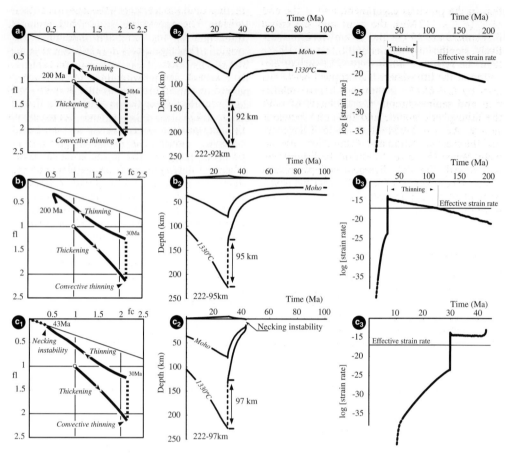

Fig. 4. Evolution of the geometry of the lithosphere following homogeneous tectonic thickening and convective thinning of the thermal boundary layer. Convergence lasts for 30 Ma, and convective thinning occurs immediately after convergence. Collapse is assumed to be accommodated by the passive displacement of the surrounding lithosphere. Three situations are considered whereby the bottom 92, 95, and 97 km are removed from the lithosphere (rows from top to bottom, respectively). The evolution of the lithospheric geometry is shown in the *fc–fl* plane (diagrams in the left column), as well as in a diagram showing the depth v. time of the main density interfaces (diagrams in the central column). Diagrams in the right column show that for each experiment significant gravity-driven flow occurs following convective thinning of the lithospheric mantle (i.e. spreading rate is higher than $10^{-17}\,\mathrm{s}^{-1}$).

is passively displaced, the gravitational force is strong enough to produce thinning in the orogenic domain at a strain rate of $c.\,10^{-14}\,\mathrm{s}^{-1}$ (Fig. 4a₃), which slowly decreases as divergent collapse proceeds. At $t_o + 33.6$ Ma the lithosphere has recovered its initial thickness but the crust is 54 km thick. At this stage, divergent collapse is still very active with a strain rate of $10^{-15}\,\mathrm{s}^{-1}$. The thickness of the lithosphere decreases to 68 km by $t_o + 71.7$ Ma, before increasing again under the action of thermal relaxation. The crust continues to thin to 32 km, and by $t_o + 80$ Ma the strain rate has dropped below $10^{-17}\,\mathrm{s}^{-1}$. A combination of erosion and thermal

relaxation slowly brings the lithosphere towards its initial geometry over a time scale of a few hundred million years. This experiment is characterized by a vigorous phase of divergent collapse. The surface elevation remains above sea level although the crust becomes a bit thinner than its initial thickness. After a few hundred of million years, the entire lithosphere will probably end up being a bit thicker than it was initially, because of a slightly thinner radiogenic crustal layer.

A similar history unfolds when the lower 95 km of the lithospheric mantle is removed (Fig. 4b). However, the collapse stage is longer that in the previous experiment, and at the end of it ($c. t_o + 112$ Ma), the crust and the entire lithosphere are 19 km and 33 km thick, respectively, significantly thinner and therefore weaker than the reference lithosphere. The surface elevation of the lithosphere has dropped below sea level by $t_o + 65$ Ma. Following thermal relaxation and sedimentation, the thickness of both the lithospheric mantle and the crust increases slowly. At $t_0 + 200$ Ma the whole lithosphere and the crust are 52 km and 22 km thick, respectively. When the lower 97 km of the lithosphere is removed (Fig. 4c), divergent collapse results in active continental rifting following the development of a necking instability at around $t_0 + 42$ Ma (Fig. 4c$_{2-3}$).

Discussion and conclusion

The simple 1-D numerical experiments presented in this paper suggests that finite lithospheric thinning and active rifting can be the result of convective thinning of the thermal boundary layer following lithospheric thickening. The fact that divergent collapse results in a lithosphere thinner than the reference lithosphere is not surprising. The gravitational force that promotes thinning following convective thinning has its origin in the large excess in gravitational potential energy stored in the thickened crust (Fig. 5). In contrast, the gravitational potential energy stored in the lithospheric mantle is quite small and promotes compression (Fleitout & Froidevaux 1982). As collapse and homogeneous thinning proceed both the crust and the mantle becomes thinner. The excess in gravitational potential energy stored in the crust decreases, whereas thinning of the lithospheric mantle produces an increasing excess of gravitational potential energy (Fig. 5), that eventually promotes further thinning. The experiments presented above show that, when enough lithospheric mantle is removed, this thinning leads towards a necking instability and the development of an active rift. The volume of lithospheric mantle that has to be removed to produce active rifting is not unrealistic. A study from Houseman & Molnar (1997) has shown that, for dry olivine, the section of the lithosphere hotter than 910–950°C is likely to be dragged into the convective mantle. For wet olivine, it is the section

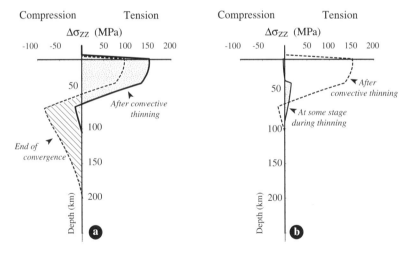

Fig. 5. The contrast in gravitational potential energy between a deformed lithospheric column and its surroundings is given by the integral of the difference in lithostatic pressure ($\Delta\sigma_{zz}$) along a deformed lithospheric column and along a column of the surrounding lithosphere. (**a**) At the end of thickening (dashed line), an excess in gravitational energy (promoting extension) is stored in the upper part of the lithosphere (mostly the crust), whereas there is a deficit in potential energy (which promotes compression) in the lower part of the lithosphere. After convective thinning of the thermal boundary layer (thick solid line), most of the deficit in potential energy is removed whereas the excess in potential energy increases significantly. The whole lithosphere is therefore under strong horizontal tensile stresses which leads to divergent collapse and thinning. (**b**) As collapse proceeds, the excess of gravitational potential energy is reduced and at one stage the crust starts to accumulate a deficit in gravitational potential energy (thin solid line). In the meantime the thinning mantle starts to store excess in gravitational potential energy. It is this excess that promotes self-enhanced extension that may lead to active continental rifting.

hotter than 750°C that can be unstable. In the numerical experiments presented above, active rifting occurs when the section of the lithosphere, hotter than 800°C, is removed.

There are many examples that suggest that divergent collapse has resulted in net crustal and lithospheric thinning. This is possibly the case in the Basin and Range Province where the crust,

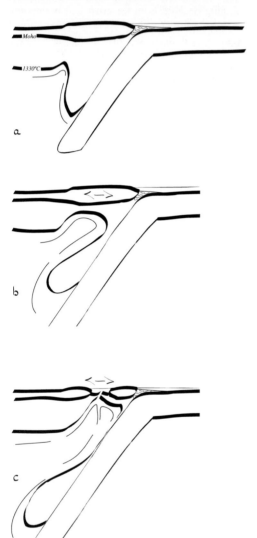

following Tertiary divergent collapse, is now 20–25 km thick and still under tensile stresses. The Carpathian Basin is another example of a low standing thin crust whose thinning has been related to post-thickening collapse. The Alboran Sea and the Aegean Sea, both the locus of very active divergent collapse during the Neogene, illustrate the case where the surface of the thinned lithosphere has dropped below sea level. In all these examples, thinned regions, some still under tensile stresses, are surrounded by regions of compressive stresses that may have inhibited, so far, active rifting.

It is possible that one of the most obvious manifestation of post-thickening active rifting occurs in the development of back-arc basins, involving the opening of an oceanic domain that detaches a piece of continental lithosphere from a thickened continental passive margin. Indeed, any mechanism that thins the sub-continental lithospheric mantle above a subduction zone will produce tensile stresses that may promote (i) failure of the overriding lithosphere; (ii) divergent collapse; (iii) active rifting; and (iv) opening of a back-arc basin. Figure 6 sketches the situation whereby the thermal boundary layer of the overriding plate is dragged into the convective mantle by the descending oceanic slab. As long as the subduction zone acts as a free boundary, the broken lithosphere segment will move counter to the subduction plate as the back-arc basin develops. Back-arc dynamics has been successfully produced in physical models (Shemenda 1993). This modelling shows that a zone of weakness in the overriding plate is necessary for back-arc extension to develop. Collapse following thickening of the overriding plate may produce this zone of weakness.

This paper has benefited from helpful and constructive reviews from O. Vanderhaeghe and P. Ryan. Many thanks to G. Houseman, S. Costa, M. Jessell and W. Schellart, for countless conversations and arguments. Thanks to S. Van Huet for having improved the English, and J. Miller for editing the text. This work was supported by ARC Large Grant No. A10017138.

Fig. 6. Back-arc extension as an example of post-convective thinning active continental rifting. This figure illustrates how the thinning of the lithospheric mantle underneath a thickened crust may induce active continental rifting, provided that the subduction zone acts as a free boundary. The force driving active continental rifting is the gravitational force related to the differential thinning of the lithospheric mantle.

References

AHNERT, F. 1970. Functional relationships between denudation, relief, and uplift in large mid-latitude drainage basins. *American Journal of Science*, **268**, 243–263.

ARTYUSHKOV, E. V. 1973. Stresses in the lithosphere caused by crustal thickness inhomogeneities. *Journal of Geophysical Research*, **78**, 7675–7708.

BOTT, M. H. P. 1993. Modelling the plate-driving mechanism. *Journal of Geological Society, London*, **150**, 941–951.

CLARK, S. P. & JAEGER, E. 1969. Denudation rates in the Alps from geochronological and heat flow data. *American Journal of Science*, **267**, 1143–1160.

CONEY, P. J. & HARMS, T. A. 1984. Cordilleran metamorphic core complexes: Cenozoic extensional relics of Mesozoic compression. *Geology*, **12**, 550–554.

COSTA, S. & REY, P. 1995. Lower crustal rejuvenation and growth during post-thickening collapse: Insights from a crustal cross section through a Varscan metamorphic core complex. *Geology*, **23**, 905–908.

DEWEY, J. F. 1988. Extensional collapse of orogens. *Tectonics*, **7**, 1123–1139.

ENGLAND, P. C. & HOUSEMAN, G. A. 1988. The mechanics of the Tibetan plateau. *Philosophical Transactions of the Royal Society London*, **326**, 301–320.

ENGLAND, P. C. & HOUSEMAN, G. A. 1989. Extension during continental convergence, with application to the Tibetan Plateau. *Journal of Geophysical Research*, **94**, 17 561–17 579.

ENGLAND, P. C. & MCKENZIE, D. P. 1982. A thin viscous sheet model for continental deformation. *Geophysical Journal Royal Astronomical Society*, **70**, 295–321.

ENGLAND, P. C. & MCKENZIE, D. P. 1983. Correction to: A thin viscous sheet model for continental deformation. *Geophysical Journal Royal Astronomical Society*, **73**, 523–532.

ENGLAND, P. C. & RICHARDSON, S. W. 1977. The influence of erosion upon mineral facies of rocks from different metamorphic environments. *Journal of the Geological Society, London*, **134**, 201–213.

FLEITOUT, L. & FROIDEVAUX, C. 1982. Tectonics and topography for a lithosphere containing density heterogeneities. *Tectonics*, **1**, 21–56.

HOUSEMAN, G. A. & ENGLAND, P. C. 1986. A dynamic model of lithosphere extension and sedimentary basin formation. *Journal of Geophysical Research*, **91**, 719–729.

HOUSEMAN, G. A. & MOLNAR, P. C. 1997. Gravitational (Rayleigh–Taylor) instability of a layer with non-linear viscosity and convective thinning of continental lithosphere. *Geophysical Journal International*, **128**, 125–150.

HOUSEMAN, G. A., MCKENZIE, D. P. & MOLNAR, P. 1981. Convective instability of a thickened boundary layer and its relevance for the thermal evolution of continental convergent belts. *Journal of Geophysical Research*, **86**, 6115–6132.

LE PICHON, X. 1983. Land-locked oceanic basins and continental collision: The eastern mediterranean as a case example. *In*: K. J. HSU (ed.) *Mountain building processes*. Academic Press, New York, 201–211.

LE PICHON, X. & ALVAREZ, F. 1984. From stretching to subduction in back-arc regions: Dynamic considerations. *Tectonophysics*, **102**, 343–357.

MARESCHAL, J. C. 1994. Thermal regime and post-orogenic extension in collision belts. *Tectonophysics*, **238**, 471–484.

MERCIER, L., LARDEAUX, J. M. & DAVY, P. 1991. On the tectonic significance of retrograde P–T–t paths in eclogites of the French Massif Central. *Tectonics*, **10**, 131–140.

MOLNAR, P. & CHEN, W. P. 1982. Seismicity and mountain building. *In*: HSU, K. J. (ed.) *Mountain building processes*. Academic Press, New York, 41–57.

MOLNAR, P. & CHEN, W. P. 1983. Focal depths and fault plane solutions earthquackes under the Tibetan plateau. *Journal of Geophysical Research*, **88**, 1180–1196.

MOLNAR, P. & LYON-CAEN, H. 1988. Some simple physical aspects of the support, structure and evolution of mountain belts. *Geological Society of America Special Paper*, **218**, 179–207.

PINET, P. & SOURIAU, M. 1988. Continental erosion and large-scale relief. *Tectonics*, **7**, 563–582.

PLATT, J. P. & ENGLAND, P. C. 1993. Convective removal of the lithosphere beneath mountain belt: thermal and mechanical consequences. *American Journal of Science*, **294**, 307–336.

RANALLI, G. 1992. Average lithospheric stresses induced by thickness change: A linear approximation. *Physics of the Earth and Planetary Interiors*, **69**, 263–269.

RANALLI, G. 1995. *Rheology of the Earth*. Chapman & Hall, London, UK.

REY, P. 1993. Seismic and tectono-metamorphic characters of the lower continental crust in Phanaerozoic areas: A consequence of post-thickening extension. *Tectonics*, **12**, 580–590.

REY, P., VANDERHEAGHE, O. & TEYSSIER, C. 2001. Gravitational collapse of continental crust: Definition, regimes, and modes. *Tectonophysics*, in press.

ROYDEN, L. & KEEN, C. E. 1980. Rifting process and thermal evolution of the continental margin of eastern canada determined from subsidence curves. *Earth and Planetary Science Letters*, **51**, 343–361.

RUXTON, B. P. & MCDOUGALL, I. 1967. Denudation rates in northeast Papua from potassium–argon dating of lavas. *American Journal of Science*, **265**, 545–561.

SANDIFORD, M. & POWELL, R. 1990. Some isostatic and thermal consequences of the vertical strain geometry in convergent orogens. *Earth and Planetary Science Letters*, **98**, 154–165.

SANDIFORD, M. & POWELL, R. 1991. Some remarks on high-temperature low-pressure metamorphism in convergent orogens. *Journal of Metamorphic Geology*, **9**, 333–340.

SANDIFORD, M., FODEN, J., ZHOU, S. & TURNER, S. 1992. Granite genesis and mechanics of convergent orogenic belts with application to the southern Adelaide fold belt. *Transactions of the Royal Society Edinburgh: Earth Sciences*, **83**, 83–93.

SHEMENDA, A. I. 1993. Subduction of the lithosphere and back-arc dynamics: Insights from physical modeling. *Journal of Geophysical Research*, **98**, 16 167–16 185.

SIBSON, R. H. 1974. Frictional constraints on thrust, wrench and normal faults. *Nature*, **249**, 542–544.

SONDER, L. J. & ENGLAND, P. C. 1986. Vertical averages of rheology of continental lithosphere: relation to thin sheet parameters. *Earth and Planetary Science Letters*, **77**, 81–90.

SONDER, L., ENGLAND, P. C., WERNICKE, B. P. &
  CHRISTIANSEN, R. L. 1987. A physical model for
  Cenozoic extension of western North America. *In*:
  COWARD, M. P., DEWEY, J. F., & HANCOCK, P. L.
  (eds) *Continental Extensional Tectonics.* Geologi-
  cal Society, London, Special Publications, **28**,
  187–201.
SUN, J. & MURRELL, A. F. 1989. On the growth
  and collapse of wide orogenic belts. *Geophysical
  Journal International,* **118**, 255–268.

TURCOTTE, D. L. & EMERMAN, S. H. 1983. Mechanisms of active and passive rifting. *Tectonophysics,*
  94, 39–50.
ZHOU, S. & SANDIFORD, M. 1992. On the stability of
  isostatically compensated mountain belts. *Journal
  of Geophysical Research,* **97**, 14 207–14 221

Episodicity during orogenesis

G. S. LISTER, M. A. FORSTER & T. J. RAWLING

Australian Crustal Research Centre, Department of Earth Sciences, Monash University, Melbourne 3800, Australia (e-mail: gordon@mail.earth.monash.edu.au)

Abstract: Deformation events and episodes of metamorphic mineral growth are usually regarded as relatively local phenomena. It is not expected that specific events and episodes within an orogenic sequence should exactly correlate over large distances. There is no obvious reason, for example, to assume that deformational and/or metamorphic events in the Western European Alps would directly correlate with events taking place in the Aegean continental crust, *c.* 1000 km distant. Yet linked episodes of deformation and metamorphism appear to take place at the same time over large distances, even in these apparently unrelated segments of the same orogenic belt. This large-scale episodic behaviour appears to be associated with switches in tectonic mode, from compressional orogenesis to extensional tectonism, and may be the result of orogenic surges and/or periods of lithospheric extension following accretion events. The effect of these switches is greatest in back-arc environments, in the over-riding plate above major subduction zones. In these environments, roll-back of the subducting lithospheric slab after individual accretion events ensures that the amount of lithospheric extension after each accretion event is large. As a result this is where coherent high-pressure metamorphic terranes formed in the preceding accretion event are exhumed, and where remnants of newly emplaced ophiolite sheets are stranded by newly formed detachment faults.

In this paper, it is suggested that the evolution of an orogen is intrinsically episodic, and that from time to time globally significant changes take place, over short time periods (e.g. <1–2 Ma). The effects of such episodes are not localized to one small part of one segment in one orogenic belt. They can be identified on a planetary scale, and there is a wide range of different phenomena associated with them (see following section).

Structural geologists and metamorphic petrologists have largely discounted the significance of episodic behaviour during the processes that lead to the formation of mountain belts (i.e. during orogenesis). From a theoretical point of view it is argued that while plate motions are smooth and continuous it should be expected that convergence across an orogenic belt would also be smooth and continuous. Events recognized locally may therefore be part of a larger continuum, and considering the larger-scale, not at all significant. Episodic behaviour inferred by field workers has been assumed to reflect the inability of geologists to see the bigger picture (Hsu 1989). Molnar & Lyon-Caen (1988) write that: 'There has been a tendency of workers in some ranges to assume that tectonic activity occurred only during short intervals ... separated by long intervals of quiescence.' However 'It seems equally probable ... that convergence was steady, possibly even at a constant rate, but with deformation concentrated in particular regions for only short times. When one region ceased being active, another presumably became active or was so already. Thus the division into separate short phases of activity may be valid only on a very small scale, but not on a regional scale.'

Diachroneity is also an important aspect of the prevailing view of orogenesis. The shape of the colliding objects is not regular. On an even larger scale, while there is general acknowledgement that different events occur at different times along an orogenic belt, these different events are not thought to be connected (in time or in space) in any particular way, except by coincidence. Without the benefit of absolute time constraints it is not easy to correlate from one part of an orogenic belt to another, so such correlations are rarely attempted.

The authors of this paper are aware that there are many factors that make it difficult to ascertain whether or not episodic behaviour does in fact take place on the large-scale. In part this circumstance arises because the separate parts of an orogenic belt have a complex evolution, and a considerable amount of data is needed for their comprehensive analysis. This difficulty has been compounded by errors in estimates for ages of different thermal and/or tectonic events, and it is

From: MILLER, J. A., HOLDSWORTH, R. E., BUICK, I. S. & HAND, M. (eds) *Continental Reactivation and Reworking.* Geological Society, London, Special Publications, **184**, 89–113. 1-86239-080-0/01/$15.00 © The Geological Society of London 2001.

only quite recently that geochronological techniques have begun to improve to the point at which more exact estimates are possible. For example, revision of the age of ultra high-pressure (UHP) metamorphism in the Dora Maira massif has now taken place and of the age of high-pressure (HP) metamorphism in the Sesia-Lanzo zone (Gebauer *et al.* 1997; Rubatto & Hermann 2000).

There is a need to overcome the argument (popular amongst structural geologists) that episodic behaviour merely reflects events that are part of a progressively evolving sequence, in particular if the pattern of movement remains the same throughout. For example, Gray & Foster (1998) consider evolution of the Palaeozoic Lachlan orogen in terms of progressive thickening of an accretionary wedge, with occasional mechanical failure producing successive thrusts. Throughout this evolution it is imagined that the overall movement picture remains constant, and that there is a smooth progressive evolution from the initial state towards the final state. In the context of this model, the breaking of a new thrust is a mere detail that should not obscure the clarity of this simple vision as to how a mountain belt evolves. Similarly Davis *et al.* (1986) imagine progressive evolution of an extensional shear zone, in its lower plate, as it is transported towards the surface from amphibolite facies to greenschist facies conditions, and finally through the brittle-ductile transformation. The movement picture might have remained the same, but nevertheless distinct episodes of movement were involved, and these took place at discrete time intervals and involved spatially distinct movement zones (ductile shear zones and/or faults). Tavarnelli & Holdsworth (1999) were able to provide a conclusive demonstration as to the importance of episodic movements in a transpressional wrench setting, since unconformities developed between different folding episodes, thus constraining the time of formation of different folds, excluding the possibility that folds had developed progressively during sedimentation. Throughout this sequence the movement picture remained more or less constant, so that fabrics were coplanar, and fold axes were colinear. Traditionally this data would have been interpreted to indicate continuous deformation over a period in excess of *c.* 30 Ma, whereas in fact movement was strongly episodic.

Another difficulty relates to the fact that it is normal to consider individual segments of an orogen independently from adjacent zones. There are some notable exceptions (e.g. Sengor 1976, 1987, 1993; Sengor *et al.* 1988) but, for example, the Himalayan collision is usually considered disjointly from the obduction of ophiolites in Papua New Guinea, and the Africa–Europe collision is

considered yet separately again. Furthermore, it is usual to explain the behaviour of particular segments of an orogen using the characteristics of individual cross-sections, whereas in fact the cause of an event may lie elsewhere.

To test the concept of global episodic behaviour is, thus, not an easy task. Many 'events' may be poorly constrained because available geochronology is not sufficiently precise, or not capable of recognizing an 'event' as a distinct entity taking place over a short time interval. The available data are often blurred (i.e. not precisely constrained in terms of absolute time). This means that even if precisely correlatable events do exist in the record, we are often unable to recognize them as such. This issue can only be resolved with the precision available to relatively modern geochronological techniques. Thus, more work is necessary before the hypothesis of globally episodic behaviour during orogenesis can be convincingly demonstrated, or set aside. Right now, however, let us cast aside these reservations, and look at a world in which a different paradigm may operate.

Global tectonic episodes during Tertiary orogenesis

The Alpine–Himalayan chain is an orogenic belt that stretches from Spain to New Zealand, over a strike-length exceeding 15 000 km (Fig. 1). It appears that during the evolution of this mountain chain there have been time periods during which seemingly catastrophic episodes of global

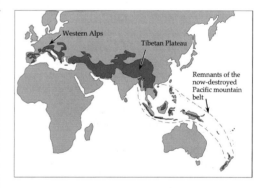

Fig. 1. The Alpine–Himalayan orogenic belt stretches from Spain to New Zealand. Along different parts of its strike length it has been dismembered as the result of relatively recent processes of extensional tectonism (e.g. in the SW Pacific). Nevertheless the identity of the orogenic belt remains unchallenged, and there is a great deal of similarity as to its tectonic evolution from one orogenic segment to another.

tectonism have taken place. The effect of these 'events' can be recognized throughout the entire extent of the orogenic belt, and correlated over large distances.

The earliest episodes are those for which least information is available (e.g. at c. 110–100 Ma or at c. 80–85 Ma). These time periods will not be discussed further in this paper. Later episodes appear to take place at c. 70–65 Ma, c. 55–50 Ma, c. 45–42 Ma, c. 35–32 Ma, c. 25 Ma, 21–19 Ma, 12–10 Ma, and at c. 5 Ma. Events at similar times can be recognized in other orogens, and some of these events appear to be globally significant.

The c. 65 Ma event

Dewey *et al.* (1989) show that at c. 66 Ma the drift of Africa relative to Europe suddenly slowed. It is suggested that this marked the onset of collision, with the first accretion of Gondwana fragments against the European margin. This first accretion event during Africa–Europe convergence was marked by an episode of high-pressure eclogite facies metamorphism (at c. 65 Ma, in the Sesia zone; Rubatto *et al.* 1999). $^{40}Ar–^{39}Ar$ step heating spectra (Venturini 1995) can be interpreted to suggest rapid exhumation shortly thereafter, as the result of an episode of continental extension.

The effects of the c. 65 Ma event are also felt throughout the African rift system, and marked by an episode of rifting and rapid erosional denudation. The rotation of crustal scale tilt-blocks, and the uplift of the Kenyan and Tanzanian rift margins produced significant topography, and a pronounced episode of denudation took place on the rift flanks. Foster & Gleadow (1992, 1996) document between 2 and 4 km of erosion that took place around this time on the flanks of the Kenya and Anza Rifts, East Africa. This was part of a more widespread denudation epoch related to the creation of topography on the sides of rifts, and erosion from the crests of domino-like crustal-scale tilt blocks (Foster & Gleadow 1996; Noble *et al.* 1997). Marked acceleration in the rate of erosion took place at c. 65 Ma (Foster & Gleadow 1996).

Accretion of arcs and ophiolite sheets to the Indian subcontinent was also taking place as India drifted north, towards its ultimate fate. Corfield *et al.* (1999) time the obduction of the Spontang Ophiolite at c. 70–65 Ma. This accretion event commenced in Late Cretaceous time, and was complete by c. 65 Ma (Corfield *et al.* 1999). It is interesting to speculate that initial collision commenced at c. 70 Ma, followed by extension at c. 65 Ma, thereby explaining the destruction and submergence of the orogenic zone. From

this point of view it is interesting to note that sillimanite facies metamorphism in the Karakoram crust took place at c. 64 Ma (Fraser *et al.* 1999; Searle *et al.* 1999). Based on the pattern of deformation and metamorphism in the Aegean crust, this Barrovian facies event in the Karakoram crust may have marked an extensional epoch taking place immediately subsequent to an accretion event.

There is probably a link between these orogenic events and the eruption of an estimated 1–2 million km^3 of flood basalts in the Deccan traps. These mafic magmas began to accumulate when India drifted over a deep-mantle-related hotspot (or megaplume) (Basu *et al.* 1993). The eruption of these magmas took place over a short time interval, covering up to 75% of the Indian subcontinent in the process. Bhattacharji *et al.* (1996) examine the western continental margin rift of the Deccan Traps and show that eruption took place between 67 and 64 Ma, with a significant peak at c. 65 Ma. The remarkable coincidence in time and space between the collision event and the eruption of these huge volumes of magma deserves further examination.

A great deal of scientific attention has been focussed on events at 65 Ma, because of the bolide impact that took place at that time (Alvarez *et al.* 1980). The significance of the c. 65 Ma event in terms of (unrelated) tectonic processes will become clearer as improved geochronology focuses attention on the rate of processes (e.g. Bhattacharji *et al.* 1996), and as more information is provided as to the exact timing of other phenomena that appear to have taken place in that period (e.g. eruption of the Kamchatka Arc in the northwest Pacific).

The c. 55 Ma event

A major episode of collisional orogeny took place at c. 55 Ma, marked by obduction of ophiolite terranes as the result of the closure of back-arc basins along the length of the Alpine–Himalayan chain. In the Himalayan segment of the orogen there is evidence for a major accretion event at c. 55 Ma. For example UHP metamorphism took place at that time in the rocks now exposed in the Tso Morari dome in the Ladakh Himalaya, NW India (de Sigoyer *et al.* 2000). It has also been suggested that the onset of the India–Asia collision took place at c. 55 Ma, based on analysis of the timing of thrust events in NW India (Corfield *et al.* 1999; Guillot *et al.* 2000). However, there is some uncertainty as to when India finally collided with Asia (see next section). The c. 55 Ma event may have marked accretion of the Dras and/or Kohlistan arcs, and preceded final collision of India with Asia (see next section).

Guillot *et al.* (1997) suggest that UHP metamorphism in the Tso Morari was the result of subduction of sediments on the Indian margin beneath Asia to *c.* 70 km. However the gneiss domes exposed throughout this region are fairly typical examples of metamorphic core complexes (e.g. Brunel *et al.* 1994). The Tso Morari dome is cored by high-pressure metamorphic rocks, and it is suggested that a metamorphic core complex formed as the result of lithospheric extension after the *c.* 55 Ma accretion event, prior to final collision with India. Widespread crustal anatexis took place in the Karakoram at *c.* 52 Ma (Searle *et al.* 1999), supporting the view that an episode of crustal extension took place subsequent to an accretion event that had resulted in a major episode of crustal shortening.

The effect of the *c.* 55 Ma event was felt globally. Cook & Crawford (1994) reported a period of rapid cooling caused by exhumation of the western metamorphic belt of the Coast orogen in southeastern Alaska, at *c.* 55 Ma. This was caused by the operation of extensional faults and shear zones at that time. Also at this time, the first of two distinct ophiolite sheets may have been emplaced in northern New Caledonia. Emplacement of the northern ophiolite mass was associated with eclogite–blueschist facies metamorphism in the underlying tectonic slices. The high-pressure sole of this ophiolite can still be observed (Rawling & Lister 1999*a*), although it has been folded and refolded on the large-scale as the result of later tectonic events. The contact zone is characterized by the formation of a 'boulder mélange' comprising high-pressure eclogite and blueschist facies exotic blocks (Rawling 1998; Rawling & Lister 1999*b*). The exact age of high-pressure metamorphism is still in doubt, although $^{40}Ar/^{39}Ar$ apparent age spectra, reported by Ghent *et al.* (1994), from the associated Diahot terrane to support emplacement at *c.* 55–50 Ma is a suggested interpretation.

The 45–42 Ma event

Le Pichon *et al.* (1992) point out that there was a major deceleration and some reorientation of the motion between India and Asia between *c.* 56 Ma and *c.* 46 Ma. More detailed information comes from the kinematics of the Central Indian Ocean Ridge, which were directly affected by this collision. Patriat & Achache (1984) reported a major slow-down of the opening rate sometime between 51 and 49 Ma, but as noted by Le Pichon *et al.*

(1992) 'the only significant reorganisation of spreading occurs between ... 46 and 43 Ma'. Le Pichon *et al.* (1992) thus support the conclusion of Dewey *et al.* (1989) that the main collision occurred near *c.* 45 Ma.

These events were marked by eclogite facies metamorphism in the Kaghan eclogites in North Pakistan (at *c.* 44 Ma; Spencer & Gebauer 1996), and by a second episode of sillimanite facies metamorphism in the Karakoram crust (also at *c.* 44 Ma; Fraser *et al.* 1999; Searle *et al.* 1999). Searle *et al.* (1999) attribute these events to the docking of the Kohlistan arc complex. Based on the pattern observed during the Aegean collision, it is suggested that high-pressure metamorphism marked the accretion event, and high-temperature metamorphism took place at the start of the subsequent extensional epoch.

In the Western European Alps, there is evidence for an accretion event that took place at the same time. This is marked by (UHP) metamorphism and the production of *c.* 44 Ma eclogites in the Zermatt–Saas zone (Rubatto *et al.* 1998). In the Aegean Alps, the third (M_{1C}) episode of high-pressure metamorphism recognized by Forster & Lister (1999*a*) probably also took place at this time. This episode of porphyroblastic growth of white mica, garnet, glaucophane and zoisite took place under blueschist facies conditions, and overprinted the *c.* 55 Ma eclogites that had formed in an earlier period of collision. These (M_{1C}) high-pressure metamorphic rocks had been exhumed by *c.* 42 Ma (Lister & Raouzaios 1996), based on $^{40}Ar/^{39}Ar$ apparent age spectra obtained by Wijbrans *et al.* (1990) from Sifnos.

At the other end of the Alpine–Himalayan chain, in New Caledonia, high-pressure eclogite facies rocks that formed in sheets at the base of the first overthrust ophiolite sheet were also significantly exhumed in this event, at *c.* 40–42 Ma (Rawling 1998; Rawling & Lister 2000). Global reorganization of plate kinematics was recorded by the change in the trajectory of the Hawaiian–Emperor seamount chain at *c.* 43 Ma (Clague & Dalrymple 1989). This implies that *c.* 1–2 Ma passed before the effect of the India–Asia collisions had become sufficiently marked as to cause large-scale reorganization of plate kinematics. A problem exists with this interpretation however. Norton (1995) suggests that the 43 Ma 'event' does not exist. He interprets the data to show that no changes in relative plate motions occurred at that time, at least for Pacific-area plates. Norton (1995) further notes that 'a large 43 Ma direction change of the Pacific plate would have to be transferred through surrounding subduction zones and should be manifested as changes in subduction direction as well as tectonic events on the surrounding continental margins'. Exhumation of the HP eclogite–blueschist belt in northern New Caledonia at $c.$ 40–42 Ma may be an effect of such reorganization, resolving some of Norton's difficulties. The 'event' at $c.$ 43 Ma may be more significant than previously recorded.

The 35 Ma event

A less well-recognized event seems to have taken place at $c.$ 35 Ma, throughout the Alpine Chain. Dewey *et al.* (1989) showed that a small change in the motion of Africa relative to Europe took place at $c.$ 35 Ma, and this was correlated with 'geologic or tectonic events'. In the Himalayan segment of the orogen, movement on the Red River fault started at this time (Harrison *et al.* 1992), and this can be interpreted as indicating full engagement of the Indian subcontinent in the collision process. Another episode of widespread crustal anatexis took place in the Karakoram at $c.$ 35 Ma (Searle *et al.* 1999) related to crustal melting as the result of mantle-related mafic underplating, or possibly the result of another extension event subsequent to a period of crustal shortening.

In the Western European Alps, a third episode of high-pressure metamorphism took place at 35 Ma (Gebauer *et al.* 1997; Rubatto & Hermann 2000), this time involving UHP metamorphism in the Dora Maira massif. These rocks were then rapidly exhumed, reaching greenschist facies conditions by $c.$ 32 Ma (Rubatto & Hermann 2000). This required movement at rates exceeding $c.$ 50 km Ma^{-1}, based on simple geometrical considerations. At the same time, in the Lepontine culmination, further to the north, a medium-pressure high-temperature Barrovian facies metamorphism took place, again followed by rapid exhumation (Schlunegger & Willett 1999). Ductile shear zones associated with back-thrusting formed at $c.$ 34 Ma (Freeman *et al.* 1997). Carmignani *et al.* (1994) indicate that crustal shortening in the adjacent Apennines ceased at $c.$ 35 Ma. Yet at the same time, brittle faulting in the Sesia Lanzo zone led to an episode of exhumation, dated at $c.$ 35 Ma by Hurford & Hunziker (1985) using fission-track zircon ages. Calc-alkaline volcanism and magmatism took place between 32 and 30 Ma in this zone (Hunziker *et al.* 1989).

In the crustal roots of the Aegean Alps, ^{40}Ar/^{39}Ar apparent age spectra suggest a major period of exhumation at $c.$ 36–35 Ma. A distinct cluster of apparent ages can be observed at this time, and these appear to correlate with ductile shear zones that accomplished considerable decrease in pressure (Lister & Raouzaios 1996). A fourth episode (M_{1D}) of high-pressure metamorphism (Forster & Lister 1999a) appears to have taken place thereafter. The M_{1D} rocks appear to have been rapidly exhumed in the period up to $c.$ 32 Ma (Lister & Raouzaios 1996).

At the other extremity of the Alpine chain in Vietnam, Jolivet *et al.* (1999) recorded $c.$ 35 Ma ^{40}Ar/^{39}Ar apparent ages in the Oligo-Miocene Bu Khang extensional gneiss dome. In New Caledonia, earlier formed high-pressure metamorphic rocks were overprinted by an episode of middle-greenschist metamorphic mineral growth, and then rapidly exhumed. This was the second of two regional extension events (Rawling & Lister 2000). At the same time the main mass of the New Caledonia ophiolite was emplaced (Cluzel 1998; Cluzel *et al.* 1998).

The $c.$ 35 Ma event was marked by significant episodes that took place in other orogenic belts, related to changes in the stress state, with or without significant deformation. The Oligocene ignimbrite flareup took place in central California (Ingersoll 1997), and this may have been caused by roll-back of the subducting East Pacific lithosphere. In parts of the Andes there may have been a change in the stress state at this time (from overall compressional orogenesis to local extension). This would have enabled large igneous bodies to intrude to relatively shallow levels (Moran-Zenteno *et al.* 1996) and once there, to dump their metal content. The largest copper porphyry so far discovered on Earth was formed at this time. This is the Chuquicamata orebody, formed in and around a granitoid intruded at $c.$ 35 Ma (McInnes *et al.* 1999). Reynolds *et al.* (1998) show that hydrothermal activity took place between $c.$ 34 and 33 Ma, followed by rapid cooling during exhumation. A second episode of hydrothermal alteration occurred at relatively shallow levels, producing quartz–sericite alteration, at $c.$ 31 Ma. Thus there is a coincidence with the broad pattern of global tectonic activity between $c.$ 35 and 36 Ma and 30 Ma, although local plate tectonic configurations may have also changed at this time.

The 21–19 Ma event

Dewey *et al.* (1989) showed that a significant change in the motion of Africa relative to Europe took place at $c.$ 20 Ma, and again this was correlated with tectonic events in the western Mediterranean. However, a local event appears to have been globally significant. During the period 21–19 Ma, a considerable extent of the Alpine–Himalayan orogenic belt suffered episodes of bimodal magmatism related to melting of the middle and lower crust and upper mantle. In many cases this appears to be the result of extensional tectonism induced by an epoch of slab retreat (i.e. roll-back of the flexure of the subducting slab; Royden 1993a, b).

Vergés & Sàbat (1999) show that crustal shortening in the Pyrenees ended at c. 24 Ma and extensional tectonism commenced. In the period 24–21 Ma rapid cooling then took place in the Magrehebides. In the Gulf of Lion, a 30° rotational opening took place in this period (Séranne 1999). In the Alboran Sea, extension took place between c. 23 and 20 Ma.

Extensional basins at formed c. 25 Ma in the Aegean (Jolivet et al. 1998). In the Cyclades early plutonism and migmatite formation took place between c. 17 and 21 Ma in the Naxos metamorphic core complex (Keay 1998; Keay et al. 2000). This was the start of a major period of Barrovian metamorphism and anatexis, followed by rapid exhumation. Similarly in the southern Aegean, immediately adjacent to the retreating Hellenic slab, exhumation of the high-pressure metamorphic rocks took place. $^{40}Ar/^{39}Ar$ apparent age spectra from the phyllite–quartzite nappe on Crete (Seidel et al. 1982) suggest that exhumation took place between c. 22 and 19 Ma (Jolivet et al. 1996).

Intense activity took place during this period in the Himalayan segment of the orogen (Burchfiel et al. 1992; Harrison et al. 1992; Hodges et al. 1993; Coleman & Hodges 1998). The episodic nature of these events has been recognized by Harrison & Yin (1999) and Yin et al. (1999). Crustal melting and leucogranite emplacement took place between 24 and 18 Ma (Harrison et al. 1992; Vannay & Hodges 1996; Vance & Harris 1999) as well as rapid exhumation of metamorphic rocks as the result of extensional shear zones and detachment faults. Harrison et al.

(1992) point out that this was accompanied by accelerated denudation rates between 20 and 18 Ma, and motion on the Main Central Thrust increased significantly thereafter. Schärer et al. (1990) recognize two distinct pulses of plutonism in the Karakoram Batholith, at c. 25 Ma and c. 21 Ma. Searle et al. (1999) discuss the Karakoram fault zone and refer to a period of leucogranite formation at 21–18 Ma, during which time the South Tibetan Detachment accomplished considerable exhumation of the higher Himalaya. Crustal shortening took place in SE Tibet between 30 and 24 Ma (Yin et al. 1999).

The effects of the 21–19 Ma event were felt at the other end of the Alpine–Himalayan belt, because coincidentally (?) collision of the Phillipines microplate with the Australian margin took place at c. 22 Ma (Baker & Malaihollo 1994). Bimodal igneous activity took place in the next few million years, during this period, in the eastern part of the Indonesian archipelago. Many different phenomena took place along the Pacific rim at this time. For example, the Trans Mojave–Sierran shear zone facilitated collapse and extension of the western North American Cordillera between 21 and 18 Ma (Dokka et al. 1998). Major extension of California and Arizona took place during period 20–18 Ma, including the formation of metamorphic core complexes. Onset of rapid extension in the southern Basin and Range province took place at c. 22–21 Ma (Foster & John 1999). Rifting and sea-floor spreading initiated in the Sea of Japan (Jolivet et al. 1992) and back-arc basins opened throughout the Pacific.

Orogeny and global sealevel variation

Collisional orogeny thickens the continental crust over large regions, and closes back-arc

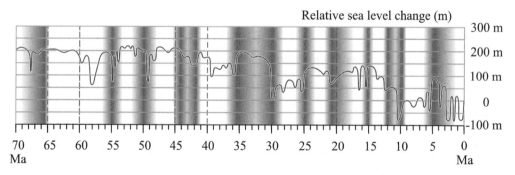

Fig. 2. Part of the global eustatic record (modified from Haq et al. 1987). Relative sea level is shown. Orogenic episodes recognized in the Alpine–Himalyan chain are shown shaded. Collisional orogeny and subsequent extensional tectonism modify the global hypsometric distribution, and thereby affect sea level. A lag between the effect of collisional events, and subsequent extensional tectonism is expected.

basins during obduction events. Large volumes of crust are affected. Orogeny can thus change the global hypsometric distribution, and directly affect the relative areal extent of the ocean basins and their average depth. In the context of the above discussion, it would not be unreasonable to expect that there would be some correlation between the orogenic events mentioned in this paper and global sea level variation (Fig. 2). However examination of the Haq *et al.* (1987) curves reveals considerable variation that does not reflect itself in terms of these known episodes of orogenesis. This is not entirely a surprise, first because significant 'events' may have been missed and second because different events may have greater or lesser global impact. It is also clear that the growth or the diminishment of polar ice sheets has a more direct influence on global sea level, and therefore a 1:1 correlation with orogeny should not be expected.

What causes global orogenic episodes?

The preceding section was not intended as an exhaustive analysis of the evolution of Earth over the past 100 Ma. There are events that have not been mentioned, some of which significant. The purpose was to demonstrate that orogenic processes over the past 100 Ma have not been smooth and continuous, and that there have been several episodes of tectonism that were global in their extent. An attempt to illustrate the magnitude of the effects associated with particular events believed to be dominant during the evolution of the Alpine–Himalyan chain was made.

The intention in the remaining sections of this paper is to explain why episodes of tectonism should correlate along the length of a 15 000 km orogenic belt. In particular, emphasis is placed on a single hypothesis that has arisen as the result of a number of independent studies undertaken at different times (in different places) during the past two decades.

Episodic behaviour in an orogenic zone can be recognized because deformation fabrics display overprinting relationships. Significant deformation episodes affect the bulk of the rock mass, and widespread overprinting of penetratively developed fabrics is usually the result of a significant change in the movement picture throughout an orogenic zone. For example, the new pattern of movement may throw a previously formed fabric into the shortening field of the later deformation. In this case the earlier formed fabric will be refolded, and a new fabric may develop axial plane to the folded older fabric. In relation to particular fabrics, the relevant question to answer concerns the origin of the change in movement picture, and ultimately their geodynamic significance.

In several of the collisional settings that have been examined, episodic behaviour is apparently associated with changes from overall crustal shortening to overall crustal extension (within the crustal volume considered). These changes are described as switches in tectonic mode. There are many potential examples of such behaviour, although the mechanism is still a matter for debate. This phenomenon has now been recognized in New Caledonia (Rawling & Lister 1999*a*), in the Cyclades, Aegean Sea, Greece and in the Otago Schist, New Zealand (Forster 1999). Other authors document similar behaviour. For example, Hodges *et al.* (1996) document rapid changes from crustal extension to crustal shortening and back to crustal extension in the central Annapurna Range in Nepal, during the period 22–18 Ma.

It is also the case that episodes of metamorphism appear to correlate with switches in tectonic mode in these settings. This phenomenon is so prevalent that it is suggested to be of general application to orogenesis in convergent zones, and a characteristic of deformation and metamorphism at depth in the crustal roots of orogenic belts in collisional settings.

Metamorphic episodes and mode switches

In each case studied, the periods in which recumbent (or upright) folds were formed were times during which the crustal volume considered was horizontally shortened. The crust was subject to large-scale thrusting, and thickening (e.g. as the result of large-scale recumbent or upright folding). In several cases the period of recumbent (and/or upright) folding ended at the same time as an episode of (static) metamorphic mineral growth. Subsequently (but while the porphyroblasts were still growing), the folds were transected by crustal-scale extensional ductile shear zones (up to 1–4 km thick). The operation of these extensional shear zones continued well past the termination of the episode of metamorphic mineral growth, producing strong alignments of metamorphic minerals in fabrics and lineations, and often accomplishing many kilobars of decompression.

A question of particular interest concerns why a period of static mineral growth should take place at the same time as a switch in tectonic mode. The answer may be that the episode of metamorphism marks the onset of collapse after accretion. However, the period of metamorphic mineral growth does not usually involve significant deformation. Porphyroblasts overgrow external foliations without disturbance. The question is then, why metamorphism should involve static mineral growth. The answer is beyond the scope of this paper, even if it was known what it might happen to be. It may be that the metamorphic event takes place with such rapidity as to appear as though it occurred under 'static' conditions, whereas really it did not. Thermal events associated with metamorphism and porphyroblastic mineral growth in the Aegean are of short duration, based on $^{40}Ar/^{39}Ar$ measurements (Baldwin & Lister 1998).

The features that need to be explained in each individual case are: (a) why switches in tectonic mode take place?; (b) why (short-lived?) pulses of metamorphic mineral growth took place at approximately the same time that the switch in tectonic mode take place?; (c) why operation of extensional shear zones should commence at the same time as the orogen collapses?; and (d) why the timing of the shear zone initiation is such that movement commences towards the end of the episode of metamorphic mineral growth? The answers to these questions are not immediately apparent, although some observations still need to be made.

Geodynamic models

Switches in tectonic mode, as above, can be caused by a variety of different geodynamic conditions. Three different types of models can be considered. Each of these provide mechanisms that have the capacity to allow explanation of an episode of extensional tectonism subsequent to a prolonged period of crustal shortening, while convergence across the orogenic zone continues throughout. However, not all of these models can explain the rapidity of the processes involved.

Effect of lithosphere drop-off or slab-tear

Several authors have considered the effect of convective removal of a thermal boundary layer when the lithosphere has been overthickened as the result of collisional orogeny (e.g. Houseman *et al.* 1981; England & Houseman 1988, 1989; Platt & England 1993). Thinning the lithosphere as the result of 'dropoff' of an overthickened thermal boundary layer substantially increases the gravitational potential energy of the overlying continental crust. This occurs at the same time as conductive warming commences, or as

heat is advected as the result of the intrusion of magmas. The strength of the orogen is therefore reduced, at the same time as gravitational potential energy is increased. As a result the orogen may be thrown into extension, as the uplifted material attempts to spread laterally.

This mechanism has been applied to several tectonic settings to explain collapse of an orogenic belt 20–30 Ma after the onset of initial collision. However, convective removal of the base of the thickened lithosphere is likely to be difficult if a subducting lithospheric slab acts as a barrier. In addition, it is difficult to explain switches in tectonic mode that take place very soon after a collision or crustal thickening event. If collapse begins within 1–3 Ma of crustal thickening, too short a time is involved to allow this sort of effect to be invoked. These models require a long lead-time before a convective removal or 'dropoff' event takes place (e.g. *c.* 20–30 Ma). Hence, it is difficult to explain switches in tectonic mode during convergence using this mechanism.

The effect of slab breakoff or slab tear are likely to provide more impetus for change. In this way von Blanckenburg & Davies (1995) were able to explain syn-collisional magmatism and the increase in temperature during exhumation of Alpine eclogites. The effects associated with slab breakoff or slab tear (for example, ingress of relatively hot asthenosphere) would be not unlike those predicted by Houseman *et al.* (1981), or Platt & England (1993), considering the effect of convective thinning of overthickened lithosphere.

Accelerated retreat of a subduction zone

Most subduction zones are not fixed, in the sense that the hinge line of the flexure of the subducting slab either retreats or is pushed seawards (Dewey 1988; Royden 1993*a, b*). This means that the overriding plate is either subject to extensional tectonism (e.g. as typified by Marianas-type subduction zones), or compressional orogenesis (e.g. as typified by Andean-type subduction). Any change in the rate that the flexure in a subducting lithospheric slab migrates is thus likely to cause a change in stress state in the adjacent (overriding) plate. The adjacent orogen may be thrown into extension if a period of accelerated retreat takes place, even if the rate of overall convergence remains the same. The overriding plate can readily switch between tectonic modes, and under certain circumstances, several oscillations of tectonic mode are to be expected.

This point is illustrated for Andean-style subduction (Fig. 3). The Andean mountain chain has formed in the over-riding plate, where a continent

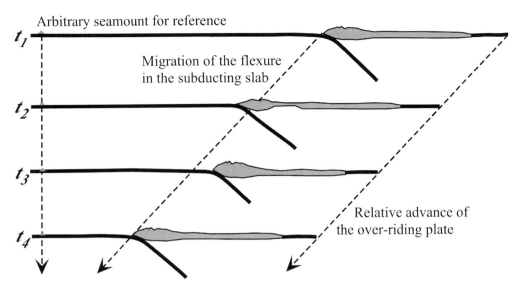

Fig. 3. Switches in tectonic mode may be a natural consequence of Andean-style convergence, as well as a feature of the style of tectonism that characterizes the SW Pacific. Here smooth convergence is shown (dashed lines t_1–t_4). The hinge of the subducting slab is pushed back during periods of Andean style convergence (t_1), whereas in an intervening period (t_2), accelerated roll-back of the hinge of the subducting lithosphere leads to a period of extensional tectonism in the opposing orogen. The switch in tectonic mode is readily reversed (t_3–t_4).

faces actively subducting oceanic lithosphere. All that is required to produce orogenesis is the slow advance of the continental mass relative to the position of the flexure of the subducting oceanic lithosphere. The flexure is forced to retreat as the result of this slow advance (i.e. it is pushed back). Smooth convergence of the opposing plates may take place, whereas the interface (defined by the orogen) may be subject to oscillatory behaviour. Note that orogenic surges (see next section) may coincide with periods of accelerated roll-back of the hinge of the subduction zone, but this may be a consequence, not a causal feature.

A switch in tectonic mode will result if there is a change in the rate at which the slab retreats (Fig. 3). At time t_2, accelerated roll-back of the subducting slab has taken place, and this causes a change from compressional orogenesis to extensional tectonism in the adjacent continental mass. This change is simply the result of the slab now retreating faster than the adjacent mass advances. Figure 4 illustrates how the geometry of slab tear can help drive a period of continental extension associated with accelerated roll back, in an Andean setting.

Where an orogen faces subducting oceanic lithosphere (for example, in the Aegean region) the 'roll-back' mechanism provides a ready explanation for a switch from crustal shortening to crustal extension (e.g. Dewey 1980; Dewey

et al. 1993). Such a switch in tectonic mode can take place at any stage during the history of orogenesis. There is no requirement that the effects of this mechanism should be limited to that of late stage orogenic collapse, as is currently considered to be the case. There is also no reason why several switches in tectonic mode should not occur throughout the history of orogenesis.

Orogenic surges

Several authors have considered the possibility that the continental crust may become over-thickened as the result of ongoing convergence. Eventually the thickened continental welt will become unstable and the orogen will then collapse (e.g. Platt 1986, 1987, 1993; Dewey *et al.* 1993). The mechanics of such a collapse event deserve some consideration, particularly in view of the data that has now emerged from the Himalayan segment of the Alpine–Himalayan orogen.

In the Himalayas, it can be argued that starting at about *c.* 21 Ma, extensional detachment to the north of the highest peaks was accompanied by accelerated motion on the Main Central Thrust to the south. This motion accomplished an apparent southward 'extrusion' of a gently-dipping wedge of metamorphic and igneous rocks (Burchfiel & Royden 1985; Burchfiel *et al.* 1992; Hodges *et al.*

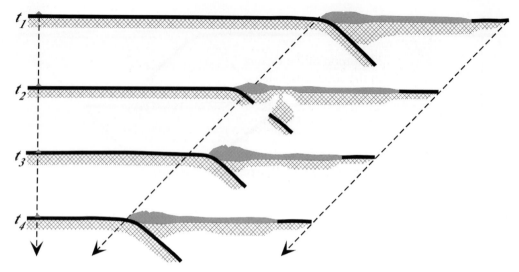

Fig. 4. Houseman *et al.* (1981) and Platt & England (1993) consider the effect of convective removal of a thermal boundary layer when the lithosphere is overthickened. However, convective removal of the base of the thickened lithosphere is likely to be difficult if there is a subducting lithospheric slab that acts as a barrier. The geometry of slab tear can produce much the same results (von Blanckenburg & Davies 1995).

1992, 1993, 1996; Grujic *et al.* 1996; Hodges 2000). Hodges *et al.* (1992, 1996) discuss the rapid exhumation and cooling of newly created metamorphic rocks beneath the South Tibetan Detachment. This occurs at the same time as these rocks are rapidly cooled above the underlying Main Central Thrust (at *c.* 19–20 Ma; Harrison *et al.* 1992). Harrison *et al.* (1992) points out that this movement was accompanied by accelerated denudation rates between 20 and 18 Ma,

and that motion on the Main Central Thrust increased significantly thereafter.

The geometry described in the Himalayas has previously been described as the result of 'extrusion'. In a geometrical sense this is a correct description, but the mechanics of extrusion usually involve forceful ejection of material, for example as the result of buoyancy (Chemenda *et al.* 1995; Hacker & Peacock 1995; Hacker 1996; Fig. 5; Wijbrans *et al.* 1993). The movement of

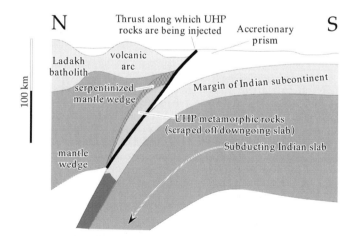

Fig. 5. Buoyancy-driven exhumation of the UHP rocks of the Tso Morari dome facilitated by serpentinite derived from the mantle wedge (modified from Guillot *et al.* 2000).

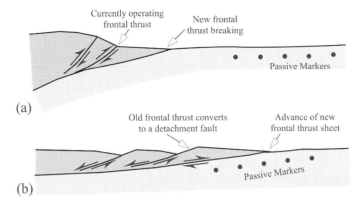

Fig. 6. Schematic illustration of the geometry of an orogenic surge. (**a**) The orogenic welt has overthickened. A new thrust has broken. (**b**) A switch in tectonic mode has taken place in the over-riding plate, and the belt has collapsed over the adjacent foreland.

material into a wedge-shaped geometry can also lead to a considerable rise in the mean stress (Mancktelow 1995), but one would suspect this is true only under constrained conditions (e.g. when the boundaries of the wedge are strong, as is the case during extrusion from a die in a metallurgical process).

The buoyancy mechanism works well for diapirs, where relative vertical motion is assured, but it is mechanically unsuitable to explain structures where the fabrics are gently dipping or horizontal. The mechanics of orogenesis do not seem compatible with the concept of forceful ejection of a gently-dipping frontal wedge out from beneath the surrounding rock. An alternative mechanical explanation is required. This might be found by considering the geometry of a 'collapse' event after a period of crustal thickening.

A period of crustal thickening will lead to an instability because the crust eventually exceeds a critical thickness, and then begins to collapse. The orogen surges over its foreland, with the same mechanics applying as does to overthickened ice sheets and glaciers. The orogen is thrown into overall extension, in this case accommodated by the operation of low-angle normal faults and ductile shear zones in the over-riding plate. The frontal slice appears to be 'extruded', but this is a consequence of the geometry of the switch into overall extensional tectonism at the same time as the orogenic welt collapses and 'surges' over the adjacent foreland (Fig. 6). An 'orogenic surge' will lead to rapid advance on the basal thrust at the same time as detachment faults are operating in the overlying slices. The 'orogenic surge' model thus readily explains the geometry of 'extrusion'. It also provides the basis for an alternative

mechanical explanation, one that is not based upon buoyancy. A switch from crustal shortening to crustal extension accompanies the 'orogenic surge', while convergence continues throughout.

An 'orogenic surge' involves an overall switch in tectonic mode in the crust overlying the basal detachment. Conversion of mechanical energy to heat as the result of pervasive strain through the rock mass as the orogen collapses may produce a thermal and/or fluid fluxing event (cf. Molnar & England 1990; Godin *et al.* 1998). This in turn may produce an episode of metamorphism, and/ or widespread crustal anatexis, as the orogen collapses. Rheological parameters are thereby changed so that the formation of ductile shear zones is favoured. Strain-softening in these movement zones then facilitates further collapse of the orogen.

As already noted in the previous section, an 'orogenic surge' can take place at the same time as an episode of accelerated roll-back of a subduction zone. These two models in conjunction thus have the capability to explain the basis for large-scale episodicity during orogenesis.

The geometry of an orogenic surge suggests that it is not possible to exhume high-pressure metamorphic rocks by this mechanism, at least during one episode of movement. The relative movement involved on the detachment faults will be too small to accomplish significant exhumation. However, this limitation will not exist if the effects of 'roll-back' are superimposed. An orogenic surge at the same time as roll-back takes place would ensure that considerably larger relative displacements are involved on individual detachments.

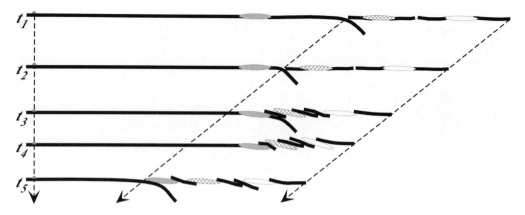

Fig. 7. Accretion events during continuous convergence in a SW Pacific-type setting, as shown by the displacement of marker points (dashed lines). Ribbons of continental crust are shown, initially in a back-arc setting (t_1–t_2). Slab roll-back eventually leads to collision (t_3). High-pressure metamorphism takes place because the crust in the back-arc region is rapidly thickened once the subduction zone can no longer absorb the necessary relative displacement (t_3–t_4). Once a new thrust transfers the newly arrived continental mass into the overriding plate, the hinge of the subduction zone is once again able to retreat, and the high-pressure metamorphics are rapidly exhumed from beneath the overthrust sheets (t_5). Ophiolite masses often represent remnants of overthrust sheets stranded by extensional detachments formed at this time. Accretion events thus lead to episodes of extensional tectonism after an episode of high-pressure metamorphism.

Migration of a thrust-detachment couple

In the example illustrated (Fig. 6), it is imagined that the onset of collapse takes place as the same time as a new frontal thrust breaks (Fig. 6a). The detachment fault shown has cut down through the approximate locus of the former frontal thrust (Fig. 6b). This leads to accelerated movement on the newly formed frontal thrust at the same time as a detachment in the hangingwall reverses the sense-of-movement on the previously active frontal thrust. The 'orogenic surge' model predicts an increase in the rate of movement on the basal thrust at the same time as detachment faulting initiates, accompanying the switch to overall crustal extension in the overriding plate.

Figure 6 shows that during orogeny, migration of the frontal thrust can be accompanied by sympathetic migration of trailing detachment faults. A potential surficial example of these processes at work comes from the Tyrrhenian–Apennine system (Lavecchia 1988). The extensional zone associated with the Tyrrhenian Sea gradually expanded, but at all times it was bounded by a thrust front that moved episodically eastwards, with a trailing extensional detachment. At the same time as a new thrust formed, the existing frontal thrust ceased operation, and converted into the trailing extensional detachment. In this case the sequence illustrated in Figure 6 has been repeated several times.

Accretion in SW Pacific settings

Tectonic mode switches that can occur in an (India–Tibet) Himalayan-type setting, and tectonic mode switches that can occur in an Andean-type setting have already been discussed. Now it is worth considering what would transpire during convergence in a SW Pacific-type setting.

The arrival of continental crust at subduction zones inhibits their operation, although overall convergence across the orogenic zone may continue. If the subduction zone cannot accommodate the necessary relative displacement, an alternative mechanism for convergence is necessary! The back-arc zone is mechanically weak, because the lithosphere is thin. This is so because either ocean-floor spreading has produced new plates,

Fig. 8. Formation and exhumation of UHP rocks of the Tso Morari dome. (**b**) A lithospheric dislocation has reversed shear sense after an accretion event (**a**). The act of accretion saw a new thrust formed, and this detached the Gondwana fragment from the down-going slab. The original thrust then reversed, and a detachment fault juxtaposed low pressure rocks against medium pressure rocks. Roll-back of the oceanic lithosphere once again commenced, and the back-arc region is subject to extreme extension (**c**). The detachment fault does not exactly follow the exact trajectory of the older thrust, so serpentinite lenses containing high-pressure boulders are preserved along the reversed thrust.

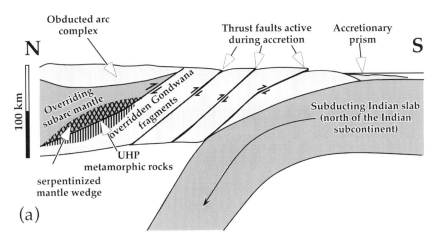

(a)

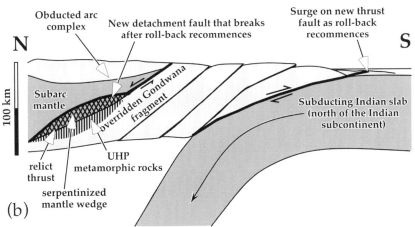

(b)

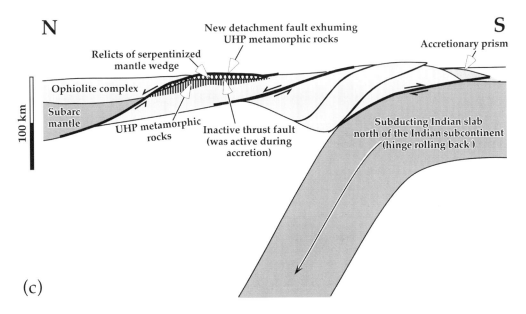

(c)

or the existing lithosphere has been stretched as the result of extension due to slab retreat. Collision, thus (initially), forces the closure of marginal basins in the over-riding plate, and this can be the cause of a major epoch of 'constructional' orogenesis along the length of a convergent zone.

This process is illustrated in Figure 7 for a SW Pacific tectonic setting. Three individual continental ribbons are shown. The hinge of the subducting slab is shown continually retreating, until eventually, further retreat is hindered by the arrival of a slice of continental crust at the subduction zone. Convergence continues, but now it is accommodated by basin inversion, and then large-scale overthrusting in the back-arc region. Metamorphism in the crustal roots of the orogenic welt thus produced may reflect the effects of a period of continental subduction (e.g. Rubatto *et al.* 1998), or simply be the result of crustal thickening due to the accretion event (Rawling & Lister 2000).

Formation and exhumation of high-pressure metamorphic rocks

The accretion model (see previous section) predicts that crustal thickening during an accretion event will eventually be terminated because new

frontal thrusts will break. These faults will separate the newly accreted crust from the down-going slab and thus allow subduction zone roll-back to once again recommence (Rawling & Lister 2000). Accelerated roll-back of the subducting slab will lead to an abrupt change in tectonic mode, and this will drive a major episode of extensional tectonism. This may take place in conjunction with an 'orogenic surge' as the overthickened crust in front of the subduction zone collapses. Figure 7 shows convergence continuing at a smooth rate throughout this process.

The 'roll-back after accretion' model explains why periods of crustal shortening are terminated by episodes of high-pressure metamorphism. The model also explains why immediate exhumation of these newly created high-pressure metamorphic rocks also takes place. It is predicted that exhumation episodes will follow each individual accretion event, although the duration of the extension event may be relatively short in comparison to the preceding period of crustal shortening.

More detail of this accretion model is provided in Figure 8 illustrating a possible model for the formation and then the exhumation of UHP rocks in the Tso Morari dome, in the Ladakh Himalaya, NW India. It is proposed that during the convergence of India toward

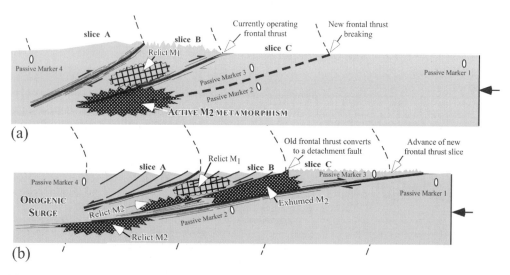

Fig. 9. Orogenic surges after accretion events can lead to a distinctive pattern of metamorphism and ^{40}Ar/^{39}Ar cooling/crystallisation ages within an orogen. (**a**) The situation is shown after a period of convergence has produced an instability. An earlier episode of high-pressure metamorphism (M_1) has already taken place, and a second period of high-pressure metamorphic mineral growth is underway. A new frontal thrust is breaking, and is about to commence operation. After the orogenic surge (**b**), the new frontal thrust has led to an advance of the orogen over its adjacent foreland, on the newly broken thrust, while the former frontal thrust in the hinterland has ceased operation, and during the orogenic surge, has acted as a detachment fault. This leads to rapid cooling and exhumation of the newly formed metamorphic rocks.

Asia, a subduction zone encountered a fragment of Gondwana that had the form of continental ribbon (e.g. as evident in SW Pacific settings; Lister & Etheridge 1989). The arrival of this fragment at a retreating subduction zone led to closure of the associated back-arc basin, and high-pressure metamorphism as the result of large-scale overthrusting (Fig. 8a). A new thrust formed, separating the fragment from its underlying mantle lithosphere, accreting the Gondwana fragment to the over-riding plate (Fig. 8b). The new thrust allows slab retreat once again to recommence, and as a result, extreme extension takes place in the overriding plate (Fig. 8c). This results in the exhumation of a coherent slice of UHP metamorphic rocks from beneath an overriding ultramafic nappe.

This is a typical occurrence of high-pressure rocks in back arc settings, and it is evident in many tectonic environments (e.g. the occurrence of high-pressure metamorphic rocks beneath the peridotite nappe in western Turkey). Conversion of mantle rocks to serpentinite may facilitate the exhumation process (Guillot et al. 2000). Nevertheless, extension driven by slab retreat is the mechanism that leads to uplift. This model is offered as an alternative to the buoyancy driven model involving exhumation of a metamorphic diapir (Fig. 5), as advocated by Chemenda et al. (1995), Hacker & Peacock 1995, or Guillot et al. (2000).

The accretion model and its effects

Orogenic surges after accretion events produce a distinctive pattern of metamorphism and ^{40}Ar/^{39}Ar cooling/crystallization ages within an orogen. These effects will be amplified if extension induced by roll-back increases the total displacement on the individual faults and shear zones. Figure 9 shows the situation after a period of convergence has produced an instability. An earlier episode of high-pressure metamorphism (M_1) has already taken place, and a second period of high-pressure metamorphic mineral growth is underway. A new frontal thrust is breaking, and is about to commence operation (Fig. 9a). The switch in tectonic mode takes place because a new frontal thrust breaks and the collapsing orogen 'surges' over its foreland.

Rapid exhumation and cooling of metamorphic rocks in the frontal slice then takes place (Fig. 9b). The accelerated movement on the frontal thrust leads to cooling because the rocks are juxtaposed against a relatively cold overridden plate. At the same time the detachment fault higher in the sequence juxtaposes

deeper level rocks against a colder upper-plate. This geometry is conducive to rapid cooling of the entire 'extruded' frontal slice.

The accretion model thus allows accurate prediction as to the distribution of ^{40}Ar/^{39}Ar apparent ages within an orogenic belt subjected to repeated episodes of crustal shortening, metamorphism, and extensional tectonism. The pattern of preservation of the metamorphic rocks is also predicted, with the oldest episodes of high-pressure metamorphism best surviving in structurally higher slices. These predictions can be obtained by assuming that the tectonic sequence illustrated in Figure 9 is been repeated several times. Each tectonic slice will have its own distinct thermal history (revealed by ^{40}Ar/^{39}Ar thermochronology). In addition different metamorphic events will dominate fabrics and microstructures in the individual slices.

Comparison of the Western Alps and the Aegean Sea, Greece

The Western European Alps and the Aegean sea, Greece, is now discussed in more detail, to illustrate the relation between switches in tectonic mode and metamorphism, and as a test of the accretion model advocated in the previous section. There is now great interest in the processes that lead to the exhumation of UHP and HP rocks, and the Aegean and the Western Alps are usually treated as quite distinct in this regard (Michard et al. 1993).

Extension in the Western Alps

Evidence for switches from crustal shortening to extensional tectonism exists in several parts of the Western Alps. Platt et al. (1987) describe intense deformation which produced well-developed c. NW trending glaucophane mineral lineations subsequent to high-pressure metamorphism in the Vanoise. They suggest that initial crustal shortening produced few penetrative fabrics because thrusting was localized and temperatures remained low. High-pressure metamorphism then, under static conditions, took place. The early fabrics, including the intensely developed c. N–S trending glaucophane lineations, may have been caused by early extensional tectonism, which exhumed the high-pressure basement rocks beneath a ductilely stretched platform cover.

In the Vanoise, a switch in tectonic mode took place subsequent to this period of extension. Crustal shortening produced km-scale upright folds and then a new generation of recumbent

folds. Regional greenschist facies metamorphism then occurred, involving widespread growth
of albite porphyroblasts under static conditions. Subsequently crustal-scale ductile shear
zones formed, with ESE- and WNW-directed
movement.

There is some argument in support of an
orogenic surge model applying to the Western
Alps at this time. Michard *et al.* (1993) note that
the 'syn-greenschist decompression of the Dora
Maira pile can be interpreted as the result of the
collapse (Dewey 1988) of the internal Penninic
wedge, thickened by the MesoAlpine collision,
which was coeval with the outward progression
of thrusts in the more external parts of the
western Alps'. Platt *et al.* (1987) note that movement on Oligocene shear zones in the internal
zone of the arcuate orogenic belt of the western
Alps is radial to the strike, and parallel to the
forethrusting direction in the immediately adjacent parts of the external zone. They note that
this suggests 'gravitational body forces dominated this period of thrusting', which is consistent
with the hypothesis that the internal zones may
have been simultaneously subjected to Oligocene
extensional tectonism. This geometry is similar to
that implied as the result of an orogenic surge
after a period of crustal shortening, as has been
illustrated. The difficulty is that (as already noted)
an orogenic surge model would not readily
exhume UHP metamorphic rocks. The displacements involved on individual faults would be too
small to exhume rocks from 80–100 km depth in a
single period of movement.

Extension in the Aegean crust

It is well-recognized that the Aegean terrane
'switched' to extensional tectonism from *c.* 25 Ma
onwards. This aspect of Aegean geology is usually
presented as the type-example of the collapse of
an orogen after the main phase of compressional
orogenesis. This switch was associated with the
development of an arcuate form to the orogenic
belt (e.g. Kissel & Laj 1988) and the extension
process was driven by 'roll-back' of the hinge
of the subducting African lithosphere. However,
as pointed out by several authors (Wijbrans &
McDougall 1986, 1988; Avigad 1998) the high-
pressure metamorphic tectonites that characterize the Cyclades were, at least partially, exhumed
considerably earlier than could be explained
by the effects of the Miocene roll-back. These
observations suggests that extensional tectonism
operated in the Aegean continental crust during
the initial stages of the Alpine orogenic history.
The effects of this early extensional tectonism

can still be discerned. Older tectonic slices can
still be recognized, although as noted by Forster
& Lister (1999*b*) the terrane has been dissected
by several generations of low-angle as well as
high-angle normal faults. It is now evident that
the structural geometries expressed in the Western Alps are not as different as might be expected
when compared to those that can be inferred for
the early Aegean continental crust.

Metamorphism in the Aegean continental crust

The Aegean crust records a sequence of four
episodes of high-pressure metamorphism (Forster & Lister 1999*a*), some linked to deformation
events that outlast the individual periods of metamorphic mineral growth. The individual events
are best-preserved in different tectonic slices, with
older metamorphic assemblages better preserved
in structurally higher slices. There is a poor record
of the first event (M_{1A}), and its existence is most
readily inferred from relict mineral assemblages
preserved as inclusions in garnet porphyroblasts
grown during later events. M_{1A} assemblages may
be more widespread, however (see below). The
second event (M_{1B}) was peak metamorphism,
with widespread formation of omphacite–jadeite
bearing eclogite-facies rocks. Note also that most
garnet that grew in these rocks formed during
later metamorphic events. There is a suggestion,
therefore, that higher pressures were involved in
this episode than currently thought to be the case.
There is no assurance that M_{1B} assemblages replaced all M_{1A} assemblages. There are some cases
where later M_{1C} blueschist assemblages directly
overprint what may be M_{1A} assemblages.

The third event (M_{1C}) was associated with
widespread development of blueschist facies mineral assemblages, commonly expressed as porphyroblasts of garnet, glaucophane, white mica
and zoisite. A major deformation commenced
toward the end of this period of porphyroblastic mineral growth (Lister & Raouzaios 1996).
These fabrics were then overprinted by a fourth
high-pressure event (M_{1D}), this time under
epidote–albite–glaucophane facies conditions
(after Evans 1990).

Based on published $^{40}Ar/^{39}Ar$ thermo-chronology, it is estimated that exhumation events in the
Aegean took place at *c.* 50–55 Ma, *c.* 45–42 Ma,
c. 36–35 Ma and *c.* 32–30 Ma. These ages are
similar to the age of eclogite facies metamorphic
events reported by Duchêne *et al.* (1997), Gebauer
et al. 1997, Rubatto *et al.* (1998), and Rubatto
& Hermann (2000) in the Western Alps. Each
period of exhumation in the Aegean took place

subsequent to one of the episodes of high-pressure metamorphism recognized by Forster & Lister (1999a). In addition each metamorphic event may have marked a switch from overall crustal shortening to overall crustal extension, although these switches in tectonic mode all took place while convergence continued.

Metamorphism in the Western Alps

Rubatto et al. (1998) demonstrate several different ages of eclogites in the Western Alps and suggest that some slices had been exhumed while others had not yet been thrust down to great depth. High-pressure eclogites of the Zermatt–Saas zone dated by Rubatto et al. (1998) crystallized at c. 44 Ma. Reddy et al. (1999) argue that these eclogites were exhumed beneath the Combin ophiolite by a crustal-scale extensional shear zone, from 45–36 Ma.

A similar circumstance may be inferred for the Dora Maira massif. Avigad (1992), Avigad et al. (1993), and Michard et al. (1993) showed that extensional shear zones accomplished exhumation of the tectonic slice bearing relicts of the UHP metamorphic event. The most recent dating (Duchêne et al. 1997; Gebauer et al. 1997; Rubatto & Hermann 2000) points to c. 35 Ma as the time of UHP metamorphism. Greenschist facies biotite $^{40}Ar/^{39}Ar$ apparent ages (Monié & Chopin 1991) are consistent with U–Pb measurements and point to retrogression at 32–30 Ma, associated with the operation of extensional shear zones.

The data from both segments of the orogenic belt can be explained if high-pressure metamorphism is related to individual accretion events, and if each accretion event is followed by an epoch of extensional tectonism, immediately exhuming the newly formed metamorphic rocks. In each case, episodes of metamorphism appear to mark switches in tectonic mode, and the beginning of periods of crustal extension during which major (c. 1–4 km thick) extensional ductile shear zones formed, and occasionally low-angle normal (detachment) faults.

The effect of multiple accretion events?

The accretion model predicted that it should be able to recognize multiple episodes of high-pressure metamorphism associated with accretion of individual continental ribbons. Each accretion event should be immediately followed by an epoch of extensional tectonism during which the newly created high-pressure rocks are exhumed. If multiple accretion events take place the process should be repeated several times.

In both the Western Alps and the Aegean, individual tectonic slices are dominated by different metamorphic assemblages, and structurally higher slices tend to record the effects of older metamorphism, e.g. the Mon Viso ophiolites that overlie the c. 35 Ma UHP rocks of the Dora Maira display blueschist and eclogite assemblages dated at c. 50 Ma (Monié & Philippot 1989). These Eoalpine units had already been exhumed by the time the Dora Maira unit was thrust over the Sanfront-Pinerolo unit (Michard et al. 1993). In both the Western Alps and the Aegean, tectonic slices may have first been juxtaposed by thrust- and fold-nappes, and then distended by extensional shear zones and related detachments. This process may have been repeated several times.

In the Aegean crust, in the Cyclades, individual metamorphic events are best preserved in different tectonic slices. $^{40}Ar/^{39}Ar$ data suggests that the final juxtaposition of the different tectonic slices was accomplished by the operation of extensional detachments and/or shear zones. However, the relation between different tectonic slices is more complex than that which could be produced by simple extension. In some circumstances the older metamorphic assemblages have been reheated in footwall slices, and this suggests earlier thrusting. Adjacent slices often appear as though they had been first juxtaposed by thrusting, and subsequently the lower slice was exhumed (from beneath the same overthrust slice), as the result of subsequent extensional tectonism (Fig. 9).

Extensional shear zones are often difficult to recognize for what they are because they interact with previously formed recumbent folds in a way that can be misleading. The later formed shear zone can be misinterpreted as the attenuated limb of a fold nappe, and therefore mistakenly identified as part of the preceding history of crustal shortening. Fortunately, an episode of static metamorphism often terminates the period of recumbent folding, and separates the two events. Such metamorphism is often marked by the growth of porphyroblasts. It can be shown that shear zone formation commences towards the end of this period of growth, allowing accurate timing of the switch in deformation style. The nature of the shear zone can be determined using $^{40}Ar/^{39}Ar$ thermochronology. However, variation in apparent age across a shear zone can allow distinction between rocks that have already cooled, and those that have been exhumed by the operation of the shear zone.

Conclusion

In this paper, the reasons why episodicity during orogenesis was such a widespread phenomenon, and why the episodes are 'in phase' over considerable distances along a mountain belt were investigated. It is concluded that the process of orogenesis is intrinsically episodic, and that many of the significant episodes result from switches from crustal shortening to crustal extension, with an period of (static) metamorphism taking place at about the same time as the switch in tectonic mode occurs. The switch in tectonic mode results because an orogenic surge takes place (with mechanics akin to that of an over-thickened glacier). It can also result because roll-back of lithospheric slabs recommences after an accretion event.

In terms of the overall energy budget, once collapse begins during an orogenic surge, gravitational potential is converted into mechanical work and thus to heat. Thermomechanical coupling ensures a 'runaway' effect, but only until the driving force is removed, which of course occurs as the result of the orogenic surge itself. Any apparent periodicity is simply the result of how long it takes to build up sufficient crustal thickness to drive the next runaway event.

Episodic behaviour on a global-scale can be explained as the response to an accretion event in an assemblage of stiff lithospheric plates separated (in part) by orogenic zones. If the individual orogenic zones have had sufficient time to achieve a critical state, a single accretion event may trigger a number of (apparently unrelated?) orogenic surges or collapse events. Geodynamic considerations suggest that episodic behaviour should be evident on a planetary scale during orogenesis, because torque balance on individual plate margins must be maintained.

The arrival of a continental mass at a subduction zone thus does not prevent ongoing convergence. However, because convergence can no longer be accommodated by subduction, the over-riding plate is subject to crustal shortening. This can result in the closure of marginal basins in the over-riding plate, and the emplacement of ophiolite sheets. High pressure (blueschist-eclogite facies) metamorphism may be the consequence, as coherent slices of the over-riding plate are overthrust, and sent to great depths in vastly overthickened crust. Eventually, large scale re-organization of plate kinematics takes place after such a collision, but not before ongoing convergence results a major epoch of orogenesis.

Where continent meets continent, and convergence continues, collapse of the orogen involves 'surges', as in glaciers. After an accretion event,

however, where the newly formed mountain chain faces an actively subducting oceanic slab it will be 'torn apart' as the result of a major epoch of lithosphere-scale extensional tectonism. This will be induced by seawards retreat of the flexure of the subducting slab (i.e. 'roll back'). As a result, the most extended regions on Earth appear to form in the over-riding plate, adjacent to retreating slabs.

Evidence for global episodes of tectonism appears to be present in the geological record. The effects of these events appears to be wider in extent than currently acknowledged. Accretion of continental fragments during orogeny may be the cause. Events at similar times can be recognized in other orogens, and these events thus appear to be globally significant. The effects of these events can be in part correlated with global sea level variation. Accretion events can be defined by determining the ages of metamorphic mineral growth in high-pressure terranes.

The work in this paper forms part of the 'Mountains and Metals' initiative of the Australian Crustal Research Centre. It was supported by a grant from the Australian Research Council.

References

ALVAREZ, L. W., ALVAREZ, W., ASARO, F. & MICHEL, H. V. 1980. Extraterrestial cause for the Cretaceous–Tertiary extinction. *Science*, **208**, 1095–1108.

AVIGAD, D. 1992. Exhumation of coesite-bearing rocks in the Dora-Maira massif (western Alps, Italy). *Geology*, **20**, 947–950.

AVIGAD, D. 1998. High-pressure metamorphism and cooling on SE Naxos. *European Journal of Mineralogy*, **10**, 1–11.

AVIGAD, D., CHOPIN, C., GOFFÉ, B. & MICHARD, A. 1993. Tectonic model for the evolution of the western Alps. *Geology*, **21**, 659–662.

BAK, P. 1996. *How nature works: the science of self-organised criticality*. Oxford University Press, Oxford.

BAKER, S. & MALAIHOLLO, J. 1994. Dating of Neogene igneous rocks in the Halmahera region; arc initiation and development. *In*: HALL, R. & BLUNDELL, D. J. (eds) *Tectonic evolution of Southeast Asia*. Geological Society, London, Special Publications, **106**, 499–509.

BALDWIN, S. L. B. & LISTER, G. S. 1998. Thermo-chronology of the South Cyclades Shear Zone, Ios, Greece: the effects of ductile shear in the argon partial retention zone (PRZ). *Journal of Geophysical Research*, **103**, 7315–7336.

BASU, A. R., RENNE, P. R., DASGUPTA, D. K., TEICH-MANN, F. & POREDA, R. J. 1993. Early and late alkali igneous pulses and a high 3He plume origin for the Deccan flood basalts. *Science*, **261**, 902–906.

BHATTACHARJI, S., CHATTERJEE, N., WAMPLER, J. M., NAYAK, P. N. & DESHMUKH, S. S. 1996. Indian intraplate and continental margin rifting, lithospheric extension and mantle upwelling in Deccan flood basalt volcanism near K/T boundary: evidence from mafic dike swarms. *Journal of Geology*, **104**, 379–398.

BRUNEL, M., ARNAUD, N., TAPPONIER, P., PAN, Y. & WANG, Y. 1994. Kongur Shan normal fault: type example of mountain building assisted by extension (Karakoram fault, eastern Pamir). *Geology*, **22**, 707–710.

BURCHFIEL, B. C. & ROYDEN, L. H. 1985. North–south extension within the convergent Himalayan region. *Geology*, **13**, 679–682.

BURCHFIEL, B. C., CHEN, Z., HODGES, K. V., LIU, Y., ROYDEN, L. H., DENG, C. & XU, J. 1992. *The South Tibetan detachment system, Himalayan Orogen; extension contemporaneous with and parallel to shortening in a collisional mountain belt.* Geological Society of America, Special Paper **269**.

CARMIGNANI, L., DECANDIA, F. A., FANTOZZI, P. L., LAZZAROTTO, A., LIOTTA, D. & MECCHERI, M. 1994. Tertiary extensional tectonics in Tuscany (Northern Apennines, Italy). *In*: SERANNE, M. & MALAVIEILLE, J. (eds) *Late orogenic extension.* Tectonophysics, **238**, 295–315.

CHEMENDA, A. I., MATTAUER, M., MALAVIEILLE, J. & BOKUN, A. N. 1995. A mechanism for syn-collisional rock exhumation and associated normal faulting: results from physical modelling. *Earth and Planetary Science Letters*, **132**, 225–232.

CLAGUE, D. A. & DALRYMPLE, G. B. 1989. Tectonics, geochronology and origin of the Hawaii–Emperor chain. *In*: WINTERER, E. L., HUSSONH, D. M. & DECKER, R. W. (eds) *The Geology of North America N, The Eastern Pacific ocean and Hawaii.* Geological Society of America, Boulder, Colorado, 188–217.

CLUZEL, D. 1998. The post-obduction flysch of Nepoui, a transported basin – inference on the age and setting of the Tertiary obduction in New Caledonia (Southwest Pacific). *Comptes Rendus de l'Academie des Sciences Serie II Fascicule A-Sciences de la Terre et des Planetes*, **327**, 419–424.

CLUZEL, D., CHIRON, D. & COURME, M. D. 1998. Upper Eocene unconformity and pre-obduction events in New Caledonia. *Comptes Rendus de l'Academie des Sciences Serie II Fascicule A-Sciences de la Terre et des Planetes*, **327**, 485–491.

COBLENZ, D. D., SANDIFORD, M., RICHARDSON, R. M., ZHOU, S. & HILLIS, R. 1995. The origins of the intraplate stress field in continental Australia. *Earth and Planetary Science Letters*, **133**, 299–309.

COLEMAN, M. E. & HODGES, K. V. 1998. Contrasting Oligocene and Miocene thermal histories from the hanging wall and footwall of the South Tibetan detachment in the central Himalaya from $^{40}Ar/^{39}Ar$ thermochronology, Marsyandi Valley, central Nepal. *Tectonics*, **17**, 726–740.

COOK, R. D. & CRAWFORD, M. L. 1994. Exhumation and tilting of the western metamorphic belt of the Coast orogen in southeastern Alaska. *Tectonics*, **13**, 528–537.

CORFIELD, R. I., SEARLE, M. P. & GREEN, O. R. 1999. Photang thrust sheet: an accretionary complex structurally below the Spontang ophiolite constraining timing and tectonic environment of ophiolite obduction, Ladakh Himalaya, NW India. *Journal of the Geological Society, London*, **156**, 1031–1044.

DAVIS, G. A., LISTER, G. S. & REYNOLDS, S. J. 1986. Structural evolution of the Whipple and South mountains shear zones, southwestern United States. *Geology*, **14**, 7–10.

DE SIGOYER, J., CHAVAGNAC, V., BLICHERT-TOFT, J., GUILLOT, S., LUAIS, B., COSCA, M., MASCLE, G. & VILLA, I. 2000. Dating continental subduction and collisional thickening in NW Himalaya: Multi-chronometry of the Tso Morari eclogites. *Geology*, **28**, 487–490.

DEWEY, J. F. 1980. Episodicity, sequence and style at convergent plate boundaries. *In*: STRANGWAY, D. W. (ed.) *The continental crust and its mineral deposits.* Geological Association of Canada, Special Paper, **20**, 553–573.

DEWEY, J. F. 1988. Extensional collapse of orogens. *Tectonics*, **7**, 1123–1139.

DEWEY, J. F., RYAN, P. D. & ANDERSEN, T. B. 1993. Orogenic uplift and collapse, crustal thickness, fabrics and metamorphic phase changes: the role of eclogites. *In*: PRITCHARD, H. M., ALABASTER, T., HARRIS, N. B. W. & NEARY, C. R. (eds) *Magmatic Processes and Plate Tectonics.* Geological Society, London, Special Publications, **76**, 325–343.

DEWEY, J. F., HELMAN, M. L., TURCO, E., HUTTON, D. H. W. & KNOTT, S. D. 1989. Kinematics of the Western mediterranean. *In*: Geological Society, London, Special Publications, **45**, 265–283.

DOKKA, R. K., ROSS, T. M. & LU, G. 1998. The Trans Mojave-Sierran shear zone and its role in early Miocene collapse of southwestern North America. *In*: HOLDSWORTH, R. E., STRACHAN, R. A. & DEWEY, J. F. (eds) *Continental transpressional and transtensional tectonics.* Geological Society, London, Special Publications, **135**, 183–202.

DUCHÈNE, S., BILCHERT-TOFT, J., LUALS, B., TELOUK, P., LARDEAUX, J.-M. & ALVAREDE, F. 1997. The Lu–Hf dating of garnets and the ages of the Alpine high-pressure metamorphism. *Nature*, **387**, 586–589.

ENGLAND, P. & HOUSEMAN, G. A. 1988. The mechanics of the Tibetan Plateau. *Philosophical Transactions of the Royal Society of London, Series A: Mathematical and Physical Sciences*, **326**, 301–320.

ENGLAND, P. & HOUSEMAN, G. A. 1989. Extension during continental convergence, with application to the Tibetan plateau. *Journal of Geophyical Research*, **94**, 17 561–17 579.

EVANS, B. W. 1990. Phase relations of epidote-blueschists. *In*: OKRUSCH, M. (ed.) *Third international eclogite conference.* Lithos, **25**, 3–23.

FRASER, J., SEARLE, M. P., PARISH, R. & NOBLE, S. 1999. U–Pb geochronology on the timing of metamorphism and magmatism in the Hunza Karakoram. *Terra Nova*, **99**, 45–46.

FORSTER, M. A. 1999. Shortening and extensional structures in a collisional setting: Otago Schist, New Zealand. *Geological Society of Australia Abstracts*, **53**, 67–68.

FOSTER, D. A. & GLEADOW, A. J. W. 1992. The morphotectonics evolution of rift-margin mountains in central kenya: constraints from apatite fission-track thermochronology. *Earth and Planetary Science Letters*, **113**, 157–171.

FOSTER, D. A. & GLEADOW, A. J. W. 1996. Structural framework and denudation history of the flanks of the Kenya and Anza Rifts, East Africa. *Tectonics*, **15**, 258–271.

FORSTER, M. A. & LISTER, G. S. 1999*a*. Separate episodes of eclogite and blueschist facies metamorphism in the Aegean core complex of Ios, Cyclades, Greece. *In*: NIOCAILL, C. & RYAN, P. D. (eds) *Continental Tectonics*. Geological Society, London, Special Publications, **164**, 157–178.

FORSTER, M. A. & LISTER, G. S. 1999*b*. Detachment faults in the Aegean core complex of Ios, Cyclades, Greece. *In*: RING, U., BRANDON, M. T., LISTER, G. S. & WILLETT, S. D. (eds) *Exhumation processes; normal faulting, ductile flow and erosion*. Geological Society, London, Special Publications, **154**, 305–324.

FOSTER, D. A. & JOHN, B. 1999. Quantifying tectonic exhumation in an extensional orogen with thermochronology; examples from the southern Basin and Range Province. *In*: RING, U., BRANDON, M. T., LISTER, G. S. & WILLETT, S. D. (eds) *Exhumation processes; normal faulting, ductile flow and erosion*. Geological Society, London, Special Publications, **154**, 343–364.

FREEMAN, S. R., INGER, S., BUTLER, R. W. H. & CLIFF, R. A. 1997. Dating deformation using Rb–Sr in white mica: greenschist facies deformation ages from the Entrelor shear zone, Italian Alps. *Tectonics*, **16**, 57–76.

GEBAUER, D., SCHERTL, H. P., BRIX, M. & SCHREYER, W. 1997. 35 Ma old ultrahigh-pressure metamorphism and evidence for very rapid exhumation in the Dora Maira Massif, Western Alps. *Lithos*, **41**, 5–24.

GHENT, E. D., RODDICK, J. C. & BLACK, P. M. 1994. $^{40}Ar/^{39}Ar$ dating of white micas from the epidote to the omphacite zones, northern New Caledonia; tectonic implications. *Canadian Journal of Earth Sciences*, **31**, 995–1001.

GODIN, L., BROWN, R. L., PARRISH, R. & HODGES, K. V. 1998. Crustal thickening, melt production and exhumation of the Himalayan metamorphic core of central Nepal. *Geological Society of America Abstracts*, **30**, 269.

GRAY, D. R. & FOSTER, D. A. 1998. Character and kinematics of faults within the turbidite-dominated Lachlan orogen: implications for tectonic evolution of eastern Australia. *Journal of Structural Geology*, **20**, 1691–1720.

GRUJIC, D., CASEY, M., DAVIDSON, C., HOLLISTER, L. S., KUENDIG, R., PAVLIS, T. L. & SCHMID, S. M. 1996. Ductile extrusion of the Higher Himalayan

Crystalline in Bhutan; evidence from quartz microfabrics. *Tectonophysics*, **260**, 21–43.

GUILLOT, S., DE SIGOYER, J. & LARDEAUX, J. M. 1997. Eclogitic metasediments from the Tso Morari area (Ladakh, Himalaya): evidence for continental subduction during India-Asia convergence. *Contributions to Mineralogy and Petrology*, **128** 197–212.

GUILLOT, S., HATORI, K. & SIGOYER, J. 2000. Mantle wedge serpentinization and exhumation of eclogites: insights from eastern Ladakh, northeast Himalaya. *Geology*, **28** 199–202.

HACKER, B. R. 1996. Eclogite formation and the rheology, buoyancy, seismicity, and H_2O content of oceanic crust. *In*: BEBOUT, G. E., SCHOLL, D. W., KIRBY, S. H. & PLATT, J. P. (eds) *Subduction top to bottom*. Geophysical Monograph, **96**, 337–346.

HACKER, B. & PEACOCK, S. M. 1995. Creation, preservation, and exhumation of UHPM rocks. *In*: COLEMAN, R. G. & WANG, X. (eds) *Ultrahigh pressure metamorphism*. Cambridge University Press, Cambridge, 159–181.

HAQ, B. U., HARDENBOL, J. & VAIL, P. R. 1987. Chronology of fluctuating sea levels since the Triassic. *Science*, **235**, 1156–1167.

HARRISON, T. M. & YIN, A. 1999. Episodic tectonics during continuous Indo–Asian convergence. European union of Geosciences conference abstracts; EUG10. *Journal of Conference Abstracts*, **4**, 42.

HARRISON, T. M., COPELAND, P., KIDD, W. S. F. & YIN, A. 1992. Raising Tibet. *Science*, **255**, 163–170.

HODGES, K. V. 1997. Rates of orogenic processes and the role of catastrophe. *Geological Society of America Abstracts*, **29**, 352.

HODGES, K. V. 1998*a*. Self-organization and the metamorphic evolution of mountain ranges. *In*: TRELOAR, P. J. & O'BRIEN, P. (conveners), *Proceedings; What drives metamorphism and metamorphic reactions; heat production, heat transfer, deformation and kinetics? Extended abstracts*. Electronic Geology, **2**.

HODGES, K. V. 1998*b*. The thermodynamics of Himalayan orogenesis. *In*: TRELOAR, P. J. & O'BRIEN, P. J. (eds) *What drives metamorphism and metamorphic relations?* Geological Society, London, Special Publications, **138**, 7–22.

HODGES, K. V. 2000. Tectonics of the Himalaya and southern Tibet from two perspectives. *In*: GEISSMAN, J. W. & GLAZNER, A. F. (eds) *Special focus on the Himalaya*. Geological Society of America, Bulletin, **112**, 324–350.

HODGES, K. V., PARRISH, R. R. & SEARLE, M. P. 1996. Tectonic evolution of the central Annapurna Range, Nepalese Himalayas. *Tectonics*, **15**, 1264–1291.

HODGES, K. V., BURCHFIEL, B. C., ROYDEN, L. H., CHEN, Z. & LIU, Y. 1993. The metamorphic signature of contemporaneous extension and shortening in the central Himalayan Orogen; data from the Nyalam transect, southern Tibet. *Journal of Metamorphic Geology*, **11**, 721–737.

HODGES, K. V., PARRISH, R., HOUSH, T., LUX, D., BURCHFIEL, B. C., ROYDEN, L. & CHEN, Z. 1992.

Simultaneous Miocene extension and shortening in the Himalayan orogen. *Science*, **258**, 1466–1470.

HOUSEMAN, G. A., MCKENZIE, D. P. & MOLNAR, P. 1981. Convective instability of a thickened boundary layer and its relevance for the thermal evolution of continental convergent belts. *Journal of Geophysical Research*, **86**(B7), 6115–6132.

HSU, K. J. 1989. Time and place in Alpine orogenesis; the Fermor Lecture. *In*: COWARD, M. P., DIETRICH, D., PARK, R. G. (eds) Geological Society, London, Special Publications, **45**, 421–443.

HUNZIKER, J. C., DESMONS, J. & MARTINOTTI, G. 1989. Alpine thermal evolution in the central and the western Alps. *In*: COWARD, M. D., DIETRICH, D. & PARK, R. G. (eds) Geological Society, London, Special Publications, **45**, 353–367.

HURFORD, A. J. & HUNZIKER, J. C. 1985. Alpine cooling history of the Monte Mucrone eclogites (Sesia-Lanzo Zone); fission track evidence. *Schweiserische Mineralogische und Petro-graphische Mitteilungen*, **65**, 325–334.

INGERSOLL, R. V. 1997. Phanerozoic tectonic evolution of central California and environs. *International Geology Review*, **39**, 957–972.

JOLIVET, L. M., FOURNIER, M. & HUCHON, P. 1992. Cenozoic intracontinental dextral motion in the Okhotsk-Japan Sea region. *Tectonics*, **11**, 968–977.

JOLIVET, L., GOFFÉ, B., BOUSQUET, R., OBERHANSLI, R. & MICHARD, A. 1998. Detachments in high-pressure mountain belts, Tethyan examples. *Earth and Planetary Science Letters*, **160**, 31–47.

JOLIVET, L., GOFFÉ, B., MONIÉ, P., TRUFFERT-LUXEY, C., PATRIAT, M. & BONNEAU, M. 1996. Miocene detachment in Crete and exhumation P-T-t paths of high-pressure metamorphic rocks. *Tectonics*, **15**, 1129–1153.

JOLIVET, L., MALUSKI, H., BEYSSAC, O., GOFFE, B., LEPRVRIER, C., PHAN, T. T. & NGUYEN, V. V. 1999. Oligocene–Miocene BuKhang extensional gneiss dome in Vietnam; geodynamic implications. *Geology*, **27**, 67–70.

KEAY, S. 1998. *The geological evolution of the Cyclades, Greece: constraints from SHRIMP U–Pb geochronology*. PhD Thesis, Australian National University.

KEAY, S., LISTER, G. S. & BUICK, I. 2000. The timing of partial melting, Barrovian metamorphism and granite intrusion in the Naxos metamorphic core complex, Cyclades, Aegean Sea, Greece. *Tectonophysics*, in press.

KISSEL, C. & LAJ, C. 1988. The Tertiary geodynamical evolution of the Aegean arc; a paleomagnetic reconstruction. *Tectonophysics*, **146**(1–4), 183–201.

LAVECCHIA, G. 1988. The Tyrrhenian–Apennines system: structural setting and seismo-tectonogenesis. *Tectonophysics*, **147**, 263–296.

LE PICHON, X., FOURNIER, M. & JOLIVET, L. 1992. Kinematics, topography, shortening, and extrusion in the India–Asia collision. *Tectonics*, **11**, 1085–1098.

LISTER, G. S. & ETHERIDGE, M. A. 1989. Detachment models for uplift and volcanism in the eastern highlands, and their application to the origin of passive margin mountains. *In*: JOHNSON, R. W., KNUTSON, J. & TAYLOR, S. R. (eds) *Intraplate volcanism in eastern Australia and New Zealand*, Cambridge University Press, Sydney, 297–313.

LISTER, G. S. & RAOUZAIOS, A. 1996. The tectonic significance of a porphyroblastic blueschist facies overprint during Alpine orogenesis; Sifnos, Aegean Sea, Greece. *Journal of Structural Geology*, **18**, 1417–1435.

MANCKTELOW, N. S. 1995. Nonlithostatic pressure during sediment subduction and the development and exhumation of high-pressure metamorphic rocks. *Journal of Geophysical Research, B Solid Earth and Planets*, **100**, 571–583.

MCINNES, B. I. A., FARLEY, K. A., SILLITOE, R. H. & KOHN, B. P. 1999. Application of apatite (U–Th)/He thermochronometry to the determination of the sense and amount of vertical fault displacement at the Chuquicamata porphyry copper deposit, Chile. *Economic Geology and the Bulletin of the Society of Economic Geologists*, **94**, 937–947.

MICHARD, A., CHOPIN, C. & HENRY, C. 1993. Compression versus extension in the exhumation of the Dora Maira coesite-bearing unit, Western Alps, Italy. *Tectonophysics*, **221**, 173–193.

MOLNAR, P. & ENGLAND, P. 1990. Temperatures, heat flux and frictional stress near major thrust faults. *Journal of Geophysical Research*, **95**, 4833–4856.

MOLNAR, P. & LYON-CAEN, H. 1988. Some simple physical aspects of the support, structure, and evolution of mountain belts. *Geological Society of America, Special Paper*, **218**, 179–207.

MONIÉ, P. & CHOPIN, C. 1991. 40Ar/39Ar dating in coesite-bearing and associated units of the Dora-Maira massif, Western Alps. *European Journal of Mineralogy*, **3**, 239–262.

MONIÉ, P. & PHILIPPOT, P. 1989. Mise en évidence de l'âge éocène moyen du métamorphisme de haute pression dans la nappe ophiolitique du MonViso (Alpes occidentales) par la methode 40Ar–39Ar. *Comptes Rendus de l'Academie des Sciences*, **309**, 254–251.

MORAN-ZENTENO, D. J., CORONA-CHAVEZ, P. & TOLSON, G. 1996. Uplift and subduction erosion in southwestern Mexico since the Oligocene; pluton geobarometry constraints. *Earth and Planetary Science Letters*, **14**(1), 51–65.

NOBLE, W. P., FOSTER, D. A. & GLEADOW, A. J. W. 1997. The post-Pan-African thermal and extensional history of crystalline basement rocks in eastern Tanzania. *Tectonophysics*, **275**, 331–350.

NORTON, I. O. 1995. Plate motions in the North Pacific: the 43 Ma nonevent. *Tectonics*, **14**, 1080–1094.

NUR, A. & BEN-AVRAHAM, Z. 1982. Oceanic plateaus, the fragmentation of continents, and mountain building. Special issue on accretion tectonics. *Journal of Geophysical Research*, **B 87**, 3644–3661.

PATRIAT, P. & ACHACHE, J. 1984. India–Eurasia collision chronology has implications for crustal shortening and driving mechanism of plates. *Nature*, **311**, 615–621.

PLATT, J. P. 1986. Dynamics of orogenic wedges and the uplift of high-pressure metamorphic rocks. *Geological Society of America Bulletin*, **97**, 1037–1053.

PLATT, J. P. 1987. The uplift of high-pressure low-temperature metamorphic rocks. *Philosophical Transactions of the Royal Society of London, Series A: Mathematical and Physical Sciences,* **321,** 87–103.

PLATT, J. P. 1993. Exhumation of high-pressure rocks: a review of concepts and processes. *Terra Nova,* **5,** 119–133.

PLATT, J. P. & ENGLAND, P. C. 1993. Convective removal of lithosphere beneath mountain belts: Thermal and mechanical consequences. *American Journal of Science,* **293,** 307–336.

PLATT, J. P., LISTER, G. S., CUNNINGHAM, P., WESTON, P., PEEL, F., BAUDIN, T. & DONDEY, H. 1987. Thrusting and backthrusting in the Briançonnais domain of the Western Alps. *In:* Geological Society, Special Publications, **45,** 135–152.

RAWLING, T. J. 1998. *Oscillating orogenesis and exhumation of high-pressure rocks in New Caledonia, SW Pacific.* PhD Thesis, Monash University.

RAWLING, T. J. & LISTER, G. S. 1999a. The high-pressure sole of the New Caledonia ophiolite belt. *In: Specialist Group for Tectonics and Structural Geology's Last Conference for the Millennium, Halls Gap, February 14–19.* Geological Society of Australia Abstracts, **53.**

RAWLING, T. J. & LISTER, G. S. 1999b. Oscillating modes of orogeny in the Southwest Pacific and the tectonic evolution of New Caledonia. *In:* RING, U., BRANDON, M. T., LISTER, G. S. & WILLETT, S. D. (eds) *Exhumation processes: Normal faulting, Ductile Flow and Erosion.* Geological Society, London, Special Publications, **154,** 109–128.

RAWLING, T. J. & LISTER, G. S. 2000. Large scale structure of the high-pressure Pam-Paniė region of New Caledonia. *Journal of Structural Geology,* in press.

REDDY, S. M., WHEELER, J. & CLIFF, R. A. 1999. The geometry and timing of orogenic extension: an example from the Western Italian Alps. *Journal of Metamorphic Geology,* **17,** 573–589

REYNOLDS, P., RAVENHURST, C., ZENTILLI, M. & LINDSAY, D. 1998. High precision $^{40}Ar/^{39}Ar$ dating of two consecutive hydrothermal events in the Chuquicamata porphyry copper system. *Chemical Geology,* **148,** 45–60.

ROYDEN, L. H. 1993a. Evolution of retreating subduction boundaries formed during continental collision. *Tectonics,* **12,** 629–638.

ROYDEN, L. H. 1993b. The tectonic expression slab pull at continental convergent boundaries. *Tectonics,* **12,** 303–325.

RUBATTO, D. & HERMANN, J. 2000. Precise pressure-temperature-time path documents fast exhumation of deeply subducted rocks. *Geology,* **29,** 3–6.

RUBATTO, D., GEBAUER, D. & COMPAGNONI, R. 1999. Dating of eclogite-facies zircons: the age of Alpine metamorphism in the sesia-Lanzo zone (Western Alps). *Earth and Planetary Science Letters,* **167,** 141–158.

RUBATTO, D., GEBAUER, D. & FANNING, M. 1998. Jurassic formation and Eocene subduction of the Zermatt-Saas-Fee ophiolites: implications for the geodynamic evolution of the Central and Western

Alps. *Contributions to Mineralogy and Petrology,* **132,** 269–287.

RYAN, P. D. & DEWEY, J. F. 1997. Continental eclogites and the Wilson Cycle. *Journal of the Geological Society, London,* **154,** 437–442.

SCHÄRER, U., COPELAND, P., HARRISON, T. M. & SEARLE, M. 1990. Age, cooling history, and origin of post-collisional leucogranites in the Kara-koram batholith; a multi-system isotope study. *Journal of Geology,* **98,** 233–251.

SCHLUNEGGER, F. & WILLETT, S. D. 1999. Spatial and temporal variations in exhumation of the central Swiss Alps and implications for exhumation mech-anisms. *In:* RING, U., BRANDON, M. T., LISTER, G. S., WILLETT, S. D. (eds) *Exhumation processes; normal faulting, ductile flow and erosion.* Geological Society, London, Special Publications, **154,** 157–179.

SEARLE, M. P., KHAN, M. A., FRASER, J. E. & GOUGH, S. J. 1999. The tectonic evolution of the Kohlistan-Karakoram collision belt along the Karakoram highway transect, north Pakistan. *Tectonics,* **18,** 929–949.

SEIDEL, E. H., KREUZER, H. & HARRE, W. 1982. The late Oligocene/early Miocene high pressure in the external Hellenides. *Geologisches Jahrbuch. Reihe E: Geophysik,* **23,** 165–206.

SENGOR, A. M. C. 1976. Collision of irregular continen-tal margins; implications for foreland deforma-tion of Alpine-type orogens. *Geology,* **4,** 779–782.

SENGOR, A. M. C. 1987. Plate tectonics and orogenic research after 25 years; synopsis of a tethyan perspective. *Tectonophysics,* **187,** 315–344.

SENGOR, A. M. C. 1993. Some current problems on the tectonic evolution of the Mediterranean during the Cainozoic. *In:* BOSCHI, E., MANTOVANI, E. & MORELLI, A. (eds) *Recent evolution and seismicity of the Mediterranean region.* NATO ASI Series. Series C: Mathematical and Physical Sciences, **402,** 1–51.

SENGOR, A. M. C., ALTINER, D., CIN, A., USTAOMER, T. & HSU, K. J. 1988. Origin and assembly of the Tethyside orogenic collage at the expense of Gondwanaland. *In:* AUDLEY-CHARLES, M. G. & HALLAM, A. (eds) *Gondwana and Tethys.* Geologi-cal Society, London, Special Publications, **37,** 119–181.

SÉRANNE, M. 1999. The Gulf of Lion continental margin (NW Mediterranean) revisited by IBS; an overview. *In:* DURAND, B., JOLIVET, L., HORVATH, F. & SERANNE, M. (eds) *The Mediterranean basins; Tertiary extension within the Alpine Orogen.* Geological Society, London, Special Publications, **156,** 15–36.

SPENCER, D. & GEBAUER, D. 1996. SHRIMP evidence for a Permian protolith age and a 44 Ma meta-morphic age for the Himalyan eclogites (upper Kaghan, Pakistan): implications for the subduction of Tethys and the subduction terminology of the NW Himalaya. *In: 11th Himalayan-Karakoram-Tibet Workshop Abstract volume.* Flagstaff, Ari-zona, 147–150.

TAVARNELLI, E. & HOLDSWORTH, R. E. 1999. How long do structures take to form in transpression zones?

A cautionary tale from California. *Geology*, **27**, 1063–1066.

TRÜMPY, R. 1973. The timing of orogenic events in the Central Alps. *In*: DE JONG, K. & SCHOLTEN, R. (eds) *Gravity and Tectonics*. Wiley-Interscience, New York, 229–252.

TRÜMPY, R. 1980. *An outline of the geology of Switzerland*. Wept & Co., Basel.

TURCOTTE, D. L. 1992. Fractals, chaos, self-organised criticality and tectonics. *Terra Review*, **4**, 4–12.

VANCE, D. & HARRIS, N. 1999. Timing of prograde metamorphism in the Zanskar Himalaya. *Geology*, **27**, 395–398.

VANDENBERG, L. C. & LISTER, G. S. 1996. Structural analysis of basement tectonites from the Aegean metamorphic core complex of Ios, Cyclades, Greece. *Journal of Structural Geology*, **18**, 1437–1454.

VANNAY, J. C. & HODGES, K. V. 1996. Tectonometamorphic evolution of the Himalayan metamorphic core between the Annapurna and Dhaulagiri, central Nepal. *Journal of Metamorphic Geology*, **14**, 635–656.

VENTURINI, G. 1995. Geology, geochemistry and geochronology of the inner central Sesia Zone (Western Alps – Italy). *Mémoires de Géologie (Lausanne)*, **25**, 126.

VERGÉS, J. & SÀBAT, F. 1999. Constraints on the Neogene Mediterranean kinematic evolution along a 1000 km transect from Iberia to Africa. *In*: DURAND, B., JOLIVET, L., HORVATH, F. & SERANNE, M. (eds) *The Mediterranean basins;* Tertiary extension within the Alpine Orogen. Geological Society, London, Special Publications, **156**, 63–80.

VON BLANCKENBURG, F. & DAVIES, H. J. 1995. Slab breakoff: a model for syncollisional magmatism and tectonics in the Alps. *Tectonics*, **14**, 120–131.

WIJBRANS, J. R. & MCDOUGALL, I. 1986. $^{40}Ar/^{39}Ar$ dating of white micas from an Alpine high pressure metamorphic belt on Naxos (Greece). *Contributions to Mineralogy and Petrology*, **93**, 187–194.

WIJBRANS, J. R. & MCDOUGALL, I. 1988. Metamorphic evolution of the Attic Cycladic metamorphic belt on Naxos (Cyclades, Greece) utilising $^{40}Ar/^{39}Ar$ age spectrum measurements. *Journal of Metamorphic Geology*, **6**, 571–594.

WIJBRANS, J. R., SCHLIESTEDT, M. & YORK, D. 1990. Single grain argon laser probe dating of phengites from the blueschist to greenschist transition on Sifnos (Cyclades, Greece). *Contributions to Mineralogy and Petrology*, **104**, 582–593.

WIJBRANS, J. R., VAN WEES, J. D., STEPHENSON, R. A. & CLOETINGH, S. A. P. L. 1993. Pressure–temperature–time evolution of the high-pressure metamorphic complex of Sifnos, Greece. *Geology*, **21**, 443–446.

YIN, A., HARRISON, T. M., MURPHY, M. A., GROVE, M., NIE, S., RYERSON, F. J., WANG, X. & CHEN, Z. 1999. Tertiary deformation history of southeastern and southwestern Tibet during the Indo-Asian collision. *Geological Society of America Bulletin*, **111**, 1644–1664.

The structure and rheological evolution of reactivated continental fault zones: a review and case study

R. E. HOLDSWORTH[1], M. STEWART[1,2] J. IMBER[1,3] & R. A. STRACHAN[4]

[1] *Reactivation Research Group, Department of Geological Sciences, University of Durham, Durham DH1 3LE, UK (e-mail: r.e.holdsworth@durham.ac.uk)*
[2] *Present address: LIG, Liverpool John Moores University, Rodney Street, Liverpool L3 5UX, UK*
[3] *Present address: Fault Analysis Group, Department of Geology, University College Dublin, Belfield, Dublin 4, Ireland*
[4] *Department of Geology, Oxford Brookes University, Gypsy Lane, Headington, Oxford OX3 0BP, UK*

Abstract: Repeated reactivation of structures and reworking of crustal volumes are characteristic, though not ubiquitous, features of continental deformation. Reactivated faults and shear zones exposed in the deeply exhumed parts of ancient orogenic belts present opportunities to study processes that influence the mechanical properties of long-lived fault zones at different palaeo-depths. Ancient basement fault systems typically comprise heterogeneous, superimposed assemblages of fault rocks formed at different times and depths for which down-temperature thermal histories are most common. Several lithological and environmental factors influence the evolution of fault rock fabrics and rheology, but most fault/shear zone arrays appear to develop as self-organized deformation systems. Once mature, the kinematic and mechanical evolution of the system is strongly influenced by the rheological behaviour of the interconnected fault/shear zone network.

A case study from the crustal-scale Great Glen Fault Zone (GGFZ), Scotland, reveals a complex evolution of mid- to upper-crustal deformation textures formed adjacent to the frictional-viscous transition. Fluid influx in the mid-crust has led to reaction softening of the rock aggregate as strong pre-existing phases such as feldspar are replaced by fine-grained, strongly aligned aggregates of weak phyllosilicates. In addition, a grainsize-controlled switch to fluid-assisted diffusional creep occurs in the highest strain regions of the fault zone. It is proposed that this led to a shallowing and narrowing of the frictional-viscous transition and to long-term overall weakening of the fault zone relative to the surrounding wall-rocks. Cataclasis is particularly important in the deeper part of the frictional regime as it helps to promote retrograde metamorphism and changes in deformation regime, by both reducing grainsize and promoting pervasive fluid influx along fault strands due to grain-scale dilatancy. Equivalent processes are likely to occur along many other long-lived, crustal-scale fault zones.

Reactivation of structures occurs when displacements are repeatedly focused along well-defined, pre-existing features such as faults, shear zones or lithological contacts (Holdsworth *et al.* 1997). There is a wealth of geological and geophysical evidence to suggest that reactivation often occurs in preference to the formation of new structures during deformation in the continental lithosphere (e.g. Sutton & Watson 1986; Butler *et al.* 1997 and references therein). As a result, pre-existing structures commonly influence the location and architecture of lithosphere deformation zones in all tectonic environments, including major transcurrent fault systems, orogenic belts, and rifted basins in both intracontinental and continental margin settings. Many long-lived structures are also known to act as channelways for the migration of hydrous fluids and magmas (e.g. see Kerrich 1986; Hutton 1988; McCaig 1997). Thus, continents are thought to carry an in-built structural inheritance that is largely absent in oceanic regions. This feature, together with the presence of a buoyant quartzofeldspathic crust that is weak relative to its oceanic equivalent, is thought to explain major differences in the large-scale deformational response

From: MILLER, J. A., HOLDSWORTH, R. E., BUICK, I. S. & HAND, M. (eds) *Continental Reactivation and Reworking.*
Geological Society, London, Special Publications, **184**, 115–137. 1-86239-080-0/01/$15.00 © The Geological
Society of London 2001.

of the lithosphere in continental and oceanic regions (see reviews by Molnar 1988, 1992; Thatcher 1995).

This paper focuses mainly on the nature and significance of reactivated structures in the continental crust. Such features are potentially very long-lived because the crust is not normally subducted during convergence and collision (Sutton & Watson 1986). Fault rock assemblages preserved along ancient, deeply exhumed reactivated basement fault zones give important insights into the causes of reactivation on various scales and at different depths in the crust. A case study from the Great Glen Fault in Scotland is presented to illustrate the importance of fluid-related fault zone processes in promoting long-term changes in rheological behaviour.

It is important to remember, however, that reactivation is *not* always ubiquitous: numerous ancient displacement zones remain quiescent subsequent to their formation, whilst more recently formed faults may not necessarily reactivate pre-existing structures (Roberts & Holdsworth 1999). Well-defined lines of geological and geophysical evidence are therefore required in order to demonstrate conclusively that individual structures or sets of structures are reactivated; some published criteria are unreliable (e.g. geometric similarity, see Holdsworth *et al.* 1997 for discussion).

Continental deformation and reactivation

Theoretical strength profiles derived from the results of experimental rock deformation studies can be constructed for an idealized, average continental lithosphere with realistic and representative compositional layering, geothermal gradient and strain rate (Fig. 1; see Kohlstedt *et al.* 1995 and references therein). For 'average' continental lithosphere, such profiles predict that there are two relatively strong, load-bearing regions – one in the upper mantle and a second in the mid-crust. The rheological behaviour and evolution of rocks in these two regions are likely to play a major role in determining the overall mechanical response of the crust and lithosphere to deformation (Fig. 1).

The location of the primary load-bearing region in the upper mantle suggests that ultimately the large-scale deformation behaviour of the continents will be mainly determined and driven by ductile shearing in the olivine-rich lithospheric mantle (e.g. Molnar 1992; Teyssier & Tikoff 1998). This suggestion seems to be borne out by the apparent close correspondence on a global scale between patterns of crustal deformation and seismic anisotropy that is thought to

track the orientation of deformation fabrics in the upper mantle (see Silver 1996 and references therein). Thus, a clearer understanding of the rheology of typical continental peridotites deformed under upper mantle conditions (e.g. Karato *et al.* 1986; Kohlstedt *et al.* 1995) will tell us much about those factors that ultimately control the large-scale strain response of the continental lithosphere. The repeated reworking of crustal volumes observed in many continental settings may, in part, reflect long-term relative weakness in underlying upper mantle shear zones.

In the upper part of the crust, it is generally agreed that deformation involves frictional flow in which the deformation mechanisms involve brittle fracture and dilatancy that depend mainly on the effective pressure (Byerlee 1978; Paterson 1978; Sibson 1983). At greater depths, the regime changes to one of viscous flow in which a range of non-frictional thermally activated deformation mechanisms are involved to produce crystal plasticity and diffusional creep (Sibson 1977; Tullis & Yund 1977; Schmid & Handy 1991). The mid-crustal region separating these two rheological regimes – the frictional–viscous (or brittle–ductile) transition (Schmid & Handy 1991) – is typically thought to lie between 10 and 15 km depth, coinciding with the secondary load-bearing region of the lithosphere (Fig. 1). Since most crustal deformation is partitioned along faults and shear zones, it is likely that the mechanical behaviour of these structures as they cut through the region of highest strength mid-crust will be particularly important in determining the location and distribution of displacements (Holdsworth *et al.* 1997; Imber *et al.* 1997; Stewart *et al.* 2000). This is likely to be a key factor in determining whether or not reactivation of crustal-scale structures occurs during a given regional deformation event.

The foregoing discussion (and Fig. 1) applies to an idealized continental lithosphere with an *average* geothermal gradient and crustal thickness. The large-scale rheological response and distribution of strength with depth in the lithosphere is particularly sensitive to the thermal state of the lithospheric section (Sonder & England 1986; Molnar 1992). Thus, the distribution of heat-producing elements within the crust and processes that thicken or thin the upper mantle can profoundly influence the large-scale mechanical response of the lithosphere to deformation (e.g. see Sandiford 1999 and references therein).

In general, if pre-existing structures are large enough to form significant regions of mechanical anisotropy in the continental lithosphere, then they are likely to localize deformation, particularly if they occur in the mechanically strong

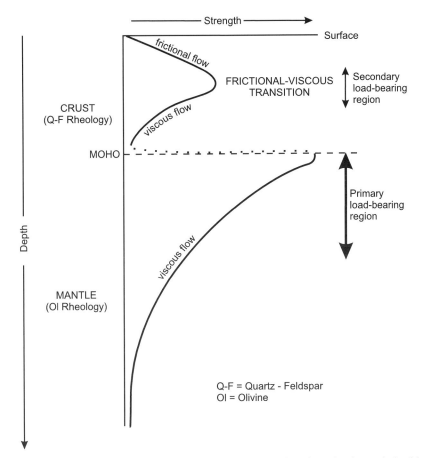

Fig. 1. Schematic strength v. depth profile for 'average' continental lithosphere showing main load-bearing regions in upper mantle and mid-crust (after Molnar 1988).

upper mantle or mid-crust. Once deformation has localized, the internal rheological behaviour of the pre-existing structures may then determine whether or not reworking or reactivation occurs. Compared to upper mantle rocks, crustal-scale fault zones of various ages and degrees of exhumation are abundant in most continental regions and are relatively accessible for direct study.

Reactivated crustal-scale fault zones

Fault rocks and exhumation

Reactivated crustal-scale structures fundamentally influence the position and architecture of a broad range of features found throughout the geological record, including the location of many economic resources and seismic hazards. Many major reactivated faults presently exposed cutting metamorphic basement have experienced a long history of exhumation and preserve fault-hosted deformation products, or 'fault rocks', that formed at different depths and times in the crust (Grocott 1977; Handy 1990b). The microstructures of these fault rocks can give an insight into the evolution of deformation mechanisms and the rheological behaviour of fault zones under a wide range of pressure and temperature conditions (e.g. Handy 1989; Snoke et al. 1998 and references therein).

Thermal history and general structure

The great majority of reactivated basement fault zones display down-temperature, retrograde metamorphic histories, often over very long time periods (e.g. Grocott 1977; Sibson 1977; Obee & White 1985; 1986; Grønlie &

Roberts 1989; Butler *et al.* 1995; Fliervoet *et al.* 1997; Imber *et al.* 1997; Stewart *et al.* 1999, 2000). As a result, early higher temperature fault rock textures and mineral assemblages are usually overprinted by lower temperature assemblages. Thus, in general, the earliest deformation products are typically mylonitic, having formed in the viscous regime, whilst the final displacement events are typically cataclastic and frictional in origin. This evidence, together with the predicted change in rheological behaviour of the crust with depth (e.g. Fig. 1), has led to the traditional model for crustal scale fault zones (Fig. 2; Sibson 1977). In this model, an interlinked network of brittle faults and cataclastic fault rocks connect directly at depth into a broader, anastomosing system of viscous mylonitic shear zones. Many basement faults display fault rocks that preserve 'semi-brittle' deformation textures (e.g. Carter & Kirby 1978; White & White 1983; Handy *et al.* 1999; Stewart *et al.* 1999) and it is generally agreed that such assemblages record deformation at depths

corresponding to the frictional viscous transition (Schmid & Handy 1991). However, there are a number of reasons why the situation is not as straightforward as the idealized fault zone shown in Figure 2 implies:

(a) Ancient fault systems often preserve *superimposed* assemblages of fault rocks formed at different times and depths (Grocott 1977). Thus, earlier structures and deformation textures may be partially or wholly overprinted by the effects of later deformation and metamorphism. Furthermore, fault rocks may be the product of very different kinematic events that were widely separated in time;

(b) rocks are compositionally heterogeneous on all scales, with adjacent rock units responding in very different ways to deformation under a given set of environmental conditions (e.g. Handy 1990*a* and references therein);

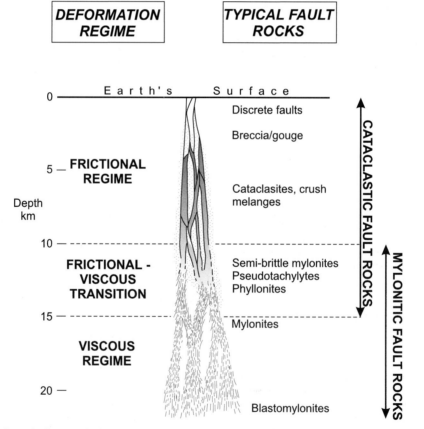

Fig. 2. Schematic diagram showing depth, fault rock distribution and deformation regimes along a vertical, crustal-scale fault zone (after Sibson 1977).

(c) there is good experimental evidence to suggest that frictional behaviour in the crust can result from several different slip mechanisms, particularly in the presence of aqueous fluids (e.g. Blanpied *et al.* 1995; Chester 1995). These mechanisms involve combined cataclasis and fluid-assisted diffusional processes (pressure solution), and are dependent on both temperature and strain rate, in addition to pressure; and

(d) deformational and metamorphic processes often profoundly modify fault rock mineralogy and microstructure in ways that lead to significant changes in rheological behaviour and mechanical strength as the fault rock evolves (e.g. Handy 1989 and references therein). This is particularly significant in the load-bearing mid-crustal region, especially if faulting occurs at these depths in the presence of a chemically active fluid phase (Imber *et al.* 1997; Handy *et al.* 1999; Stewart *et al.* 2000). As a result, the location and character of the frictional viscous transition can evolve with time in any given fault zone (Schmid & Handy 1991; Stewart *et al.* 2000).

Faults or shear zones that have acted as conduits for the passage of hydrothermal fluids or granitoid magmas at some stage in their history may display complex thermal histories (e.g. Strong & Hanmer 1981; Hollister & Crawford 1986; D'Lemos *et al.* 1997; Hanmer 1997; McCaffrey 1997). In many such cases, the intensity of high temperature metamorphism and shear localization associated with magma intrusion may totally obscure any evidence for the presence of a pre-existing structure that then has to be inferred on regional grounds (e.g. Holdsworth 1994). Following intrusion of fluids and/ or magma, such faults and shear zones often exhibit long histories of subsequent reactivation that must result in part from thermal softening of the lithospheric channelway (e.g. Holdsworth 1994; Tommasi *et al.* 1994; Saint Blanquat *et al.* 1998).

Controls on fault rock evolution

Localization, weakening and reactivation

The partitioning of shear strain into faults and shear zones is recognized at every scale in all regions of crustal deformation (e.g. Ramsay 1980 and references therein). Such *localization* behaviour is known to occur widely in both strain softening and strain hardening deformation regimes (e.g. Griggs & Handin 1960; Cobbold

1977; Aydin 1978). It is incorrect, therefore, to always equate localization behaviour with strain softening or weakening (Hobbs *et al.* 1990). It is also generally assumed that reactivation, a process that usually involves localization of shear strain, reflects the long-term mechanical weakness of pre-existing faults and shear zones relative to their surrounding wall-rocks (e.g. Watterson 1975; White *et al.* 1986; Holdsworth *et al.* 1997). This assertion is probably correct during the early stages of reactivation since the stress-strain behaviour of the previously inactive system is likely to be predominantly determined by the *constitutive* or *material* behaviour of the fault rocks, i.e. whether they are intrinsically strain softening, strain hardening, etc. However, once deformation is re-established, the overall *system* behaviour will additionally be influenced and even changed by the prevailing kinematic boundary conditions (Hobbs *et al.* 1990). The influence of such boundary conditions is largely determined by the deformation regime. In particular, materials deforming by mechanisms that involve significant amounts of dilatancy (i.e. frictional flow, hydrofracture veining events) will be particularly sensitive to kinematic boundary conditions (Edmond & Paterson 1972). In most geological environments, the imposed conditions will lead to strain hardening behaviour of the system, irrespective of the constitutive properties of the fault rocks (Hobbs *et al.* 1990). In contrast, materials deforming by pressure-insensitive, non-dilational mechanisms (e.g. crystal plasticity) are less sensitive to kinematic boundary conditions and, in most cases, localization is associated with strain induced weakening (Hobbs *et al.* 1990). In the case of reactivated structures, this means that once displacement is re-established, the stress-strain behaviour will be significantly influenced by the mechanisms that prevail during ongoing deformation. Some reactivation events involving brittle deformation and significant dilatancy may lead to the onset of strain hardening behaviour.

Fault rock controls and the evolution of the fault system

The textural evolution and distribution of deformation within most fault zones can be determined by the operation of up to six interrelated factors that can be conveniently subdivided into lithological and environmental controls (Fig. 3; cf. Knipe 1989). The importance of the six factors will change in both space and time as the fault system evolves and accumulates displacement, leading to a heterogeneous

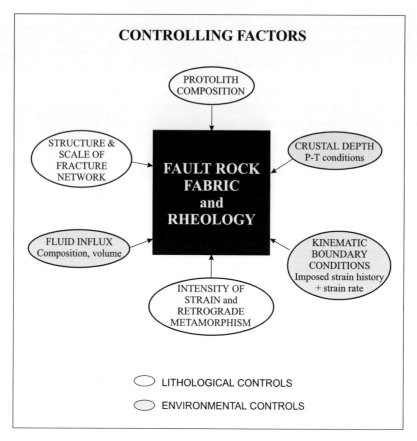

Fig. 3. Lithological (unshaded) and environmental (grey shading) factors controlling the fault-rock fabrics and rheological evolution in major fault zones. The structure and scale of the fault zone-related fracture network includes features such as scaling characteristics, degree of interconnectivity, permeability, etc. (e.g. Cowie *et al.* 1996 and references therein).

distribution of fault rocks, textures and deformation histories (e.g. see reviews by Handy 1989; Schmid & Handy 1991). However, faults and shear zones tend to develop as self-organizing deformation systems on all scales (e.g. Sornette *et al.* 1990; Handy 1990*a*, 1994; Sornette *et al.* 1993). Strains are increasingly localized to form interconnected, narrow displacement zones (faults, shear zones) that surround elongate lenses of less highly deformed material (e.g. Figs 2, 6, 8 and 9). This configuration is mechanically stable and is thought to allow rock-systems to deform over long time-scales by heterogeneous steady-state flow in which the strain response of the entire system will be controlled by the kinematic behaviour of the interconnected fault/shear zone network. Therefore, in mature fault/shear zone systems, it is likely that the rheological properties and evolution of the fault rocks in the inter-connected, highest strain fault/shear zone strands will ultimately control the behaviour of the whole system (Handy 1990*a*).

A case study of a reactivated fault system: the Great Glen Fault, Scotland

Introduction

Many recent investigations into how faulting processes may alter the mechanical strength of natural, large-scale fault zones have focused on fault rocks formed in the upper crust (e.g. Zoback & Lachenbruch 1992 and references therein). In these cases, localization of deformation along major fault strands has been linked to weakening caused by either the presence of anomalously low-friction clay gouges (Wang 1984; Morrow *et al.* 1992) and/or to the development of high fluid pressures in excess of hydrostatic pressures (Byerlee 1990; Rice 1992; Chester *et al.* 1993). If, however, the long-term mechanical properties of crustal fault zones are determined mainly in the region where faults cut through the strong mid-crust, then it is necessary to study the deeper parts of fault zones. Deeply-exhumed major faults or shear zones like the Great Glen Fault Zone in Scotland permit direct examination at the surface of fault rock assemblages and deformation processes that occurred in the region close to the ancient frictional-viscous transition. In common with many other examples, the Great Glen Fault Zone has a long history of movement spanning hundreds of millions of years. This may indicate that such faults possess geometric and/or internal mechanical properties that make them susceptible to persistent activity over long timescales. They are therefore an ideal natural laboratory in which to investigate whether mid-crustal internal fault zone processes are one of the possible causes of reactivation behaviour.

Regional geology

The Great Glen Fault Zone (GGFZ) is a sub-vertical, major fault that cuts NE-SW across the Ordovician–Silurian Caledonian orogenic belt in the NW British Isles (Fig. 4a; Kennedy 1946; Flinn 1961). Geophysical studies demonstrate that the GGFZ is coincident with a sub-vertical discontinuity seen in deep seismic reflection profiles down to at least 40 km depth (Hall *et al.* 1984). Regional structural studies suggest that the earliest, main phase of sinistral movements along the GGFZ occurred during the later stages of the Caledonian orogeny between *c.* 428 Ma and 390 Ma (Stewart 1997; Stewart *et al.* 1997, 1999). Based on regional, tectonostratigraphic considerations and offsets of mantle reflectors, displacements of at least 200 km are likely (see Soper & Hutton 1984; Snyder & Flack 1990). Structural analysis and correlation of offset geological markers across the GGFZ suggest reactivation events involving 20 km of dextral shear during the Devonian–Carboniferous (Rogers *et al.* 1989) and 10 km of additional dextral movement, probably during the Tertiary (Underhill & Brodie 1993). Historical records of earthquake activity and the preservation of supposed seismically induced slump structures in Quaternary glacial deposits adjacent to the GGFZ provide apparent evidence of more recent seismic activity (Musson 1994).

On the Scottish mainland, different structural levels are preserved adjacent to the GGFZ as a result of differential displacement and exhumation across the structure (Stewart 1997; Stewart *et al.* 1999, 2000). A central fault core *c.* 300 m wide is thought to contain the principal displacement surfaces and separates the two regional crustal blocks that define the wall-rocks to the NW and SE of the GGFZ, respectively, the Northwest Highland and Grampian crustal blocks (Fig. 4b). In addition, two narrow, highly deformed fault-bounded slivers – the Torcastle and Aberchalder blocks – are entrained within the fault zone (Fig. 4b).

To the northwest of the GGFZ, the NW Highland block is dominated by Neoproterozoic metasedimentary rocks of the Moine Supergroup (Holdsworth *et al.* 1994 and references therein). The narrow, fault-bounded Torcastle block appears to also be derived from a Moinian protolith (Stewart *et al.* 2000). To the southeast of the GGFZ, Neoproterozoic to early Cambrian metasedimentary rocks of the Dalradian Supergroup dominate the Grampian block (Harris *et al.* 1994 and references therein). In the northeastern part of the Grampian Block, rocks thought to form the lower parts of the Dalradian Supergroup are underlain by migmatites which may, at least in part, correlate with the Moine Supergroup (Highton *et al.* 1999). Sheared rocks within the Aberchalder block have been interpreted as derived from these migmatitic protoliths (Stewart *et al.* 1999).

Distribution of fault-related deformation and metamorphism

The main phase of sinistral shearing along the GGFZ produced a zone of brittle to semi-brittle deformation textures and associated retrograde metamorphism up to 3 km wide (Stewart 1997; Stewart *et al.* 1999, 2000). The effects of faulting and associated retrograde metamorphism overprint predominantly metasedimentary rocks that underwent regional deformation under greenschist- to amphibolite-facies metamorphic conditions. Psammitic lithologies are dominant, with lesser amounts of semi-pelite, pelite and granitic gneiss. Overlying the metamorphic rocks either side of the Great Glen are sequences of continental sedimentary rocks of probable Devonian age (Old Red Sandstone; Fig. 4b). Small outliers of conglomerate and sandstone occur within the GGFZ, either unconformably overlying the metamorphic rocks, or, more typically as narrow fault-bounded slivers (Stoker 1982;

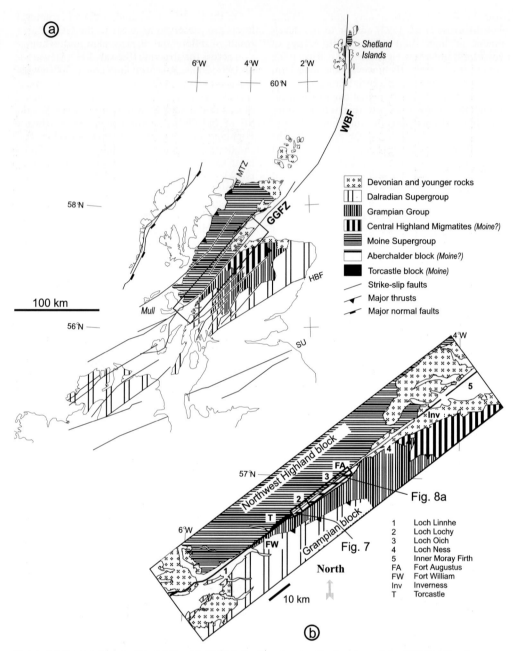

Fig. 4. (a) Regional map of the UK and Irish Caledonides showing the location of the GGFZ. Box shows location of Figure 4b. GGFZ, Great Glen Fault Zone; WBF, Walls Boundary Fault Zone; MTZ, Moine Thrust Zone; HBF, Highland Boundary Fault; SU, Southern Uplands Fault. (b) The main structural components of the GGFZ, together with localities discussed in the text and areas shown in Figures 7 and 8. Note that the main key also applies to this map.

May & Highton 1992). These Old Red Sandstone (ORS) rocks are an important time marker in the fault zone as they allow the effects of pre- and post-ORS deformation to be distinguished.

Detailed mapping by Stewart (1997) of all exposures in the GGFZ in the area between Fort Augustus and Loch Linnhe (Fig. 4b) allows a series of schematic cross-sections through

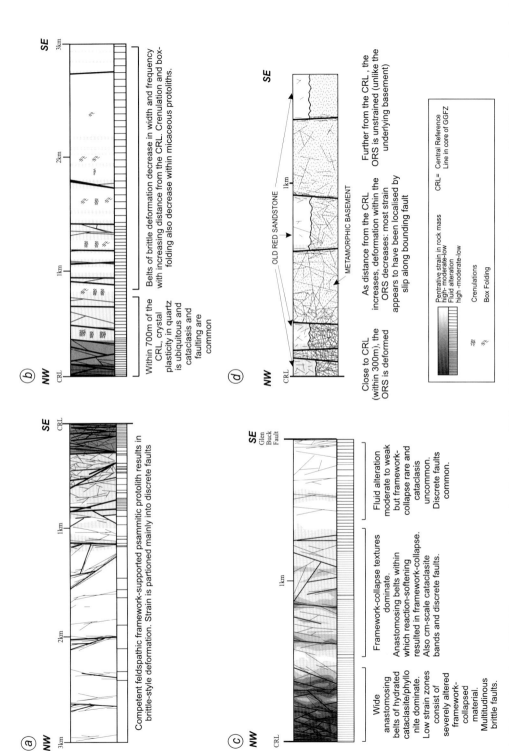

Fig. 5. Schematic regional NW–SE cross-sections through: (**a**) northwest Highland block; (**b**) Grampian block; (**c**) Aberchalder block; (**d**) ORS deposits and associated basement in region immediately SE of GGFZ. The fault networks, brittle fold structures, GGFZ-related deformation intensity and degree of low-grade alteration are also indicated.

the NW Highland, Grampian and Aberchalder blocks to be constructed (Figs 5a–c). For comparative purposes, a schematic section showing the distribution and deformation of the ORS deposits found immediately southeast of the core region of the GGFZ is also presented (Fig. 5d). A detailed account of the textures and rheological significance of rocks from the Torcastle block is given elsewhere (Stewart *et al.* 2000).

Importantly, fault-related strain in the outliers of Devonian ORS is much less intense compared to the adjacent, underlying units of metamorphic basement in the GGFZ. Deformation is restricted to discrete brittle microfracturing and faulting immediately adjacent to major bounding faults (see Fig. 5d). This suggests that the ORS was deposited unconformably upon pre-existing fault rocks in the core of the GGFZ, confirming the pre-ORS age of the fault rock assemblages discussed here.

Although the effects of deformation associated with the main phase of sinstral movement are highly heterogeneous, the schematic sections (Figs 5a–d) show that a general increase in strain intensity occurs toward the centre of the fault zone. In the kilometre nearest to the fault core, pre-existing regional banding and deformation fabrics are subvertical, striking sub-parallel to the main fault zone. In a relatively complete section through the northwest margin of the Grampian Block exposed along the River Spean (for location see Fig. 4b), progressive reorientation of the regional structures into parallelism with the GGFZ is observed and appears to have occurred by rotations of fault-bounded blocks (see Stewart *et al.* 1999). In all sections, deformation is predominantly brittle involving the development of fracture networks, cataclasite seams, breccias, minor brittle kink and box folds and crenulation fabrics (Figs 5a–d). As the intensity of fracturing increases, there is a general broadening and interlinking of cataclasite and breccia seams that reflects increasing strain. At the same time, increasing amounts of retrograde alteration occur involving the growth of sericitic white mica and chlorite. In much of the fault zone, these minerals are apparently late, post-kinematic alteration products after feld-spar and biotite, respectively, although in regions close to the fault core, there is evidence for widespread syntectonic growth of phyllosilicates synchronous with shearing (see below). Collectively, these features suggest that the fracture systems formed during shearing along the GGFZ have acted as channelways for the passage of a chemically active, hydrous fluid phase; there is a clear spatial association between the intensity of fracturing and the intensity of retrograde alteration.

The style of deformation is significantly influenced by the lithology of the metamorphic protoliths. Fault-related strain was preferentially focused into micaceous lithologies (pelite, semipelite) in the lithologically heterogeneous rocks of the Grampian and Torcastle blocks (e.g. Fig. 5b; Stewart *et al.* 1999, 2000). In contrast, the homogeneous feldspathic Moine psammites and granitic gneiss in the NW Highland block show very little lithologically controlled partitioning of strain intensity and appear to have behaved in a much more brittle manner suggesting that they represent relatively strong mechanical units compared to other parts of the fault zone. In addition, as we show in the following section, syntectonic fluid-related alteration has led to a marked focusing of shearing in much of the Aberchalder Block and in a zone several hundred metres southeast of the core of the GGFZ in the Grampian block. The development of these rocks is thought to have important mechanical consequences and so they are now discussed in more detail.

Fault rocks from the GGFZ core: Grampian and Aberchalder blocks

Within the Grampian block, two localities occur close to the core of the GGFZ on the southern shoreline of Loch Linnhe (Figs 4b and 6) and along stream sections in the Dochanassie area (Figs 4b and 7). A third core zone locality lies within the Aberchalder block near Loch Oich (Figs 4b and 9).

Comparison of deformation fabrics at these sites is particularly important, because they represent deformation in different protolith types. The Grampian Group comprises psammites, quartzites and units of semi-pelite/pelite interbanded on a tens of centimetre to tens of metre scale, whilst the Aberchalder block comprises monotonous psammites.

Loch Linnhe shoreline (Grampian block): Good quality exposures of GGFZ-related fault-rocks occur along the shore of Loch Linnhe approximately 4 km SW of Fort William (NN 0732 7051 to NN 0760 7085; Fig. 6) and 750 m from the CRL. Subvertical, interbanded units (0.5–5 m thick) of psammite and semi-pelite display pervasive brittle fracturing and offsets along discrete systems of faults. At all scales, strain appears to have been localized into NE-trending, centimetre- to tens of centimetre-wide phyllonitic horizons that appear to be largely derived from

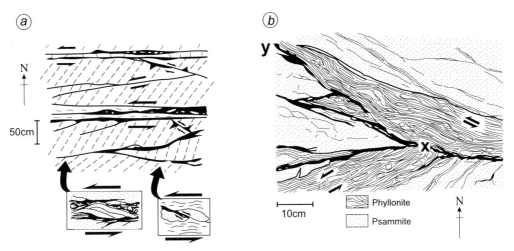

Fig. 6. (**a**) Plan of NE trending, mainly sinistral shears related to the GGFZ on the SE shoreline of Loch Linnhe (NN 0732 7051 to 0760 7085). (**b**) Detailed plan view of sinistral phyllonitic shear zone (at NN 0742 7054) offset (X to Y) by left-lateral brittle fault. After Stewart *et al.* (1999).

pre-existing layers of semi-pelite and pelite. Anastomosing networks of dark-green, mm-scale phyllonitic, mainly sinistral shear zones are developed in both psammitic and semi-pelitic/pelitic hosts (Fig. 6). These higher strain shears separate lenses of mainly less deformed psammite and comprise fine-grained chlorite and white mica. Faults associated with these phyllonites display tens of cm-scale offsets and bands of cataclastically deformed material are common at the shear zone margins and along the fault planes. A sparse network of still narrower units (μm to mm wide) of fine-grained, darker-coloured phyllosilicate-rich fault rocks are also present and appear to form the most deformed regions.

In thin section, brittle deformation textures dominate (Fig. 9a). Psammitic protoliths are pervaded by trans- and intracrystalline fractures in feldspar, quartz and mica aggregates. Quartz grains show limited flattening and incipient dynamic recrystallization suggesting minor amounts of low-temperature crystal plasticity. Calcite is commonly precipitated along fractures and feldspars are widely altered to fine-grained sericite. The phyllonitic shear-zones preserve three domains with differing finite strain intensities and textures. Volumetrically, 20–40% of the shear zones comprise low strain domains that comprise elongate discontinuous bands of psammite in which feldspars have experienced severe sericitisation where the primary grain shape is preserved (e.g. LS in Fig. 9a). The high strain, anastomosing network of dark-coloured phyllosilicate-rich shears (HS in Fig. 9a) comprise ultrafine-grained foliated aggregates of sericite and chlorite enveloping small subangular to rounded clasts of quartz and sericitised feldspar. Intermediate between these two domain types are transitional areas in which rotated blocks of severely altered psammite are pervaded by poorly developed, discontinuous networks of fine phyllosilicate-rich shears (IS on Fig. 9a). These domains are highly elongate, typically 0.5–2.0 mm wide (although occasionally several centimetres wide) and are aligned to define the main macroscopic element of the foliation in the phyllonites.

At high magnifications, the highest-strain, phyllosilicate-rich layers preserve dark-coloured/black solution seams defined by concentrations of opaques and insoluble phases (Fig. 9b), whilst quartz and chlorite 'beards' are common in the pressure shadows of larger porphyroclasts (e.g. Fig. 9c, d). The aligned solution seams and fibrous overgrowths define the main foliation, together with highly elongate domains of ultra-fine-grained, monomineralic sericite and rare flattened and aligned quartz clasts. Solution seams are particularly abundant at the boundary between the lower strain domains and through-going shears (e.g. Fig. 9b). These textures indicate the operation of fluid-assisted diffusive mass transfer (DMT) deformation mechanisms, at least within the high-strain phyllosilicate-rich domains. In some samples, calcite is precipitated along fractures or can be disseminated throughout the fine matrix in the finer-grained areas of fault rock (Fig. 9a).

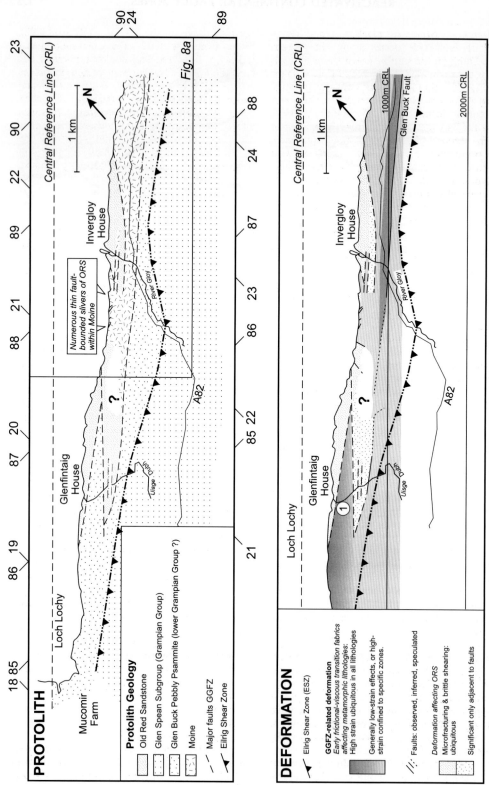

Fig. 7. Maps showing (**a**) protolith geology and (**b**) GGFZ-related deformation intensity for the Dochanassie area. Location of area shown in Figure 4b. Locality 1 is discussed in the text. The area includes the SW end of the Aberchalder block (Moine) and therefore overlaps with the region shown in Figure 8a (box).

Dochanassie (Grampian block): Good exposures of fault-rock occur along the stream Uisge-Dubh on the south-eastern shore of Loch Lochy, approximately 17 km NE of Fort William (Fig. 7), and within 1 km of the CRL. Large fault-bounded slivers of ORS conglomerates and sandstones form areas of higher ground, whilst the basement rocks are mainly exposed in stream sections. Heavily fractured quartzites and psammites predominate, together with numerous units of phyllonites. Two very different types of phyllonite are preserved: amphibolite-facies phyllonites formed in a pre-GGFZ regional ductile thrust zone that runs through this area (the Eilrig shear zone; ESZ; Phillips *et al.* 1993; Stewart 1997) and finer-grained, lower temperature phyllonites related to the GGFZ. Where GGFZ-related deformation intensities are low, the earlier regional phyllonites are coarse grained and statically recrystallized with strong relict S–C fabrics. In contrast, GGFZ-related phyllonite tends to be ultra-fine grained with a dark green/grey colour in which streaks of lighter, buff-coloured material occur. In thin section, these phyllonites comprise ultrafine-grained sericite and chlorite, whilst lighter areas are formed by highly elongate domains of sericitised feldspar and quartz. These phyllonites usually lie close to or within zones of cataclastically deformed psammite in which feldspars display varying degrees of retrograde sericitisation. Overall, the style of deformation at Dochanassie is comparable to that at Loch Linnhe in that phyllonites are preferentially developed in units of micaceous protolith.

Several units of quartzite (e.g. at Loc 1 Fig. 7) exhibit strongly flattened quartz ribbons surrounded by a fine-grained, dynamically recrystallized matrix. These are quartz mylonites and they are pervaded by numerous micaceous shears in which muscovite preserves textures indicative of dynamic recrystallization. These are clearly related to deformation associated with the GGFZ as they lack the coarse, statically recrystallized textures typically associated with the regional ESZ.

Loch Oich (Aberchalder block): Homogeneous, NE-trending, subvertical psammitic gneisses are locally overlain unconformably by ORS sandstones and conglomerates that also occur as fault-bounded slivers in the area around Loch Oich (Figs 4b and 8a). The best section of fault rocks occurs in a man-made dry gully *c.* 250 m northeast of the Allt na Criche (locality 1 Fig. 8a; NN 3090 9915), providing continuous good quality exposure along a 170 m section orthogonal to the GGFZ (Fig. 8b).

At the top of the section (NN 3103 9892), 430 m from the CRL, the psammitic gneiss is heavily fractured and cut by numerous discrete fault planes and cm-scale cataclasite bands. As the CRL is approached, the frequency and width of cataclasite zones increases (Fig. 8b). The intervening psammites becomes increasingly altered due to intense sericitisation of feldspars. This eventually leads to 'framework collapse' of the pre-existing, strong feldspar grains to form elongate aggregates of strongly aligned, fine-grained sericite that surround largely undeformed grains of quartz. In the 100 m of the section nearest to the CRL, the most altered psammites become increasingly sheared to form centimetre- to tens of centimetre-scale bands of phyllonite that superficially resemble those described at Loch Linnhe. Meanwhile, the intervening cataclasites become increasing hydrated as the CRL is approached and display spectacular evidence for the onset of bulk ductile deformation (Fig. 9e, f). Millimetre to metre-scale angular to rounded blocks of psammite supported in a dark cataclasite matrix become increasingly deformed in an apparently ductile manner. The long axes of the flattened clasts become aligned sub-parallel to the foliation in the surrounding ultrafine-grained, foliated cataclasite matrix composed of sericite, chlorite and interspersed quartz porhyroclasts (Fig. 9e). A crude discontinuous foliation is also defined by solution seams and the alignment of elongate quartz grains (Fig. 9f). The feldspar grains within the psammite clasts are severely sericitised, often with an increase in the degree of framework collapse from the centre of the clast to the margin. The collapse of the feldspar grains appears to facilitate the ductile shape change of the pre-existing polycrystalline clasts.

Elsewhere in the Aberchalder block, some quartz-rich units preserve fine aggregates of equigranular polygonal quartz grains surrounding larger, deformed quartz crystals displaying internal undulose extinction and deformation bands (localities 2 & 3 Fig. 8a). Feldspar grains are undeformed and muscovite grains display mica-fish shapes. These fabrics are similar to annealed, semi-brittle protomylonitic textures described in the Torcastle block along strike and to the SW (Stewart *et al.* 2000).

Discussion

In the metamorphic basement, the restricted range of rock types and lack of lithologies suitable for the growth of widespread metamorphic index minerals make it difficult to assess

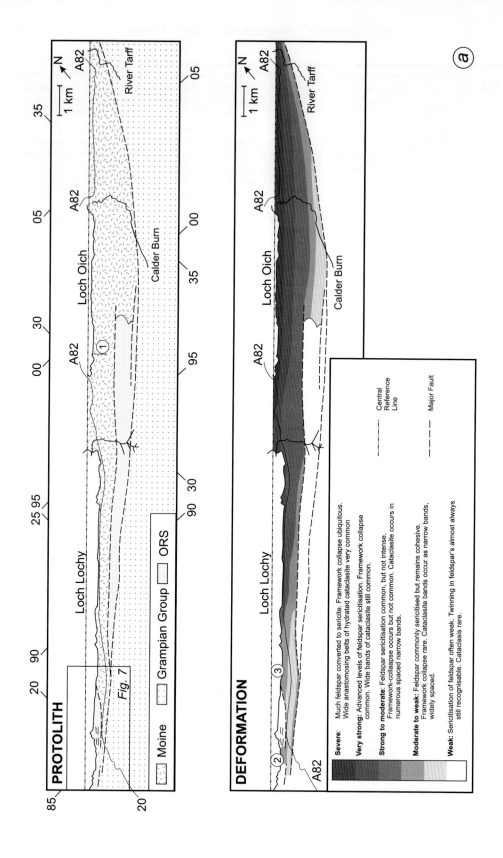

PROTOLITH

Loch Lochy Loch Oich River Tarff Calder Burn

A82 A82 A82

Fig. 7

1 km N ↑

Moine Grampian Group ORS

85 20 90 25 95 00 05 35

20 30 90 95 00 35 05

DEFORMATION

Loch Lochy Loch Oich River Tarff Calder Burn

A82 A82 A82

1 km N ↑

Central Reference Line

Major Fault

Severe: Much feldspar converted to sericite. Framework collapse ubiquitous. Wide anastomosing belts of hydrated cataclasite very common

Very strong: Advanced levels of feldspar sericitisation. Framework collapse common. Wide bands of cataclasite still common.

Strong to moderate: Feldspar sericitisation common, but not intense. Framework-collapse occurs but not common. Cataclasite occurs in numerous spaced narrow bands.

Moderate to weak: Feldspar commonly sericitised but remains cohesive. Framework collapse rare. Cataclasite bands occur as narrow bands, widely spaced.

Weak: Sericitisation of feldspar often weak. Twinning in feldspar's almost always still recognisable. Cataclasis rare.

a

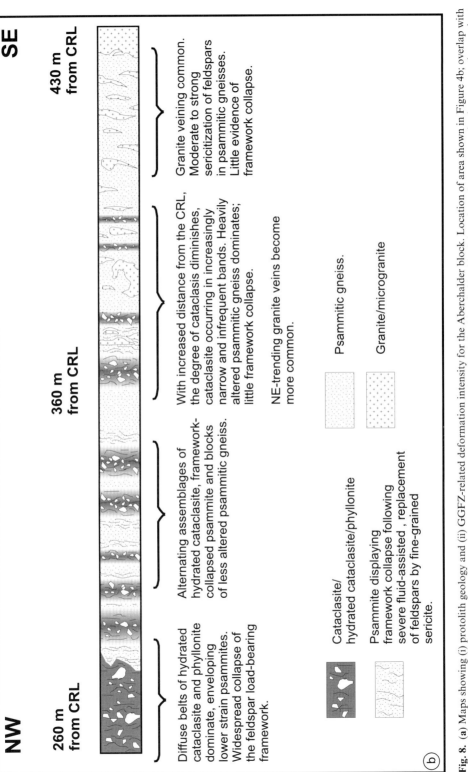

Fig. 8. (a) Maps showing (i) protolith geology and (ii) GGFZ-related deformation intensity for the Aberchalder block. Location of area shown in Figure 4b; overlap with Figure 7 also indicated (box). Localities 1–3 are discussed in text. (b) Section through part of the Aberchalder block based on the exposures in the man-made trench at NN 3090 9915 to 3124 9882 (locality 1 in Fig. 8a). After Stewart *et al.* (1999).

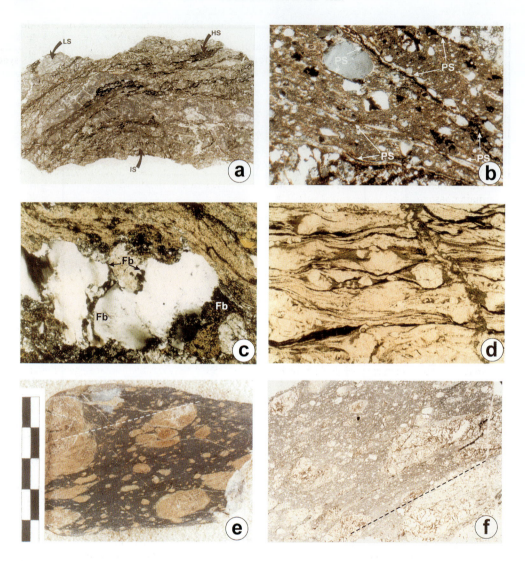

Fig. 9. Hydrated fault rocks from the core regions of the GGFZ. (**a**) Low power ppl view of typical phyllonite, Loch Linnhe (NN 0760 7085). Field of view 4.2 cm across. HS, High strain domain, IS, Intermediate strain domain; LS, Low strain domain. See text for details. (**b**) Higher power crossed polars view of high strain domain in phyllonite, Loch Linnhe (NN 0732 7051). Field of view 0.6 cm across. PS, Pressure solution seams; note that the two margins of the high strain domain are shown at the top right and bottom of the photo. (**c**) Higher power crossed polars view of fibrous qtz–chlorite overgrowths (labelled Fb) on pulled-apart quartz porphyroclasts, Loch Linnhe phyllonite (NN 0732 7051). Field of view 0.16 cm across. (**d**) Higher power ppl view of pressure solution seams (dark areas) and chlorite-white mica fibrous overgrowths on clasts within phyllonite, Loch Linnhe, NN 0732 7051. Field of view 0.4 cm. (**e**) Polished slab showing hydrated, partially collapsed cataclasite breccia – note alignment of elliptical Moine psammite clasts with foliation in matrix (dashed line). Aberchalder block, Loch Oich, NN 3102 9940. Scale bar in cm. (**f**) Low power ppl view of hydrated cataclasite breccia showing framework collapse producing aligned, deformed clasts and foliation (dashed line) in ultrafine grained matrix with fine dark pressure solution seams. Aberchalder block, Loch Oich, NN 3212 9940. Field of view 3.2 cm.

the physical conditions under which GGFZ-related deformation occurred. Away from the core region of the fault zone, the fault fabrics preserved are almost entirely the product of frictional processes involving mechanical dis-aggregation and comminution of all component minerals as the main strain accommodating mechanisms. Quartz often preserves textures indicative of low-temperature plasticity and strain hardening behaviour, suggesting that temperatures were insufficient to facilitate dislocation climb (i.e. <300°C). Assuming a normal geothermal gradient (30°C km^{-1}), this corresponds to depths of 10 km or less. The exposed sections of the GGFZ closest to the fault core in the Grampian and Aberchalder blocks display fabrics that differ from the rest of the fault zone in two ways:

(1) *Semi-brittle fabrics are locally preserved.* Quartz-rich fault rocks associated with the main phase of sinistral shearing display textures indicative of crystal plasticity in quartz, whilst feldspar deformation was cataclastic, i.e. semi-brittle deformation (Carter & Kirby 1978; White & White 1983; Schmid & Handy 1991; Handy *et al.* 1999). Similar fabrics are also found in the Torcastle Block along strike to the SW (Stewart *et al.* 2000) and suggest deformation at temperatures that are noticeably higher than the other parts of the GGFZ (e.g. 250–450°C). Such a proposal is consistent with the ubiquitous growth of white mica and chlorite synchronous with shearing under low greenschist-facies metamorphic conditions. Assuming a normal geothermal gradient (30°C km^{-1}), this corresponds to a depth range of 8–15 km, and it is suggested that shearing occurred in the region of the frictional-viscous transition and the deeper part of the frictional flow regime. Thus either the fault core and associated slivers have been uplifted and more deeply exhumed compared to the fault margins or significant amounts of frictional shear heating must have occurred along the fault zone. The widespread retrograde alteration consistent with significant influx of hydrous fluids would seem to mitigate against the second possibility.

(2) *Syn-tectonic hydration and framework collapse of fault rocks.* Fluid influx has led to considerable retrograde alteration of both pre-existing, regionally metamorphosed country rocks and later brittle cataclasites/breccias to produce an assemblage of fine-grained, foliated chlorite-white mica-carbonate fault rocks ranging from hydrated cataclasites to phyllonites (e.g. Fig. 9a–f). The cataclasites and breccias, in particular, preserve spectacular evidence for framework collapse due to the replacement of originally strong feldspar grains with fine-grained aggregates of weak phyllosilicate. The most highly sheared hydrated fault rocks are fine-grained phyllonites derived from cataclasitic protoliths and they consistently preserve widespread textural evidence for the operation of fluid-assisted diffusional creep mechanisms, including the development of fine dissolution seams and fibrous overgrowths on clasts (e.g. Figs 9a–d and f). Further detailed studies, for example of grainsize distributions and grain shapes in ultrafine-grained cataclasites, are required in order to assess whether processes such as grain boundary sliding are occurring in these rocks.

The most hydrated and altered regions consistently correspond to the most highly sheared parts of the fault zone exposed in the core regions of the GGFZ. This strongly indicates that, once the hydrated fault rocks began to form, they increasingly became the focus of fault-related shearing. Hydration occurs in two quite distinct rock types: (a) pre-existing mica-rich protoliths (semi-pelites/pelites, pre-GGFZ phyllonites), e.g. Loch Linnhe, Dochanassie; and (b) GGFZ-related cataclasites, e.g. Loch Oich. It is suggested that subsequent to hydration, shearing localizes due to: (a) reaction softening due to the growth of fine-grained phyllosilicate phases at the expense of stronger pre-existing phases; and (b) a grainsize controlled switch in deformation regime from cataclasis to one involving fluid-assisted diffusional creep in the finer-grained cataclastic fault rocks. The structural collapse of the pre-existing load-bearing framework of feldspar grains would also rapidly lead to a strong, tectonically induced preferential alignment of the fine-grained phyllosilicate grains which may still further weaken and focus strain into the hydrated fault rocks (Shea & Kronenberg 1993; Wintsch *et al.* 1995).

The pressure solution seams and fibrous mineral overgrowths recognized in the high strain core regions of the GGFZ could conceivably indicate multimechanism frictional behaviour with the simultaneous operation of cataclasis and pressure solution, as envisaged by the experimentally-based models of Chester (1995). We consider this unlikely based on the close association between diffusional microstructures and those associated with reaction softening and framework collapse, particularly within the Aberchalder block. In our view, the flattening out of clasts observed in the hydrated cataclasites (e.g. Fig. 9e, f) cannot be attributed to frictional flow processes involving solution creep. Nevertheless, we cannot eliminate the possibility that multimechanism frictional behaviour has played an important role in determining the rheological behaviour of the GGFZ at some time in its history.

In conclusion, the distribution of deformation and associated retrograde hydration associated with the GGFZ are strongly influenced by protolith composition. In the lithologically heterogeneous rocks of the Grampian block, cataclasis, hydration and subsequent viscous shearing are localized principally within pre-existing micaceous units, leading to a highly partitioned pattern of fault related deformation (Figs 5b and 6). In the compositionally homogeneous, feldspathic rocks of the Aberchalder block, however, the effects of cataclasis, hydration and subsequent viscous shearing are more diffuse (e.g. Fig. 8b). Here the intensity of hydration is determined by the proximity of rocks to individual GGFZ-related fault zones or belts of cataclasite breccias. In both scenarios, however, the focusing of shearing into the most altered regions leads to the establishment of an interlinked network of narrow high strain zones. The high degree of interconnectivity means that any effects on the internal mechanical strength will affect most of the fault zone on a regional scale.

Conclusions

Whilst the deformation behaviour of the entire lithosphere may ultimately be determined in the upper mantle, much of the strain accommodated by the continental crust appears to be preferentially focused into large-scale fault and shear zone networks that extend across a broad depth range. Many of these structures are reactivated over timescales of tens to hundreds of millions of years and appear to have accommodated displacements in preference to the formation of new structures. This is likely to occur because such tectonic discontinuities are large-scale anisotropies with an appropriate geometry (orientation relative to regional stresses, size, interconnectivity) to efficiently accommodate further displacement. It may, in addition, reflect long-term changes in the internal mechanical properties of the fault zones themselves that make them relatively weak compared to surrounding unfaulted wall-rocks.

The distribution and textural evolution of fault rocks preserved in the margins and high strain core of the GGFZ appear to be typical of the mid-crustal section of many crustal-scale reactivated structures (e.g. for another example, see Imber *et al.* 1997). The case study presented here shows that under the P–T conditions typically associated with crustal depths between 8 and 15 km, the syn-tectonic influx of hydrous

fluids can bring about fundamental changes in the internal deformation behaviour of fault zones. Cataclasis may be particularly important in the deeper part of the frictional regime as it could promote pervasive fluid influx along fault strands due to grain-scale dilatancy. This would help to bring about widespread retrograde reaction softening of the rock aggregate as strong, pre-existing phases such as feldspar are replaced by fine-grained, strongly aligned weak phyllosilicate assemblages (Wintsch *et al.* 1995). In addition, following cataclasis, a grainsize-controlled switch to diffusional creep could occur in the highest strain parts of the fault zone.

A speculative model that has general application to all crustal-scale fault zones is proposed (Fig. 10). Cataclastic deformation in the deeper part of the frictional regime and in the frictional–viscous transition zone leads to pervasive influx of hydrous fluids. If influx is rapid, leading to the development of supra-hydrostatic fluid pressures, the frictional–viscous transition will be transiently shifted to greater depths (e.g. Handy 1989; Handy *et al.* 1999). On a longer timescale, however, we propose that the affects of grain-scale fluid-rock interactions begin to dominate the rheological behaviour, i.e. the onset of reaction softening and fluid-assisted DMT following cataclasis and the ubiquitous growth of fine-grained, retrograde phyllosilicate (see also examples described by White & Knipe 1978; Mitra 1984; Brodie & Rutter 1985). Following the arguments of Schmid & Handy (1991), the reaction softening and change to diffusional creep could lead to a progressive *shallowing* of the domain of viscous flow in the most altered, finest-grained, high strain parts of the fault zone (arbitrarily shown in the centre of the vertical fault zone modelled in Fig. 10). If diffusional creep is the dominant mechanism in the high strain regions, they are likely to be relatively weak compared to the adjacent rock units that must still deform by frictional or crystal plastic mechanisms (e.g. Schmid *et al.* 1977; Schmid 1982; Handy 1989). As a result, the interconnected fault network should increasingly focus displacement, leading to an additional vertical *narrowing* of the frictional–viscous transition zone in the vicinity of the fault zone (Fig. 10). As this depth range corresponds to the main load-bearing region of the crust, these processes may also have profound mechanical consequences. The strength curve shown for the core of the fault zone in Figure 10 is loosely based on experimentally-derived data for the deformation of phyllosilicate-rich rocks and single mica grains (e.g. Janecke & Evans 1988; Mares & Kronenberg 1993; Wintsch *et al.* 1995). If it is

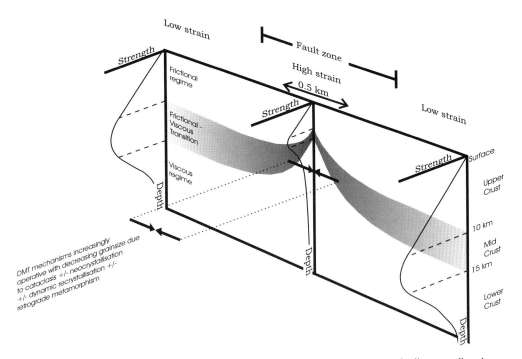

Fig. 10. Schematic 3-D strength profiles through a vertical, crustal-scale fault zone and adjacent wall rocks. As strain localizes increasingly into the high-strain fault strands, the grain size-controlled switch from frictional cataclasis to diffusional viscous creep leads to a shallowing and narrowing of the frictional–viscous transition within the fault zone. Schematic stength v. depth curves are shown for the wall rocks and centre of the fault zone to illustrate the proposed long-term weakening affect – see text for details. In this hypothetical example, the weakest region is shown in the centre of the fault zone – in real examples, there may be several of these regions corresponding to major fault strands, including those at the fault zone margin(s).

appropriate to apply such curves to the fault zone as a whole, then the fault zone processes will ultimately lead to the establishment of a large-scale, long-term zone of weakness.

This case study has emphasized the roles of cataclasis and retrogression in the presence of a fluid phase, but, in the region of the frictional–viscous transition, there are several other deformational and metamorphic processes that may lead to marked grainsize reduction and the onset of diffusional creep (Fig. 10). These include dynamic recrystallization of quartz (e.g. White *et al.* 1980) and fluid-assisted neocrystallization of feldspar (Fitz Gerald & Stünitz 1993; Stünitz & Fitz Gerald 1993). Variables such as temperature, protolith composition, the presence and composition of a hydrous fluid phase and the degree of fracture and shear zone interconnectivity on several scales are likely to determine which of these processes are dominant in a given fault zone. In many natural fault zones, more than one grainsize-reduction mechanism may operate simultaneously at different depths and/ or at the same depth.

The speculative model presented here predicts changes in internal fault zone strength that may, in part, account for the long-term reactivation of faults. In the future we need to address a more difficult problem: why it is that some crustal faults fail to reactivate? Some examples never reactivate, whilst others, like the GGFZ, fail to move during regionally important episodes such as the Mesozoic opening of the nearby Moray Firth Basin (Fig. 4a; Underhill & Brodie 1993) despite having shown a previous history of repeated movement. Still other faults reactivate only after very long periods of quiescence. To some extent, such behaviour may be determined by geometric factors such as incorrect orientation relative to the regional stress field or by changes in the rheology of the underlying mantle lithosphere. It is also pertinent to ask what processes might act upon a fault zone to bring about strengthening of the fault relative to its surrounding wall rocks. Obvious strengthening processes could include pervasive post-kinematic grain-growth ('annealing') and mineralization associated with the emplacement of fault-controlled plutons or hydrothermal vein systems. There is little evidence for post-kinematic annealing or mineralization of fault rocks in the GGFZ fault core sections examined during the present study.

In conclusion, the present investigation illustrates that fault rocks in reactivated basement fault zones can tell us a great deal about the distribution, nature and possible interaction of contemporaneous and different deformation mechanisms across a broad range of spatial and temporal scales. Such qualitative observations give a unique and direct insight into long-term weakening mechanisms that operate in natural fault zones at depth in the load-bearing regions of the crust. They illustrate the fundamental rheological importance of fluid-rock interactions in determining the long-term strain response of the crust during large-scale deformation of the continental lithosphere.

REH would like to thank Statoil UK for financial support of the Reactivation Research Group. MS and JI acknowledge receipt of studentships from Oxford Brookes University and NERC respectively. Lorraine Beacom, Randy Parrish, Mike Searle, Tony Dore and Erik Lundin are thanked for discussions of fault zone examples and concepts. Jurgen Streit, Bob Hatcher and Jodie Miller provided detailed and helpful reviews that have led to a much improved paper.

References

AYDIN, A. 1978. Small faults formed as deformation bands in sandstone. *Pure and Applied Geophysics*, **116**, 913–930.

BLANPIED, M. L., LOCKNER, D. A. & BYERLEE, J. D. 1995. Frictional slip of granite at hydrothermal conditions. *Journal of Geophysical Research*, **100**, 13,045–13,064.

BRODIE, K. H. & RUTTER, E. H. 1985. On the relationship between deformation and metamorphism, with special reference to the behaviour of basic rocks. *In*: THOMPSON, A. B. & RUBIE, D. (eds) *Kinetics, textures and deformation. Advances in Physical Geochemistry*. Springer, New York, **4**, 138–179.

BUTLER, C. A., HOLDSWORTH, R. E. & STRACHAN, R. A. 1995. Evidence for Caledonian sinistral strike-slip motion and associated fault zone weakening, Outer Hebrides Fault Zone, NW Scotland. *Journal of the Geological Society, London*, **152**, 743–746.

BUTLER, R. W. H., HOLDSWORTH, R. E. & LLOYD, G. E. 1997. The role of basement reactivation in continental deformation. *Journal of the Geological Society, London*, **154**, 69–71.

BYERLEE, J. D. 1978. Friction of rocks. *Pure and Applied Geophysics*, **116**, 615–626.

BYERLEE, J. D. 1990. Friction, overpressure and fault normal compression. *Geophysics Research Letters*, **17**, 2109–2112.

CARTER, N. L. & KIRBY, S. H. 1978. Transient creep and semibrittle behaviour of crystalline rocks. *Pure and Applied Geophysics*, **116**, 807–839.

CHESTER, F. M. 1995. A rheologic model for wet crust applied to strike-slip faults. *Journal of Geophysical Research*, **100**, 13 033–13 044.

CHESTER, F. M., EVANS, J. P. & BIEGEL, R. L. 1993. Internal structure and weakening mechanisms of the San Andreas Fault. *Journal of Geophysical Research*, **98**, 771–786.

COBBOLD, P. R. 1977. Description and origin of banded deformation structures. II Rheology and growth of banded perturbations. *Canadian Journal of Earth Sciences*, **14**, 2510–2523.

COWIE, P. A., KNIPE, R. J. & MAIN, I. G. 1996. Introduction to the Special Issue. *Journal of Structural Geology*, **18**, v–xi.

D'LEMOS, R. D., SCHOFIELD, D. I., HOLDSWORTH, R. E. & KING, T. R. 1997. Deep crustal and local rheological controls on the siting and reactivation of fault and shear zones, northeastern Newfoundland. *Journal of the Geological Society*, **154**, 117–122.

EDMOND, J. M. & PATERSON, M. S. 1972. Volume changes during the deformation of rocks at high pressures. *International Journal of Rock Mechanics and Mining Sciences*, **9**, 161–182.

FITZ GERALD, J. D. & STÜNITZ, H. 1993. Deformation of granitoids at low metamorphic grade I: Reactions and grain size reduction. *Tectonophysics*, **221**, 269–297.

FLIERVOET, T. F., WHITE, S. H. & DRURY, M. R. 1997. Evidence for dominant grain-boundary sliding deformation in greenschist- and amphibolite-grade polymineralic ultramylonites from the Redbank Deformed Zone, Central Australia. *Journal of Structural Geology*, **19**, 1495–1520.

FLINN, D. 1961. Continuation of the Great Glen Fault beyond the Moray Firth. *Nature*, **191**, 589–591.

GRIGGS, D. T. & HANDIN, J. 1960. Observations on fracture and a hypothesis of earthquakes. *In*: GRIGGS, D. & HANDIN, J. (eds) *Rock Deformation*. Geological Society of America Memoir, **79**, 347–364.

GROCOTT, J. 1977. The relationship between Precambrian shear belts and modern fault systems. *Journal of the Geological Society of London*, **133**, 257–261.

GRØNLIE, A. & ROBERTS, D. 1989. Resurgent strike-slip duplex development and the Hitra-Snåsa and Verran Faults, Møre–Trøndelag Fault Zone, Central Norway. *Journal of Structural Geology*, **11**, 295–305.

HALL, J., BREWER, J. A., MATTHEWS, D. H. & WARNER, M. 1984. Crustal structure across the Caledonides from the WINCH seismic profile: Influences on the evolution of the Midland Valley of Scotland. *Transactions of the Royal Society, Edinburgh, Earth Sciences*, **75**, 97–109.

HANDY, M. R. 1989. Deformation regimes and the rheological evolution of fault zones in the lithosphere: the effects of pressure, temperature, grain-size and time. *Tectonophysics*, **163**, 119–152.

HANDY, M. R. 1990a. The solid-state flow of poly-mineralic rocks. *Journal of Geophysical Research*, **95**, 8647–8661.

HANDY, M. R. 1990b. The exhumation of cross-sections of the continental crust: structure, kinematics and rheology. *In*: SALISBURY, M. H. & FOUNTAIN, D. M. (eds) *Exposed Cross-Sections of the Continental Crust*. Kluwer Academic Publishers, Netherlands, 485–507.

HANDY, M. R. 1994. The energetics of steady state heterogeneous shear in mylonitic rock. *Material Science and Engineering*, **A175**, 261–272.

HANDY, M. R., WISSING, S. B. & STREIT, J. E. 1999. Frictional-viscous flow in mylonite with varied bimineralic composition and its effect on lithospheric strength. *Tectonophysics*, **303**, 175–191.

HANMER, S. 1997. Shear zone reactivation at granulite facies: the importance of plutons in the localization of viscous flow. *Journal of the Geological Society, London*, **154**, 111–116.

HARRIS, A. L., HASELOCK, P. J., KENNEDY, M. J. & MENDUM, J. R. 1994. The Dalradian Supergroup in Scotland, Shetland and Ireland. *In*: GIBBONS, W. & HARRIS, A. L. (eds) *A revised correlation of Precambrian rocks in the British Isles*. Geological Society, London, Special Report, **22**, 33–53.

HIGHTON, A. J., HYSLOP, E. K. & NOBLE, S. R. 1999. U–Pb zircon geochronology of migmatization in the northern Central Highlands: evidence for pre-Caledonian (Neoproterozoic) tectonometamorphism in the Grampian Block, Scotland. *Journal of the Geological Society, London*, **156**, 1195–1204.

HOBBS, B. E., MÜHLHAUS, H.-B. & ORD, A. 1990. Instability, softening and localization of deformation. *In*: KNIPE, R. J. & RUTTER, E. H. (eds) *Deformation Mechanisms, Rheology and Tectonics*. Geological Society, London, Special Publications, **54**, 143–165.

HOLDSWORTH, R. E. 1994. Structural evolution of the Gander–Avalon terrane boundary: a reactivated transpression zone in the NE Newfoundland Appalachians. *Journal of the Geological Society, London*, **151**, 629–646.

HOLDSWORTH, R. E., STRACHAN, R. A. & HARRIS, A. L. 1994. Precambrian rocks in northern Scotland: the Moine Supergroup. *In*: GIBBONS, W. & HARRIS, A. L. (eds), *A revised correlation of Precambrian rocks in the British Isles*. Geological Society, London, Special Report, **22**, 23–32.

HOLDSWORTH, R. E., BUTLER, C. A. & ROBERTS, A. M. 1997. The recognition of reactivation during continental deformation. *Journal of the Geological Society, London*, **154**, 73–78.

HOLLISTER, L. S. & CRAWFORD, M. L. 1986. Melt-enhanced deformation: a major tectonic process. *Geology*, **14**, 558–561.

HUTTON, D. H. W. 1988. Granite emplacement mechanisms and tectonic controls: inferences from deformation studies. *Transactions of the Royal Society of Edinburgh: Earth Sciences*, **79**, 245–255.

IMBER, J., HOLDSWORTH, R. E., BUTLER, C. A. & LLOYD, G. E. 1997. Fault zone weakening pro-cesses along the reactivated Outer Hebrides Fault Zone, Scotland. *Journal of the Geological Society, London*, **154**, 105–109.

JANECKE, S. U. & EVANS, J. P. 1988. Feldspar-influenced rock rheologies. *Geology*, **16**, 1064–1067.

KARATO, S.-I., PATERSON, M. S. & FITZGERALD, J. D. 1986. Rheology of synthetic olivine aggregates: influence of grain size and water. *Journal of Geophysical Research*, **91**, 8151–8176.

KENNEDY, W. Q. 1946. The Great Glen Fault. *Quarterly Journal of the Geological Society of London*, **102**, 41–76.

KERRICH, R. 1986. Fluid transport in lineaments. *Philosophical Transactions of the Royal Society, London*, **A317**, 219–251.

KOHLSTEDT, D. L., EVANS, B. & MACKWELL, S. J. 1995. Strength of the lithosphere: Constraints imposed by laboratory experiments. *Journal of Geophysical Research*, **100**, 17 587–17 602.

KNIPE, R. J. 1989. Deformation mechanisms – recognition from natural tectonites. *Journal of Structural Geology*, **11**, 127–146.

MCCAFFREY, K. J. W. 1997. Controls on the reactivation of a major fault zone: the Fair Head–Clew Bay line in Ireland. *Journal of the Geological Society, London*, **154**, 129–133.

MCCAIG, A. M. 1997. The geochemistry of volatile fluid flow in shear zones. *In*: HOLNESS, M. B. (ed.) *Deformation-enhanced Fluid Transport in the Earth's Crust and Mantle*. Chapman & Hall, London, 227–266.

MARES, V. M. & KRONENBERG, A. K. 1993. Experimental deformation of muscovite. *Journal of Structural Geology*, **15**, 1061–1075.

MAY, F. & HIGHTON, A. J. 1997. *Geology of the Invermoriston District*. Memoir of the British Geological Survey Sheet **73**(W), Scotland.

MITRA, G. 1984. Brittle to ductile transition due to large strains along the White Rock Thrust, Wind River mountains, Wyoming. *Journal of Structural Geology*, **6**, 51–61.

MOLNAR, P. 1988. Continental tectonics in the aftermath of plate tectonics. *Nature*, **335**, 131–137.

MOLNAR, P. 1992. Brace–Goetze strength profiles, the partitioning of strike-slip and thrust faulting at zones of oblique convergence and the stress-heat flow paradox of the San Andreas Fault. *In*: EVANS, B. & WONG, T.-F. (eds) *Fault Mechanics and Transport Properties in Rocks*. Academic Press, London, 359–435.

MORROW, C., RADNEY, B & BYERLEE, J. 1992. Frictional strength and the effective pressure law of Montmorillonite and Illite clays. *In*: EVANS, B. & WONG, T.-F. (eds) *Fault Mechanics and Transport Properties in Rocks*. Academic Press, London, 69–88.

MUSSON, R. M. W. 1994. *A catalogue of British earthquakes AD 684–1993*. Technical Report of the British Geological Survey, Edinburgh, **WL-94 04**.

OBEE, H. K. & WHITE, S. H. 1985. Faults and associated fault rocks of the Southern Arunta block, Alice Springs, Central Australia. *Journal of Structural Geology*, **7**, 701–712.

OBEE, H. K. & WHITE, S. H. 1986. Microstructural and fabric heterogeneities in fault rocks associated with a fundamental fault. *Philosophical Transactions of the Royal Society, London*, A317, 99–109.

PATERSON, M. S. 1978. *Experimental Rock Deformation – The Brittle Field.* Springer, Berlin.

PHILLIPS, E. R., CLARK, G. C. & SMITH, D. I. 1993. Mineralogy, petrology and microfabric analysis of the Eilrig Shear Zone, Fort Augustus, Scotland. *Scottish Journal of Geology*, 29, 143–158.

RAMSAY, J. G. 1980. Shear zone geometry: A review. *Journal of Structural Geology*, 2, 83–99.

RICE, J. R. 1992. Fault stress states, pore pressure distributions and the weakness of the San Andreas Fault. *In*: EVANS, B. & WONG, T.-F. (eds) *Fault Mechanics and Transport Properties in Rocks*. Academic Press, London, 475–503.

ROBERTS, A. M. & HOLDSWORTH, R. E. 1999. Linking onshore and offshore structures: Mesozoic extension in the Scottish Highlands. *Journal of the Geological Society, London*, 156, 1061–1064.

ROGERS, D. A., MARSHALL, J. E. A. & ASTIN, T. R. 1989. Devonian and later movements on the Great Glen fault system, Scotland. *Journal of the Geological Society, London*, 146, 369–372.

SAINT BLANQUAT, M., TIKOFF, B., TEYSSIER, C. & VIGNERESSE, J. L. 1998. Transpressional kinematics and magmatic arcs. *In*: HOLDSWORTH, R. E., STRACHAN, R. A. & DEWEY, J. F. (eds) *Continental Transpressional and Transtensional Tectonics.* Geological Society, London, Special Publications, 135, 327–340.

SANDIFORD, M. 1999. Mechanics of basin inversion. *Tectonophysics*, 305, 109–120.

SCHMID, S. M. 1982. Microfabric studies as indicators of deformation mechanisms and flow laws operative in mountain building. *In*: HSÜ, K. J. (ed.) *Mountain Building Processes*. Academic Press, London, 95–110.

SCHMID, S. M. & HANDY, M. R. 1991. Towards a genetic classification of fault rocks: Geological usage and tectonophysical implications. *In*: MÜLLER, D. W., MCKENZIE , J. A. & WEISSERT, H. (eds) *Controversies in Modern Geology*. Academic Press, London, 339–361.

SCHMID, S. M., BOLAND, J. N. & PATERSON, M. S. 1977. Superplastic flow in fine-grained limestone. *Tectonophysics*, 43, 257–291.

SHEA, W. T. JR & KRONENBERG, A. K. 1993. Strength and anisotropy of foliated rocks with varied mica contents. *Journal of Structural Geology*, 15, 1097–1122.

SIBSON, R. H. 1977. Fault rocks and fault mechanisms. *Journal of the Geological Society, London*, 133, 191–213.

SIBSON, R. H. 1983. Continental fault structure and the shallow earthquake source. *Journal of the Geological Society, London*, 140, 741–767.

SILVER, P. G. 1996. Seismic anisotropy beneath the continents: probing the depths of geology. *Annual Reviews of Earth and Planetary Science*, 24, 385–432.

SNOKE, A. W., TULLIS, J. & TODD, V. R. 1998. *Fault-related rocks: a photographic atlas*. Princeton University Press, New Jersey.

SNYDER, D. B. & FLACK, C. A. 1990. A Caledonian age for reflectors within the mantle lithosphere north and west of Scotland. *Tectonics*, 9, 903–922.

SONDER, L. & ENGLAND, P. 1986. Vertical averages of rheology of the continental lithosphere: relation to thin sheet parameters. *Earth and Planetary Science Letters*, 77, 81–90.

SOPER, N. J. & HUTTON, D. H. W. 1984. Late Caledonian Sinistral displacements in Britain: Implications for a three-plate collision model. *Tectonics*, 3, 781–794.

SORNETTE, A., DAVY, P. & SORNETTE, D. 1993. Fault growth in brittle-ductile experiments and the mechanics of continental collisions. *Journal of Geophysical Research*, 98, 12 111–12 139.

SORNETTE, D., DAVY, P. & SORNETTE, A. 1990. Structuration of the lithosphere in plate tectonics as a self-organised critical phenomenon. *Journal of Geophysical Research*, 95, 17 353–17 361.

STEWART, M. 1997. *Kinematic evolution of the Great Glen Fault Zone, Scotland*. Unpublished PhD Thesis, Oxford Brookes University.

STEWART, M., STRACHAN, R. A. & HOLDSWORTH, R. E. 1997. Direct field evidence for sinistral displacements along the Great Glen Fault Zone: late Caledonian reactivation of a regional basement structure? *Journal of the Geological Society, London*, 154, 135–139.

STEWART, M., STRACHAN, R. A. & HOLDSWORTH, R. E. 1999. Structure and early kinematic history of the Great Glen Fault Zone, Scotland. *Tectonics*, 18, 326–342.

STEWART, M., HOLDSWORTH, R. E. & STRACHAN, R. A. 2000. Deformation processes and weakening mechanisms within the frictional-viscous transition zone of major crustal-scale faults: insights from the Great Glen Fault Zone, Scotland. *Journal of Structural Geology*, 22, 543–560.

STOKER, M. S. 1982. Old Red Sandstone sedimentation and deformation in the Great Glen Fault Zone NW of Loch Linnhe. *Scottish Journal of Geology*, 18, 147–156.

STRONG, D. F. & HANMER, S. K. 1981. The leucogranites of southern Brittany: Origin by faulting, frictional heating, fluid flux and fractional melting. *Canadian Mineralogist*, 19, 163–176.

STÜNITZ, H. & FITZ GERALD, J. D. 1993. Deformation of granitoids at low metamorphic grade II: Granular flow in albite-rich mylonites. *Tectonophysics*, 221, 299–324.

SUTTON, J. & WATSON, J. V. 1986. Architecture of the continental lithosphere. *Philosophical Transactions of the Royal Society, London*, A317, 5–12.

TEYSSIER, C. & TIKOFF, B. 1998. Strike-slip partitioned transpression of the San Andreas Fault system: a lithospheric scale approach. *In*: HOLDSWORTH, R. E., STRACHAN, R. A. & DEWEY, J. F. (eds) *Continental Transpressional and Transtensional Tectonics*. Geological Society, London, Special Publications, 135, 143–158.

THATCHER, W. 1995. Microplate versus continuum descriptions of active tectonic deformation. *Journal of Geophysical Research*, **100**, 3885–3894.

TOMMASI, A., VAUCHEZ, A., FERNANDES, L. A. D. & PORCHER, C. C. 1994. Magma-assisted strain localisation in an orogen-parallel transcurrent zone of southern Brazil. *Tectonics*, **13**, 421–437.

TULLIS, J. & YUND, R. A. 1977. Experimental deformation of dry Westerly granite. *Journal of Geophysical Research*, **82**, 5705–5718.

UNDERHILL, J. R. & BRODIE, J. A. 1993. Structural geology of Easter Ross, Scotland: Implications for movement on the Great Glen Fault. *Journal of the Geological Society, London*, **150**, 515–527.

WANG, C-Y. 1984. On the constitution of the San Anreas Fault zone in central California. *Journal of Geophysical Research*, **89**, 5858–5866.

WATTERSON, J. 1975. Mechanism for the persistence of tectonic lineaments. *Nature*, **253**, 520–522

WHITE, J. C. & WHITE, S. H. 1983. Semi-brittle deformation within the Alpine fault zone, New Zealand. *Journal of Structural Geology*, **5**, 579–589.

WHITE, S. H. & KNIPE, R. J. 1978. Transformation- and reaction-enhanced ductility in rocks. *Journal of the Geological Society, London*, **135**, 513–516.

WHITE, S. H., BRETAN, P. G. & RUTTER, E. H. 1986. Fault-zone reactivation: kinematics and mechanisms. *Philosophical Transactions of the Royal Society, London*, **A317**, 81–97.

WHITE, S. H., BURROWS, S. E., CARRERAS, J., SHAW, N. D. & HUMPHREYS, F. J. 1980. On mylonites in ductile shear zones. *Journal of Structural Geology*, **2**, 175–187.

WINTSCH, R. P., CHRISTOFFERSON, R. & KRONENBERG, A. K. 1995. Fluid- rockreaction weakening of fault zones. *Journal of Geophysical Research*, **100**, 13021–13032.

ZOBACK, M. D. & LACHENBRUCH, A. H. 1992. Introduction to the special section on the Cajon Pass scientific drilling project. *Journal of Geophysical Research*, **97**, 4991–4994.

Geodynamics of central Australia during the intraplate Alice Springs Orogeny: thin viscous sheet models

EMILY A. ROBERTS[1] & GREGORY A. HOUSEMAN[2,3]

[1] née Neil, Australian Geodynamics Cooperative Research Centre, VIEPS Department of
Earth Sciences, Monash University, Melbourne, Victoria 3168, Australia
[2] VIEPS Department of Earth Sciences, Monash University, Melbourne,
Victoria 3168, Australia
[3] Present address: School of Earth Sciences, University of Leeds, Leeds LS2 9JT, UK

Abstract: This study investigates possible mechanisms that can account for the intraplate deformation in central Australia and the Canning Basin during the Devo-Carboniferous Alice Springs Orogeny. The intraplate orogeny in central Australia seems to have occurred without the association of a significant collisional orogenic event at the plate boundary. In contrast, the present-day Tian Shan may be viewed as a consequence of the plate boundary collision which has produced the Himalayas and the Tibetan Plateau. The experiments presented in this paper examine a mechanism that produces intraplate thickening and thinning of the crust but leaves the boundary relatively undeformed. A thin viscous sheet approximation of continental lithosphere is used to demonstrate that a clockwise rotational northern boundary acting on a lithospheric sheet with an internal weak zone may produce crustal thickening in the region representing central Australia, and thinning in the region representing the Canning Basin. In this model a clockwise rotation of the northern boundary may be induced either by an eastward shear traction or by a clockwise bending moment. The relation between rotation of the boundary and maximum crustal thickening factor is, to first order, independent of the way in which deformation is driven. It depends primarily on the relative dimensions of the intracratonic weak zone. For plausible estimates of thin viscous sheet geometry and variation of lithospheric strength within the sheet, it is inferred that clockwise rotation of the northern Australian block of order $20-25°$ is required to produce a maximum crustal thickening factor in central Australia of order 1.67. These calculations indicate that the depth-averaged strength of the lithosphere in central Australia prior to the Alice Springs Orogeny was of order $B_0 = 0.8 \times 10^{13}$ Pa s$^{1/3}$, assuming plausible estimates of plate boundary force of 5×10^{12} Nm^{-1} and orogenic time span of 100 Ma. Based on a simplified approximate model for lithospheric strength this strength coefficient corresponds to a Moho temperature in central Australia of order 520°C. The concentration of deformation in this relatively narrow zone that stretches E–W across the continent, implies that the blocks to the N and S are much stronger, a difference which might be explained by a decrease in Moho temperature of order 60°C. If Moho temperatures prior to the Alice Springs orogeny were higher than those estimated above, the required deformation may have been compressed into a shorter period than 100 Ma, or may have been episodic rather than continuous.

Central Australia consists of a number of intracratonic basins separated by areas of exposed basement crust (Figs 1 and 2). The basins, namely the Officer, Amadeus and Ngalia basins, have depositional histories that date back to the Late Proterozoic (Wells & Moss 1983). The basement regions that separate these basins, the Arunta Block to the north and the Musgrave Block in the south, consist mainly of metamorphosed and deformed igneous and sedimentary rocks of Lower and Middle Proterozoic ages (Stewart et al. 1984).

The region of Central Australia has been a focus of tectonic activity since the Early Proterozoic. A combination of Rb–Sr, ^{39}Ar/^{40}Ar and U–Pb zircon ages from the Arunta Block in central Australia reveal a series of tectonic events that occurred in the region (e.g. Shaw & Black 1991; Black & Shaw 1995; Collins & Williams 1995; Dunlap & Teyssier 1995). These events have been documented by Collins & Shaw (1995) as the Strangways Orogeny 1780–1730 Ma, the Argilke thermal event 1680–1660 Ma, the Chewings Orogeny c. 1600 Ma, the Anmatjira uplift event 1500–1400 Ma, the Teapot event c. 1200–1150 Ma and the Alice Springs Orogeny 400–300 Ma. Further to the south, in the Musgrave Block the Petermann Ranges

From: MILLER, J. A., HOLDSWORTH, R. E., BUICK, I. S. & HAND, M. (eds) *Continental Reactivation and Reworking*. Geological Society, London, Special Publications, **184**, 139–164. 1-86239-080-0/01/$15.00 © The Geological Society of London 2001.

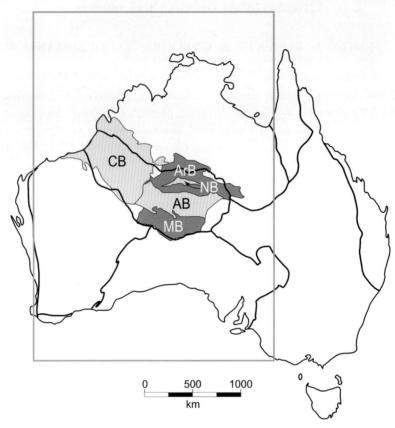

Fig. 1. Sketch of Australia showing the crustal mega-elements defined by Shaw *et al.* (1996) using geophysical terrain boundaries. The Central mega-element incorporates the Canning Basin (CB), the Amadeus Basin (AB), the Ngalia Basin (NB), the Arunta Block (ArB) and the Musgrave Block (MB), and corresponds roughly to the postulated weak zone defined for the thin viscous sheet models presented later in this paper. Corresponding thin viscous sheet dimensions are shown by grey rectangular boundary.

Orogeny 600–500 Ma was also recorded. The Alice Springs Orogeny was the final episode of tectonic activity to affect central Australia (Shaw *et al.* 1984; Stewart *et al.* 1984), and was responsible for defining the present day shape of the Arunta Block and Amadeus Basin. The Amadeus Basin saw no further sedimentation after the Alice Springs Orogeny. The youngest sediments in the basin are Late Devonian age and are believed to be the result of weathering and erosion of the topography associated with the Alice Springs Orogeny. There is a general lack of Carboniferous sediments over most of continental Australia. Powell (1984) interprets this observation as a result of Australia's movement to southern polar latitudes during the Carboniferous, implying that Australia may, therefore, have been under a continental ice sheet similar to that present in Antarctica today.

The Alice Springs Orogeny has been described as an intraplate orogenic event by several authors (e.g. Forman & Shaw 1973; Duff & Langworthy 1974; Shaw *et al.* 1984; Teyssier 1985) for the following reasons;

(1) The sediments within the Amadeus, Ngalia, Georgina and Officer basins of central and northern Australia are characterized by interconnected, correlatable sequences and the depocenters of these sequences are at least several hundred kilometres from any present-day, Palaeozoic or late Proterozoic continental margin (Wells & Moss 1983; Shaw *et al.* 1984). If there had been much separation between the regions of northern and southern Australia before the Alice Springs Orogeny it would be unlikely that

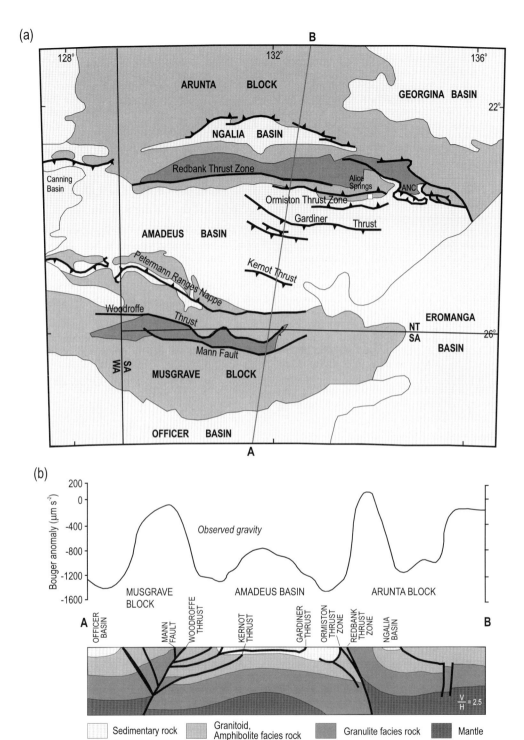

Fig. 2. (**a**) Central Australia, regional geology (ANC, Arltunga Nappe Complex). (**b**) Schematic cross-section of the regional geology showing displacement of Moho by Redbank Thrust Zone (RTZ) and Woodroffe Thrust. Gravity profile over the geological cross-section shows the large gravity anomalies associated with these structures (Figure simplified from Shaw *et al.* 1991, fig. 10).

· the pre-Devonian sediments in these basins could be correlated.

(2) Middle Cambrian palaeomagnetic poles from central and southern Australia (Klootwijk 1980) plot very close together, suggesting that these regions were adjacent in Mid Cambrian time. Palaeopoles determined for the Canning Basin and Central Australia during the Late Devonian are also very similar (Hurley & Van der Voo 1987; Chen et al. 1993). While the interpretation of no net relative movement is not conclusive because of the data uncertainty, it suggests that these regions have been closely associated at least since the early Palaeozoic, participating together in the drift of Gondwanaland.

(3) No relict ophiolite or oceanic crust has been discovered in the region, whereas in modern regions of intercratonic continental collision, for example in the Himalayas (Jan & Windley 1990; Arif & Jan 1993), and in the Alps (Cartwright & Barnicoat 1999) ophiolite has been exhumed and preserved. Other features indicative of plate margin tectonism such as sequences of acid volcanics and intrusives, and high-pressure low-temperature metamorphism are also absent in central Australia.

These lines of evidence suggest that the Alice Springs Orogeny was indeed an intraplate orogenic event, rather than a plate boundary collision between a northern and a southern Australian plate.

Effect of the Alice Springs Orogeny in central Australia

Although tectonism of Alice Springs Orogeny (ASO) age has been detected in many parts of Australia, including Mt. Isa (Spikings et al. 1996) and in the Flinders Ranges (Mitchell et al. 1997), the orogeny is primarily recognized in the Arunta Block and Amadeus Basin of central Australia. In central Australia, the Alice Springs Orogeny produced a regime of N–S shortening that extended from the Arunta Block southward to the Amadeus Basin, a distance of 300–400 km (Fig. 2a). Within this region most of the deformation associated with the Alice Springs Orogeny was concentrated in the Arunta Block along the Redbank Thrust Zone (Goleby et al. 1989). The movement on the Redbank was associated with uplift of the Arunta Block (Lambeck et al. 1988; Shaw et al. 1991) and reactivated older E–W trending mobile belts of the mid-Early Proterozoic (Plumb 1979; Shaw et al. 1984; Shaw et al. 1991).

From detailed structural mapping, Teyssier (1985) estimated that the shortening associated with the orogeny must be between 50 and 100 km minimum. Reconstructions based on thermal history and geochronology combined with structural mapping caused Dunlap et al. (1995) to suggest that c. 85 km of shortening had occurred in the Arltunga Nappe Complex (ANC, Fig. 2) during the ASO.

Late Devonian – Early Carboniferous tectonics of the Canning Basin

During the earlier part of the Alice Springs Orogeny, when central Australia was undergoing shortening and compression, the region of the Canning Basin in north-western Australia (Fig. 3) was experiencing (NE–SW) extensional movement which occurred from Late Devonian to Early Carboniferous (Shaw et al. 1994). A clear change in the trend of magnetic and gravity lineations marks the transition from the extensional domain of the north-west to the convergent domain of central Australia. Braun et al. (1991) suggested that the transition could be

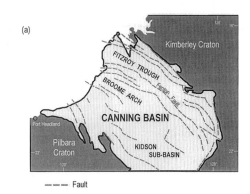

(a)

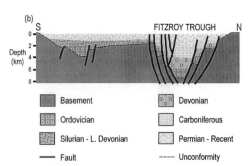

(b)

Fig. 3. (a) Regional geology of the Canning Basin. **(b)** Generalized cross-section of the Canning Basin from S–N, showing the increased thickness of the sediment pile in the Fitzroy Trough (Figure modified from Purcell & Poll 1984).

caused by a major sinistral shear zone, the Lasseter shear zone, though there is no clear evidence of major displacement.

Several periods of extension are recorded in the Canning Basin between 400 and 300 Ma, the time window of the ASO. Work by Kennard et al. (1994) separates the extension into a 'pre-rift' phase in the Middle Devonian that may be related to extension in the lower crust and mantle, and a rift phase in the Late Devonian known as the Pillara Extension (initiated at 375 Ma). The sag phase of the Pillara Extension was interrupted by at least two periods of renewed extension and fault-block tilting, the Van Emmerick extension in the latest Frasnian (c. 360 Ma) and the Red Bluffs Extension at the beginning of the Tournaisian (c. 354 Ma) (Kennard et al. 1994; Shaw et al. 1994). During these periods of extension in the Canning Basin the sediments of the Devonian to Early Carboniferous Megasequence were deposited. (Shaw et al. 1984, 1994; Kennard et al. 1994). The thickness of this Megasequence varies from 1 km in the middle of the Canning Basin (in the Kidson Sub-basin, Fig. 3) to >6 km in the Fitzroy Trough (Drummond et al. 1991).

The Fitzroy Trough occurs on the northern margin of the Canning Basin (Fig. 3) and consists of a series of NW–SE trending half-grabens that formed in the Middle to Late Devonian. Drummond et al. (1991) have estimated the extension in the Fitzroy Trough to be c. 20–25 km in the Late Devonian, 10 km in the Carboniferous and 5 km in the Permian. The present day thickness of the basement crust beneath the sediments of the Fitzroy Trough is c. 21 km, with more than 85% of this crustal thinning occurring during the Late Devonian–Early Carboniferous (Drummond et al. 1991). While 30–35 km of extension occurred across the Fitzroy Trough during the Devo-Carboniferous, it is likely that extension also occurred in other regions of the Basin and that the total amount of extension within the Canning Basin during this time was somewhat greater than 35 km.

Global position of Australia during the Devo-Carboniferous

While central Australia was experiencing the compression and shortening associated with the Alice Springs Orogeny, the Canning Basin was experiencing periods of extension, subsidence and sedimentation. The question therefore arises: what kind of tectonic environment is required to allow intraplate compression in one region alongside intraplate extension in an adjacent region?

In order to understand the tectonic environment in which the Alice Springs Orogeny occurred we now look at the geographic and tectonic context of Australia during the Devonian to Carboniferous. Throughout that time, Australia was situated on the north-east margin of Gondwanaland with Antarctica to the south and India to the south-west. Scotese & McKerrow (1990) have published palaeogeographic reconstructions of the continents during the Carboniferous, showing that during the Namurian (333–315 Ma) the latitude of central Australia was c. 35°S. Australia then migrated south to c. 50°S in the Westphalian (315–296 Ma) as Gondwanaland rotated clockwise relative to Laurentia, causing the north-western margin of Gondwanaland to collide with northern Europe and Laurentia (Scotese & McKerrow 1990), producing progressively the Hercynian, Variscan, Appalachian/ Mauritanide and Ouachita orogenic collision fronts (Scotese & McKerrow 1990).

Tectonic activity was also occurring on the northern margin of Gondwanaland closer to the Australian continent. During the Early Palaeozoic several continental blocks (North China, South China, Tarim, Indochina, and Qaidam) were close by to the west (Metcalfe 1996). The correlatable distribution of Devonian vertebrate faunas present across these blocks and other parts of north-eastern Gondwanaland suggests that these blocks were in close proximity to Gondwanaland up until the late Devonian (Long & Burrett 1989; Young 1990) before separating and moving north to become part of present day Asia. Metcalfe (1996) suggests a Late Devonian clockwise rotation of these Asian crustal blocks away from northern Gondwanaland. Such a separation is consistent with the penecontemporaneous anti-clockwise rotation of Gondwanaland proposed by Chen et al. (1993) during the Late Devonian–Early Carboniferous.

In addition to the activity on the north African and Australian margin of Gondwanaland, the opposite margin was experiencing a sequence of orogenic movements resulting from a long maintained subduction zone that extended from SE Australia across Antarctica to South America (Coney 1992; Chen et al. 1993). This subduction zone was instrumental in the accretion, deformation and assembly of southeastern Australia and contributed to the tectonics of the Lachlan and New England orogens of south-eastern Australia.

Rifting of East Asian blocks from the north Australian margin of Gondwanaland, the collision of the north African margin with Laurentia, and long maintained subduction on the South American–Antarctic–East Australian margin, were the main tectonic events to affect Gondwanaland during the Devo-Carboniferous. Each of these events may therefore have contributed plate boundary stresses which drove deformation in the intraplate Alice Springs Orogeny in central Australia.

Veevers *et al.* (1994) suggested that the stresses produced in the collision of Laurentia and Gondwanaland were transmitted through Pangea to cause the deformation associated with the Alice Springs Orogeny. That collision front was, however, roughly 15 000 km from the site of intraplate deformation in Central Australia. It therefore seems unlikely that intraplate stresses associated with this collision could have been sufficiently focused to cause the Alice Springs Orogeny. In particular it seems unlikely that an event at such a distance could produce the localized compression and extension seen in central Australia and the Canning Basin on a lengthscale much smaller than that over which the stresses would have been transmitted.

The Lachlan Orogeny lasted from the Mid- to Late Cambrian to the Early Carboniferous, affecting most of Victoria and south-eastern New South Wales (Gray 1988). It was followed by the New England Orogeny which lasted from Mid Carboniferous to Early Mesozoic affecting north-eastern New South Wales and the east coast of Queensland (Murray *et al.* 1987). While the development of the Lachlan and New England orogens was roughly contemporaneous with the Alice Springs Orogeny, and plate deformation models which incorporate both events have been developed (Braun & Shaw 1999, 2001), it seems likely that these orogens developed independently. There is no evidence that subduction along the Lachlan–New England margin of Australia had any direct link with the mechanism driving intraplate orogeny in central Australia. While this subduction zone may have had some influence on the tectonics of central Australia, it is unlikely to be the driving mechanism.

The rotation and rifting of Asian continental blocks off the north Australian margin of Gondwanaland occurred roughly 1500 km from the central Australian region. The proximity of this event suggests that it may have had a much greater influence on Australian tectonics than could be produced by the Laurentian collision ten times further away. Palaeomagnetic data indicate that these blocks rotated clockwise as they moved to the north (Metcalf 1996) while Gondwanaland rotated anticlockwise penecontemporaneously (Chen *et al.* 1993). Such a relative rotation on the northern margin of Gondwanaland could conceivably induce an eastward variation of stress along the Australian margin from extensional to compressional, which in turn, might explain a distribution of deformation in which compression in central Australia was synchronous with extension in the Canning Basin.

Variation of lithospheric strength within the continent

Whatever the source of the boundary stresses that produced the Alice Springs Orogeny, the fact that deformation was focused in the central Australian and Canning Basin regions must reflect on the lithospheric strength of those regions. If stress is applied to a heterogeneous continent, those regions that are weaker will deform more easily than surrounding terrains and will show most of the strain associated with plate boundary stresses. If a sequence of tectonic events occurred preferentially in central Australia because of a weak lithosphere, it appears that the lithosphere remained weak after each event, or at least was weakened again prior to each subsequent event. It appears from the concentration and intensity of deformation in the Arunta and Musgrave Blocks throughout the Proterozoic and Palaeozoic that the lithosphere beneath the central Australian region has been weak relative to neighbouring regions.

Kusznir & Park (1984) have shown that, where the lithosphere is weak, it can flow in response to externally applied boundary stress. The stress may therefore be focussed by ductile deformation in the lower plate, leaving the more brittle upper plate under even greater stress. This type of deformation is termed 'thick-skinned' type tectonism, or alternatively 'whole lithosphere failure'. Following their interpretation the primary surface deformation associated with the Alice Springs Orogeny, shearing and reactivation of the crustal scale Redbank Thrust Zone, may simply indicate that the lithosphere underlying the region was weak.

Several hypotheses have been presented to explain how such a zone of lithospheric weakness could occur. Shaw *et al.* (1984) propose the existence of a long-lasting elongate plume in the mantle underlying the Arunta Block establishing a zone of weakness, which acted as a focus of tectonic activity for the next 1500 Ma. Etheridge *et al.* (1987) have proposed that the earliest orogenesis in central Australia, between 1920 and 1820 Ma, was the result of delamination of the underlying lithosphere. Upwelling of asthenospheric material after such an event, to replace the delaminated lithosphere, can cause an increase in the local geothermal gradient heating the remaining lithosphere and lower crust, and thus weakening the lithosphere. Subsequent orogenic cycles occurring in the region could have repeatedly reactivated the same weakened lithosphere. The actively deforming regions of central Asia show variations in the intensity of deformation which appear to correlate with regions that were most recently active, suggesting that prior tectonic activity may serve to weaken the lithosphere (Molnar & Tapponnier 1981) rendering it more liable to remobilization (Krabbendam 2001).

Using standard rheological models, the strength of the lithosphere may be parameterized in terms of the temperatures at the base of the crust (England 1983; Sonder & England 1986). Any mechanism that causes an increase in Moho temperature will presumably cause weakening of the lithosphere. Sandiford & Hand (1998) have proposed that heating and weakening of the lithosphere prior to both the Petermann Ranges Orogeny and the Alice Springs Orogeny occurred as a result of the deposition of thick sediment sequences in the Amadeus Basin. They proposed that these sediments insulated the high heat producing basement of central Australia, heating the crust and elevating the Moho temperature. The 7–8 km of sediment deposited in the northern and southern regions of the Amadeus Basin prior to the Alice Springs Orogeny, could increase the Moho temperature by $c.\,110°C$, significantly reducing lithospheric strength in that region (Sandiford & Hand 1998).

A band of weak lithosphere extending beneath central Australia and the Canning Basin may explain why tectonic activity was focused in these two regions. The postulated weak zone corresponds roughly to the Central Australian mega-element defined by Shaw et al. (1996) as a geophysical terrain (Fig. 1). In this study the thin viscous sheet model is used (Bird & Piper 1980; England & McKenzie 1982) to investigate the tectonic environment associated with this intraplate deformation and the nature of the lithosphere beneath central Australia and the Canning Basin. Within the model, a lateral viscosity variation in which viscosity is greater in the northern and southern parts of the lithospheric sheet but decreases to a minimum in the middle of the weak region (Fig. 4) is defined. The weak zone represents the regions of central Australia and the Canning Basin. These regions are assumed to have been underlain by weak lithosphere because of their prior history.

Thin viscous sheet methods

The continental lithosphere is represented as a thin sheet of viscous fluid, assuming a power-law

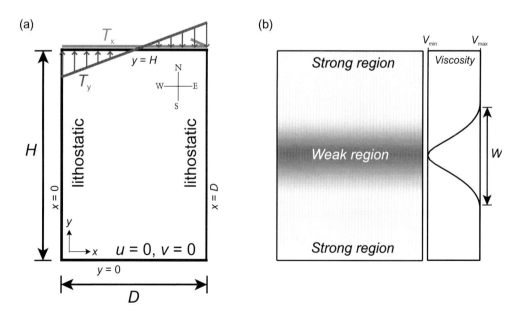

Fig. 4. (a) Schematic representation of boundary conditions used in the thin viscous sheet calculations. On the northern boundary ($y = H$) two independent traction conditions are applied. The E–W directed shear traction T_x is represented by a grey horizontal arrow, and the N–S directed traction $T_y(x)$ is represented by grey arrows in the y-direction. On the southern boundary $u = v = 0$. (b) Map, in greyscale, of the variation of the strength parameter B' (Eqn 11), with $W = H/2$, $V_{min} = 1.0$ and $V_{max} = 5.0$. Adjacent profile shows the N–S profile of B' within the sheet.

viscosity and incompressible deformation. In this approximation the horizontal components of velocity (u, v) within the lithosphere are assumed independent of depth, and horizontal gradients of the vertical shear stress components are neglected. The following constitutive equation is used (England & McKenzie 1982):

$$\dot{\varepsilon}_{ij} = B^{-n} \Theta^{n-1} \bar{\tau}_{ij} \qquad (1)$$

where $\bar{\tau}_{ij}$ and $\dot{\varepsilon}_{ij}$ are the vertically averaged components of the deviatoric stress and strain-rate tensors, respectively; n is a constant referred to as the rheological exponent; and Θ is the second invariant of the deviatoric stress tensor. The factor B includes the dependence of the vertically averaged viscosity of the lithosphere on temperature and composition, as summarized by England & McKenzie (1982). In order to investigate the effects of a heterogeneous strength distribution, a spatially varying distribution of the parameter B within the horizontal plane (x, y) is assumed (Fig. 4).

For the experiments presented in this paper, values of 1 and 3 were used for the rheological exponent n. A rheological exponent of $n = 1$ describes a thin sheet with Newtonian viscosity. A rheological exponent of $n = 3$ is appropriate for a lithosphere whose vertically averaged rheology is dominated by the power law creep of olivine (Karato et al. 1986).

The stress-balance equations describe conservation of momentum in a viscous creeping material. In dimensionless variables these equations are (England & McKenzie 1982)

$$\frac{\partial \bar{\tau}_{ij}}{\partial x_j} - \frac{\partial \bar{\tau}_{zz}}{\partial x_i} = \frac{Ar}{2} \frac{\partial S^2}{\partial x_i} \qquad (2)$$

where S is the crustal thickness (normalized by the length scale L), and Ar is the Argand number, defined by England & McKenzie (1982) for the case of an Airy-type isostatic model. The Argand number is a dimensionless measure of the magnitude of buoyancy forces caused by gradients in crustal thickness relative to the internal viscous stresses associated with deformation:

$$Ar = \frac{gL\rho_c(1 - \rho_c/\rho_m)}{B_0(U_0/L)^{1/n}} \qquad (3)$$

where g is the acceleration due to gravity; L is the lithospheric thickness; ρ_c and ρ_m are the density of crust and mantle, respectively; and U_0 is the velocity scale. For the experiments described in this paper Ar is set to zero appropriate for situations where viscous stress is much greater than buoyancy stress. The buoyancy stress acting on the system depends partly

on the topographic relief produced by the shortening. Since the Alice Springs Orogeny occurred over a period of as much as 100 Ma, the effects of erosion would continually reduce the impact of the buoyancy force, so the assumption that viscous stress is much greater than buoyancy stress $(Ar = 0)$ may be a valid approximation for the Alice Springs Orogeny. Since the actual crustal thickness in central Australia prior to the Alice Springs Orogeny is unknown, an initial crustal thickness of 35 km throughout the modelled region, typical for continental lithosphere, is used (Christensen & Mooney 1995).

The finite element method as described by Houseman & England (1986) is used to solve equation (2). The resulting distribution of velocity can then be used to determine the rate of crustal thickening in terms of the vertical strain rate, $\dot{\varepsilon}_{zz}$, at each location in the indented region.

$$\frac{1}{S} \frac{\partial S}{\partial t} = \dot{\varepsilon}_{zz} = -\left(\frac{\partial u}{\partial x} + \frac{\partial v}{\partial y}\right) \qquad (4)$$

Equation (4) is integrated in time to compute crustal thickness distribution as the region is progressively deformed. In these experiments erosion is not included, so crustal volume is conserved. The calculations presented here all use dimensionless variables for time, stress, strain-rate, velocity etc. Dimensionless variables, distinguished by a 'prime' are defined by:

$$x = x'D \qquad (5)$$
$$y = y'D \qquad (6)$$
$$t = t't_0 \qquad (7)$$
$$u = u'\left(\frac{D}{t_0}\right) = u'U_0 \qquad (8)$$
$$\dot{\varepsilon} = \frac{1}{t_0} \dot{\varepsilon}' \qquad (9)$$
$$\tau = \Sigma\tau' \qquad (10)$$
$$B = B_0 B' \qquad (11)$$

where D is the width of the thin sheet. Dimensionalizing factors are: t_0 for time; U_0 for velocity; Σ for stress; and B_0 for the lithospheric strength parameter. The value of B_0 represents the value of the lithospheric strength parameter in the middle of the weak zone, where $B' = 1.0$.

Boundary conditions – rotational northern boundary

The region of the Australian lithospheric plate represented by the thin sheet is shown schematically in Figure 1 (grey rectangular boundary). We use a thin viscous sheet with a width to length ratio of $D:H = 1:1.4$ to represent central Australia and surrounding regions. The sheet has nominal dimensions of width ($D = 2500$ km), length ($H = 3500$ km) and thickness ($L = 100$ km). To simplify description the cardinal directions E and N are used to refer to the directions x and y respectively. On the southern boundary ($y = 0$) velocity in both x and y directions is set to zero ($u = 0$, $v = 0$) (Fig. 4a), representing the idea that this part of the sheet is rigidly embedded in the adjoining super-continent.

In order to test the hypothesis that deformation in the interior of the plate was caused by stresses acting on the boundary of the plate we explore the effect of a rotational type condition in which stresses are applied to the northern margin of the sheet, causing the boundary to rotate in a clockwise direction. This condition provides a simple representation of the tectonic environment proposed by Chen *et al.* (1993) and Metcalf (1996), where blocks of present day China rotated clockwise relative to the northern margin of Gondwanaland as they rifted away from the super continent.

The rotation of this boundary may occur in response to either of two separate tractions which are imposed on the northern boundary (Fig. 4a): an east-directed traction T'_x (initially equivalent to tangential stress σ_{xy}), constant on $y = H$ causes the upper boundary to move eastwards, causing dextral shearing between the northern boundary and the rigid southern boundary. A north-directed traction which varies linearly from $+T'_y$ ($x = 0$) to $-T'_y$ ($x = D$) across the northern boundary is also considered. This applied normal traction therefore induces a bending moment on the sheet producing compression in the east and extension in the west (Fig. 4a). The magnitudes of the two tractions are considered variable but are invariant in any particular experiment.

On the two side boundaries ($x = 0$; $x = D$) a lithostatic-stress boundary condition described by Houseman & England (1993) is imposed. The lithostatic boundary condition assumes that the tangential shear stress is zero, and assumes that the normal stress is constant, (and equal to the initial confining stress), everywhere along the boundary. The lithostatic condition represents the situation of an external weak region which permits movement normal to the boundary.

Strength distribution within the sheet

A spatially varying strength parameter $B(x, y)$ (equation 1) with a reference value of B_0 is assumed. In the series of experiments presented here, B is initially constant in x but varies with y;

it is a material constant advected by the flow. At the beginning of each experiment B is maximum at the northern and southern margins of the sheet but follows a harmonic cosine dependence on y with a minimum in the middle of the sheet (Fig. 4b):

$$B(y) = \left(\frac{B_0}{2}\right)\left((V_{max} + V_{min})\right.$$

$$\left. - (V_{max} - V_{min})\cos\left[\frac{2\pi}{W}\left(y - \frac{H}{2}\right)\right]\right)$$

$$\text{for } \left(\frac{H - W}{2}\right) \le y \le \left(\frac{H + W}{2}\right) \quad (12)$$

$$B(y) = B_0 V_{max}$$

$$\text{for } y < \frac{H - W}{2} \quad \text{or} \quad y > \frac{H + W}{2}$$

where H (Fig. 4a) is the length of the sheet; and W is the width of the weak zone (Fig. 4b); V_{max} is the maximum value of B' (equation 11), which occurs on the southern and northern parts of the solution region; and V_{min} is the minimum value of B' ($V_{min} = 1$ at $y = H/2$). The distributions of boundary stress and viscosity within the sheet were chosen in order to avoid large deformation at the boundary of the sheet. Other indenter type models for intraplate orogeny (Neil & Houseman 1997) produce extensive crustal thickening at the plate boundary and there is no evidence for such a plate boundary event to the north of Australia at the time of the Alice Springs Orogeny.

Numerical experiments

Experiments with Newtonian viscosity ($n = 1$) are described first in order to gain a broad understanding of the parameter space. Further experiments with non-Newtonian viscosity ($n = 3$) are also discussed because $n = 3$ probably provides a better approximation of lithospheric behaviour (e.g. England & Houseman 1986; Sonder & England 1986). In order to quantify the effect of the two independent boundary tractions, T'_x and T'_y (Fig. 4a), on the model continent with the central weak zone, we measured for each experiment, three key parameters (Fig. 5) as a function of increasing time:

(1) the rotation angle (θ) of the northern boundary (measured in degrees clockwise from east).
(2) the displacement, in the x direction, of the top right hand corner of the sheet (d', a measure of total strain in x direction).
(3) the maximum crustal thickening factor produced in the sheet ($\beta_{max} = S_{max}/S_0$).

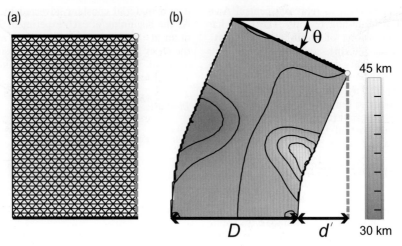

Fig. 5. (a) Undeformed finite element mesh used to calculate deformation within the thin viscous sheet.
(b) Deformed sheet showing the three methods of recording strain. (1) Angle of rotation of the northern boundary
(θ). (2) E–W displacement of the northeast corner of the sheet (d'). (3) Crustal thickness distribution (grey scale
with contours at 2 km intervals) determining the maximum crustal thickening factor produced (β_{max}).

Results of experiments with Newtonian viscosity (n = 1)

In the initial experiments either T'_y or T'_x was set
to zero while the other was varied. Figure 6 shows
how the thin viscous sheet deforms under the
action of a bending moment with $T'_y = 1.0, n = 1$,
$V_{max} = 5$ and $W = H$. The bending moment on
the northern boundary of the thin viscous sheet
causes that boundary to rotate $c.\ 35°$ before $t' = 2$
has elapsed. This calculation shows that this type
of viscosity distribution in conjunction with a
bending moment T'_y on the northern boundary,
produces deformation concentrated in the central
weak zone with shortening in the east and extension in the west. The convergent zone corresponds
to central Australia and the extensional region
corresponding to the Canning Basin. By the time
that the northern boundary has rotated $c.\ 35°$, a
maximum crustal thickness of $c.\ 42$ km (central
Australia) and a minimum crustal thickness of
$c.\ 30$ km (Canning region) have been achieved.

Varying T'_y shows how the rotation of the
northern margin (θ), the total eastward displacement (d') and the maximum crustal thickening factor (β_{max}) depend on T'_y. Figure 7a shows
the relationship between d' and dimensionless
time (t') for $T'_y = \frac{1}{4}, \frac{1}{2}, 1.0, 2.0$ and 4.0 (T'_x is zero).
Figure 7a shows that d' is approximately linearly
dependent on time until displacements of order d'
$c.\ 0.35$ are achieved. Lines of best fit were calculated for the initial linear phase of these graphs
for each value of T'_y. During this linear phase of

deformation the displacement depends on dimensionless time t' according to;

$$d' = k_d t' \tag{13}$$

where k_d is the slope of the line. Plotting k_d v. T'_y
(Fig. 8a) then yields a straight line that shows
the linear dependence of the deformation rate on
the magnitude of the bending moment:

$$k_d = C_2 T'_y \tag{14}$$

Combining equations (13) and (14) gives the
approximately linear dependence of d' on T'_y and
t' as;

$$d' = C_2 T'_y t' \tag{15}$$

Similar calculations were carried out to determine the dependence of the angle of rotation (θ)
on time and traction. Figure 7b shows the dependence of θ on t' and once again the dependence
is approximately linear for rotations up to about
$50°$, and the rate of rotation is simply proportional to the applied stress T'_y (Fig. 8c), hence:

$$\theta = C_4 T'_y t' \tag{16}$$

For the experiments summarized in Figure 8a, c
$C_2 = 0.21$ and $C_4 = 16.6$ (or $C_4 = 0.29$ for θ in
radians). These results show that θ and d', both
measures of strain, are linearly dependent on t'
(for strains of order 20%) and that both rates of
strain are linearly dependent on T'_y. If this is true
for the N–S traction T'_y, how does deformation
depend on T'_x?

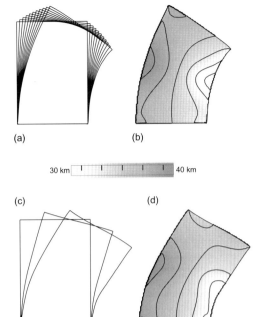

30 km   40 km

Fig. 6. (**a**) Effect of the bending moment applied to the northern boundary of the thin viscous sheet (at intervals of $t' = 0.2$), for an experiment with $n = 1$ $W = H$, $T'_y = 1.0$ and $T'_x = 0.0$ for conditions specified in (**a**). (**b**) The distribution of crustal thickness at $t' = 2.0$ when the northern margin has rotated $c.\,34°$. (**c**) Effect of an east directed traction applied to the northern boundary of the thin viscous sheet (at intervals of $t' = 0.2$), for an experiment with $n = 1$, $W = H$, $T'_x = 1.0$ and $T'_y = 0.0$ for conditions specified in (**a**). (**d**) The distribution of crustal thickness at $t' = 0.4$ when the northern margin has rotated $c.\,32°$ for conditions specified in (**c**). Crustal thickness contours for (**b**) and (**d**) are at 2 km intervals and other boundary conditions are as specified in Figure 4.

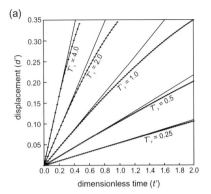

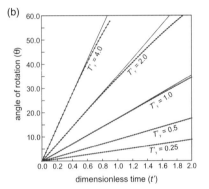

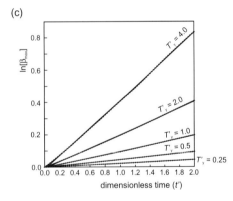

Fig. 7. Time dependence of the three principal deformation indicators: (**a**) displacement d'; (**b**) rotation, θ; and (**c**) logarithm of crustal thickening factor, $\ln[\beta_{max}]$, for experiments with $n = 1$, $T'_x = 0$ and $T'_y = 0.25$, 0.5, 1.0, 2.0 and 4.0. In each case the approximately linear relationship between the deformation indicator and t' is described by a best-fit straight line with slope (a) k_d, (b) k_θ, and (c) k_β.

A series of experiments was then conducted in order to determine the dependence of d' and θ on dimensionless time (t') when the deformation is driven only by the eastward directed shear T'_x. For $T'_x = \frac{1}{4}, \frac{1}{2}, 1.0, 2.0$ and 4.0 $(T'_y = 0)$ the deformation parameters d' and θ show a similar approximate linear dependence on time to that shown in Figure 7a, b. In this case both d' and θ initially follow a linear dependence on time with slope that increases in proportion to T'_x (Fig. 8b, d).

$$d' = C_1 T'_x t' \qquad (17)$$

and

$$\theta = C_3 T'_x t' \qquad (18)$$

For the experiments in which $T'_x = 0$ the best fit values of $C_1 = 1.4$ and $C_3 = 77.3$ (or $C_3 = 1.3$ for θ in radians) are obtained.

Having established that the slopes of the θ v. t' and d' v. t' graphs are linearly dependent on

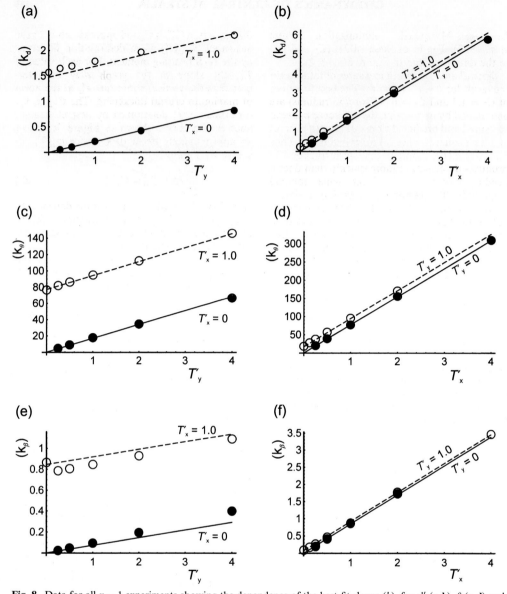

Fig. 8. Data for all $n = 1$ experiments showing the dependence of the best-fit slopes (k), for d' **(a, b)**, θ **(c, d)** and $\ln[\beta_{\max}]$ **(e, f)** v. time t' plots, on both T'_x and T'_y. The best fit straight lines are described by the numerical values in Table 1 with Equations (19), (20) and (24).

both T'_y and T'_x separately, it can be anticipated that, if both types of traction are present, the effects on θ and d' add together linearly:

$$d' = (C_1 T'_x + C_2 T'_y)t' \qquad (19)$$

and

$$\theta = (C_3 T'_x + C_4 T'_y)t' \qquad (20)$$

To test this idea a series of experiments, firstly with $T'_x = 1.0$ and $T'_y = 0, \frac{1}{4}, \frac{1}{2}, 1.0, 2.0$ and 4.0

and then with $T'_y = 1.0$ and $T'_x = 0, \frac{1}{8}, \frac{1}{4}, \frac{1}{2}, 1.0$ and 2.0, were run. Once again the dependence of d' on t' for the five different values of T'_y, was approximately linear for displacements $d' \lesssim 0.35$.

The best fit values of $C_1 = 1.5$ and $C_2 = 0.22$ for all of our measurements were determined by minimizing the difference between measured and predicted slopes of the $d(t')$ curves in a least squares sense for all experiments using

Levenberg–Marquardt minimization (Press et al. 1992). The fit of these values (C_1 and C_2) to the data is shown in Figure 8(a, b).

Since d' and θ are both measures of total strain the result for θ v. t' is similar. The best fit values of $C_3 = 1.3$ and $C_4 = 0.30$ (for θ in radians) are determined by minimizing the difference between measured and predicted slopes of the $\theta(t')$ curves in a least squares sense for all experiments. Thus $\theta(t')$, like $d'(t')$, shows that the shear traction T'_x) produces deformation more quickly than does a bending moment (T'_y) of the same nominal magnitude. The factor C_3/C_4 (≈ 4.5) is significantly less than C_1/C_2 (≈ 6.9) (Table 1), however indicating that the x-directed traction produces a greater ratio of displacement to rotation than does the y-directed traction. The accuracy of the fit to the data of equation (20) is shown in Figure 8c, d using the above best fit parameters for C_3 and C_4.

Both these measures of total strain (d' and θ) are approximately linearly dependent on dimensionless time (t'). Thus the strain-rates within the system are approximately constant in time, at least for sufficiently small t'. Furthermore, the measures of deformation rate $\dot{\theta}$ and $\dot{d}$ are both linearly dependent on the driving tractions T'_y and T'_x. The linearity of the deformation clearly depends on the use of a Newtonian viscosity for the thin viscous sheet, and implies that the non-linear influence of the continuously changing boundary geometry can be neglected for $d' \lesssim 0.35$ and $\theta \lesssim 60°$.

Another measure of the deformation produced by the rotation and displacement of the northern boundary is the crustal thickening factor that occurs within the sheet. The parameter β_{max} describes the maximum crustal thickening factor within the sheet at any time t'. In general the crustal thickening factor β_{max} will depend on time according to:

$$\frac{1}{\beta}\frac{\partial\beta}{\partial t'} = \dot{\varepsilon}'_{zz} \quad (21)$$

where $\dot{\varepsilon}'_{zz}$ is the vertical strain rate. If that strain-rate is constant, then equation (21) has the following solution:

$$\beta = e^{\dot{\varepsilon}'_{zz}t'} \quad (22)$$

since $\beta = 1$ when $t = 0$. The approximately linear dependence of d' and θ on t' shown above for $n = 1$ indicates that strain-rates are approximately constant with time. Equation (22) therefore suggests that a plot of $\ln[\beta_{max}]$ v. t' should yield a straight line. The results presented in Figure 7c for $T'_y = \frac{1}{4}, \frac{1}{2}, 1.0, 2.0$ and 4.0 and $T'_x = 0$, show

that equation (22) is a good approximation to the measured $\beta_{max}(t')$ when deformation is driven by the N–S bending moment. For each value of T'_y, the slope of the graph may be interpreted as the vertical strain rate $\dot{\varepsilon}'_{zz}$ at the point of maximum crustal thickening. The slopes, k_β, (or strain rates) determined by best fit straight lines to the curves shown in Figure 7c follow an approximately linear dependence on T'_y, as shown by Figure 8e:

$$\ln[\beta_{max}] = C_6 T'_y \quad (23)$$

A similar relationship occurs for the dependence of β_{max} on t' and T'_x when deformation is driven only by the east directed shear stress acting on the northern boundary (Fig. 8f).

The dependence of β_{max} on t' for $T'_y = 1$ and $T'_x = \frac{1}{8}, \frac{1}{4}, \frac{1}{2}, 1.0$ and 2.0 and $T'_x = 1$ and $T'_y = \frac{1}{4}, \frac{1}{2}, 1.0, 2.0$ and 4.0 (Fig. 8(e, f) open circles) were also examined. In each of these experiments, the dependence of $\ln[\beta_{max}]$ on t' is nearly linear, as in the experiments with $T'_x = 0$ or $T'_y = 0$ (Fig. 8(e, f) filled circles). In each case the slope of the graph was measured to determine an empirical strain-rate, and we sought a relationship of the form:

$$\ln[\beta_{max}] = C_5 T'_x + C_6 T'_y \quad (24)$$

A least squares fit to all of the measured slopes gives $C_5 = 0.85$ and $C_6 = 0.073$, and the fit of equation (24) to the individual measurements is shown in Figure 8e, f.

The empirical coefficients that describe the dependence of d', θ and β_{max} on t', T'_x and T'_y are summarized in Table 1. Note that these coefficients have been determined specifically for the geometry $H/D = 1.4$ and the viscosity distribution with $W = H$. For all three measures of the deformation, (d', θ, β_{max}) a shear traction T'_x of a given magnitude produces crustal deformation at a faster rate than a bending moment (T'_y) with the same maximum magnitude. If we compare one experiment with $T'_x = 1$ and $T'_y = 0$ with a second experiment in which $T'_x = 0$ and $T'_y = 6.9$, we see similar d' in each case but the bending moment produces greater rotation while the eastward shear traction produces greater crustal thickening.

Results of experiments with non-Newtonian viscosity (n = 3)

To determine how the deformation of the thin viscous sheet proceeds when a non-Newtonian

Table 1. T'_x and T'_y coefficients for $n = 1$ experiments

$n = 1$	T'_x coefficient	T'_y coefficient	Ratio of coefficients
d'	$C_1 = 1.5$	$C_2 = 0.22$	$C_1/C_2 = 6.9$
θ	$C_3 = 77$	$C_4 = 17$	$C_3/C_4 = 4.5$
β_{max}	$C_5 = 0.85$	$C_6 = 0.073$	$C_5/C_6 = 11$

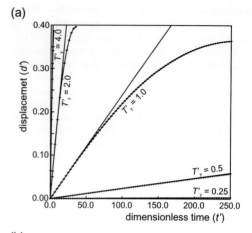

(a)

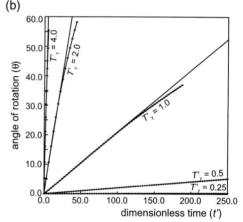

(b)

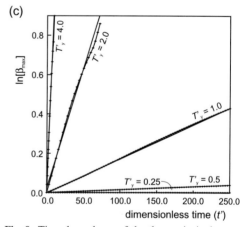

(c)

($n = 3$) viscosity is assumed, the dependence of d', θ and β_{max} on t' with varying northern boundary tractions T'_y and T'_x is measured again. Initially the effect of varying T'_y is investigated for $T'_y = \frac{1}{4}$, $\frac{1}{2}$, 1.0, 2.0 and 4.0 and $T'_x = 0$ (Fig. 9). The distribution of the strength parameter B is again defined by Figure 4 with $W = H$ and $V_{max} = 5$ but note that, for $n = 3$, a given increase in B has a much greater effect in suppressing deformation than is the case for $n = 1$.

The dependence of d', θ and $\ln[\beta_{max}]$ on time is still approximately linear (Fig. 9) for sufficiently small strain in spite of the non-Newtonian viscosity. The dependence of slope (k) on T'_y is, however, non-linear for each of d', θ and $\ln[\beta_{max}]$. In the thin viscous sheet approximation the strain rate varies as stress to the power n (equation 1). Since d', θ and $\ln[\beta_{max}]$ are all measures of strain, the rate of increase with time is expected to be proportional to strain rate and therefore proportional to stress to the power n. This hypothesis is tested in Figure 10(a, c, e filled circles) by plotting $k^{1/3}$ v. T'_y where k in each case is measured from plots of d', θ and $\ln[\beta_{max}]$ v. t' (e.g. Fig. 9).

For each of the deformation measures an approximate linear dependence of ($k^{1/3}$) on T'_y is found, and the lines of best fit determined in Figure 10(a, c, e solid lines) imply the following dependence of d', θ and β_{max} on t' and T'_y when $T'_x = 0$:

$$d' = G_2 T'^3_y t' \tag{25}$$

$$\theta = G_4 T'^3_y t' \tag{26}$$

$$\beta_{max} = e^{G_6(T'_y)^3 t'} \tag{27}$$

A similar method was used to determine the dependence of d, θ and β_{max} on t' and T'_x when $T'_y = 0$. Figure 10(b, d, f, filled circles) shows the dependence of d', θ and $\ln[\beta_{max}]$ on t' for $T'_y = 0$ and $T'_x = \frac{1}{4}, \frac{1}{2}$, 1.0, 2.0 and 4.0. In each case the dependence is initially linear in t' and the slopes (k) were determined from these experiments.

Fig. 9. Time dependence of the three principal deformation indicators for experiments using non-Newtonian viscosity ($n = 3$), boundary conditions as shown in Figure 4, and strength distribution defined by $W = H$, $V_{max} = 5.0$, $V_{min} = 1.0$, for $T'_y = 0.25$, 0.5, 1.0, 2.0 and 4.0, and $T'_x = 0.0$. (**a**) d' v. t'. (**b**) θ v. t'; and (**c**) $\ln[\beta_{max}]$ v. t'.

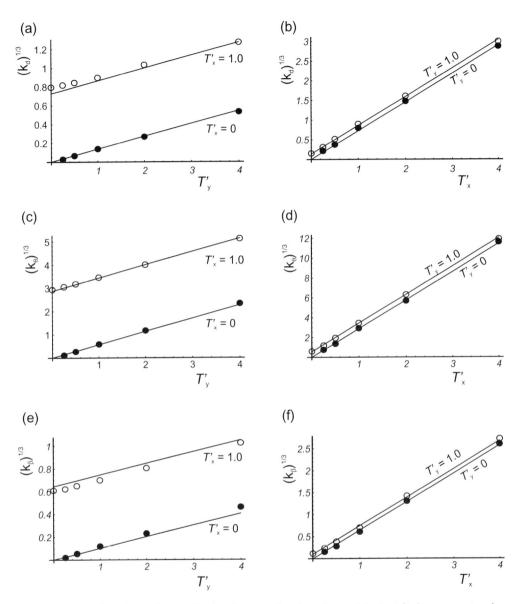

Fig. 10. Data for all $n = 3$ experiments showing the power-law dependence of the best-fit slope parameters k_d (**a**, **b**), k_θ (**c**, **d**) and k_β (**e**, **f**) T'_x and T'_y, with power-law index 1/3. Best-fit straight lines use Equations (32–34) with constants G_i from Table 2.

Plots of the cube root of slope ($k^{1/3}$) v. T'_x are linear in every case, indicating the following dependencies of d', θ and β_{max} on t' and T'_x when $T'_y = 0$:

$$d' = G_1 T'^3_x t' \qquad (28)$$

$$\theta = G_3 T'^3_x t' \qquad (29)$$

$$\beta_{max} = e^{G_5 (T'_x)^3 t'} \qquad (30)$$

The linear dependence of $k^{1/3}$ on T'_x and T'_y for each of these variables separately suggests that we should examine a combined functional relation of the form:

$$k_d^{1/3} = G_1 T'_x + G_2 T'_y \qquad (31)$$

and similar relations for θ and $\ln[\beta_{max}]$. In that case the dependence of d', θ and β_{max} on t', T'_y

Table 2. T'_x and T'_y coefficients for $n=3$, $H/D=1.4$ and $W/H=1.0$

$n=3$	T'_x coefficient	T'_y coefficient	Ratio of coefficients
d'	$G_1 = 0.72$	$G_2 = 0.14$	$G_1/G_2 = 5.11$
θ	$G_3 = 2.85$	$G_4 = 0.57$	$G_3/G_4 = 5.01$
β_{max}	$G_5 = 0.65$	$G_6 = 0.086$	$G_5/G_6 = 7.56$

and T'_x is parameterized as follows:

$$d' = (G_1 T'_x + G_2 T'_y)^3 t' \qquad (32)$$

$$\theta = (G_3 T'_x + G_4 T'_y)^3 t' \qquad (33)$$

$$\beta_{max} = e^{(G_5 T'_x + G_6 T'_y)^3 t'} \qquad (34)$$

Further experiments with $n=3$, $T'_y = 1.0$ and $T'_x = \frac{1}{4}, \frac{1}{2}$, 1.0, 2.0 and 4.0 and also $T'_x = 1.0$ and $T'_y = \frac{1}{4}, \frac{1}{2}$, 1.0, 2.0 and 4.0, again show the approximately linear dependence of d', θ and $\ln[\beta_{max}]$ on t' for sufficiently small strain ($\theta <$ c. 50°). A least squares fit of the $k^{1/3}$ values measured for all $n=3$ experiments was used to determine the constants G_1–G_6 shown in Table 2.

The fit to the experimentally determined slopes k_d, k_θ, k_β of equations (32–34) with the constants given in Table 2 is shown in Figure 10.

Experiments with non-Newtonian viscosity ($n=3$) and wider geometry

Two other factors which should influence the dependence of d', θ and β_{max} on t' are the length to width ratio of the thin viscous sheet, and the width of the low viscosity zone. In order to investigate the effect of these two parameters a further series of numerical experiments with $n=3$ are conducted. Firstly, the width of the thin sheet was increased so that the ratio of width to height was $D:H=2.0:1.4$ in place of $D:H=1.0:1.4$ used in the previous calculations. Secondly, (and separately) the effect of reducing the lateral extent of the low viscosity zone was investigated by using $W = H/2$ (equation 12) in place of $W = H$ as used in the previous experiments. The regions of the sheet to the north and the south of the low viscosity zone are set to a strength coefficient $V_{max} \times B_0$, and so remain relatively rigid.

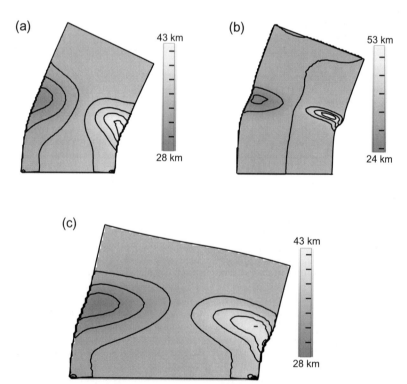

Fig. 11. Crustal thickness distributions for experiments with $n = 3$, $T'_x = 1$ and $T'_y = 1$ for (a) original geometry and $t' = 0.4$ with crustal thickness contours at 2 km intervals; (b) with narrow weak zone ($W = H/2$), and $t' = 0.6$ (contours at 5 km intervals). (c) with double width geometry ($D:H=2.0:1.4$), and $t' = 2.0$ (contours at 2 km intervals).

Table 3. *Coefficients of* T'_x *and* T'_y *for* n = 3, H/D = 0.7 *and* W/H = 1.0

n = 3	T'_x coefficient	T'_y coefficient	Ratio of coefficients
d'	$G_1 = 0.47$	$G_2 = 0.088$	$G_1/G_2 = 5.4$
θ	$G_3 = 1.5$	$G_4 = 0.38$	$G_3/G_4 = 3.9$
β_{max}	$G_5 = 0.46$	$G_6 = 0.070$	$G_5/G_6 = 6.6$

Table 4. *Coefficients of* T'_x *and* T'_y *for* n = 3, H/D = 1.4 *and* W/H = 0.5

n = 3	T'_x coefficient	T'_y coefficient	Ratio of coefficients
d'	$G_1 = 0.81$	$G_2 = 0.13$	$G_1/G_2 = 6.4$
θ	$G_3 = 3.1$	$G_4 = 0.57$	$G_3/G_4 = 5.5$
β_{max}	$G_5 = 0.86$	$G_6 = 0.18$	$G_5/G_6 = 4.7$

Graphs of d', θ and $\ln[\beta_{max}]$, v. t' for the experiments with $D:H = 2:1.4$ (e.g. Fig. 11c) again show a dependence on t', T'_x and T'_y described by equations (32–34), though the values of the coefficients G_i, $i = 1, 6$ are changed. As in the earlier analysis with $D:H = 1:1.4$, the numerical coefficients of equations (32–34) were determined for each of d', θ and $\ln[\beta_{max}]$ by measuring the gradient of the linear part of each curve and determining the least squares best fit values of the coefficients G_i in equations (32–34).

Although doubling the width of the sheet does not affect the nature of the power-law dependence, it decreases the values of the G_i coefficients (compare Table 2 and Table 3) by a factor of approximately 1/2 to 2/3. Thus, if the original width of the sheet is doubled, the boundary tractions required to produce the same strain indicators (d', θ and β_{max}) are increased by between 50 and 100%. Alternatively if the boundary traction is held at the same level, the rates of strain are decreased by a factor of order 1/8 to 1/3 when the width is doubled.

Experiments with non-Newtonian viscosity (n = 3) and a narrow weak zone, (W = H/2)

The deformation for experiments with the narrow low viscosity zone (e.g. Fig. 11b) was analysed by the same techniques. As for the preceding series of experiments, the linear dependence on time and the power-law dependence on T'_y and T'_x are found for each of d', θ and β_{max}, but the coefficients of G_i are again changed. The measured gradients, $k^{1/3}$, were again fit to equations of the form of equations (32–34) and best fit coefficients were determined (Table 4). The coefficients G_i in Table 4 are generally greater than the coefficients in Table 2 though the difference varies from near zero for G_2 and G_4 to around 100% for G_6. Thus, for $n = 3$ the same boundary tractions applied to a sheet in which the weak zone is narrower will produce strain more quickly, or alternatively the same degree of strain can be produced by lesser boundary tractions. Concentrating the deformation in a narrower zone implies greater strain-rates in that zone but the

local increase in strain rate is here achieved without increasing driving traction on the northern boundary.

Figure 11 shows the crustal thickness distribution for the three different geometries here investigated using non-Newtonian ($n = 3$) viscosity. In each case conditions at the northern boundary have produced crustal thickening in the region corresponding to central Australia and crustal thinning in the region corresponding to the Canning Basin. Figure 11(b) shows that a narrow weak zone causes concentration of deformation into a narrower area within which greater extremes of crustal thickness are achieved.

Relation between crustal thickening (β_{max}) and rotation of the northern region (θ)

In order to implement a rotational type boundary condition on the northern boundary we considered two possible contributions to the boundary forces and treated them separately. These two components may be described as an x-directed traction, initially a shear, and a y-directed bending moment. The effects of these two tractions on the stress field in the interior of the plate are simply additive, and since both types of boundary traction produce similar deformation fields, it would, therefore, be difficult to distinguish between the two in a real geological example. The two types of traction produce a similar deformation distribution although the x directed traction T'_x appears to deform the sheet more quickly and therefore produces a higher strain-rate for the same maximum traction. Since other factors such as the strength contrast and width of the weak zone also affect the strain-rate, it generally would be impossible to determine the relative contribution of each of these two tractions from geological data. More significant than the relative contributions of the two types of traction, however, is the total amount of rotation (θ) that may be produced by either. It appears that the resulting crustal thickness distribution depends strongly on the amount of rotation of the northern block but is insensitive to the type

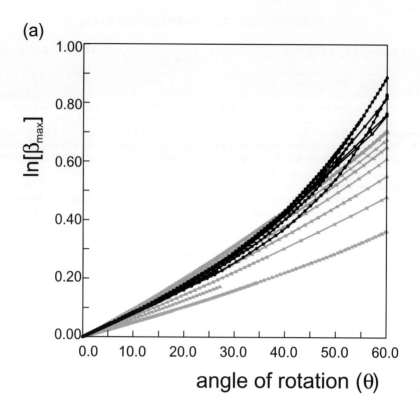

(a)

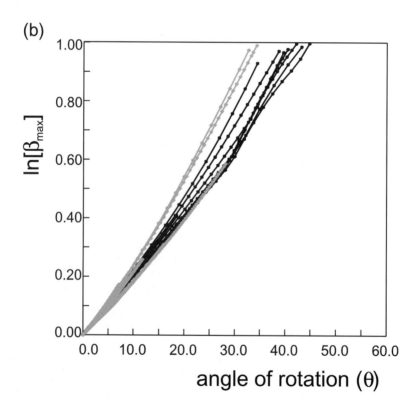

(b)

of boundary traction, T'_x or T'_y, that produces the rotation.

Figure 12 shows, for all the experiments conducted in this study, the dependence of $\ln[\beta_{max}]$ on θ where β_{max} is the maximum crustal thickening factor achieved in the weak zone, and θ is the angle of rotation of the northern boundary. The Newtonian experiments and the non-Newtonian experiments with the same geometry ($W/H = 1.0$ and $H/D = 1.4$) produce a very similar dependence of β_{max} on θ (Fig. 12a). The $n = 1$ experiments produce a more nearly linear relationship and show a greater spread of gradients than do the $n = 3$ experiments but the ratio of maximum crustal thickening to angle of rotation is similar in both cases. The observation that the relationship between maximum crustal thickening factor in the weak zone and angle of rotation of the northern region is broadly independent of the rheological exponent (n) is a remarkable result.

Comparing Figure 12a and 12b it is seen that the dependence of β_{max} on θ is affected by the initial geometry of both the sheet and the weak zone. Figure 12b shows the relationship between β_{max} and θ for experiments, with $n = 3$ where the width of the sheet (D) has been doubled (Fig. 12b, grey lines) and experiments where the width of the sheet has remained the same, but the width of the weak zone (W) has been halved (Fig. 12b, black lines). In both cases the ratio of $\ln[\beta_{max}] : \theta$ is increased by a factor of about two, relative to those experiments with the original geometry (Fig. 12a). The similar dependence of β_{max} on θ for both sets of experiments in Figure 12b can be explained by realizing that when the weak zone is narrowed the remaining regions of the sheet, top and bottom, remain relatively rigid, with a strength coefficient of $V_{max} \times B_0$. Tractions applied to the upper boundary, therefore, are transferred through the rigid part of the sheet to the weak region with little or no strain, so that the tractions effectively act on the northern boundary of the weak zone rather than the northern boundary of the sheet. The geometric ratio that affects the relation between shortening and rotation is therefore not $D : H$ but rather $D : W$ where D is the E–W width of the sheet and W is the N–S width of the weak zone. In these experiments either D has been doubled or W has been halved, so in either case the ratio of $D : W$ has doubled. From Figure 12b it appears that doubling D/W causes the gradient of the $\ln[\beta_{max}]$ v. θ graph to be

approximately doubled. If the weak zone is narrow, greater crustal thicknesses are produced by smaller rotations of the northern block. The thickening factor β_{max} is thus primarily determined by the width of the weak zone relative to the length of the boundary on which the traction applies, and by θ the magnitude of rotation of the northern boundary. In all experiments the strain indicators $\ln[\beta_{max}]$, θ and d' appear to increase almost linearly with time for a constant applied boundary traction.

To produce a crustal thickening factor of $\beta_{max} = 1.66$ in the weak zone of the model with $H/D = 1.4$ and $H/W = 1$ requires a clockwise rotation of the northern boundary of $c. 45°$ (Fig. 12a). If however the geometry is altered so that the width of the weak zone is halved or the E–W width of the system doubled, then a smaller rotation angle of $c. 20–25°$ of the northern boundary will produce the same degree of crustal thickening.

Discussion

Application of experimental results to the Alice Springs Orogeny

To simplify the analysis of the problem, the results that have been presented so far have been described using dimensionless variables. In order to discuss the relevance of these results to central Australia, actual geological constraints must be applied to the model. We consider:

(1) the timing and duration of the orogeny;
(2) the length-scale and shape of the lithospheric regions involved;
(3) the shortening associated with the Alice Springs Orogeny;
(4) the post-orogenic width of the shortened zone; and
(5) the amount of crustal thickening produced.

Various geochronological studies of the central Australian region indicate that the Alice Springs Orogeny occurred over a time window of approximately 100 Ma between 400 and 300 Ma (e.g. Dunlap et al. 1995; Foden et al. 1995 (365–308 Ma); Mawby et al. 1995). It is less clear, however, whether deformation occurred as a continuous process or as discrete events during that time (Dunlap et al. 1995). The magnitude of strain rate depends primarily on the magnitude

Fig. 12. The dependence of $\ln[\beta_{max}]$ on θ for all the experiments conducted in this study. (**a**) shows $n = 1$ experiments (grey) and $n = 3$ experiments (black) with boundary conditions as specified in Figure 4, $H/D = 1.4$ and $W = H$. (**b**) $n = 3$ experiments with $H/D = 0.7$ (grey) and $n = 3$ experiments where $W = H/2$ (black).

of the plate boundary driving tractions and on the lithospheric strength coefficient of the weak zone. Variation in time of the plate boundary stress could produce an episodic strain history. Here it is assumed, for simplicity, that deformation was continuous but the relationship between rotation of the northern lithospheric block and shortening or extensional strain in the central weak zone depends only on the cumulative strain, and is in-sensitive to the variation of strain-rate with time.

In comparing the rectangular thin viscous sheet model to the region of Australia which we are considering (Fig. 1), a dimensionless length unit of 1.0 corresponds to $D = 2500$ km. The thin viscous sheet thus has dimensions of 2500 km by 3200 km for $H/D = 1.4$. Shortening of c. 100 km associated with the Alice Springs Orogeny (ASO) appears to have been concentrated in the central and southern Arunta Block and in the central and northern Amadeus Basin (Teyssier 1985; Dunlap *et al.* 1995). From the middle of the Arunta Block to the middle of the Amadeus Basin is a distance of c. 300 km. If we assume that the orogeny produced 100 km of shortening, and that the width of the shortened zone is 300 km after deformation, an estimate of the lithospheric thickening that has occurred can then be made.

Assuming that shortening is uniformly distributed, is approximately plane-strain in the vertical

section, and that no mass is lost due to erosion, the area A_a (Fig. 13a) must be equal to the area A_b (Fig. 13b) by conservation of volume, such that the maximum crustal thickening factor is:

$$\beta_{max} = \frac{S}{S_0} \approx 1.33 \qquad (35)$$

where S_0 is the original crustal thickness and S is the post-orogenic crustal thickness. It is, however, unlikely that crustal thickening would have been uniform across the orogenic zone. Suppose therefore that the perturbation to lithospheric thickness follows a cosine dependence of wavelength 300 km such that the lithosphere is thickest in the middle of the shortened zone and decreases to its pre-orogenic thickness at the edges of the zone (Fig. 13c). The maximum crustal thickening may then be estimated by integrating the thickness function to obtain the cross-sectional area A_c (Fig. 13c) from which the maximum crustal thickening factor is:

$$\beta_{max} = \frac{S}{S_0} \approx 1.66 \qquad (36)$$

These two estimates of maximum crustal thickening factor provide plausible upper and lower limits on β_{max}. These estimates of the crustal thickening factor are based on the estimates for crustal shortening and deformation distribution relevant to the Alice Springs Orogeny and are not affected by the redistribution of crustal mass by erosional processes. Ignoring the effect of erosion is consistent with the assumption of the numerical experiments ($Ar = 0$).

The lateral extent of the orogeny also provides some constraint on the W/H ratio and therefore on the angle of rotation θ since the angle of rotation required in the model to produce $\beta_{max} = 1.66$ is a function of the ratio W/H. For $W/H = 1$ the lateral extent of crustal thickening in the weak zone is of order 800 km (Fig. 11a, c). A weak zone of $W/H = 0.5$, however, produces a zone of crustal thickening with a lateral extent of order 300 km (Fig. 11b). Assuming that the lateral extent of crustal thickening in the Alice Springs Orogeny was of order 300 km, it seems that a ratio of $W/H = 0.5$ and consequently a rotation (θ) of order 20° to 25°, are the parameter values that best represent the deformation associated with the Alice Springs Orogeny. If such a rotation occurred between northern and southern Australia during the Alice Springs Orogeny it could potentially be measured using palaeomagnetic data. Future work in collecting Devonian and Carboniferous palaeomagnetic poles for both north and south Australia could be of value in showing whether such a rotation has occurred.

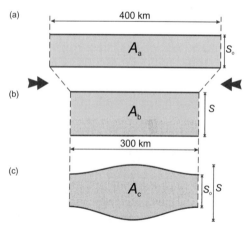

Fig. 13. Schematic model (shown in vertical section) used to estimate the lithospheric thickening factor produced by convergence in central Australia. An initial rectangular block of length 400 km (**a**) is shortened to 300 km, either by (**b**) uniform shortening, giving $\beta_{max} = 1.33$, or by (**c**) shortening which obeys a cosine dependence on x giving $\beta_{max} = 1.66$. Isostatic adjustment of the crustal columns must occur but is not shown here.

Scaling of plate boundary force and lithospheric strength

Within the model the parameter B_0 (equations 1, 11) describes the strength of the lithosphere in the middle of the weak zone. We estimate the apparent value of B_0 from the relationships between the dimensionless variables used in the calculations ($\dot{\varepsilon}'$, t', τ') and the dimensional parameters that apply to the earth ($\dot{\varepsilon}$, t, τ).

Substituting equations (9) and (11) into equation (1) and inverting, gives the following:

$$\tau = B_0 B' \left[\left(\frac{1}{t_0}\right)^{1/n} (\dot{E}'^{(1/n-1)} \dot{\varepsilon}') \right] \qquad (37)$$

where $\dot{E}'$ is the second invariant of the dimensionless strain-rate tensor, and B' is the dimensionless value of lithospheric strength (V_{min}).

Since dimensionless stress and strain-rate are also related by:

$$\tau' = B' E'^{\dot{Y}(1/n-1)} \varepsilon^{\dot{Y}} \qquad (38)$$

it then follows from equation (10) that the scaling factor for stress Σ is related to the scaling factors for time t_0, and strength B_0 by:

$$\Sigma = B_0 (t_0)^{-(1/n)}. \qquad (39)$$

Rearranging equation (39), and noting that t_0 is the ratio of actual time scale to dimensionless time scale, the parameter B_0 may also be expressed as:

$$B_0 = \Sigma \left(\frac{t}{t'}\right)^{1/n} \qquad (40)$$

where t is the dimensional time taken to reach a crustal thickening factor of β_{max} (the duration time of the orogeny). Since a relationship between t' and $\ln[\beta_{max}]$ has already been determined in equation (34), B_0 can be expressed in terms of t and $\ln[\beta_{max}]$:

$$B_0 = \frac{t^{1/n}(G_5 T_x + G_6 T_y)}{(\ln[\beta_{max}])^{1/n}} \qquad (41)$$

where T_x and T_y are now the dimensionalized tractions acting on the boundary. For example assuming that the duration of the orogeny (t) is 100 Ma, β_{max} is 1.66, $n = 3$, G_5 from Table 4 and the depth averaged traction T_x acting at the plate boundary is 50 MPa, the value of lithospheric strength in the weak zone (B_0) is:

$$B_0 = 0.8 \times 10^{13} \, \text{Pa} \, \text{s}^{1/3}. \qquad (42)$$

The estimated values of B_0 are simply proportional to the assumed magnitude of the plate boundary driving stress, but vary only as the third power of the assumed time scale. Although this estimate of B_0 is as poorly constrained as the plate boundary traction, it appears consistent with the value of $B = 1.7 \times 10^{13} \, \text{Pa} \, \text{s}^{1/3}$ calculated by Houseman & Molnar (1997) for vertically averaged, continental lithosphere involved in the India–Asia collision. The India–Asia collision of course is on a much larger scale and has progressed much faster apparently than the Alice Springs Orogeny.

Since the parameter B is a measure of lithospheric strength it is influenced by the thermal structure of the lithosphere. England (1983) showed that, for lithosphere in which the integrated strength resides primarily in the mantle, the force per unit length acting on a vertical section of lithosphere which is shortening under conditions of plane strain and pure shear, may be expressed approximately as:

$$F \approx 2^{(1-n)/2n} \left(\frac{\dot{\varepsilon}}{A}\right)^{1/n} \frac{nRT_m^2}{Q\gamma} \exp\left[\frac{Q}{nRT_m}\right] \qquad (43)$$

where F is the force per unit length; T_m is the temperature at the Moho; γ is the geothermal gradient within the mantle lithosphere (assumed constant); R is the universal gas constant; and A and Q are material constants and all other parameters are as defined previously. Under the same deformation conditions, we may write $\tau = 2^{(1-n)/2n} B \dot{\varepsilon}^{1/n}$. Since $F = L\tau$, where L is the thickness of the lithosphere, and τ is the plate boundary stress (relative to the lithostatic stress level) comparison with equation (43) allows the strength coefficient B to be expressed approximately as a function of T_m as:

$$B = \frac{nRT_m^2}{A^{1/n}LQ\gamma} \exp\left[\frac{Q}{nRT_m}\right]. \qquad (44)$$

By equating the two expressions for the lithospheric strength coefficient (equations 41 and 44) we can describe the dependence of maximum crustal thickening factor β_{max} on the plate boundary force per unit length $F_x = LT_x$, or $F_y = LT_y$, the time span of the orogeny (t), the Moho temperature (T_m), and other rheological parameters:

$$\beta_{max} = \exp\left[\left(\frac{Q\gamma(G_5 F_x + G_6 F_y)}{nRT_m^2}\right)^n \right.$$
$$\left. \times At \exp\left[\frac{-Q}{RT_m}\right]\right] \qquad (45)$$

The values of Q and A are specific to the material constituting the lithosphere which is predominantly olivine. From experiments on the rheology of olivine, Karato *et al.* (1986) determined that wet olivine had a rheological exponent of $n = 3$ and values of $Q = 420 \, \mathrm{kJ \, mol}^{-1}$ and $A = 1.9 \times 10^3 \, \mathrm{s}^{-1} \, \mathrm{MPa}^{-3}$. If the maximum crustal thickening factor is known, equation (45) may be inverted in various ways, so that any of the following parameters may be expressed as a function of the other parameters, considered known: force per unit length (F_x or F_y), duration of the orogeny t, Moho temperature T_m, and maximum crustal thickening factor β_{max}. Substituting the above values of Q and A and $\gamma = 8°\mathrm{C/km}$ into equation (45) allows several different graphical representations of that equation (Figs 14–16).

While the timing of the orogeny and the crustal thickening factor produced in the orogeny are somewhat constrained there is very little constraint on the magnitude of the boundary stresses or on the strength of the weak lithosphere in central Australia. Plausible values for plate boundary force are suggested from studies of other orogens. An upper limit for the magnitude of plate boundary forces was provided by England & Houseman (1986) who showed that the deformation associated with the Himalayan front and the Tibetan plateau could be supported by a vertically integrated plate boundary force of order $2 \times 10^{13} \, \mathrm{Nm}^{-1}$. In a study of the magnitude of boundary forces acting on highly heterogeneous continental lithosphere of northeastern Brazil, Tommasi *et al.* (1995) found that plate boundary forces of between $1.2 \times 10^{12} \, \mathrm{Nm}^{-1}$ and $9.0 \times 10^{12} \, \mathrm{Nm}^{-1}$ were required to effectively model the deformation in that region. Bird (1978) calculated that the force exerted by a subducting slab on the overriding plate was of order $4 \times 10^{12} \, \mathrm{Nm}^{-1}$. In the following discussion a reference value of $5 \times 10^{12} \, \mathrm{Nm}^{-1}$ (corresponding to an average boundary stress of 50 MPa acting on a 100 km thick lithosphere) is used as a plausible estimate of the magnitude of forces that may have been acting at the plate boundary during the Alice Springs Orogeny.

Assuming that crustal thickening of $\beta_{\mathrm{max}} = 1.66$ occurred in a specific time interval t, Figure 14 shows the required plate boundary force F_x for a range of Moho temperatures in the zone of deformation. If the applied boundary force per unit length causing the orogeny was of order $5 \times 10^{12} \, \mathrm{Nm}^{-1}$ then, for the orogeny to occur over a period of 100 Ma requires a Moho temperature of $c. 520°\mathrm{C}$ (Fig. 14). If, for the same boundary force, the orogeny occurred over a much shorter time period, then a weaker and

hotter lithosphere is required. For the same deformation to occur within 5 Ma, for example, a Moho temperature of $c. 565°\mathrm{C}$ is implied. As would be expected, larger boundary stresses allow the required deformation to occur in a cooler, more rigid lithosphere, while smaller boundary stresses require a hotter lithosphere to produce the same deformation. For the calculations presented here a value of $V_{\mathrm{max}} = 5$ has been used to represent the strong lithosphere near the plate boundaries. While no external constraints have been determined for the value of V_{max} (these zones appear relatively rigid in the experiments), lithosphere that is five times stronger than the lithosphere in the weak zone implies a Moho temperature on the order of $50°$ to $70°\mathrm{C}$ colder than that of the weak zone (Neil & Houseman 1997; fig. 6).

Also based on equation (45), Figure 15 shows how the time period of the orogeny (t) depends on the Moho temperature in the weak zone (T_m) for (a) different values of the applied boundary force (F_x) and (b) different crustal thickening factors β_{max}. Figure 15 shows that, regardless of the magnitude of the force per unit length applied at the boundary, or assumed crustal thickening factor, the required duration of orogenesis decreases, almost exponentially with increasing Moho temperature. If the crustal thickening factor increases, or if the plate boundary force decreases, the required duration of the orogeny increases.

Figure 16 shows how the maximum crustal thickening factor (β_{max}) depends on Moho temperature for the reference plate boundary force

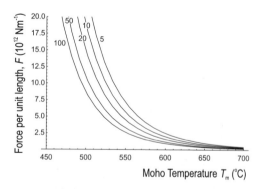

Fig. 14. The dependence of plate boundary force per unit length F_x on Moho temperature T_m, using a non-Newtonian viscosity ($n = 3$) with rheological parameters for wet olivine, as specified in the text. The force required to produce maximum crustal thickening of $\beta_{\mathrm{max}} = 1.66$ is shown for a range of orogenic durations t. Each curve is labeled with the t value (in Ma). This graph was calculated assuming that deformation was driven only by T'_x with $T'_y = 0$.

(a)

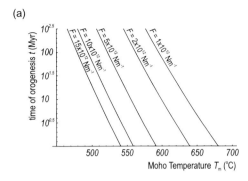

(b)

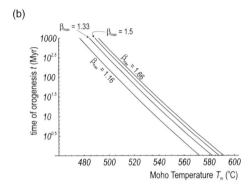

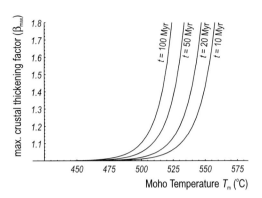

Fig. 16. Maximum crustal thickening factor ($\beta_{\max}$) v. Moho temperature in the weak zone (T_m) for non-Newtonian viscosity ($n = 3$), and an applied boundary force of 5×10^{12} Nm^{-1}. Each line shows the dependence for a different value of orogenic duration t (10 Ma to 100 Ma). This graph was calculated assuming that deformation was driven by T'_x and that $T'_y = 0$.

Fig. 15. (a) Dependence of time period of the orogeny (t) on the Moho temperature in the weak zone (T_m), for different values of boundary force between $F_x = 1 \times 10^{12}$ Nm^{-1} and $F_x = 15 \times 10^{12}$ Nm^{-1}, where $\beta_{\max} = 1.66$. **(b)** Dependence of t on T_m for different degrees of crustal thickening ($\beta_{\max}$ from 1.16 to 1.66) for calculations with a boundary force of $F_x = 5 \times 10^{12}$ Nm^{-1}. Both graphs are based on Equation (45), assuming a wet olivine constitutive law ($n = 3$) with a driving stress provided by T'_x with $T'_y = 0$.

of 5×10^{12} Nm^{-1} and a range of orogenic time intervals. The strong non-linearity of equation (45) implies that small changes in Moho temperature can lead to large differences in the maximum crustal thickening factor for a constant boundary force. If Moho temperatures in central Australia during the Alice Springs Orogeny were $> c.\ 525°C$ then these calculations imply that the duration of the orogeny was <100 Ma and therefore that deformation might have occurred in smaller, discrete time intervals rather than continuously. Of course Moho temperature estimates depend on the assumed magnitude of the plate boundary force. Moreover, the simplified rheological model used here neglects the contribution to lithospheric strength of the crustal layer. Thus it leads perhaps to an overstatement of the tem-

perature dependence of lithospheric strength. This model provides a guide to the effect of relative changes in Moho temperature rather than a robust estimate of actual Moho temperature.

Conclusions

The numerical experiments presented here demonstrate that clockwise rotation of the northern boundary of a lithospheric sheet with an internal weak zone may produce crustal thickening in the region representing central Australia and thinning in the region representing the Canning Basin. In this model a clockwise rotation of the northern boundary may be induced either by an eastward shear traction or by a clockwise bending moment. It appears that, for $n = 3$ calculations, the actual stress regime that produces the rotation does not influence the dependence of maximum crustal thickening $\beta_{\max}$ on boundary rotation θ since, for all combinations of T'_x and T'_y, tested for a specific geometry, the same, almost linear dependence of $\beta_{\max}$ on θ is seen. Changing the geometry of the sheet, however, does affect the slope of the $\beta_{\max}$ v. θ graph. If the E–W width of the continental sheet is doubled, or if the N–S width of the weak zone is halved, then the slope of the $\beta_{\max}$ v. θ dependence is effectively doubled and the same crustal thickening factor is predicted for a clockwise rotation only half as great. For plausible estimates of thin viscous sheet geometry and variation of lithospheric strength within the sheet, we infer that clockwise rotation of the northern Australian block of order 20–25° is required to produce crustal thickening in central Australia of order $\beta_{max} = 1.67$. If a clockwise rotation of this magnitude did occur between northern and southern Australia during the Alice Springs Orogeny it could, in principle, be measured by a change in paleomagnetic declination.

These calculations indicate that for an orogenic time span of 100 Ma and a plate boundary force of order $5 \times 10^{12}\,\text{Nm}^{-1}$, the depth-averaged strength of the lithosphere in central Australia prior to the Alice Springs orogeny was of order $B_0 = 0.8 \times 10^{13}\,\text{Pa}\,\text{s}^{1/3}$. Using a simplified model of lithospheric strength (England 1983) this strength coefficient (B_0) corresponds approximately to a pre-orogenic Moho temperature in central Australia of $c.\,520°\text{C}$. The concentration of deformation in this relatively narrow zone that stretches E–W across the continent, implies that the blocks to the North and South are stronger than the central region. Such a strength difference could be explained by a decrease in Moho temperature of $c.\,60°\text{C}$ from central Australia to the North and South. If Moho temperatures prior to the Alice Springs orogeny were higher than those estimated above, the lithosphere would be weaker and the same amount of deformation may have occurred in a time period of <100 Ma. Shorter periods of deformation would imply that the deformation associated with the orogeny was episodic rather than continuous.

References

ARIF, M. & JAN, M. Q. 1993. Chemistry of chromite and associated phases from Shangla ultramafic body in the Indus suture zone of Pakistan. *In*: TRELOAR, P. J. & SEARLE, M. P. (eds) *Himalayan Tectonics*. Geological Society, London, Special Publications, **74**, 101–112.

BIRD, P. 1978. Stress and temperature in subduction shear zones: Tonga and Mariana. *Geophysical Journal of the Royal Astronomical Society*, **55**, 411–434.

BIRD, P. & PIPER, K. 1980. Plane-stress finite element models of tectonic flow in southern California. *Physics of the Earth and Planetary Interiors*, **21**, 158–175.

BLACK, L. P. & SHAW, R. D. 1995. An assessment, based on U–Pb zircon data, of Rb–Sr dating in the Arunta Inlier, central Australia. *In*: COLLINS, W. J. & SHAW, R. D. (eds) *Time limits on Tectonic Events and Crustal Evolution Using Geochronology: Some Australian Examples*. Precambrian Research, **71**, 3–15.

BRAUN, J., MCQUEEN, H. & ETHERIDGE, M. 1991. A fresh look at the late Paleozoic tectonic history

of west central Australia. *Exploration Geophysics*, **22**, 49–54.

BRAUN, J. & SHAW, R. 1999. A continental-scale tectonic model for the reactivation of Proterozoic sutures in the Mid to Late Paleozoic. *Geological Society of Australia Abstracts*, **54**, 8.

BRAUN, J. & SHAW, R. 2001. A thin-plate model of Palaeozoic deformation of the Australian lithosphere: Implications for understanding the process of cartonization, orogenesis, reactivation and reworking. *In*: MILLER, J. A., HOLDSWORTH, R. E., BUICK, I. S. & HAND, M. (eds) *Continental Reactivation and Reworking*. Geological Society, London, Special Publications, **184**, 165–193.

CARTWRIGHT, I. & BARNICOAT, A. C. 1999. Stable isotope geochemistry of Alpine ophiolites: a window to ocean-floor hydrothermal alteration and constraints on fluid-rock interaction during high-pressure metamorphism. *International Journal of Earth Sciences*, **88**, 219–235.

CARTWRIGHT, J. & BUICK, I. S. 1999. The flow of surface-derived fluids through Alice Springs age middle-crustal shear zones, Reynolds Range, central Australia. *Journal of Metamorphic Geology*, **17**, 397–414.

CHEN, Z., LI, Z. X., POWELL, C. MCA. & BALME, B. E. 1993. Palaeomagnetism of the Brewer Conglomerate in central Australia, and fast movement of Gondwanaland during the Late Devonian. *Geophysical Journal International*, **115**, 564–574.

CHRISTENSEN, N. I. & MOONEY, W. D. 1995. Seismic velocity structure and composition of the continental crust: A global view. *Journal of Geophysical Research*, **100**, 9761–9788.

COLLINS, W. J. & SHAW, R. D. 1995. Geochronology constraints on orogenic events in the Arunta Inlier: a review. *In*: COLLINS, W. J. & SHAW, R. D. (eds) *Time limits on Tectonic Events and Crustal Evolution Using Geochronology: Some Australian Examples*. Precambrian Research, **71**, 315–346.

COLLINS, W. J. & WILLIAMS, I. S. 1995. SHRIMP ion-probe dating of short-lived Proterozoic tectonic cycles in the northern Arunta Inlier, central Australia. *In*: COLLINS, W. J. & SHAW, R. D. (eds) *Time limits on Tectonic Events and Crustal Evolution Using Geochronology: Some Australian Examples*. Precambrian Research, **71**, 69–89.

CONEY, P. J. 1992. The Lachlan belt of eastern Australia and Circum-Pacific tectonic evolution. *In*: FERGUSSON, C. L. & GLEN, R. A. (eds) *The Palaeozoic Eastern Margin of Gondwanaland: Tectonics of the Lachlan Fold Belt, Southeastern Australia and Related Orogens*. Tectonophysics, **214**, 1–25.

DRUMMOND, B. J., SEXTON, M. J., BARTON, T. J. & SHAW, R. D. 1991. The nature of faulting along the margins of the Fitzroy Trough, Canning Basin, and implications for the tectonic development of the Trough. *Exploration Geophysics*, **22**, 111–116.

DUNLAP, W. J. & TEYSSIER, C. 1995. Paleozoic deformation and isotopic disturbance in the southern Arunta Block, central Australia. *In*: COLLINS, W. J. & SHAW, R. D. (eds) *Time limits on Tectonic Events and Crustal Evolution Using*

Geochronology: Some Australian Examples. Precambrian Research, **71**, 229–250.

DUNLAP, W. J., TEYSSIER, C., MCDOUGALL, I. & BALDWIN, S. 1995. Thermal and structural evolution of the intracratonic Arltunga Nappe Complex, central Australia. *Tectonics*, **14**, 1182–1204.

ENGLAND, P. C. 1983. Constraints on extension of continental lithosphere. *Journal of Geophysical Research*, **88**, 1145–1152.

ENGLAND, P. C. & HOUSEMAN, G. 1986. Finite strain calculations of continental deformation, II: Comparison with the India–Asia collision zone. *Journal of Geophysical Research*, **91**, 3664–3676.

ENGLAND, P. C. & MCKENZIE, D. P. 1982. A thin viscous sheet model for continental deformation. *Geophysical Journal of the Royal Astronomical Society*, **70**, 295–321. (Erratum, 1983. *Geophysical Journal of the Royal Astronomical Society*, **73**, 523–532.)

ETHERIDGE, M. A., RUTLAND, R. W. R. & WYBORN, L. A. I. 1987. Orogenesis and tectonic process in the Early to Middle Proterozoic of northern Australia. *In*: KRÖNER, A. (ed.) *Precambrian Lithospheric Evolution*. American Geophysical Union, Geodynamics Series, **17**, 131–147.

FODEN, J., MAWBY, J., KELLY, S., TURNER, S. & BRUCE, D. 1995. Metamorphic events in the eastern Arunta Inlier, Part 2. Nd–Sr–Ar isotopic constraints. *Precambrian Research*, **71**, 207–227.

FORMAN, D. J. & SHAW, R. D. 1973. Deformation of the crust and mantle in central Australia. *Bureau of Mineral Resources, Australia Bulletin*, **144**, 20.

GOLEBY, B. R., SHAW, R. D., WRIGHT, C., KENNETT, B. L. N. & LAMBECK, K. 1989. Geophysical evidence for 'thick-skinned' crustal deformation in central Australia. *Nature*, **337**, 325–330.

GRAY, D. R. 1988. Structure and Tectonics. *In*: DOUGLAS, J. D. & FERGUSON, J. A. (eds) *Geology of Victoria*. Geological Society of Australia, Melbourne, 1–36.

HOUSEMAN, G. A. & ENGLAND, P. C. 1986. Finite strain calculations of continental deformation, I: Method and general results for convergent zones. *Journal of Geophysical Research*, **91**, 3651–3663.

HOUSEMAN, G. A. & ENGLAND, P. C. 1993. Crustal thickening versus lateral expulsion in the India–Asia continental collision. *Journal of Geophysical Research*, **98**, 12 233–12 249.

HOUSEMAN, G. A. & MOLNAR, P. 1997. Gravitational (Rayleigh–Taylor) instability of a layer with non-linear viscosity and convective thinning of continental lithosphere. *Geophysical Journal International*, **128**, 125–150.

HURLEY, N. F. & VAN DER VOO, R. 1987. Paleomagnetism of upper Devonian reefal limestones, Canning Basin, Western Australia. *Geological Society of America Bulletin*, **98**, 138–146.

JAN, M. Q. & WINDLEY, B. F. 1990. Cr–spinel–silicate chemistry in ultramafic rocks of the Jijal complex, NW Pakistan. *Journal of Petrology*, **31**, 667–715.

KARATO, S., PATERSON, M. S. & FITZGERALD, J. D. 1986. Rheology of synthetic olivine aggregates:

Influence of grainsize and water. *Journal of Geophysical Research*, **91**, 8151–8176.

KENNARD, J. M., JACKSON, M. J., ROMINE, K. K., SHAW, R. D. & SOUTHGATE, P. N. 1994. Depositional sequences and associated petroleum systems of the Canning Basin, W.A. *In*: PURCELL, P. G. & PURCELL, R. R. (eds) *The sedimentary basins of Western Australia*. Proceedings of the Petroleum Exploration Society of Australia Symposium, Perth, 657–676.

KLOOTWIJK, C. T. 1980. Early Palaeozoic Palaeomagnetism in Australia. *Tectonophysics*, **64**, 249–332.

KRABBENDAM, M. 2001. When the Wilson Cycle breaks down: How orogens can produce strong lithosphere and inhibit their future reworking. *In*: MILLER, J. A., HOLDSWORTH, R. E., BUICK, I. S. & HAND, M. (eds) *Continental Reactivation and Reworking*. Geological Society, London, Special Publications, **184**, 57–75.

KUSZNIR, N. J. & PARK, R. G. 1984. Intraplate lithosphere deformation and the strength of the lithosphere. *Geophysical Journal of the Royal Astronomical Society*, **79**, 513–538.

LAMBECK, K., BURGESS, G. & SHAW, R. D. 1988. Teleseismic travel-time anomalies and deep crustal structure in central Australia. *Geophysical Journal of the Royal Astronomical Society*, **94**, 105–124.

LONG, J. & BURRETT, C. 1989. Fish from the Upper Devonian of the Shan-Thai Terrane indicate proximity to East Gondwana and South China terranes. *Geology*, **17**, 811–813.

MAWBY, J., FODEN, J., KELLEY, S. & MCDOUGALL, I. 1995. Chronological constraints on the thermal evolution of the Harts Range, Arunta Inlier. *Geological Society of Australia Abstracts*, **40**, 102–103.

METCALF, I. 1996. Gondwanaland dispersion, Asian accretion and evolution of eastern Tethys. *Australian Journal of Earth Sciences*, **43**, 605–623.

MITCHELL, M. M., KOHN, B. P. & FOSTER, D. A. 1997. Apatite Fission Track Thermochronology of Eastern South Australia. *Abstracts, Geodynamics and Ore Deposits Conference*, Australian Geodynamics Cooperative Research Centre, Ballarat, Victoria, 125.

MOLNAR, P. & TAPPONNIER, P. 1981. A possible dependence of tectonic strength on the age of the crust in Asia. *Earth and Planetary Science Letters*, **52**, 107–114.

MURRAY, C. G., FERGUSSON, C. H., FLOOD, P. G., WHITAKER, W. G. & KORSCH, R. J. 1987. Plate tectonic model for the Carboniferous evolution of the New England Fold Belt. *Australian Journal of Earth Sciences*, **34**, 213–236.

NEIL, E. A. & HOUSEMAN, G. A. 1997. Geodynamics of the Tarim Basin and the Tian Shan in central Asia. *Tectonics*, **16**, 571–584.

PLUMB, K. A. 1979. Tectonic evolution of Australia. *Earth Science Reviews*, **14**, 205–249.

POWELL, C. MCA. 1984. Terminal fold-belt deformation: Relationship of mid-Carboniferous megakinks in the Tasman fold belt to coeval thrusts in cratonic Australia. *Geology*, **12**, 546–549.

PRESS, W. H., TEUKOLSKY, S. A., VETTERLING, W. T. & FLANNERY, B. P. 1992. *Numerical Recipes in Fortran, The Art of Scientific Computing, 2nd edition.* Cambridge University Press, New York.

PURCELL, P. G. & POLL, J. 1984. The seismic definition of the main structural elements of the Canning Basin. *In*: PURCELL, P. G. (ed.) *The Canning Basin, W.A.* Geological Society of Australia, Incorporated and the Petrolium Exploration Society of Australia Ltd., Perth, 75–84.

SANDIFORD, M. & HAND, M. 1998. Controls on the locus of intraplate deformation in central Australia. *Earth and Planetary Science Letters*, **162**, 97–110.

SCOTESE, C. R. & MCKERROW, W. S. 1990. Revised world maps and introduction. *In*: MCKERROW, W. S. & SCOTESE, C. R. (eds) *Palaeozoic Palaeogeography and Biography.* Geological Society Memoir, **12**, 1–21.

SHAW, R. D. & BLACK, L. P. 1991. The history and tectonic implications of the Redbank Thrust Zone, central Australia, based on structural, Metamorphic and Rb–Sr isotopic evidence. *Australian Journal of Earth Sciences*, **38**, 307–332.

SHAW, R. D., KORSCH, R. J., WRIGHT, C. & GOLEBY, B. R. 1991. Seismic interpretation and thrust tectonics of the Amadeus Basin, central Australia, along the BMR regional seismic line. *In*: KORSCH, R. J. & KENNARD, J. M. (eds) *Geological and geophysical studies in the Amadeus Basin, central Australia.* Bureau of Mineral Resources, Australia, Bulletin of Geology and Geophysics, **236**, 385–408.

SHAW, R. D., SEXTON, M. J. & ZEILINGER, I. 1994. *The tectonic framework of the Canning Basin, W.A., including the 1:2 million structural element map of the Canning Basin.* Australian Geological Survey Organization Record **1994/48**.

SHAW, R. D., STEWART, A. J. & BLACK, L. P. 1984. The Arunta Inlier: A complex ensialic mobile belt in central Australia. Part 2: Tectonic history. *Australian Journal of Earth Sciences*, **31**, 457–484.

SHAW, R. D., WELLMAN, P., GUNN, P., WHITAKER, A. J., TARLOWSKI, C. & MORSE, M. 1996. *Guide to using the Australian Crustal Elements Map.* Australian Geological Survey Organization Record **1996/30**.

SONDER, L. J. & ENGLAND, P. C. 1986. Vertical averages of rheology of the continental lithosphere: Relation to thin-sheet parameters. *Earth and Planetary Science Letters*, **77**, 81–90.

SPIKINGS, R. A., FOSTER, D. A., GLEADOW, A. J. W. & KOHN, B. P. 1996. Phanerozoic and denudational history of the Mt. Isa Inlier, Queensland. *Geological Society of Australia Abstracts*, **41**, 414.

STEWART, A. J., SHAW, R. D. & BLACK, L. P. 1984. The Arunta Inlier: A complex ensialic mobile belt in central Australia. Part 1: Stratigraphy, correlation and origin. *Australian Journal of Earth Sciences*, **31**, 445–455.

TEYSSIER, C. 1985. A crustal thrust system in an intracratonic tectonic environment. *Journal of Structural Geology*, **7**, 689–700.

TOMMASI, A., VAUCHEZ, A. & DAUDRÈ, B. 1995. Initiation and propagation of shear zones in a heterogeneous continental lithosphere. *Journal of Geophysical Research*, **100**, 22 083–22 101.

VEEVERS, J. J., CLARE, A. & WOPFNER, H. 1994. Neocratonic magmatic-sedimentary basins of post-Variscan Europe and post-Kanimblan eastern Australia generated by right lateral transtension of Permo-Carboniferous Pangea. *Basin Research*, **6**, 141–157.

WELLS, A. T. & MOSS, J. F. 1983. The Ngalia Basin, Northern Territory: Stratigraphy and structure. *Bureau of Mineral Resources, Australia, Bulletin* **212**, 88.

YOUNG, G. C. 1990. Devonian vertebrate distribution patterns and cladistic analysis of palaeogeographic hypotheses. *In*: MCKERROW, W. S. & SCOTESE, C. R. (eds) *Palaeozoic Palaeogeography and Biography.* Geological Society Memoir, **12**, 243–255.

A thin-plate model of Palaeozoic deformation of the Australian lithosphere: implications for understanding the dynamics of intracratonic deformation

JEAN BRAUN & RUSSELL SHAW

Research School of Earth Sciences, The Australian National University
Canberra ACT 0200, Australia (e-mail: jean.braun@anu.edu.au)

Abstract: Presented here are the results of a thin-plate model of the continental lithosphere in which deformation is driven by velocities specified along the plate boundaries. The geometry of the model, the strength of each lithospheric block, and the boundary conditions have been chosen to reproduce the major tectonic episodes experienced by the Australian continent during a 200 Ma time period starting in the Ordovician (i.e. 470 Ma). The model's focus is on the reactivation and/or reworking of zones of weakness within the continent that have either been set *a priori* or developed in response to previous tectonic regimes. Using the tectonic history of the Australian continent as a natural laboratory in which hypotheses on the nature and style of intracratonic deformation can be tested, the following conclusions can be made: (i) intracratonic deformation results from the concentration of strain into regions of decreased lithospheric strength; these weak zones are often caused by previous intracratonic deformation and/or develop at the interface between regions of contrasting strength; (ii) repeated deformation episodes lead to strain localization; (iii) localized deformation may also take place as the result of the constructive interaction between two tectonic regimes originating on separate margins; and (iv) there are mechanisms that operate within the lithosphere by which deformation leads to local strengthening.

According to the theory of Plate Tectonics, orogenesis can be regarded as the product of the interaction of quasi-rigid lithospheric plates forced to move along the Earth's surface under the action of tectonic forces originating in the underlying, viscous mantle. Consequently, the theory can explain why orogenesis commonly takes place along plate margins; it is also likely to affect mostly continents, because continental plates are much weaker than their oceanic counterparts. In many instances, however, orogens develop in a continent's interior, away from active plate boundaries. This implies that either (a) the forces driving continental deformation originate in the mantle beneath the continents, or (b) the continental lithosphere (or part of it) is acting as an effective stress guide, capable of transmitting substantial horizontal stresses over large distances; in both cases, there must exist mechanisms through which deformation is allowed to localize, either by stress concentration or due to a local reduction in lithospheric strength.

A well-documented example of intracratonic deformation is the late Devonian to mid-Carboniferous Alice Springs Orogeny of central Australia (Forman & Shaw 1973) which has left major lithospheric-scale scars in the interior of the continent (Lambeck *et al.* 1988; Goleby *et al.* 1989), as evidenced by the presence of large Bouguer gravity anomalies (Wellman 1976). These anomalies result from large offsets in the crust-mantle boundary caused by N–S shortening in the late Palaeozoic (Goleby *et al.* 1989). Interestingly, at the same time, the northwestern part of the Australian continent was subjected to strong NNE–SSW extension which led to lithospheric stretching, subsidence and the accumulation of 10 km of mostly marine sediments in the Fitzroy Trough (Forman & Wales 1981; Braun *et al.* 1991; Braun & Shaw 1998*b*). That concomitant events of opposite character took place in very close proximity to each other suggests that intracontinental deformation can be a complex process.

There is, at present, no consensus on the nature of the mechanism(s) driving intracontinental deformation. As mentioned earlier, two end-member hypotheses have been proposed: remote in-plane forces originating along the plate boundaries (at subduction zones or mid-ocean ridges) (Lambeck 1983*a*; Thomas & Gibbs 1985), v. local vertical forces originating in the mantle directly beneath the continent (Middleton 1989; Houseman & Molnar 1997; Neil & Houseman 1999). For example, the Alice Springs Orogeny

From: MILLER, J. A., HOLDSWORTH, R. E., BUICK, I. S. & HAND, M. (eds) *Continental Reactivation and Reworking.*
Geological Society, London, Special Publications, **184**, 165–193. 1-86239-080-0/01/$15.00 © The Geological
Society of London 2001.

has been regarded by many (Lambeck 1983*b*; Shaw & Black 1991) as the classical example of reactivation of a pre-existing structure (the Redbank Thrust Zone along the southern margin of the Arunta Block) by in-plane forces originating along the northern or southern margin of the continent, while others (Braun & Shaw 1998*a*) prefer a mechanism that invokes mantle delamination or gravitational instabilities at or near the base of the thermo-mechanical lithosphere. It is also possible that both mechanisms were at play during the Alice Springs Orogeny.

In this study, the first hypothesis is followed which assumes that in-plane forces play a dominant role in deforming continent interiors. The circumstances under which these forces are likely to reactivate internal structures inherited from previous tectonism such as, in the case of the Australian continent, Proterozoic accretionary events (Myers *et al.* 1996), are investigated. Here, focus is placed on a set of relatively well-defined orogenic events which took place within the interior of the Australian plate during a 200 Ma time interval (from the early Ordovician to the late Carboniferous), these are then used as 'natural experiments' to investigate the dynamic behaviour of the continental lithosphere.

To achieve this, a sophisticated thermo-mechanical numerical model is used. The model's behaviour is based on what is believed to be a reasonable representation of the mechanical properties of the continental lithosphere, i.e. the strength distribution with depth within the lithosphere, and how it is affected by the temperature distribution. Temperature is, in turn, controlled by the evolving vertical distribution of heat producing radioactive elements (Ranalli 1997), the advection of heat by rock uplift/subsidence and thermal conduction. The model's behaviour is also dictated by (a) the presence of mechanical heterogeneities arbitrarily introduced at the start of computation and (b) a set of velocity boundary conditions imposed along the continent margins.

Little is known, however, about the thermal state of the Australian continent or the magnitude and direction of the relative movement between lithospheric plates in the early Palaeozoic. Therefore, the geometry of the mechanical heterogeneities introduced in the model and the velocity conditions imposed along the boundaries are constrained only to first-order by the geological record. It is shown, however, how the model can be used to put further constraints on the assumed initial strength distribution and the boundary conditions, as a result of the feedback between deformation and strength (resulting from the strong thermo-mechanical coupling).

The model is then used to explore the following questions regarding the behaviour of the interior of continental plates during orogenic events:

(a) Can a complex tectonic history, like that of the Australian continent in the Palaeozoic, be understood by invoking solely in-plane driving forces?
(b) What complexities can develop from the interplay between a set of interacting strong and weak zones and different sets of imposed boundary conditions?
(c) How is the internal strength of the continental interior affected by progressive and repeated lithospheric-scale deformation events?

Tectonic evolution of Australia in the middle Palaeozoic

In this section, a brief outline of the various large-scale orogenic events that have shaped the Australian continent during the middle Palaeozoic, more precisely from the middle Ordovician to the late Carboniferous is given. This period has been selected for two resaons: firstly, its tectonic evolution is relatively well constrained by geological observations; secondly, many of the large–scale geological features of the present-day Australian craton have formed during this time period.

Unnamed continental-wide Ordovician extension (475–450 Ma)

From 475 Ma to 455 Ma, extension-driven subsidence led to the deposition of a thick sedimentary sequence over a broad area of eastern Australia (Fergusson & Coney 1992); the most likely driving mechanisms are back-arc spreading or platform sedimentation (Mitrovica *et al.* 1989; Gallagher & Lambeck 1989; de Caritat & Braun 1992) associated with a subduction zone along the eastern seaboard.

During a similar time period (i.e. from 470 to 450 Ma), extension and deposition in the Willara and Kidson sub-basins of the Canning Basin during the Samphire Marsh Movement (Shaw *et al.* 1994) suggest that extension/sea-floor spreading was taking place along the northern margin of the continent. Similarly, in central Australia, deep-seated extension in the lower crust and mantle can explain (a) a rapid increase in subsidence rate during deposition of the Ordovician Stairway Formation (Shaw *et al.* 1991, fig. 5), (b) palaeogeographic reconstructions of changes in the distribution of Ordovician deposition (Walley *et al.* 1991), and (c) high-grade metamorphism in subsequently exhumed basement (Harts Range) during the so-called Larapinta Event at 475–450 Ma (Hand *et al.* 1999*a*, *b*; Buick *et al.* 2001).

Benambran Orogeny (440–420 Ma)

At 440 Ma, subduction along the eastern seaboard changed character (potentially due to a younging of the subducting plate or the subduction of a triple junction) (Collins & Vernon 1992) and led to compression of the overriding continent. This caused shortening of the recently deposited turbidite sequence in the western and central Lachlan Fold Belt (from 440 to 425 Ma) during the Benambran Orogeny (Collins & Vernon 1992; Gray *et al.* 1997). A similar event is proposed to explain Silurian deformation and regional metamorphism in the Coen Inlier and Georgetown province (cf. Bain and Draper 1997).

A second compressive pulse is recorded along the margins of the Wagga-Omeo Metamorphic Belt, leading to deformation in the eastern Lachlan Fold Belt during the Quidongan Orogeny (430–420 Ma) and the more restricted Bindian Orogeny (410–405 Ma) (Foster *et al.* 1998), potentially corresponding to the closure of an ocean basin and the accretion of the Wagga-Omeo Metamorphic Belt (Gray *et al.* 1997) to the Australian continent.

Tabberaberan Orogeny (410–370 Ma)

Final accretion of the eastern Lachlan Fold Belt (during the Tabberabberan Orogeny) corresponds to a widespread phase of compressional deformation in the Lachlan Fold Belt (Gray *et al.* 1997). In the Early to Mid-Devonian (410–385 Ma), extension and subsidence took place in the Drummond Basin, in central Queensland, probably driven by a phase of back-arc extension associated with the ongoing subduction along the eastern seaboard (de Caritat & Braun 1991, 1992). During a similar time interval (410–370 Ma), central Australia experienced mild compression during the Pertnjara Movement (Shaw 1991) marking the start of the Alice Springs Orogeny.

Alice Springs Orogeny (370–300 Ma)

The Late Devonian to late Carboniferous tectonic history of the Australian continent is dominated by concomitant N–S compression in central Australia and NNE–SSW extension in northwestern Australia during the well-documented Alice Springs Orogeny (Braun *et al.* 1991; Shaw & Black 1991). Little is known of the nature and location of the mechanism that is responsible for this continent-wide deformation. However, at the time, Australia was part of a larger land mass including Antarctica to the South and India to the West. For this reason, and because deformation during the Alice Springs Orogeny was partitioned between two N–S trending corridors, it is most likely that the driving forces originated along the northern margin of the continent (Braun *et al.* 1991).

Plate reconstructions (mostly based on palaeomagnetic evidence) suggest that, during this time interval, several micro-plates were actively rifting from Gondwana along what is now the northern margin of the Australian continent (see Klootwijk (1996) for a comprehensive review). The micro-plates involved in the successive rifting episodes included North and South China, the Tarim Basin and Indochina (Klootwijk 1996). The exact timing of rifting and the pre-rift location of each micro-plate are not well known and prevent us from building an accurate model for the forces that must have been active at this time along the northern margin of the continent.

The relatively well-preserved sedimentological and structural evidence from the various central and western Australian basins (Forman & Wales 1981; Shaw 1987; Shaw *et al.* 1994; Colwell & Kennard 1996) suggests, however, that rifting of the micro-plates along the northwestern margin of the continent was not always orthogonal to the plate margin but, at times, may have resulted in a torque being applied to the northern margin of the continent. It is also possible, as speculated by Klootwijk (1995) on the basis of palaeomagnetic evidence, that this extended period of rifting was interrupted by an episode of convergence between the Australian craton and the Siberian terrane of 'Altaids' during the mid-Carboniferous.

Extension in the Petrel sub-basin (the main Late Palaeozoic depocentre in the Bonaparte Gulf Basin) during the Late Devonian to mid-Carboniferous (Gunn 1998) is clearly concomitant with extension in the on-shore Fitzroy Trough (northeastern Canning Basin) (Forman & Wales 1981; Shaw *et al.* 1994). The sediment record suggests that both basins accommodated a similar amount of NE–SW extension by movement along crustal-scale listric normal faults. The only major difference between the two basins is the post-extensional thermal subsidence sag phase, which is much more clearly defined in the Petrel sub-basin than in the Fitzroy Trough. This disparity must reflect a difference in the style and/or location of extension in the underlying mantle lithosphere.

Compressional regime of the New England Orogen (380–325 Ma)

In the Late Devonian to mid-Carboniferous, the New England Fold Belt was part of an active convergent margin associated with a westerly dipping subduction zone (Murray *et al.* 1987). Apart from the well defined volcanic arc, forearc basin and subduction complex system preserved along this palaeo-margin, there is little indication of large-scale crustal shortening and/or sediment deposition in the neighbouring Drummond Basin (de Caritat & Braun 1991). This suggests that the continent interior was only subjected to mild E–W compression.

The Early Carboniferous Kanimblan Orogeny (360–320? Ma)

The last major tectonic event to affect southeastern Australia in the Palaeozoic is the Late Devonian to early Carboniferous Kanimblan Orogeny, mostly recognized in the northern part of the eastern Lachlan Fold Belt (Powell *et al.* 1977). This event is marked by mild compression of Devonian sediments following reactivation of pre-existing structures (Morand & Gray 1990). Several interpretations of this event have been presented, including (1) the last stage in the closure of a marginal sea along the eastern seaboard (Fergusson & Coney 1992), (2) the postorogenic isostatic adjustment of a region of anomalously thickened crust (Gray *et al.* 1997), or (3) the result of southward propagation of deformation from the active margin of the New England Fold Belt (Powell *et al.* 1977).

Transpressional regime of the New England Orogen (325–300 Ma)

In the mid-Carboniferous, changes in the subduction style along the eastern seaboard (Veevers *et al.* 1982) transformed the northeastern Australian margin from a convergent plate boundary to a convergent-transform plate boundary (Murray *et al.* 1987). This event led to large dextral movement along continental-scale structures (the Yarrol-Great Moreton Fault and the Peel Fault; Fig. 4 in Murray *et al.* 1987) similar in size and nature to the present-day San Andreas fault system. A substantial component of E–W

compression resulted in crustal thickening, and the formation of a foreland basin by lithospheric flexure to the west of the collision front (the lower section of the Galilee and Bowen basins in south-central Queensland) (de Caritat & Braun 1992). Mild reactivation in a transpressional dextral sense is also observed along pre-existing structures in the eastern and central Lachlan Fold Belt (Morand & Gray 1990).

Model description

Mechanical model

It is assumed that, at the scale of the continent, lithospheric deformation may be approximated by that of a thin viscous sheet in which a dynamic equilibrium exists between viscous and gravitational forces (Bird & Piper 1980; England & McKenzie 1982, 1983). The basic coupled partial differential equations governing the deformation of an incompressible thin viscous lithospheric plate of viscosity η, thickness L and crustal thickness S are (Bird & Piper 1980):

$$\frac{\partial \bar{p}}{\partial x} + \frac{\partial}{\partial x}\left[2\eta\left(\frac{\partial u}{\partial x} + \frac{\partial v}{\partial y}\right)\right]$$
$$+ \frac{\partial}{\partial y}\left[\eta\left(\frac{\partial u}{\partial y} + \frac{\partial v}{\partial x}\right)\right] = 0 \qquad (1)$$

$$\frac{\partial \bar{p}}{\partial y} + \frac{\partial}{\partial y}\left[2\eta\left(\frac{\partial u}{\partial x} + \frac{\partial v}{\partial y}\right)\right]$$
$$+ \frac{\partial}{\partial x}\left[\eta\left(\frac{\partial u}{\partial y} + \frac{\partial v}{\partial x}\right)\right] = 0 \qquad (2)$$

$$\bar{p} = \frac{1}{L}\int_0^L p\,dz = \frac{1}{L}\int_0^L \rho g z\,dz$$
$$= \frac{\rho_c g S^2}{2L}\left(1 - \frac{\rho_c}{\rho_m}\right) + \frac{\rho_m g L}{2} \qquad (3)$$

$$\frac{1}{S}\frac{\partial S}{\partial t} = -\left(\frac{\partial u}{\partial x} + \frac{\partial v}{\partial y}\right) \qquad (4)$$

where, x and y, are the horizontal spatial coordinates; and u and v, are the Eulerian velocities in the x- and y-directions, respectively. ρ_c and ρ_m are the mean crustal and mantle densities, respectively; their values are given in Table 1. $\bar{p}$ is the depth-averaged lithostatic pressure.

Under the thin sheet approximation, it is not possible to resolve features in the predicted deformation field on a scale smaller than the

Table 1. *Model parameter values*

Parameter	Symbol	Value
Granulite rheology:		
Pre-exponential constant	B_c	$4.75 \times 10^6 \,\mathrm{Pa}\, \mathrm{s}^{1/n_c}$
Stress exponent	n_c	3.1
Activation energy	Q_c	$243 \times 10^3 \,\mathrm{J}\, \mathrm{mol}^{-1}$
Olivine rheology:		
Pre-exponential constant	B_m	$2.64 \times 10^5 \,\mathrm{Pa}\, \mathrm{s}^{1/n_m}$
Stress exponent	n_m	4.5
Activation energy	Q_m	$498 \times 10^3 \,\mathrm{J}\, \mathrm{mol}^{-1}$
Crustal brittle parameter	$A_{0,c}$	In compression: $4 \times 10^4 \,\mathrm{Pa}\, \mathrm{m}^{-1}$ In extension: $1 \times 10^4 \,\mathrm{Pa}\, \mathrm{m}^{-1}$
Mantle brittle parameter	$A_{0,m}$	In compression: $4 \times 10^4 \,\mathrm{Pa}\, \mathrm{m}^{-1}$ In extension: $1 \times 10^4 \,\mathrm{Pa}\, \mathrm{m}^{-1}$
Crustal density	ρ_c	$2850 \,\mathrm{kg}\, \mathrm{m}^{-3}$
Mantle density	ρ_m	$3200 \,\mathrm{kg}\, \mathrm{m}^{-3}$
Initial lithospheric thickness	L	120 km
Initial crustal thickness	S	35 km
Thermal conductivity	k	$3 \,\mathrm{W}\, \mathrm{m}^{-1} \,°\mathrm{C}^{-1}$
Specific heat	c	$1 \times 10^3 \,\mathrm{J}\, \mathrm{kg}^{-1} \,°\mathrm{C}^{-1}$
Temperature at the base of the lithosphere	T_L	1350°C
Erosion constant	K_D	$1 \times 10^2 \,\mathrm{m}^{-2} \,\mathrm{yr}^{-1}$
Elastic thickness	t_a	50 km
Young modulus	E_a	$1 \times 10^{11} \,\mathrm{Pa}$
Poisson's ratio	ν	0.25

lithosphere thickness (a few hundreds of kilometres). This is an approximation that limits the results of this type of model to the study of lithospheric deformation at a continental scale (thousands of kilometres) and of processes that affect strength at lithospheric scale (Enlgand & McKenzie 1982). The viscosity η is assumed to be a complex, non-linear function of the local strain rate, $\dot{\epsilon}$, and temperature, T:

$$\eta = \frac{\tau}{\dot{\epsilon}}$$

$$= \frac{1}{L\dot{\epsilon}} \left[\int_0^S \min\left(B_c \dot{\epsilon}^{1/n_c} \exp\left(\frac{Q_c}{n_c RT}\right), A_c z \right) dz \right.$$

$$\left. + \int_S^L \min\left(B_m \dot{\epsilon}^{1/n_m} \exp\left(\frac{Q_m}{n_m RT}\right), A_m z \right) dz \right]$$

$$\tag{5}$$

which represents the 'mean strength' of a lithospheric column. The values of the creep parameters, B_c, B_m, n_c, n_m, Q_c and Q_m are derived from the results of laboratory experiments on rock specimens believed to be representative of crustal and mantle compositions (i.e. the Adirondak granulite rheology of Wilks & Carter (1990) and the olivine-dominated rheology of

Chopra & Paterson (1981)). The value of the parameters characterizing the brittle regime, A_c and A_m, depends on the sign of $(\partial u/\partial x + \partial v/\partial y)$, i.e. whether the system is in a local state of extension or compression. The value of the various rheological parameters is given in Table 1.

This formulation leads to a strong feedback between lithospheric deformation and lithospheric strength through the dependence of η on S and T. As demonstrated by England (1983) and Braun (1992), lithospheric thinning (or, conversely, thickening) may lead to short-term lithospheric weakening (or stengthening), but long-term lithospheric strengthening (or weakening). The short-term response results from the quasi-isothermal advection of density/rheological interfaces such as the Earth's surface, the Moho discontinuity and the lithosphere/asthenosphere boundary. The long-term response results from conductive cooling or heating. These concepts are illustrated in Figure 1. Note that, as shown by Sandiford (1999), lithospheric thinning may lead, under special circumstances, to long-term lithospheric-scale weakening, for example in situations where lithospheric extension leads to accumulation of a thick sedimentary basin characterized by rocks with a high content in radioactive elements and/or a low thermal conductivity.

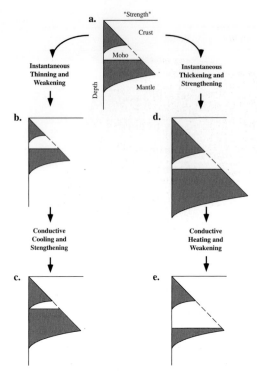

Fig. 1. Schematic representation of the coupling between deformation and strength in the continental lithosphere. (**a**) The strength of the lithosphere is characterized by two maxima separated by a minimum at the base of the crust. (**b**) Thinning leads to an instantaneous local reduction in integrated strength (i.e. the surface of the shaded area). (**c**) Conductive cooling leads to a long term increase in integrated strength. (**d**) Conversely, thickening leads to an instantaneous increase in integrated strength. (**e**) Conductive cooling leads to a long term reduction in integrated strength.

Thermal model

A vertical temperature profile, $T(z)$, is computed at each point of the model by solving, at every time step, the one-dimensional, vertical heat transfer equation which includes the effects of thermal conduction, advection and radiogenic heat production:

$$\rho_c \left(\frac{\partial T}{\partial t} + w \frac{\partial T}{\partial z} \right) = \frac{\partial}{\partial z} \left(k \frac{\partial T}{\partial z} \right) + \rho H \quad (6)$$

w, is the vertical velocity of rocks with respect to the free surface at $z = 0$; c, is the specific heat of rocks; k, is the thermal conductivity; and H, is the heat production. The value of these parameters is given in Table 1. In any given lithospheric column, the vertical advection velocity

resulting from internal deformation of the lithosphere can be approximated by a linear function of depth:

$$w = \frac{\partial S}{\partial t} \frac{z}{S} \quad (7)$$

Including the vertical advection term in the heat transfer equations ensures that we capture the first-order thermal (hence mechanical) disturbance that results from lithospheric deformation. The conduction term introduces a time-dependence to the thermal (and thus mechanical) response of the lithosphere to deformation.

The surface and basal temperatures are assumed to be constant:

$$T(z = 0) = 0 \quad (8)$$

$$T(z = L) = T_L \quad (9)$$

Strength distribution

The initial strength of the various 'cratons' and pre-Ordovician 'orogenic belts' is set by varying the thermal properties of the lithosphere (Fig. 2). Weak zones are embedded by locally increasing the concentration of radiogenic heat producing rocks in the uppermost crust (H). This leads to an increased thermal gradient that results in higher temperatures at all depths within the lithosphere, but especially near the Moho. In this way, the concept introduced by Sandiford & Hand (1998) and Hand & Sandiford (1999) is followed whereby the presence of a highly radioactive sedimentary blanket, deposited during a previous tectonic episode, strongly influences the strength of the underlying lithosphere. Note that, at the start of each model run, the temperature at every point in the model is allowed to reach thermal equilibrium. This is equivalent to assuming that the radioactive rocks have been in place for a very long time (i.e. much longer than the lithospheric conductive time scale).

Strong inclusions, corresponding to cratons, are embedded by increasing locally the thickness of the thermal lithosphere; this approach is consistent with the relatively well-accepted definition of cratons, espoused by Sengör (1999) that 'It is the geothermal gradient that determines when and where a craton will form. Cratons form above the coldest parts of the upper mantle'.

To determine the initial strength distribution of the model lithosphere, the large-scale division of the Australian continent suggested by Shaw et al. (1996) in their crustal element map is followed. Western, South and North Australia have not been substantially deformed since the

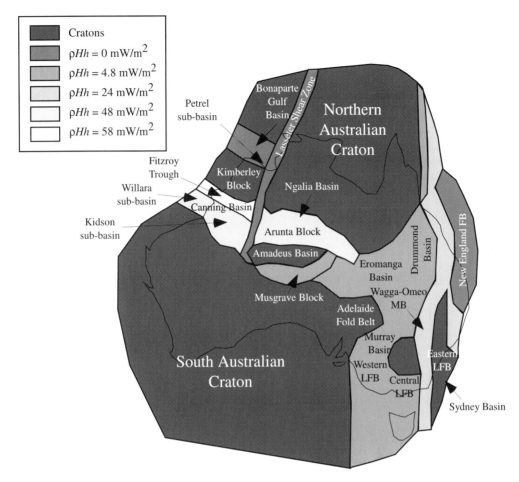

Fig. 2. Initial strength distribution assumed in the numerical model expressed as a plot of the integrated heat production in the uppermost crust. The location of the various tectonic elements described in the text is also shown.

late Proterozoic and, for the purpose of our calculations, are therefore considered as quasi-rigid cratons (i.e. characterized by a very low geothermal gradient). Central Australia, the Tasman and New England fold belts have been 'overprinted' by Palaeozoic tectonic events and must therefore be mechanically weaker. In this model, they are characterized by a more 'typical' continental geothermal gradient. Within this broad division, small fragments of continents are given different thermal, hence mechanical, initial properties. From west to east, the weaker fragments are: the Fitzroy Trough (northwestern Canning Basin), the Arunta Block and a N–S running belt along the east coast which comprises the Wagga-Omeo Metamorphic Belt in the south. Stronger fragments are: the Amadeus Basin, the crustal block beneath the southern part

of the Murray Basin and the Sydney Basin. Also imbedded are two regions of relative weakness within the western part of Northern Australia: a rectangular region which corresponds to the Bonaparte Gulf Basin and a NE–SW corridor corresponding to the Lasseter Shear Zone of Braun *et al.* (1991). The geometry of these smaller scale features has been derived from the crustal element map (Shaw *et al.* 1996).

The *absolute* strength of each continental block/fragment (determined by the assumed concentration of radioactive elements in the upper crust) cannot be obtained from the geological record. The model results are mostly affected by the *relative* strength of each component. The values shown in Figure 2 have thus been determined by trial and error: a large number of model runs were performed in which the relative strength of the various blocks was varied. For each model run, the strain distribution, surface topography, crustal thickness and amount of denudation/sedimentation were compared to what can be derived from the geological record. The results presented here correspond to the 'best-fit' model run. The results of other model runs are also shown to illustrate how sensitive the model is to the assumed initial strength distribution as well as the value of model parameters. This approach is justified, because this study does not aim at simply 'reproducing' a chapter in the geological history of the Australian continent but at using it to improve our understanding of the behaviour of the continental lithosphere. This is achieved by analysing each feature of the model output to determine which of the various physical processes built into the model controls its evolution.

It is comforting to note that the thermal anomalies introduced in the model are similar in character and magnitude to the distribution of low seismic shear wave velocity anomalies recently evidenced beneath several parts of the Australian continent (Fig. 3) (van der Hilst *et al.* 1998; Debayle & Kennett 2000). The 'cratons' in

this model (Fig. 2) correspond to regions characterized by a thick, cold lithospheric keel (Fig. 3) such as observed beneath the Western Australian Craton, the Gawler Block, the Northern Australian Craton and the Kimberley Block (Myers *et al.* 1996) and proposed to explain a marked magnetic feature beneath the Murray Basin (Thalhammer *et al.* 1998).

The regions of enhanced radiogenic heat production in the model correspond to areas that are now characterized by relatively thin lithosphere (Fig. 3), but also high surface heat flux (Cull 1982) and/or the presence of a mid-Proterozoic sedimentary cover (Hand & Sandiford 1999) such as the Adelaide Fold Belt, parts of central Australia, the Fitzroy Trough and the Halls Creek Province (marked as the Lasseter Shear Zone in Figure 2), as well as large parts of eastern Australia.

Erosion/sedimentation model

Transport of mass at the Earth surface by erosion and deposition is assumed to be proportional to local slope (Culling 1960). This

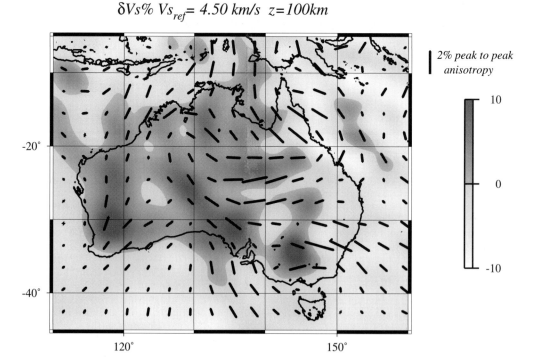

Fig. 3. Distribution of seismic surface wave anomalies along a 100 km deep horizontal section beneath the Australian continent as derived from waveform inversion of teleseismic events recorded at sites located within the Australian continent. The short segments represent the fast direction of horizontally propagating SV waves (after Debayle & Kennett 2000).

simple approach to a very complex problem involving a large number of processes acting on a range of spatial and temporal scales is justified when implemented at the continental scale. This assumption leads to a simple 'diffusion' equation governing the evolution of surface topography, h:

$$\frac{\partial h}{\partial t} = K_D \left(\frac{\partial^2 h}{\partial x^2} + \frac{\partial^2 h}{\partial y^2} \right) \quad (10)$$

K_D, is a measure of the efficiency of transport mechanisms (such as the formation and transport of soil, fluvial and glacial erosion, landsliding, etc.) in redistributing mass in response to relief production by tectonic and flexural processes. The value of the coefficient K_D is given in Table 1. The rate of change of topography is used to update the crustal thickness, S, and contributes to the vertical advection velocity, w, in the heat transfer equation (eq. 6).

From this formulation a crude estimate of raw uplift and erosion (denudation) before any isostatic adjustment is gained. Cooling resulting from denudation could potentially be related to closure temperatures recorded by ^{40}Ar–^{39}Ar spectra and Apatite Fission Track Data (AFTD). This eroded material is then available for mass transfer into depocentres where it accumulates as sediment, thereby conserving mass. The end result is that the topographic surface at any given time is progressively smoothed relative to the amount of cumulated emergence and corresponding subsidence in any given region.

Flexural isostasy

To include the effects of isostasy, the erosion/sedimentation model is coupled to a thin elastic plate model in which the vertical deflection, W, of the surface is calculated by solving the following equation (Turcotte & Shubert 1982):

$$D\nabla^4 W(x, y) + \rho_m g W = V(x, y) \quad (11)$$

The load V is estimated at each point of the model as the difference between the weight of the lithosphere and that of a reference column. This load is a function of local crustal thickness S, topography h and, to a lesser degree, to the temperature distribution $T(z)$. The flexural rigidity, D, is given by (Turcotte & Shubert 1982):

$$D = \frac{E_a t_a^3}{12(1 - \nu^2)} \quad (12)$$

where, t_a, is the effective elastic thickness; E_a is Young modulus; and ν is Poisson's ratio. The value of these parameters is given in Table 1.

The computed vertical deflection, W, is added to the surface topography, h, but does not affect the crustal thickness, S.

Numerical implementation

The mechanical (1–2) and erosion (10) equations are solved on an irregular mesh by means of the finite element method. At the end of every time step, the predicted velocity field is used to advect the numerical mesh which, in turn, is allowed to evolve dynamically via the injection of nodes in regions where the advection leads to inadequate spatial discretization. The finite element discretization is forced to be the Delaunay tessellation around the nodes using the efficient method described in Sambridge et al. (1995).

The mechanical equations are solved using six-node quadratic triangular elements similar to those employed by Bird & Piper (1980) and Houseman & England (1986). The transient 1-D heat transfer equation (6) is solved using three-node quadratic elements and a fully implicit time integration scheme. The flexure Equation (11) is solved by a spectral method (Nunn & Sleep 1984).

Boundary conditions

In the model, deformation is driven by imposed kinematic boundary conditions along the sides of the model, which represent tectonic processes acting along the margins of the continent such as oceanic subduction, continent–continent collision, arc accretion or rifting. The velocity boundary conditions imposed along the sides of the model are shown in Figure 4 and explained in Table 2. Each set of imposed velocity vectors corresponds to an orogenic phase described earlier. The remaining parts of the model boundaries are either set to zero velocity (fixed boundary condition) or zero stress (free boundary condition).

The western and southern margins of the continents are set to be fixed at all times in all model runs. This is meant to represent the presence of Antarctica to the south of Australia and the Indian sub-continent to the west of Australia during most of the Palaeozoic. Antarctica and India did not separate from Australia until the Cretaceous (c. 120 Ma and 90 Ma, respectively).

It is very difficult to estimate a priori the exact direction and magnitude of the imposed velocities from the geological data. Little information exists on the rate at which plates or plate fragments moved with respect to each other in

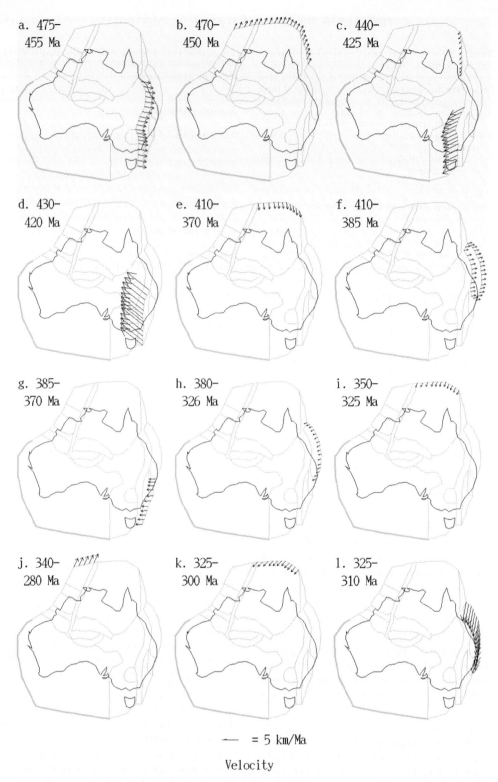

Fig. 4. Summary of the velocity boundary conditions imposed along the margins of the continent. The thick grey line indicates the part of the boundary that is fixed (no velocity boundary condition). Other parts of the boundary are free. The tectonic events corresponding to each time window (panels a to l) are given in Table 2.

Table 2. *Correspondence between the velocity boundary conditions shown in Figure 4 and the various tectonic events described earlier in the text*

Plate boundary process	Orogenic Phase	Panel in Figure 4
Back-arc spreading associated with subduction along the eastern seaboard	Unnamed Ordovician extension	a
Rifting along the northern margin of Australia	Samphire Marsh Movement	b
Subduction of a triple junction along the eastern seaboard	Benambran Orogeny	c
Closure of oceanic basin and accretion of WOMB	Quidongan Orogeny	d
Mild compression along northern margin (origin unknown)	Pertnjara Movement	e
Back-arc extension associated with subduction along eastern seaboard	Subsidence in Drummond Basin	f
Accretion of eastern LFB	Tabberaberan Orogeny	g
Subduction along central section of eastern margin	Compressional Regime of the New England Fold Belt	h
Simultaneous convergence and extension along northern margin (possibly associated with rifting of a micro-plate)	Alice Springs Orogeny	i, j & k
Change from normal to oblique convergence along eastern margin	Transpressional Regime of New England Fold Belt	l

Note that the late Devonian to Early Carboniferous Kanimblan Orogeny has not been included in the model as a set of imposed velocities; it has been assumed that this tectonic event resulted from other plate boundary processes active during that time period.

the Palaeozoic. As will be shown in the following section, the model predictions are, however, very sensitive to the value of the imposed velocities. As for the assumed initial strength distribution, the boundary conditions shown in Figure 4 were therefore determined by trial and error to optimize the fit between the predicted strain, denudation, topography and crustal thickness and the large-scale observations derived from the geological record. By presenting results from different model runs, we will demonstrate how sensitive the model is to the boundary conditions.

Two important points must be made here. Firstly, remind the reader that the purpose of this modelling exercise is not to determine the nature of the tectonic forces responsible for these orogenic events, but to determine how deformation, imposed along the margins of the continent, propagates within the continental interior and localizes in certain regions as a result of factors such as initially reduced strength or syn- and post-deformation thermal weakening/strengthening; it is therefore appropriate to impose velocity boundary conditions. Secondly, the model is not a first-order kinematic model where imposed deformation interacts simply with erosional processes, flexure and strength contrasts imposed *ab initio*. It is a dynamic model because the strength of the various units is allowed to evolve

through time in response to deformation, erosion, and other factors influencing the thermal and mechanical evolution of the lithosphere.

Model results

The results of the best-fit model run are shown in Figures 5 to 11 in terms of:

- a series of contour plots of the strain rate expressed as the second invariant of the deviatoric part of the strain rate tensor (Fig. 5);
- contour plots of the cumulative strain expressed as the second invariant of the deviatoric part of the accumulated strain (Fig. 6);
- contour plots of the crustal thickness (Fig. 7);
- contour plots of the temperature at the crust-mantle boundary (Fig. 8);
- contour plots of topography (Fig. 9);
- contour plots of cumulative denudation and sediment thickness (Fig. 10); and
- vector plots of the instantaneous velocity field (Fig. 11).

Note that the measure of strain and strain rate used in these figures does not carry information about the nature of the deformation, i.e. whether it is extensional, compressional or strike-slip. This information can be obtained from the

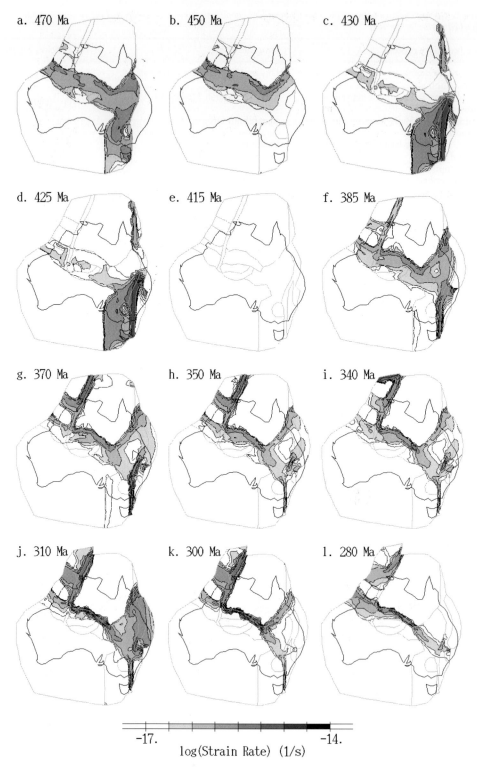

Fig. 5. Contour plots of the logarithm of the computed strain rate at 12 selected times in the model evolution.

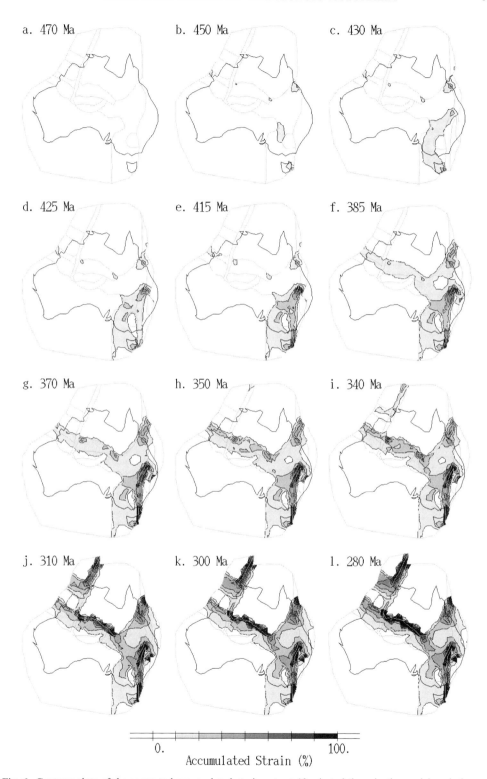

a. 470 Ma   b. 450 Ma   c. 430 Ma

d. 425 Ma   e. 415 Ma   f. 385 Ma

g. 370 Ma   h. 350 Ma   i. 340 Ma

j. 310 Ma   k. 300 Ma   l. 280 Ma

0.   100.

Accumulated Strain (%)

Fig. 6. Contour plots of the computed accumulated strain rate at 12 selected times in the model evolution.

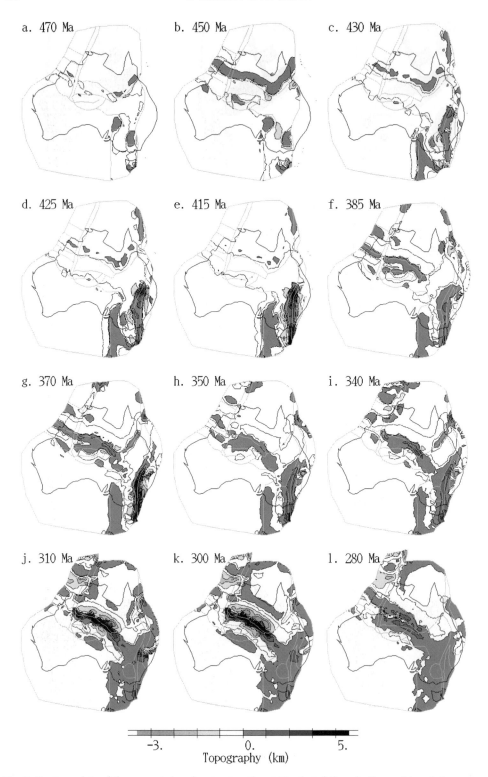

Fig. 9. Contour plots of the computed surface topography at 12 selected times in the model evolution.

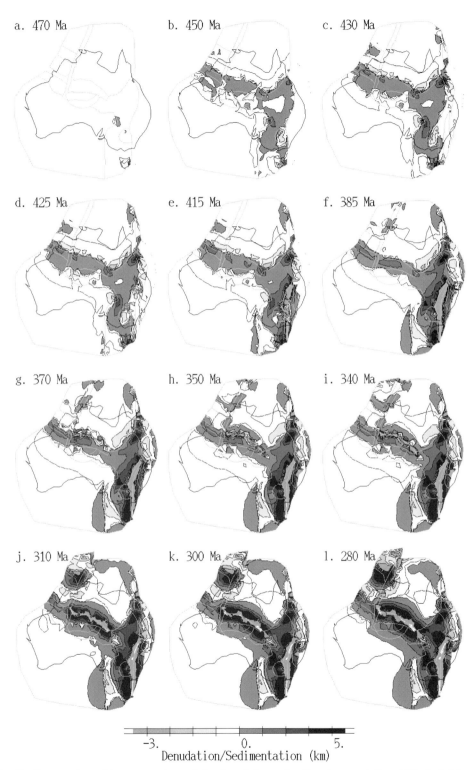

Fig. 10. Contour plots of the computed cumulative denudation and sediment accumulation at 12 selected times in the model evolution.

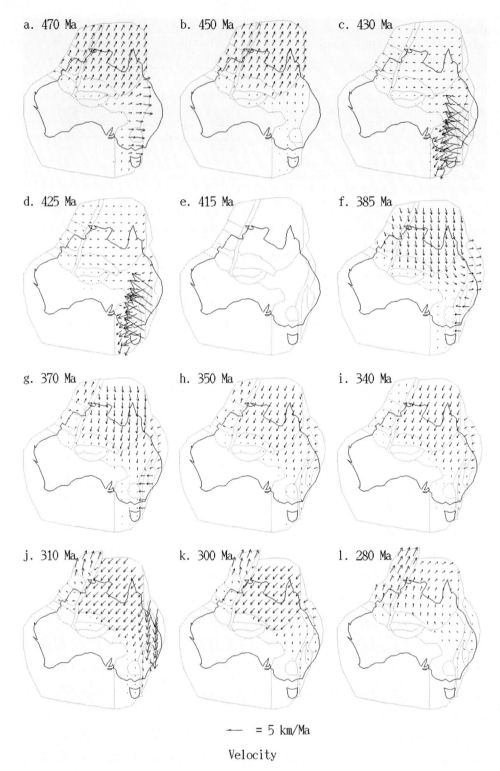

Fig. 11. Plot of the computed instantaneous velocity vectors at 12 selected times in the model evolution.

velocity vector plots. Each figure contains twelve panels which correspond to 'stills' in the evolution of the model at arbitrarily selected time steps.

General model behaviour

Before describing the model's predictions during the successive tectonic events, the behaviour of the system as a whole is considered.

Contour plots of strain rate through time (Fig. 5) clearly indicate that successive deformation episodes tend to progressively localize the deformation. This is evidenced in several parts of the model, namely central, northwestern and southeastern Australia. The first continent-wide extensional episode in the Ordovician results in distributed, divergent deformation in the Canning Basin, in the Arunta Block and Amadeus Basin and along a N–S striking corridor in the east (Fig. 5a). By contrast, deformation during the Carboniferous Alice Springs Orogeny (Fig. 5k) is restricted to the Fitzroy Trough (northern Canning) and the margins of the Arunta Block. Similarly, during the last stages of orogenic development in the Lachlan Fold Belt (i.e. the Devonian Tabberabberan Orogeny – Fig. 5f), deformation is localized along a few discrete N–S striking shear zones. This behaviour is dictated by the strongly non-linear dependence of lithospheric rheology on its thermal state and crustal thickness (Braun 1992).

The opposite observation can be made regarding crustal thickness (Fig. 7). Any anomaly in crustal thickness that is created by an orogenic event gives rise to buoyancy forces which act to reduce gradients in crustal thickness (England & McKenzie 1982). Subsequent viscous flow leads to a progressive lateral spreading of the crustal thickness anomaly into regions that have not been affected by the orogenic event. This behaviour, therefore, results from our assumption of a finite, i.e. non-zero, value for the Argand number (Ar) which 'measures the tendency of the lithosphere to strain in response to the buoyancy forces generated by crustal thickness contrasts' (England & McKenzie 1982).

Computed values of the Moho temperature (Fig. 8) show a high degree of spatial variability; this is because temperature calculations are done along a series of one-dimensional vertical profiles (i.e. at each node of the mechanical grid) and horizontal conductive heat transport is not permitted in our calculations. In general, regions that undergo compression experience a progressive heating of the Moho, whereas regions that undergo extension experience a progressive cooling. The time lag between the deformation

episode and its thermal effect at the Moho discontinuity is of the order of 5–10 Ma which is, approximately, the conductive time scale for heat transfer at the scale of the crust (L = 35 km).

Surface topography (Fig. 9) is always transient as erosion, denudation and accompanying sediment accumulation are active at all times. The rate at which topography evolves by surface processes is arbitrarily determined by the value of the transport coefficient, K_D, the value of which cannot be determined by simple experiments or observations as it applies to mass transport at very large spatial scales (continental scale) and temporal scales (tens of Ma).

As a first order approximation, 'denudation' (Fig. 10) can be interpreted as the amount of accumulated denudation and is commonly expressed by the grade of uplifted metamorphic rocks, whereas what is called 'sedimentation' in the model is equivalent to the total, time-integrated, sediment accumulation over the time interval considered in the model.

A detailed description of the model results

Time interval 470–450 Ma. The model predicts that deformation (strain rate) (Fig. 5a, b), and subsequently subsidence and deposition (Fig. 10a, b), are weakly concentrated along the regions of maximum contrast in lithospheric strength, i.e. the edges of the initially weak mobile belt to the east of the southern and northern Australian blocks. Because the Northern Australian Craton and the crust underlying the Proterozoic Kimberley Basin are assumed to be very strong, deformation localizes in the weaker mobile belts, i.e. the Canning Basin and the central Australian basins, and to a lesser degree in the Bonaparte Gulf Basin.

It is very clear that, in these early stages of model evolution, the distribution of strain is controlled by the presence of assumed strength heterogeneities. In Figure 12, the results of a model experiment in which the lithosphere is assumed to be initially uniformly strong are shown. In this situation, the results show that, as demonstrated by England (1987) and England et al. (1985), the deformation of a uniform thin viscous sheet is mostly controlled by the nature of the boundary conditions. As shown by others before us (See Tommasi et al. (1995), for example), this demonstrates that intracratonic deformation cannot be driven by plate boundary processes alone. It is the presence of mechanical weaknesses or strong inclusions within the plate interior that permits deformation to localize away from plate boundaries.

a.

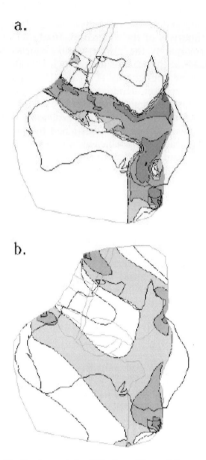

b.

Fig. 12. Contour plots of the logarithm of the computed strain rate at one selected time in the early parts of the model evolution for two model runs. (**a**) The best-fit model run. (**b**) The uniform lithosphere model run.

Further inspection of the model results (Figs 9b and 10b) shows that, in all tectonically active areas, sediment is provided by erosion of up-lifted areas along the margins of the stable cratons. This uplift of basin flanks is driven by the flexural response of the lithosphere to thin-ning (Braun & Beaumont 1989). The width and amplitude of the uplift along the margins of the cratons are determined by the assumed litho-spheric elastic thickness. Despite clear evidence that this parameter is variable within continental interiors and particularly the Australian conti-nent (Zuber et al. 1989), for simplicity of calcu-lation, in our model this parameter is assumed to be spatially uniform. In Figure 13, the computed topography at 450 Ma in a model experiment is shown, in which the assumed elastic thick-ness, t_a, is 100 km, i.e. twice the value used in the

reference model experiment shown in Figures 5 to 11. The uplifted areas are wider but the ampli-tude of the uplift is smaller in comparison with the reference experiment.

Localized extension, as in the Canning Basin, leads to thinning of the crust and post-exten-sional cooling of the uppermost mantle that will result in strengthening of the lithosphere. This process, termed 'post-extensional mantle heal-ing', has been described by Braun (1992). As will be shown later, it has important consequences on the distribution of deformation in subsequent tectonic events.

Finally, note that the eastern part of the Lachlan Fold Belt and the New England Fold Belt have not been included in the model cal-culations yet as, during this time interval, those two terranes were separated from the rest of the continent by a wide marginal sea (Collins & Vernon 1992; Gray et al. 1997).

Time interval 450–415 Ma. During this time interval, the model predicts that the successive compressional episodes driven by subduction along the eastern margin of the continent have led to deformation, uplift and erosion of the sedimentary fill that was deposited in the previous time window (i.e. the Ordovician sediments) in the eastern and central Lachlan Fold Belt (Fig. 5c, d). The model also predicts deposition in regions immediately west of the the Wagga-Omeo Metamorphic Belt. Exhumation and, hence, metamorphism are also predicted during this time period in the Wagga-Omeo Metamorphic Belt (Fig. 10d), which, in this model, is regarded as a weak zone that develops into an orogenic belt.

The model also predicts that, during this time period, oblique subduction along the northern part of the eastern margin may have caused northern Australia to move south-westwards with respect to southern Australia (Fig. 11d), causing sinistral transpression in the Canning Basin. This may be the cause for the formation of the observed major unconformity between Ordovician and Devonian sediments in the Can-ning Basin (attributed to the so-called Prices Creek movement) (Shaw et al. 1994).

The model results also suggest that this rela-tive movement of northern Australia caused deformation in central Australia. This phase of transpression may correspond to the Rodingan movement (Shaw 1991) documented in the stratigraphy of the Amadeus Basin. This tectonic phase is also clearly evidenced in palaeogeo-graphic reconstructions (Fig. 17, BMR Palaeo-geographic Group, 1990) which show that the regions of marine deposition in central Australia contracted during this time interval.

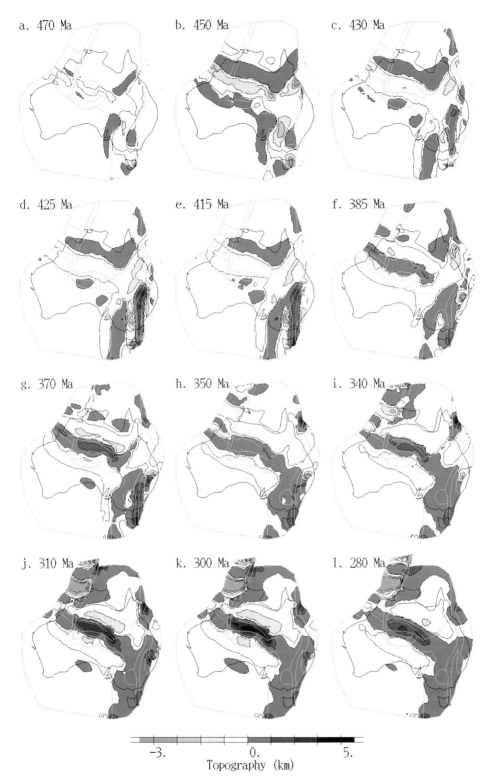

Fig. 13. Contour plots of the computed surface topography at 12 selected times for a model run in which the elastic thickness of the lithosphere is 100 km, twice as much as in the 'best-fit' model run, of which the computed topography is shown in Figure 9.

Time interval 415–370 Ma. In the southeastern part of the continent, the predicted deformation (strain rate) is localized along a relatively narrow zone (Fig. 5g) which corresponds to a zone of lithospheric weakness (i.e. high Moho temperature, Fig. 8g) inherited from previous deformation events (Fig. 5c, d). This zone corresponds quite closely to parts of Southeastern Australia that underwent east-west shortening in the Early Devonian: open folding in the Melbourne zone of central-eastern Victoria, tight chevron-folding in central Victoria (Gray 1988) and, farther north, N–S trending crenulation cleavages along the margins of the Wagga-Omeo Metamorphic Belt (Sandiford *et al.* 1988).

These predictions clearly demonstrate that, at this stage in the model evolution, patterns of deformation are not solely dictated by the imposed initial strength distribution, but are influenced by the healing or weakening effects of previous tectonic events. This is clearly demonstrated in Figure 14 which shows the results of a model experiment in which the temperature is forced to remain unchanged through time (in other words, the processes of thermal weakening and healing are suppressed). As the two models evolve, major differences develop (compare Fig. 5 and Fig. 14). When thermal feedback on rheology is neglected, the lithosphere has little memory of previous tectonic events and the distribution of deformation is mostly controlled by the assumed initial strength distribution. The effects are significant: compare the strain rate values and distribution in central Australia at the peak of the Alice Springs Orogeny (Figs 5k and 14k) and notice the lack of reactivation of Ordovician–Devonian structures in the LFB during the Carboniferous when thermal feedback is ignored (Fig. 14h–k).

The model also predicts that the extension imposed along the northeasterm margin of the continent leads to the deposition of sedimentary cover and strengthening of the lithosphere in central Queensland (Fig. 10e, f). Convergence imposed along the northern margin of the continent leads to localized deformation along the margins of the Arunta Block, which becomes emergent (Fig. 10g). This, in turn, leads to subsidence and sediment accumulation in the Amadeus and Ngalia Basins. In this model, this episode also marks the birth of the N–S striking shear zone which will accommodate a large amount of dextral movement between western and central Australia during the Alice Springs Orogeny (and corresponds to the so-called Lasseter Shear Zone of Braun *et al.* 1991).

Time interval 370–300 Ma. The northward velocity applied along the northwestern boundary of the continent leads to concomitant extension in the Fitzroy Trough and the Bonaparte Gulf Basin (compare strain rate in Fig. 5f, g). The partitioning of deformation between those two regions depends on their relative strength which, in turn is determined mostly by the amount of deformation these two regions have experienced in the past, and, to a lesser degree, by their initial strength (i.e. their initial thermal state). In particular, the amount of extension experienced during the Ordovician extensional episode is the most critical determining factor for the relative mechanical strength between the two regions. To illustrate this point, Figure 15 shows the distribution of strain rate at the peak of the Alice Springs Orogeny (i.e. 350 Ma) for the best-fit model run (Fig. 15b) and for a model run in which the amount of extension experienced in the Ordovician has been enhanced by a factor of 3 (Fig. 15a). The results demonstrate that the greater the extension is in the Ordovician, the lesser the extension will localize in the Fitzroy Trough in the Carboniferous. This is because extension in the Ordovician tends to concentrate in the Canning Basin (Fig. 5b) leading to strengthening of this region by post-extension mantle healing during the Devonian.

The Ordovician extensional episode also affects the distribution of deformation in central Australia (Fig. 15). There, the velocity boundary condition imposed along the northern edge of the continent creates stresses that are transmitted across the quasi-rigid Northern Australian craton to cause deformation within the weaker, more readily deformed Arunta block. The model predicts that as the extension in the Ordovician is enhanced, the deformation tends to migrate from the southern margin of the Arunta Block to the northern margin of the Musgrave Block. The main reason for these differences in behaviour is, once again, the tendency for the continental lithosphere to become stronger in regions that have experienced extension in a relatively distant past (in this case in the Ordovician).

Note that it is not claimed that the azimuth and magnitude of all velocity vectors imposed along the margins of the continent in the best-fit model run are uniquely determined by the modelling exercise. Because the deformation pattern predicted by the model depends on many other factors such as the initial distribution and nature of mechanical heterogeneities and the model rheological and thermal parameters, other velocity distributions could be considered. For example, Roberts & Houseman (2001) demonstrated how shearing the northern

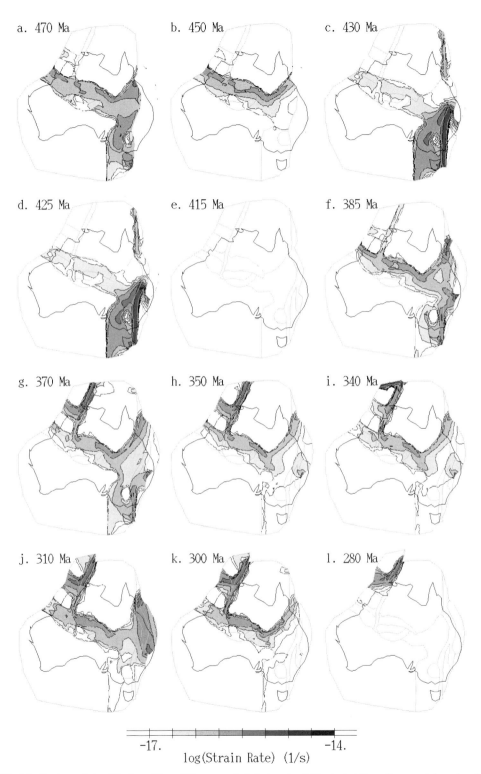

Fig. 14. Contour plots of the logarithm of the computed strain rate at 12 selected times for a model run in which the temperature field is not allowed to change in response to deformation.

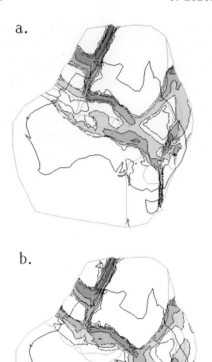

a.

b.

Fig. 15. Comparison between results of two model runs differing by the amount of assumed Ordovician extension. (**a**) Strain rate distribution at 350 Ma in a model run in which the amount of extension imposed along the northern margin of the continent in the Ordovician (panel b in Fig. 4) has been increased. (**b**) Strain rate distribution at 350 Ma in the best-fit model run.

eastern margin. We therefore propose that early Carboniferous deformation in the central and eastern Lachlan Fold Belt may be the result of the interaction between the constructive influence of the Alice Springs Orogeny in central Australia and deformation in the New England Fold Belt in the east. As noted earlier, however, late reactivation of early Palaeozoic structures in the Lachlan Fold Belt is only possible if the lithosphere has a 'memory' of past tectonic activity. In the model experiment in which thermal feedback on rheology is neglected, the southeastern corner of the continent does not experience any Carboniferous deformation (Fig. 14h–k).

Time interval 300–280 Ma. After the peak of the Alice Springs Orogeny (i.e. for times <300 Ma), the model predicts that central Australia undergoes an extensional episode (Fig. 11), as some of the movement imposed along the northern margin of the Bonaparte Gulf Basin is transferred, across the Lasseter Shear Zone, to northern Australia. There is little, if any, evidence of this late, post-Alice Springs Orogeny movement in central Australia. If it ever took place, it is likely that its imprint on the geological record will have been erased during the postorogenic evolution of the central Australian region (i.e. erosion and collapse of the uplifted cratonic areas and rebound of the basement supporting the synorogenic sedimentary basins).

Imposed transpression along the northeastern margin causes intense deformation along the boundary between the New England Fold Belt and the rest of the continent as far south as the northern margin of the Sydney Basin (Fig. 5j). This deformation leads to further subsidence and sedimentation in a foreland basin to the west (which could correspond to sedimentation in the lower Galilee Basin – de Caritat & Braun 1992) (Fig. 10j).

boundary of the continent may lead to a deformation pattern that is very similar to the one observed during the Alice Springs Orogeny.

During this time interval, the model also predicts the development of a N–S striking narrow zone of intense deformation in the southeastern part of the continent (Fig. 5f–j). Detailed analysis of the results show that this region is affected by oblique compression (Fig. 11) which is produced by the combined effects of two 'farfield' active plate boundaries: N–S compression along the northern margin of the continent and E–W compression (from 380 Ma to 340 Ma) followed by oblique compression (from 325 Ma to 300 Ma) along the northern section of the

Comparison with continental-scale datasets

Tectonic element map (derived from gravity and magnetic data)

In Figure 16, the regions of the crustal element map of Australia recently prepared by Shaw *et al.* (1996) are highlighted, which are defined as 'geophysically overprinted' in the Palaeozoic. These are zones where one set of geophysical features are progressively replaced by another. In our model predictions, these regions are all characterized by high values of accumulated strain (Fig. 6). Apart from the offshore areas which are not included in the crustal element

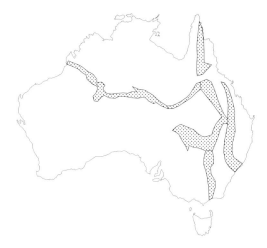

Fig. 16. Outline of the regions which, according to the tectonic element map of Australia (Shaw *et al.* 1996), have been geophysically overprinted during the Palaeozoic.

map, the most striking discrepancy is the Lasseter Shear Zone, which does not appear as tectonically reactivated (or overprinted) in the Palaeozoic but is predicted, in our model, to have accommodated large dextral strikeslip movement between central and northwestern Australia. We argue that this discrepancy arises from the lack of syn-deformational deposition and/or volcanism related to the Palaeozoic deformation and that the geophysical signature of the area has, therefore, been attributed by Shaw *et al.* (1996) to Proterozoic events.

Seismic anisotropy

The short segments superimposed on the seismic velocity anomaly map of Figure 3 represent the fast direction of horizontally propagating SV waves derived from a seismic model of the upper 100 km of the Australian lithosphere (Debayle & Kennett 2000). As suggested by Debayle & Kennett (2000), it is likely that this azimuthal anisotropy is related to the most recent local deformation event frozen in the lithosphere.

The relationship between crystallographic orientation and mantle deformation is still poorly understood. Theoretical studies (Nicolas & Poirier 1976) and observations of mantle peridotites (Ribe & Yu 1991) suggest that, at low to moderate strain levels, the fast *a*-axis of olivine crystals is parallel to the direction of maximum extension but at high strain levels and/or high temperature

and pressure (1500 K and 300 MPa) it becomes orientated with flow direction (Zhang & Karato 1995). In general, in compressional systems, the *a*-axis of olivine crystals should therefore be in the direction perpendicular to the shortening direction whereas in continental rifts, the *a*-axis should be orientated in the direction of extension. In regions of strike-slip deformation, the *a*-axis should be close to the azimuth of major faults, especially after large strain.

Comparing our model results to the distribution of anisotropy in the top 100 km of the Australian continent supports the assumption that anisotropy was set by lithospheric deformation during the most recent tectonic events. In central Australia, the direction of the fast axis is clearly E–W, perpendicular to the direction of lithospheric shortening (compare Fig. 3 to Fig. 11k). In northwestern Australia (Canning Basin and Bonaparte Gulf Basin), the fast axis is parallel to the direction of the late Palaeozoic extension (i.e. NNE–SSW). The anisotropy observed beneath the Lachlan Fold Belt is more difficult to reconciliate with the overal E–W compression experienced by the area during most of the Early to mid-Palaeozoic. It is likely, however, that seismic anisotropy in eastern Australia has been strongly affected by the more recent opening of the Tasman Sea, a deformation event that is not included in our model.

Crustal thickness

Figure 17 shows a seismically derived map of estimates of depth to Moho beneath Australia (Clitheroe *et al.* 2000), which should be compared to our model predictions of final crustal thickness (Fig. 7l). Thick crust (>40 km) characterizes central and southeastern Australia and, as suggested by our model predictions, may have resulted from crustal shortening during the mid- to Late Palaeozoic, by which time the Australian continent would have been almost completely cratonized. Thin crust (<35 km) is observed beneath the Canning Basin of northwestern Australia, which, according to our model, must be attributed to crustal thinning in the Late Devonian to mid-Carboniferous.

The relatively low crustal thickness observed in the northern Adelaide Fold Belt (28 km) and the strongly contrasting thick estimates for northern and southern Australia, are likely to be related to pre-Ordovician tectonics (Clitheroe *et al.* 2000) and cannot, therefore, be reproduced in this model, which assumes a uniform crustal thickness of 35 km at the onset of the first imposed tectonic event (475 Ma).

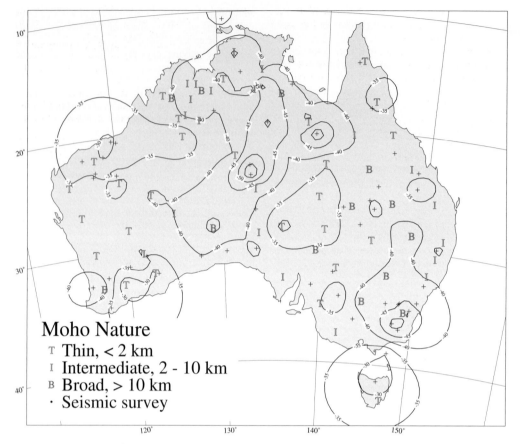

Fig. 17. Contour map of the Moho derived from inversion of teleseismic events recorded at the seismic stations shown on the map. Moho transition zone width is indicated for broad-band seismic stations by a letter symbol (from Clitheroe *et al.* 2000).

Conclusions

To study the processes that are responsible for the deformation of continental interiors, we have developed a thin viscous plate model of the continental lithosphere in which the material strength is affected by previous deformation events through a strong thermo–mechanical feedback. We have used the relatively well-documented late Palaeozoic tectonic history of the Australian continent as a case example on which our model can be applied. We have been able to find a set of velocity boundary conditions as well as an initial distribution of mechanical heterogeneities that lead to predictions of strain, sedimentation/exhumation, topography and crustal thickness that are compatible with the geological record.

The results of these computations have led to the following conclusions regarding the thermo-

mechanical evolution of a continental area subjected to intense and repeated intracratonic deformation.

- The tectonic history of the Australian continent can be reproduced by assuming that tectonic forcing originates at plate boundaries. There is no need to invoke sub-lithospheric vertical forces originating beneath the continent.

- The most likely mechanism for intracratonic deformation is the concentration of strain by transmission of horizontal stress originating at plate boundaries into regions of decreased lithospheric strength; an example is the deformation along the southern margin of the Arunta Block during the Alice Springs Orogeny.

- These weak zones are often caused by previous intracratonic deformation or develop at the interface between regions of normal or reduced strength and regions of increased strength (cratons).

- Repeated deformation episodes may lead to strain localization; an example is the progressive concentration of strain into narrow north-south corridors in the development of the Lachlan Fold Belt. This behaviour results from the assumed non-linear, thermally-activated rheology of the continental lithosphere.
- Localized deformation may also take place as the result of the constructive interaction between two sets of tectonic forces acting on separate margins.
- There are mechanisms that operate within the lithosphere by which deformation leads to local strengthening; this is particularly true of regions which have undergone extension followed by conductive cooling.

The authors wish to thank D. Gray, M. Hand and A. M. Walley for suggestions/comments made during the development of the model and its application to the Australian continent. We thank M. Sandiford, G. Houseman and an anonymous reviewer for constructive reviews of an earlier version of this manuscript. We also thank E. Debayle and G. Clitheroe for giving us permission to use their seismic models. Part of the work reported here was conducted while R. D. Shaw was at the Australian Geological Survey Organisation (AGSO) as part of the Australian Geodynamics Cooperative Research Centre (AGCRC) and is published with the permission of the CEO (AGSO) and the Director of the AGCRC.

References

BAIN, J. H. C. & DRAPER, J. J. 1997. *North Queensland Geology.* Australian Geological Survey Organisation Bulletin **240** and Queensland Department of Mines and Energy, Queensland Geology, **9**, 600.

BIRD, P. & PIPER, K. 1980. Plane-stress finite-element models of tectonic flow in Southern California. *Physics of the Earth and Planetary Interiors,* **21**, 158–175.

BRAUN, J. 1992. Postextensional mantle healing and episodic extension in the Canning Basin. *Journal of Geophysical Research,* **97**, 8927–8936.

BRAUN, J. & BEAUMONT, C. 1989. A physical explanation of the relation between flank uplifts and the breakup unconformity at rifted continental margins. *Geology,* **17**, 760–764.

BRAUN, J. & SHAW, R. D. 1998a. Contrasting styles of lithospheric deformation along the northern margin of the Amadeus Basin, central Australia. *In*: BRAUN, J., DOOLEY, J., GOLEBY, B., VAN DER HILST, R. & KLOOTWIJK, C. (eds) *Structure and Evolution of the Australian Continent.* American Geophysical Union, Geodynamics Series, **26**, 139–156.

BRAUN, J. & SHAW, R. D. 1998b. Extension in the Fitzroy Trough, Western Australia: an example of reactivation tectonics. *In*: BRAUN, J., DOOLEY, J., GOLEBY, B., VAN DER HILST, R. & KLOOTWIJK, C. (eds) *Structure and Evolution of the Australian Continent.* American Geophysical Union, Geodynamics Series, **26**, 157–174.

BRAUN, J., MCQUEEN, H. & ETHERIDGE, M. 1991. A fresh look at the late Palaeozoic tectonic history of western-central Australia. *Exploration Geophysics,* **22**, 49–54.

BUICK, I. S., MILLER, J. A., WILLIAMS, I. S. W. & CARTWRIGHT, I. 2001. Ordovician high-grade metamorphism of a newly discovered late Neo-Proterozoic terrane in the northern Harts Range, central Australia. *Journal of Metamorphic Geology,* **19**, 373–394.

CHOPRA, P. N. & PATERSON, M. S. 1981. The experimental deformation of dunite. *Tectonophysics,* **78**, 453–473.

CLITHEROE, G., GUDMUNDSSON, O. & KENNETT, B. L. N. 2000. The crustal thickness of Australia. *Journal of Geophysical Research,* **105**(B6), 13 697–13 714.

COLLINS, W. J. & VERNON, R. H. 1992. Palaeozoic arc growth, deformation and migration across the Lachlan Fold Belt, southeastern Australia. *Tectonophysics,* **214**, 381–400.

COLWELL, J. B. & KENNARD, J. M. 1996. *AGSO Petrel sub-basin Study 1995–1996.* Australian Geological Survey Organisation Record **1996/40**.

CULL, J.-P. 1982. An appraisal of Australian heat-flow data. *BMR Journal of Australian Geology and Geophysics,* **7**, 11–21.

CULLING, W. E. H. 1960. Analytical theory of erosion. *Journal of Geology,* **68**, 336–344.

DEBAYLE, E. & KENNETT, B. L. N. 2000. The Australian continental upper mantle: structure and deformation from surface waves. *Geophysical Journal International,* **105**, 25 423–25 450.

DE CARITAT, P. & BRAUN, J. 1991. Extension/collision cyclicity and the convergent margin history of outboard Gondwana (eastern Australia). *EOS Transactions of the American Geophysical Union,* **72**, 300.

DE CARITAT, P. & BRAUN, J. 1992. Cyclic development of sedimentary basins at convergent plate margins – 1. Structural and tectono-thermal evolution of some Gondwana basins of eastern Australia. *Journal of Geodynamics,* **16**, 241–282.

ENGLAND, P. 1983. Constraints on extension of continental lithosphere. *Journal of Geophysical Research,* **88**, 1145–1152.

ENGLAND, P. 1987. Diffuse continental deformation: length scales, rates and metamorphic evolution. *Philosophical Transactions of the Royal Society of London,* **A321**, 3–22.

ENGLAND, P. C. & MCKENZIE, D. P. 1982. A thin viscous sheet model for continental deformation. *Geophysical Journal of the Royal Astronomical Society,* **70**, 295–231.

ENGLAND, P. C. & MCKENZIE, D. P. 1983. Correction to: 'A thin viscous sheet model for continental deformation'. *Geophysical Journal of the Royal Astronomical Society,* **73**, 523–232.

ENGLAND, P. HOUSEMAN, G. & SONDER, L. 1985. Length scales for continental deformation in convergent, divergent and strike-slip environments: analytical and approximate solutions for a thin viscous sheet model. *Journal of Geophysical Research*, **90**, 3551–3557.

FERGUSSON, C. L. & COWEY, P. J. 1992. Convergence and intraplate deformation in the Lachlan Fold Belt of southeastern Australia. *Tectonophysics*, **214**, 417–439.

FORMAN, D. J. & SHAW, R. D. 1973. *Deformation of the crust and mantle in central Australia*. Bureau of Mineral Research Bulletin **144**.

FORMAN, D. J. & WALES, D. W. 1981. *Geological evolution of the Canning Basin, Western Australia*. Bureau of Mineral Resources Bulletin **210**.

FOSTER, D. A., GRAY, D. R., KWAK, T. A. P. & BUTCHER, M. 1998. Chronology and tectonic framework of turbidite-hosted gold deposits in the western Lachlan Fold Belt, Victoria: 40Ar–39Ar results. *Ore Geology Reviews*, **13**, 229–250.

GALLAGHER, K. & LAMBECK, K. 1989. Subsidence, sedimentation and sea-level changes in the Eromanga Basin, Australia. *Basin Research*, **2**, 115–131.

GOLEBY, B. R., SHAW, R. D., WRIGHT, C., KENNETT, B. L. N. & LAMBECK, K. 1989. Geophysical evidence for 'thick-skinned' crustal deformation in central Australia. *Nature*, **337**, 325–330.

GRAY, D. R. 1988. Structure and tectonics. *In*: DOUGLAS, J. G. & FERGUSSON, J. A. (eds) *Geology of Victoria* (second edition). Geological Society of Australia, Victorian Division, Melbourne, 1–36.

GRAY, D. R., FOSTER, D. A. & BUCHER, M. 1997. Recognition and definition of orogenic events in the Lachlan Fold Belt. *Australian Journal of Earth Sciences*, **44**, 489–501.

GUNN, P. J. 1998. Bonaparte Basin: evolution and structural framework. *In*: PURCELL, P. G. & PURCELL, R. R. (eds) *The North West Shelf, Australia*. Proceedings of the Petroleum Exploration Society Symposium, Perth, 275–285.

HAND, M. & SANDIFORD, M. 1999. Intraplate deformation in central Australia, the link between subsidence and fault reactivation. *Tectonophysics*, **305**, 121–140.

HAND, M., MAWBY, J., KINNY, P. & FODEN, J. 1999a. U–Pb ages from the Harts Range, central Australia: evidence for early Ordovician extension and constraints on Carboniferous metamorphism. *Journal of the Geological Society, London*, **156**, 715–730.

HAND, M., MAWBY, J., MILLER, J. A., BALLEVRE, M., HENSEN, B., MOELLER, A. & BUICK, I. S. 1999b. *Tectonothermal evolution of the Harts and Strangways Range region, eastern Arunta Inlier, central Australia*. Field Guide No. 4 Specialist Group in Geochemistry, Mineralogy and Petrology, Geological Society of Australia.

HOUSEMAN, G. & ENGLAND, P. 1986. Finite strain calculations of continental deformation. 1. Method and general results for convergent zones. *Journal of Geophysical Research*, **91**, 3651–3663.

HOUSEMAN, G. A. & MOLNAR, P. 1997. Gravitational (Rayleigh–Taylor) instability of a layer with non-linear viscosity and convective thinning of continental lithosphere. *Geophysical Journal International*, **128**, 125–150.

KLOOTWIJK, C. 1995. Palaeomagnetism suggests mid-Carboniferous convergence between Greater Australia and Altaids. *AGSO Research Newsletter*, **22**, 14–17.

KLOOTWIJK, C. 1996. Phanerozoic configurations of Greater Australia: Evo of the North West Shelf. Part One: Review of reconstruction models. *Australian Geological Survey Organisation Record*, **1996/51**, 106.

LAMBECK, K. 1983a. The role of compressive froces in intracratonic basin formation and mid-plate orogenies. *Geophysical Research Letters*, **10**, 845–848.

LAMBECK, K. 1983b. Structure and evolution of the intracratonic basins of central Australia. *Geophysical Journal iof the Royal Astronomical Society*, **74**, 843–886.

LAMBECK, K., BURGESS, G. & SHAW, R. D. 1988. Teleseismic travel time anomalies and deep crustal structure in central Australia. *Geophysical Journal of the Royal Astronomical Society*, **94**, 105–124.

MIDDLETON, M. F. 1989. A model for the formation of intracratonic sag basins. *Geophysical Journal of the Royal Astronomical Society*, **99**, 665–676.

MITROVICA, J. X., BEAUMONT, C. & JARVIS, G. T. 1989. Tilting of continental interiors by the dynamical effects of subduction. *Tectonics*, **8**, 1079–1094.

MORAND, V. J. & GRAY, D. R. 1990. Major fault zones related to the Omeo Metamorphic Complex, northeastern Victoria. *Australia Journal of Earth Sciences*, **38**, 203–221.

MURRAY, C. G., FERGUSSON, C. L., FLOOD, P. G., WHITAKER, W. G. & KORSCH, R. J. 1987. Plate tectonic model for the Carboniferous evolution of the New England Fold Belt. *Australian Journal of Earth Sciences*, **34**, 213–236.

MYERS, J. S., SHAW, R. D. & TYLER, I. M. 1996. Tectonic evolution of Proterozoic Australia. *Tectonics*, **15**, 1431–1446.

NEIL, E. & HOUSEMAN, G. 1999. Rayleigh–Taylor instability of the upper mantle and its role in intraplate orogeny. *Geophysical Journal International*, **138**, 89–107.

NICOLAS, A. & POIRIER, J.-P. 1976. *Crystalline Plasticity and Solid State Flow in Metamorphic Rocks*. John Wiley, New York.

NUNN, J. A. & SLEEP, N. H. 1984. Thermal contraction and flexure of intracratonic basins: A three-dimensional study of the Michigan Basin. *Geophysical Journal of the Royal Society*, **76**, 587–635.

POWELL, C. MCA., EDGECOMBE, D. R., HENRY, N. M. & JONES, J. G. 1977. Timing of regional deformation in the Hill Trough: a reassessment. *Journal of the Geological Society of Australia*, **23**, 407–422.

RANALLI, G. 1997. Rheology of the lithosphere in space and time. *In*: BURG, J.-P. & FORD, M. (eds) *Orogeny through Time*. Geological Society, London, Special Publications, **121**, 19–37.

RIBE, N. & YU, Y. 1991. A theory for the evolution of orientation textures in deformed olivine polycrystals. *Journal of Geophysical Research*, **96**, 8325–8335.

ROBERTS, E. A. & HOUSEMAN, G. A. 2001. Geodynamics of central Australia during the intraplate Alice Springs Orogeny: thin viscous sheet models. *In*: MILLER, J. A., HOLDSWORTH, R. E., BUICK, I. S. & HAND, M. (eds) *Continental Reactivation and Reworking*. Geological Society, London, Special Publications, **184**, 139–164.

SAMBRIDGE, M., BRAUN, J. & MCQUEEN, H. 1995. Geophysical parameterization and interpolation of irregular data using natural neighbours. *Geophysical Journal International*, **122**, 837–857.

SANDIFORD, M. 1999. Mechanics of basin inversion. *Tectonophysics*, **305**, 109–120.

SANDIFORD, M. & HAND, M. 1998. Controls on the locus of intraplate deformation in central Australia. *Earth and Planetary Sciences Letters*, **162**, 97–110.

SANDIFORD, M., MARTIN, N. & LOHE, E. M. 1988. Shear zone deformation in the Yackandandah granite, northeast Victoria. *Australian Journal of Earth Sciences*, **35**, 223–230.

SENGÖR, A. M. C. 1999. Continental interiors and cratons: any relation? *Tectonophysics*, **305**, 1–42.

SHAW, R. D. 1987. *Basement uplift and basin subsidence in central Australia*. Unpublished PhD Thesis, The Australian National University, Canberra, Australia.

SHAW, R. D. 1991. The tectonic development of the Amadeus Basin, central Australia. *In*: KORSCH, R. J. & KENNARD, J. M. (eds) *Geological and geophysical studies in the Amadeus Basin, central Australia*. Bureau of Mineral Research Bulletin, **236**, 429–461.

SHAW, R. D. & BLACK, L. P. 1991. The history and tectonic implications of the Redbank Thrust Zone, central Australia, based on structural, metamorphic and Rb–Sr isotopic evidence. *Australian Journal of Earth Sciences*, **38**, 307–332.

SHAW, R. D., ETHERIDGE, M. A. & LAMBECK, K. 1991. Development of the Late Proterozoic to mid-Palaeozoic, intracratonic Amadeus Basin in central Australia: a key to understanding tectonic forces in plate interiors. *Tectonics*, **10**(4), 688–721.

SHAW, R. D., SEXTON, M. J. & ZEILINGER, I. 1994. *The tectonic framework of the Canning Basin, W.A., including the 1:2 million structural elements map of the Canning Basin*. Australian Geological Survey Organisation Record **1994/48**.

SHAW, R. D., WELLMAN, P., GUNN, P., WHITAKER, A. J., TARLOWSKI, C. & MORSE, M. 1996. *Guide to using the Australian Crustal Elements map*. Australian Geological Survey Organisation Record **1996/30**.

THALHAMMER, O. A. R., STEVENS, B. P. J., GIBSON, J. & GRUM, W. 1998. Tibooburra Granodiorite, western New South Wales: emplacement history and geochemistry. *Australian Journal of Earth Sciences*, **45**, 775–789.

THOMAS, M. D. & GIBBS, R. A. 1985. Proterozoic plate subduction and collision: processes for reactivation of Archaean crust in the Churchill Province. *In*: AYES, L. D., THURSTON, P. C., CARD, K. D. & WEBSTER, W. (eds) *Evolution of Archaean Supracrustal Sequences*. Geological Association of Canada, Special Paper, **28**, 263–279.

TOMMASI, A., VAUCHEZ, A. & DAUDRÉ, B. 1995. Initiation and propagation of shear zones in a heterogeneous continental lithosphere. *Journal of Geophysical Research*, **100**, 22083–22101.

TURCOTTE, D. L. & SHUBERT, G. 1982. *Geodynamics: Applications of Continuum Physics to Geological Problems*. John Wiley & Sons, New York.

VAN DER HILST, R. D., KENNETT, B. L. K. & SHIBUTANI, T. 1998. Upper mantle structure beneath Australia from portable array deployments. *In*: BRAUN, J., DOOLEY, J., GOLEBY, B., VAN DER HILST, R. & KLOOTWIJK, C. (eds) *Structure and Evolution of the Australian Continent*. American Geophysical Union, Geodynamics Series, **26**, 39–57.

VEEVERS, J. J., JONES, J. G. & POWELL, C. MCA. 1982. Tectonic framework of Australia's sedimentary basins. *Association of Petroleum Exploration of Australia Journal*, **22**, 283–300.

WALLEY, A. M., COOK, P. J., BRADSHAW, J., BRAKEL, A. T., KENNARD, J. M., LINDSAY, J. F., NICOLL, R. S., OWEN, M., SHERGOLD, J. H., TOTTERDEL, J. M. & YOUNG, G. C. 1991. The Palaeozoic palaeogeography of the Amadeus Basin Region. *In*: KORSCH, R. J. & KENNARD, J. M. (eds) *Geological and geophysical studies in the Amadeus Basin, central Australia*. Bureau of Mineral Resources, Australia Bulletin, **236**, 155–169.

WELLMAN, P. 1976. Gravity trends and the growth of Australia – a tentative correlation. *Journal of the Geological Society of Australia*, **23**, 11–14.

WILKS, K. & CARTER, N. L. 1990. Rheology of some continental lower crustal rocks. *Tectonophysics*, **182**, 57–77.

ZHANG, S. & KARATO, S. I. 1995. Lattice preferred orientation of olivine aggregates deformed in simple shear. *Nature*, **375**, 774–777.

ZUBER, M. T., BECHTEL, T. D. & FORSYTH, D. W. 1989. Effective elastic thicknesses of the lithosphere and mechanisms of isostatic compensation in Australia. *Journal of Geophysical Research*, **94**, 9353–9367.

Tectonic feedback, intraplate orogeny and the geochemical structure of the crust: a central Australian perspective

M. SANDIFORD[1], M. HAND[2] & S. McLAREN[2]

[1] *School of Earth Sciences, University of Melbourne, Victoria, Australia*
(e-mail: m.sandiford@earthsci.unimelb.edu.au)
[2] *Department of Geology and Geophysics, University of Adelaide, South Australia*

Abstract: The geological record of intraplate deformation in central Australia implies that past tectonic activity (basin formation, deformation and erosion) has modulated the response of the lithosphere during subsequent tectonic activity. In particular, there is a correspondence between the localization of deformation during intraplate orogeny and the presence of thick sedimentary successions in the preserved remnants of a formerly widespread intracratonic basin. This behaviour can be understood as a kind of 'tectonic feedback', effected by the long-term thermal and mechanical consequences of changes in the distribution of heat producing elements induced by earlier tectonism. From a geochemical point of view, one of the most dramatic effects of intraplate orogeny in central Australia has been the exposure, in the cores of the orogens, of deep crustal rocks largely depleted in the heat producing elements. The geochemical structuring of the crust associated with the erosion of the heat-producing upper crust resulted in long-term cooling of the deep crust and upper mantle with associated lithospheric strengthening. This is illustrated here by mapping the consequences of deformation and associated tectonic responses onto the $h–q_c$ plane, where h is the characteristic length-scale for heat production distribution, and q_c is the total crustal heat production. Because rates of intraplate deformation in central Australia appear to be much slower than that typical of plate margin orogens, it is possible that the ongoing geochemical structuring of the crust has played an important role in terminating intraplate orogeny in central Australia by providing a 'thermal lock'. The diagnostic geophysical signature of this lock may be the extraordinary gravity anomalies of the central Australian intraplate orogens.

Central Australia provides an intriguing record of intraplate deformation resulting in the development of a number of extraordinary orogens that expose deep crustal rocks in their cores (Forman 1971; Teyssier 1985; Goleby *et al.* 1989). The spatial distribution of intraplate deformation during the Alice Springs and Petermann orogenies suggests that the mechanical response of this region has varied through time. Consequently, this response is important for our understanding of the processes controlling reactivation in intraplate settings. In previous contributions it has been argued that the distribution of deformation in central Australia has been strongly influenced by prior tectonic processes, such as basin formation and erosion (Sandiford & Hand 1998*a*; Hand & Sandiford 1999). This paper focusses on the way in which intraplate deformation modifies the geochemical structure of the crust in this region, in particular the distribution of heat sources. Recognizing that the distribution of heat sources provides a first-order control on the thermal structure of the crust, the conse-

quences of intraplate deformation for the long-term thermal and mechanical evolution of the crust is then explored. Our aim is to demonstrate the relevance of a novel style of 'tectonic feedback' which seems to have modulated the long-term tectonic behaviour of the central Australian region, and which may be relevant to continental interiors in general. The notion of tectonic feedback is compatible with a highly temperature dependent lithospheric rheology, as suggested by many experimental studies of natural rocks. Such feedback has profound implications for the long-term tectonic evolution of continental interiors, particularly their potential for reactivation.

The record of intraplate deformation in central Australia

Central Australia encompasses the region comprising northern South Australia, the central and southern parts of the Northern Territory, and the

From: MILLER, J. A., HOLDSWORTH, R. E., BUICK, I. S. & HAND, M. (eds) *Continental Reactivation and Reworking*. Geological Society, London, Special Publications, **184**, 195–218. 1-86239-080-0/01/$15.00 © The Geological Society of London 2001.

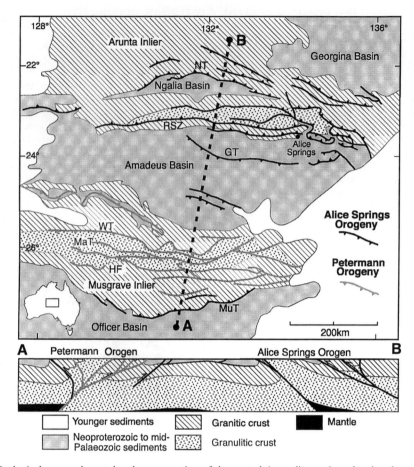

Fig. 1. Geological map and crustal scale-cross section of the central Australian region, showing the structurally remnant Neoproterozoic basins (the Officer, Amadeus, Ngalia and Georgina Basins), separated by basement inliers (the Musgrave and Arunta Inliers). Within the inliers we can distinguish two distinct types of terrane: (1) gneissic granite terranes that form the peripheral regions of the inliers and which are unconformable overlain by the Neoproterozoic sediments; (2) mafic granulite terranes that define the cores of the inliers and which are tectonically juxtaposed with the gneissic granite terranes. This juxtaposition reflects, in part, the strain associated with intraplate orogeny accumulated during the exhumation of the basement inliers from beneath a formerly more or less continuous intracratonic basin. Crustal scale cross-sections based on seismic reflection profiling (Goleby *et al.* 1988, 1989) show that the mafic granulite terranes are representative of the Central Australian lower crust, while the gneissic granite terranes are representative of the mid-upper crust. The central Australian intraplate orogens are characterised by extraordinary gravity anomalies (Mathur 1976). These gravity anomalies are amongst the largest known from the continental interiors. Moreover, they clearly relate to structures that were active during the Petermann and Alice Springs Orogenies. GT, Gardiner Thrust; HF, Hinckley Fault; NT, Napperby Thrust; RSZ, Redbank Shear Zone; UT, Uluru Thrust; WT, Woodroofe Thrust.

contiguous parts of Western Australia (Fig. 1). During the Neoproterozoic this region was largely covered by an extensive intracratonic basin. The record of intraplate deformation in this region is preserved in structures associated with the formation of this basin as well as its subsequent structural dismemberment. This record has been well documented in numerous publications by the Australian Government Survey Organization (AGSO) and its predecessor, the Bureau of Mineral Resources (BMR). In this section, a brief synopsis of this record as largely elucidated by AGSO and BMR geologists is provided, concentrating on the evolution of intraplate orogeny.

Central Australia comprises Palaeo- to Mesoproterozoic, intermediate – high grade, metamorphic complexes overlain by an extensive system of Neoproterozoic to Phanerozoic basins (Fig. 1). The main crust forming events occurred in the intervals 1900–1600 Ma and 1100–1000 Ma, by which time the region was apparently largely cratonized. From about 800 Ma, regional scale subsidence resulted in widespread sedimentation as reflected in the sequences preserved in the Officer, Amadeus, Ngalia, Georgina and Wiso Basins. These basins are now separated by inliers of older meta-morphic 'basement'. The general absence of facies changes and thickness variations in the sediments adjacent to these boundaries, and the continuity of stratigraphy between basins (e.g. Walter *et al.* 1995), provides a compelling argument for the former continuity of these basins. The formerly more or less continuous basin that is assumed to have covered almost the entire Central Australian region through the Neoproterozoic, prior to its fragmentation into the present series of structural-remnant basins, has been termed the Centralian Superbasin.

The fragmentation of the Centralian Superbasin began in the latest Neoproterozoic with the exhumation of the Musgrave Inlier in northern South Australia, which effectively isolated the Officer Basin from the rest of the Centralian Superbasin (Fig. 1). This period of exhumation accompanied the Petermann Orogeny (570–530 Ma), the first of two major intraplate orogenies to have affected the central Australian region. The second major episode, in the Devonian and Carboniferous (430–300 Ma), is known as the Alice Springs Orogeny.

The Petermann Orogen forms an east–west trending belt, with the principal locus of deformation centred on the northern part of the Musgrave Inlier in the vicinity of the Woodroffe and Mann Faults. Along the northern margin of the orogen, differential denudation across large-scale thrust faults resulted in important structuring of the crust. The major structural feature is the Woodroffe Thrust; a south-dipping mylonite zone up to 3 km thick (Edgoose *et al.* 1993; Camacho *et al.* 1995; Stewart 1995) offsetting the Moho by *c.* 20 km (Lambeck & Burgess 1992) and associated with a prominent gradient in the gravity field. South of the Woodroffe Thrust, deformation produced an imbricate thrust stack in which levels of denudation decrease northward from *c.* 40 km to 30 km in the immediate hanging wall of the Woodroffe Thrust (Scrimgeour & Close 1998). The high-pressure rocks are predominantly composed of felsic and mafic orthogneiss. North of the Woodroffe Thrust, mid-crustal rocks are dominated by porphyritic granites and granite gneisses which have been imbricated in a large-scale duplex system that

accommodated around 100 km of shortening (Flöttmann & Hand 1999). The structurally higher parts of the duplex contain basal units of the Amadeus Basin, metamorphosed at greenschist to mid-amphibolite facies conditions. Further north within the Amadeus Basin, Petermann-aged structures are defined by sub-metamorphic, open to tight folding of the Neoproterozoic sequences. The northern margin of the Petermann Orogen thus exposes a near crustal scale cross-section.

As yet there are no definitive estimates for the duration of, or the amount of shortening during, the Petermann Orogeny. Sedimentological and thermochronologic data suggest that the Petermann Orogen began around 570 Ma and continued until about 530 Ma (Maboko *et al.* 1992; Walter & Gorter 1994; Hoskins & Lemon 1995; Walter *et al.* 1995; Lindsay & Leven 1996; Camacho *et al.* 1997). However, in the southern Amadeus Basin sequences older than 570 Ma thicken toward the orogen (Wells *et al.* 1970; Hand & Sandiford 1999), suggesting that loading may have begun prior to 570 Ma, and that the orogeny may have lasted more than 40 Ma. Estimates of shortening along the northern edge of the orogen are constrained by the geometry of Amadeus Basin sequences. However, the bulk of the orogen contains no preserved cover sequences and shortening can only be indirectly inferred from the geometry of, and metamorphic offsets associated with, shear zones. The Woodroffe Thrust dips at around 30° and shows about 12 km of differential offset across it (Scrimgeour & Close 1998), implying that it has accommodated approximately 20 km of shortening. To the south of the Woodroffe Thrust, shear zones typically dip at $\geq 45°$ and the highest pressure rocks are located within a corridor some 30 km wide (Camacho *et al.* 1997). Although by no means definitive, it appears likely that the shortening associated with the Petermann Orogeny was less than <200 km.

The Alice Springs Orogeny resulted in exhumation of the Palaeo to Mesoproterozoic Arunta Inlier from beneath the northern fragment of the Centralian Superbasin isolating the Amadeus and Ngalia Basins from the Georgina and Wiso Basins to the north (Fig. 1). Shortening began possibly as early as 450 Ma (Shaw & Black 1991; Mawby *et al.* 1999), but certainly by 400 Ma, and was marked by the inversion of a late Cambrian to mid-Ordovician marine basin and the deposition of clastic sediment (Shaw *et al.* 1991*a*). Isotopic and sedimentological evidence indicates that deformation continued, at least intermittently, for at least 100 Ma, terminating in the early Permian (Shaw *et al.* 1991*a*; Wells & Moss

1983). Total shortening is estimated at 100–125 km (Teyssier 1985; Flottmann & Hand 1999).

The major Alice Springs structure is the Redbank Shear Zone; a reverse sense shear zone dipping north at *c.* 45° that offsets the Moho by at least 20 km (Goleby *et al.* 1989; Korsch *et al.* 1998). The Redbank Shear Zone is associated with one of the largest gravity anomalies (*c.* 150 mgal) known from continental interiors (Mathur 1976). Early to mid-Devonian synorogenic sediments in the northern Amadeus Basin immediately south of the Redbank Shear Zone contain clasts of the underlying sequences including basement (Jones 1972; Jones 1991) indicating that movement on the Redbank Shear Zone was initiated at or before 400–390 Ma (see also Shaw & Black 1991). As with the Woodroffe Thrust in the Petermann Orogen, deformation along the Redbank Shear Zone has produced a crustal-scale lithological subdivision. To the south of the Redbank Shear Zone, the Arunta Inlier exposes large granitic complexes, which have intruded low-pressure amphibolite to greenschist grade metasediments. Immediately north of the Redbank Shear Zone, in the core of the Arunta Inlier, medium pressure mafic and felsic granulite outcrop over a zone up to 50 km across strike.

Further north, in the central part of the Arunta Inlier, the Napperby Thrust carries a basement wedge composed mainly of granite and granite gneisses over sediments in the northern Ngalia Basin (Wells & Moss 1983). The Napperby Thrust was initiated during late Devonian to early Carboniferous (Wells & Moss 1983), at least 50 Ma after deformation began along the Redbank Shear Zone. It accommodated approximately 20 km of shortening on a surface that dips northwards at 30° (Wells & Moss 1983; Bradshaw & Evans 1988). Along the northern margin of the Arunta Inlier, granitic basement has been thrust northward over Neoproterozoic and early Palaeozoic cover sequences belonging to the Georgina and Wiso Basins.

In summary, the Petermann and Alice Springs Orogens show important similarities:

- in both cases, deformation along crustal–penetrative thrust faults has resulted in the formation of major gravity gradients;
- compared with typical plate margin orogens, the total shortening is relatively minor (of the order 100 km); and
- given the limited shortening, the interval over which deformation occurred is relatively long implying slow bulk orogenic shortening rates; in the case of the Petermann Orogeny (where absolute age constraints are relatively poor)

the orogen probably lasted at least 40 Ma, while the Alice Springs Orogeny lasted at least 100 Ma and possibly 150 Ma.

Distribution of deformation and sediment thickness in the Centralian superbasin

An intriguing aspect of the record of intraplate deformation in central Australia relates to the way in which the locus of deformation changed in space and time. The Petermann and Alice Springs Orogens appear to be more or less mirror images of each other. Both are characterized by E–W structural trends (reflecting predominant N–S shortening), therefore they must have formed in response to similarly oriented intraplate stress fields. However, it is apposite to ask why the Alice Springs aged deformation was not also localized in the vicinity of the Petermann Orogen but some 300 km further north? This question has been addressed elsewhere in a series of papers (Sandiford & Hand 1998*a*; Hand & Sandiford 1999), which were motivated by the observation that the locus of intraplate deformation appears to mimic the main sedimentary depocentres in the Centralian Superbasin (Fig. 2). In the following section these observations are briefly summarized.

Immediately prior to the Petermann Orogeny, the Centralian Superbasin was thickest (>4 km, Fig. 2a) in the vicinity of the Musgrave Inlier, the region in which deformation was subsequently localized. At the same time a relatively thin sheet (<2 km, Fig. 2a) of sediment covered the Arunta Inlier. Following the Petermann Orogeny the locus of subsidence shifted to the northern part of the Amadeus Basin in the vicinity of the Arunta Inlier, while the major Petermann Orogen structures such as the Woodroffe Thrust were stripped of their cover and exhumed from deep levels (Fig. 2b). Prior to the Alice Springs Orogen, the northern margin of the Amadeus Basin contained at least 6 km of sediment and locally as much as 8 km (Fig. 2b), with isopachs increasing northwards across the basin towards the region where the most intense Alice Springs Orogenic deformation was localized. Importantly, the major Petermann-aged structures such as the Woodroffe Thrust remained inactive during the Alice Springs Orogeny (for more details see Hand & Sandiford 1999).

It is believed that these observations imply that past tectonic activity (basin formation, deformation and erosion) has modulated the response of the central Australian lithosphere during subsequent tectonic activity. This suggests that the link is primarily thermal, with the behaviour

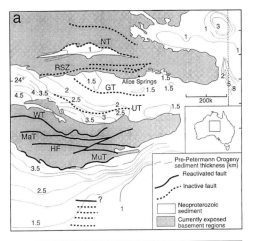

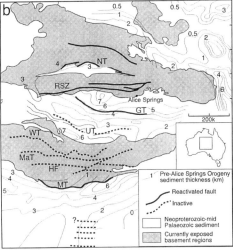

Fig. 2. Isopach maps for the Centralian Superbasin constructed for time intervals immediately prior to the onset of the Petermann Orogeny (Fig. 2a) and the Alice Springs Orogeny (Fig. 2b). Details of these isopach distributions are outlined in Hand & Sandiford (1999). It is noted that prior to the Petermann Orogeny the deepest parts of the basin were in the vicinity of the Petermann Orogen. Following the Petermann Orogeny, which resulted in the exhumation of the Musgrave Inlier, the depocentre shifted north, such that by the time of the Alice Springs Orogen the main sedimentary mass was located in the vicinity of the northern Amadeus basin, and the southern Arunta Inlier, which became the locus of Alice Springs aged deformation. HF, Hinckley Fault; MaT, Mann Thrust; MuT, Munyari Thrust; NT, Napperby Thrust; RSZ, Redbank Shear Zone; UT, Uluru Thrust; WT, Woodroofe Thrust.

reflecting an important example of 'tectonic feedback' effected by the long-term thermal consequences of changes in the distribution of heat producing elements in the crust. In order to quantify the way this feedback has helped to shape the tectonic evolution of this part of the continent, it is necessary to first understand the way in which heat production is distributed in the central Australian crust.

Heat production distributions in the central Australian region

As discussed earlier the Petermann and Alice Springs Orogens show remarkable structural similarities. This similarity extends to the petrological and geochemical character of the orogens. Each is characterized by a high-grade gneissic core granulite comprising significant proportions mafic granulites. This core is tectonically juxtaposed against essentially granitic terranes along the major structures active during the intraplate orogenies. The fact that the sediments of the Centralian Superbasin unconformably overlie these gneissic granite terranes implies that they formed the upper crust prior to basin formation. Seismic reflection and refraction profiling (Goleby *et al.* 1989) together with gravity measurements (Mathur 1976) across the Alice Springs Orogen confirms the fact that the mafic granulites exposed in the core of the orogen connect directly with, and are therefore representative of, the mid-lower crust beneath the basins, while the granitic terranes connect with the mid-upper crust (Fig. 1). By virtue of the fact that these orogens now expose terranes that were at very different crustal levels prior to deformation, it is possible to evaluate the way in which heat production is distributed in the crust. This section reviews the constraints on the distribution of heat sources in the various terranes that have been exhumed from different levels of the crust. The heat production character of the main lithostratigraphic units can be estimated using geochemical analyses and calibrated airborne radiometric data (Fig. 3). Note that heat production values listed below are quoted at 350 Ma, as appropriate to the Alice Springs Orogeny (i.e., about 6% greater than modern day values). Estimates of the characteristic thickness of the sequences derived from seismic, structural and stratigraphic studies then allows the estimation of the total thermal energy budget of the various crustal levels, which are expressed in terms of their contribution to the surface heat flow.

The crustal sections exposed in the vicinity of the Woodroffe Thrust in the Petermann

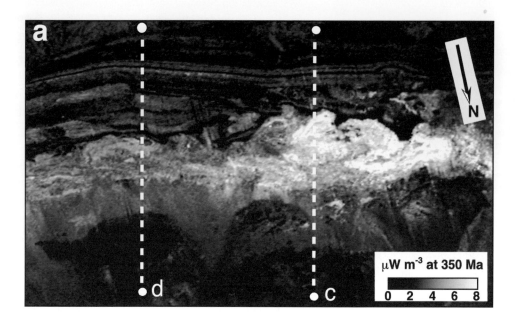

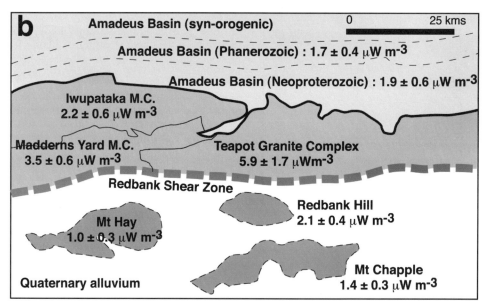

Fig. 3. (a) Image of heat production in the western Macdonnell Ranges (Hermannsburg Sheet) derived from Northern Territory Geological Survey airborne radiometrics. **(b)** shows the area averaged heat production rates for each of the main lithological components of this region. In the south (top), sediments of the Amadeus Basin are characterized by low-intermediate heat production rates. In the central region, high-heat producing Mesoproterozoic granites of the Teapot granite Complex, and intermediate heat producing metasediments and orthogneisses of the Iwupataka and Madderns Yard Metamorphic Complexes formed the basement upon which the Amadeus Basin sediments were deposited. North of the Redbank Shear Zone (bottom of figure), mafic and felsic granulites exhumed from the deep crust occur as isolated monadnocks (Mt Hay, Mt Chapple and Redbank Hill) surrounded by Quaternary alluvium which has largely been shed from the granites to the south. These deep crustal rocks are characterised by low-intermediate heat production rates. **(c)** heat production profile (black line) from the Ormiston Gorge region superposed with the heat production distribution (gray line) used to model heat production (equation 1). **(d)** heat production profile from the Ellery Creek region. Heat production rates are calculated at 350 Ma (i.e, during the Alice Springs Orogeny).

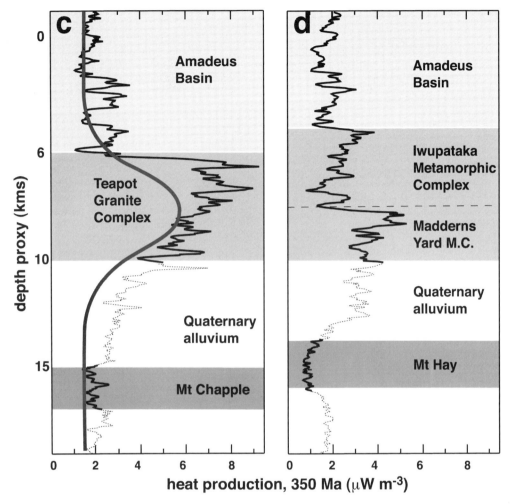

Fig. 3. (*continued*)

Orogen and the Redbank Shear Zone in the Alice Springs Orogen suggest that the central Australian crust has a relatively simple three layered structure comprising mafic to felsic lower crust, granitic and metasedimentary mid-upper crust and unmetamorphosed sedimentary upper crust (e.g. Goleby *et al.* 1989; Shaw *et al.* 1991*b*). The boundaries between these layers are generally sharp and are marked either by major shear zones or unconformities (Fig. 3).

The *mid-lower crust* is dominated by Palaeo to Mesoproterozoic mafic to felsic granulite. Modelling of gravity and seismic data suggests that this layer is between 10 and 20 km thick (Goleby *et al.* 1989; Lambeck & Burgess 1992). The heat production is typically $<2 \mu W m^{-3}$ in felsic lithologies and $c. 1 \mu W m^{-3}$ in mafic rocks (Fig. 3). In the Arunta Inlier these mid-lower

crustal units tend to outcrop as isolated monadnocks, prohibiting quantification of area-averaged heat production rates from the airborne radiometric data with any confidence. However, a conservative estimate of the average heat production of $1 \mu W m^{-3}$ (see exposures at Mount Hay, Mount Chapple and Redbank Hill, Fig. 3a and b) implies that the mid-lower crust may contribute between about 10 and $20 mW m^{-2}$ to the surface heat flow.

The *mid-upper crust* consists predominantly of Palaeo- to Mesoproterozoic metasedimentary and granitic lithologies. Gravity and seismic data, as well as baric offsets across large shear zones suggest that this mid to upper crustal layer is typically at least 10 km thick (Goleby *et al.* 1988, 1989; Warren & Hensen 1989; Lambeck & Burgess 1992). In much of the Musgrave Inlier, this layer has been removed as a consequence of the relatively deep levels of denudation associated with the Petermann Orogeny. In the footwall of the Woodroffe Thrust, where this layer has been preserved, individual granites up to 4 km thick with outcrop extents exceeding 2000 km^2 are exposed in the cores of regional-scale antiformal culminations enclosed by cover sequences of the Amadeus Basin (Flöttmann *et al.* 1999). In the Arunta Inlier, granitic and metasedimentary lithologies dominate, reflecting the generally smaller amounts of denudation compared to the Musgrave Inlier. In the central and northern parts of the Inlier, granite accounts for up to 75% of the outcrop (Offe 1978; Wells 1982; Haines *et al.* 1991; Young *et al.* 1995). Granite heat production rates are typically greater than 5 μW m^{-3}, with some individual granite bodies contributing up to 10 μW m^{-3} making them, by global standards, extremely effective heat producers (Sandiford & Hand 1998*b*). In the western Macdonnell Ranges (Fig. 3a), the Teapot Granite Complex averages 5.9 μW m^{-3}. The intervening metasedimentary complexes such as the Iwupataka Complex generate around 2–3.5 μW m^{-3}. The area averaged heat production rates for the mid-upper crustal zone shown in Figure 3a is 4.1 μW m^{-3}, suggesting it contributes at least 40 mW m^{-2} to the surface heat flow. It is probable that there are significant spatial variations in the heat contributed by the mid-upper crust (cf. Fig. 3a, b), although the nature of the amplitude and wavelengths of such variations are not yet understood.

The *upper-most crust* consists of the sediments of the Centralian Superbasin. The sedimentary sequences are characterized by intermediate to low heat production rates. Along the northern margin of the Amadeus Basin, where the entire sequence is exposed in near vertical profile along the Macdonnell homocline, heat production varies between 0.5 and 2.5 μW m^{-3} and averages *c.* 1.8 μW m^{-3} (Fig. 3a, b). Together with estimates of isopachs, these heat production rates suggest that prior to each orogeny the basin contributed, at most, between 7 and 12 mW m^{-2} to the surface heat flow.

Figures 3c and d show two heat production traverses across the Macdonnell Ranges from the northern Amadeus Basin to the hanging wall of the Redbank Shear Zone. This section effectively provides a depth slice through the three layered crustal structure, with 'steps' in the heat production occurring at the unconformity at the base of the Amadeus Basin, and at the Redbank Shear Zone. The profile shows that the heat production distribution immediately prior to the Alice Springs Orogeny was characterized by a pronounced maxima at palaeo-depths of about 8–12 km. This distribution contrasts dramatically with the common perception that crustal heat production decreases as an exponential function of depth.

The heat production data from the central Australian region suggests that, prior to each orogeny, the average crustal contribution to the heat flow was likely to have exceeded 50 mW m^{-2}. In the context of global continental heat flow, especially for Precambrian terranes, this value is high (e.g. McLennan & Taylor 1996). However, it is consistent with other heat flow-heat production data from Australia. These data define an anomalous heat flow zone that includes most, but not all, of Proterozoic Australia (Cull 1982). This zone includes the eastern part of the Gawler Craton (and the adjacent Stuart Shelf), the Willyama and Mount Painter Inliers in South Australia and New South Wales, the Mount Isa Inlier in western Queensland, and the Tennant Creek Inlier in the Northern Territory (Cull 1982). The average surface heat flow in this zone is 85 mW m^{-2} (Cull 1982). The zone is underlain by seismically fast and, by inference, cold and thick (>200 km) lithospheric mantle (Zielhuis & van der Hilst 1996; van der Hilst *et al.* 1998) implying mantle heat flows are unlikely to exceed *c.* 15 mW m^{-2}. An important implication is that the crustal contribution to the heat flow in Proterozoic Australia may be as high as 70 mW m^{-2}. This amounts to a lithospheric U, Th, K endowment approximately twice the continental average (e.g. McLennan & Taylor 1996), and is consistent with widespread U-mineralization through the region, including some of the largest known U-deposits (i.e. Olympic Dam).

Measured heat flow near Alice Springs is currently 60 mW m^{-2} (Cull 1982) suggesting that crustal sources in the vicinity of the heat flow record currently contribute *c.* 45 mW m^{-2}. At Alice Springs, denudation following the Alice Springs Orogeny has removed all the cover and some of the enriched basement, while only some 10 km further north in the hanging wall of the Redbank Shear Zone both the upper and mid-crustal section has been removed. These observations suggest that the current crustal contribution to the heat flow is significantly lower than was likely to have applied immediately prior to the Alice Springs Orogeny.

Thermal aspects of heat source distribution

The analysis presented in the previous section suggests that the central Australian crust is strongly stratified in heat producing elements, with much of the heat production concentrated in mid-upper crustal level gneissic granite terranes, where the average heat production is estimated to be at least $4 \, \mu W \, m^{-3}$. Sequences both above (basin-filling sediments) and below (lower crustal gneisses) this layer, are characterized by signifi-cantly lower heat production rates. Intraplate tectonism in central Australia displaced this heat production anomaly relative to the surface. During basin formation, the high-heat producing layer was buried by up to 8 kilometres of sedi-ment. Following intraplate orogeny, locally the layer was completely removed by erosion. While

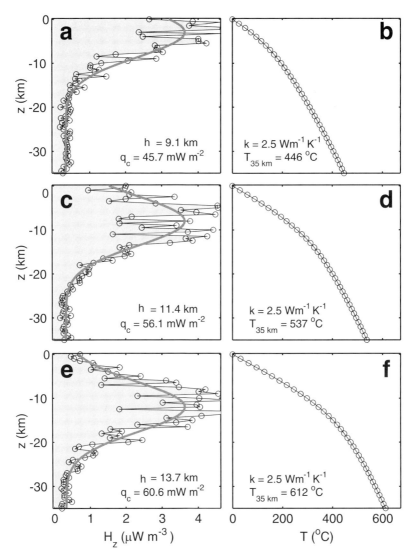

Fig. 4. Illustration of the thermal consequences of burial of an anomalous heat producing layer such as that believed to characterize the Proterozoic granitic rocks in central Australia, as shown in Figure 3. The three distributions on the left correspond to increasing depth of burial (as represented by the parameter h). Note that while the form of the heat production distribution in each is the same, and can be approximated by equation (1) as shown by the thicker solid line, the total crustal contribution to the surface heat flow q_c increases with burial due to the fact that the rocks that contribute to the burial (e.g. the sediments) add to the total heat production. The right hand side figures show the geotherms corresponding to each of the heat production distributions, and highlight the sensitivity of temperatures at depth (e.g. $T_{35\,km}$) to the depth of the heat producing layer (see Fig. 5).

the tectonic processes associated with such translation must have been associated with thermal transients it is necessary to emphasize the significant long-term thermal effects resulting from the movement of this heat source (e.g. Fig. 4). 'Long-term' implies the thermal regimes that approximate steady-state, and which applied following the dissipation of all thermal transients associated with tectonism. In so doing, we are necessarily restricting our attention to thermal consequences of processes on timescales greater than about 100 Ma. Such timescales are appropriate to the overall development of the Centralian Superbasin (which contains a depositional record spaning almost 500 Ma) and the consequences of the Petermann Orogeny during the Alice Springs Orogeny, some 100–200 Ma later.

In order to quantify the long-term thermal consequences of such vertical translation it is useful to approximate the distribution with an analytic function. Following Sandiford & Hand (1998*b*), the distribution of heat production in central Australia is approximated as an exponential function which attains a maximum H_i at a discrete level z_i within the crust (Fig. 3b):

$$H(z) = H_i \exp\left(\frac{-(z - z_i)^2}{h_r^2}\right) \quad (1)$$

In equation (1) the parameter h_r provides a measure of the spread of the heat production distribution with the heat production falling to $H_i e^{-1}$ at depths $z_i \pm h_r$. Subject to a basal heat flow, q_m, applied at a depth beneath which heat production is negligible and the heat production distribution in equation (1) the resulting temperature field is:

$$
\begin{aligned}
T(z) = -\frac{q_m z}{k} &+ \frac{H_i h_r^2}{2k}\left(\exp\left(-\frac{z_i^2}{h_r^2}\right)\right.\\
&\left. - \exp\left(-\frac{(z_i - z)^2}{h_r^2}\right)\right) + \frac{h_r H_i \sqrt{\pi}}{2k}\\
&\times \left(z \operatorname{Erf}\left(\frac{z_c - z_i}{h_r}\right) + (z_i - z)\operatorname{Erf}\left(\frac{z - z_i}{h_r}\right)\right.\\
&\left. + z_i \operatorname{Erf}\left(\frac{z_i}{h_r}\right)\right) \quad (2)
\end{aligned}
$$

where k, is the thermal conductivity assumed to be both temperature and depth independent. Note that the 1-D approximation in equation (2) implies no lateral variations in heat production parameters. Figure 3 clearly shows significant lateral variation in heat production in the high-heat producing granitic layer. In choosing parameters relevant to central Australia it is

therefore important to use average, rather than extreme, parameters. In the following, heat production parameters contribute a total of $45 \, \text{mW m}^{-2}$ to the surface heat flow with a maximum heat production of $4 \, \mu\text{W m}^{-3}$, thus constraining the characteristic length scale h_r to be *c*. 7 km. In view of our earlier discussion this is a rather conservative estimate of the contribution made by crustal sources in this region.

In the following section the long-term thermal consequences of the burial of such a heat-producing layer, by calculating the changes in the temperature at deep crustal- upper mantle depths accompanying erosion and/or sedimentation are explored. Figure 5a shows that the erosional removal of the sedimentary cover from this heat producing layer will lead to long-term cooling of the Moho by about *c*. 18°C per kilometre of denudation for parameter ranges appropriate to central Australia. Note that this estimate is sensitive to the value of the mantle heat flux and the thermal conductivity, which together determine the thermal gradient in the deeper part of the lithosphere, both of which are relatively poorly constrained. Approximately half the Moho cooling is due to shallowing of the Moho, while the remainder is due to deep crustal cooling in response to the change in the depth of the heat producing layer.

The evaluation of the long-term thermal effects of the burial of such a sequence beneath an accumulating sedimentary pile requires an understanding of the factors that contribute to subsidence (e.g. Sandiford 1999). If the pre-existing crust is simply buried beneath the sedimentary pile such that the whole crust including the Moho is displaced, then the long-term Moho temperature will be the opposite of that induced by erosion (i.e. for $q_c = 45 \, \text{mW m}^{-2}$, Moho temperatures increase by about 18°C km^{-1} of basin-fill). However, more typically basin formation is induced by crustal stretching and the Moho displacement will not be directly coupled with the thickness of the sedimentary basin. Moreover, crustal stretching will attenuate pre-existing crustal heat production. As discussed by Sandiford (1999), the thermal consequences of this style of basin formation will reflect a competition between burial of the pre-existing heat production, which tends to heat the crust, and the attenuation of the pre-existing heat production, which tends to cool the crust. Figure 5b shows that for the case where the total subsidence is sufficient to preserve the Moho at its pre-stretching depth, then Moho temperatures will rise about 9°C for each kilometre of subsidence. In a more typical case, where the Moho shows long-term shallowing as a consequence

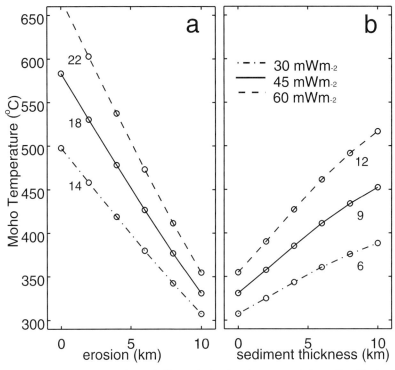

Fig. 5. Illustration of the long-term thermal effects of (**a**) exhuming a central Australian like heat production distribution from beneath a thick sedimentary succession such as the Centralian Superbasin, and (**b**) burying it beneath a sedimentary succession, where the subsidence is related to crustal stretching (see text for discussion). Three different curves represent three different total abundances of heat producing elements, i.e. $q_c = 30$, 45, and $60\,\mathrm{mW\,m^{-2}}$. For Central Australia, we estimate q_c is on average at least $45\,\mathrm{mW\,m^{-2}}$. The numbers on each curve indicate the magnitude of the change in Moho temperature per kilometre of denudation/burial. Other model parameters are $q_m = 20\,\mathrm{mW\,m^{-2}}$, $k = 3\,\mathrm{W\,m^{-1}\,K^{-1}}$. Increasing q_m and/or decreasing k will tend to increase the magnitude of the thermal response to denudation/burial.

of rifting, Moho temperatures may in fact be reduced (e.g. Sandiford 1999). However, even for cases where the Moho is cooled because of such shallowing, the temperature at any given depth will inevitably increase due to the thermal consequences of the burial of the heat-producing layer (see below).

A generalized model for heat production changes during intraplate deformation

In the previous section it was demonstrated that changes in the heat production distribution have a profound effect on the thermal structure in the deep crust and upper mantle. This can be generalized using a simple parameterization of crustal heat production involving just two parameters: the vertically-integrated crustal heat production, q_c, and the characteristic length

scale for heat production, h. Formally, q_c and h are defined as follows:

$$q_c = \int_0^{z_c} H(z)\,dz \qquad (3)$$

$$h = \frac{1}{q_c} \int_0^{z_c} (H(z)z)\,dz \qquad (4)$$

Importantly, this definition of h makes no explicit assumption about the form of the heat production distribution. h, can be thought of as the *effective depth* of the heat production distribution (Fig. 6). Physically, it can be understood as the depth at which concentrating all q_c heat sources (i.e., as a delta spike) does not alter the thermal structure at or beneath the base of the crust. The parameterization of the crustal heat production distribution in terms of h and q_c has a number of benefits. Firstly, it allows

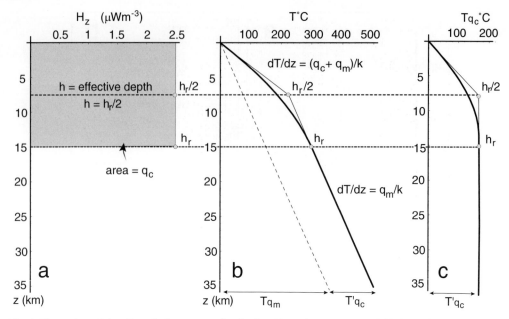

Fig. 6. Illustration of the effect of a heat source distribution (**a**) on the temperature field shown by the thick solid line in (**b**). The dashed line in (**b**) represents the contribution of mantle heat flow. The difference between the actual temperature field and the mantle contribution is a measure of the crustal heat source contribution, T_{q_c}, as shown in (**c**). T_{q_c} attains a maximum value T'_{q_c} $(= hq_c/k)$ at the base of the heat producing layer (i.e. depth h_r). The parameter h represents the *effective depth* of the heat production. For the heat source distribution consisting of a layer of thickness h_r with uniform heat production, the effective depth $h = h_r/2$. Concentrating all q_c heat sources as a delta spike at depth h does not affect the value of T'_{q_c} (as shown by the thin solid line in (**b**) & (**c**)) which shows the temperature field for such a distribution.

the effects of tectonic processes to be quantitatively portrayed on the h–q_c plane. Secondly, the contribution of crustal heat sources to the temperature field at or beneath the base of the heat producing parts of the lithosphere (i.e. the Moho) can be expressed succinctly as a linear function of both h and q_c:

$$T'_{q_c} = \frac{q_c h}{k} \qquad (5)$$

T'_{q_c}, can be considered as the thermal contribution caused by the heat source distribution relative to an otherwise identical lithosphere with no heat production (Fig. 6). The long-term consequences of tectonic processes on the heat production parameters of the crust can be illustrated using a simple scenario appropriate to crustal stretching and basin formation (Fig. 7). Normally continental extension is accompanied by basin formation with associated subsidence continuing for up to *c.* 100 Ma as thermal transients induced by the extension are dissipated. In order to provide a clear picture of the combined effects of crustal stretching and basin

formation on the heat production distribution we treat them sequentially. Figure 7a shows a crust initially 30 km thick, with a simple heat source distribution consisting of an upper crustal layer some 15 km thick characterized by an uniform heat production ($3\,\mu W\,m^{-3}$) beneath which heat production is assumed to be negligible. This crust is therefore characterized by a total (vertically-integrated) heat production of $45\,mW\,m^{-2}$. For this distribution, the appropriate length scale that conforms to equation (4) is given by the mean depth of the heat production, i.e. $h = 7.5\,km$ (see Appendix 1 for further details). From equation (5), $T'_{q_c} = 112\,°C$ ($k = 3\,W\,m^{-1}\,K^{-1}$). A stretching event which homogeneously attenuates the crust to 2/3 its original thickness (Fig. 7b) reduces q_c to $30\,mW\,m^{-2}$, h to 5 km and T'_{q_c} to 50°C. If allowed to thermally equilibrate temperatures at any depth below the heat producing layer would cool by 62°C as the direct consequence of the changes in the heat production distribution. Note that because of the 10 km reduction of the Moho depth inherent in this scenario there would be significant additional cooling of the Moho or any other

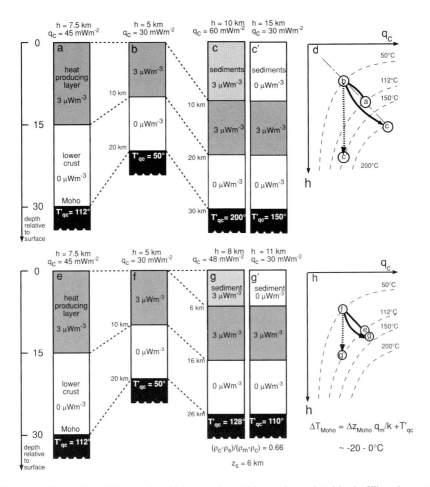

Fig. 7. Schematic illustration of the way in which crustal stretching and associated basin filling change the heat production parameters in the crust (a more detailed account of the long-term thermal consequences of basin formation is given in Sandiford 1999). Column (**a**) illustrates the reference state prior to crustal stretching, and the numbers along side the Moho in each column, represent the contribution of heat sources to the temperature field at that depth (i.e. T'_{q_c}), if the column were left to thermally equilibrate as illustrated. The long-term changes in the temperature of the Moho contributed by changing heat source distributions are given by the change in T'_{q_c} following a tectonic process. For this scenario, the combined effects of crustal stretching (**b**) and basin filling with heat-producing sediments (**c**) result in a long-term heating of the Moho by 88°C. The thermal effects are dependent on thermal properties of the basin filling sediments. However, even in the case where basin filling sediments contain no heat production (**c′**) there is a long-term increase in Moho temperature of 38°C. (**d**) shows that the various scenarios can be represented on the h–q_c plane (see also Fig. 8). Figure 7c applies when the density of the basin-fill is similar to the pre-existing crust. However, even when sediments are less dense, and the basin therefore less deep $\Delta T'_{q_c}$ may be positive (Fig. 7e–f). Note however, that in this case the long-term shallowing of the Moho will mean that $\Delta T_{Moho} < \Delta T'_{q_c}$ (see Sandiford 1999).

material point beneath the base of the heat producing layer. The effect of filling the accommodation space created by extension with sediments depends on the properties of the sediments. If the sediments have heat production rates comparable to the pre-existing upper crust (i.e. $3\,\mu W\,m^{-3}$) and are sufficiently dense that in filling the basin the Moho is displaced back to its

original depth (Fig. 7c), the resulting heat producing layer would be some 20 km thick (comprising a 10 km thick basin together with the 10 km thick attenuated pre-existing heat producing layer) and would contribute $60\,mW\,m^{-2}$ to the surface heat low. In this scenario (h is 10 km, T'_{q_c} is 200°C), the combined effects of stretching and sedimentation yield a change in T'_{q_c} of *c.* 88°C, or about 9°C/km of sediment-fill (cf. Fig. 5b). While the long-term thermal response is sensitive to the heat production contributed by the sediments, there maybe a long term increase, albeit small, in T'_{q_c} even when the sediment-fill contains no heat production (Fig. 7c). This highlights the fact that the long term thermal response to extensional basin formation reflects a competition between attenuation (leading to cooling) and burial (leading to heating) of the pre-existing heat production. Where the density of the basin-filling sediments is somewhat less than the crustal average, isostasy dictates a long-term shallowing of the Moho, possibly by as much as 4 km for the stretching scenario outlined

above depending on the density of the sediment, with a correspondingly smaller T'_{q_c} (Fig. 6e–g). Moreover, the shallowing of the Moho in this scenario causes additional Moho cooling (by approximately $q_m \Delta z/k$, where Δz, is the amount of Moho shallowing), implying that even though $\Delta T'_{q_c}$ may be positive (i.e. long-term heating at a specific depth), ΔT_{Moho} may be negative (i.e. long-term cooling of the Moho). A more detailed analysis of the long-thermal consequences of rift basin formation is given by Sandiford (1999).

The above analysis shows that the long-term thermal consequences of tectonic processes relate directly to the way these processes change h and q_c. Figure 8 uses these two parameters to illustrate the thermal consequences of basin formation (dashed lines) induced by a 10% crustal stretch for the case where the basin filling sediments (*c.* 3 km) have only half the heat production of the upper crust. Also shown, in solid lines is the effect of a 10% homogeneous shortening of the crust, followed by erosion such that crustal thickness returns to its original (pre-shortening) value. Although a 10% strain at the crustal scale represents only a very mild deformation, Figure 8 shows that, when coupled with an associated surface response such as erosion, such deformations can result in very substantial redistribution of crustal heat sources. This in turn has profound consequences for the long-term thermal state of the lithosphere. In general, crustal stretching and basin formation will lead to long-term heating at deep crustal and upper mantle levels, while crustal shortening and associated erosion leads to long-term cooling (note that the Moho may experience additional changes in temperature if its depth is also changed). For a given crustal strain, the most dramatic changes in temperature are expected when the crustal heat production distribution is already strongly differentiated, and unusually enriched (i.e. low h and high q_c), as appears to be the case for much of the Australian Proterozoic crust including central Australia. In contrast, the processes are rather less effective when operating on relatively undifferentiated crust (i.e. large h).

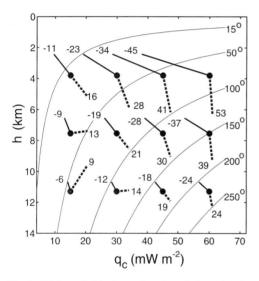

q_c (mW m^{-2})

Fig. 8. With a suitable chosen length scale (equation 3, Appendix 1), the long term thermal consequences of a tectonic process can be predicted from the associated changes in h and q_c, allowing portrayal on the h–q_c plane. The bold lines show the effects of the imposition of a 10% stretch with associated basin filling (*c.* 3.5 km) on q_c and h. The thin lines show the effects of a 10% crustal shortening, followed by erosion (*c.* 3.5 km) back to the pre-shortening crustal thickness. The initial geometry, indicated by the bullet points, consists of an upper crustal layer of thickness $2h$ characterized by a uniform heat production of magnitude $q_c/(2h)$. Note that for a given crustal strain the most dramatic changes in temperature are expected when the crust is already strongly stratified in heat producing elements, and unusually enriched (i.e. low h and high q_c), as appears to be the case for the central Australian crust. In contrast, the processes are rather less effective when operating on crust which shows no substantial differentiation of heat producing elements (e.g. large h). In this modelling, basin-filling sediments are assumed to contribute $2 \mu\mathrm{W\,m}^{-3}$, while the conductivity is assumed to be $3\,\mathrm{Wm}^{-1}\,\mathrm{K}^{-1}$.

Mechanical consequences of heat source redistribution

Emphasis has been placed on the long-term thermal consequences of burial and denudation in central Australia, as reflected in thermal regime in the deep crust and upper mantle. Primarily because the upper mantle temperatures are likely to provide a useful proxy for lithospheric strength (e.g. Sonder & England 1986). In this section an attempt to quantify how such long-term thermal changes might change the strength parameters of the lithosphere using a 'Brace-Goetze' rheology in which the lithosphere is assumed to deform by a combination of frictional sliding and ductile creep (e.g. Brace & Kohlstedt 1980). The application of the 'strength envelope' approach implicit in this 'Brace-Goetze' rheology has become widespread in geodynamic problems, and is now very familiar. However, such calculations are highly uncertain (e.g. Paterson 1987), because:

- the extrapolation of rheological flow laws from laboratory conditions and timescales to the geological realm involves many orders of magnitude;
- our imprecise understanding of the compositional and mineralogical structure of the lithosphere translates to uncertainties in knowledge of what particular flow law applies in any given part of the lithosphere; and
- our imprecise knowledge of the thermal property structure of the lithosphere (particularly thermal conductivity) results in considerable uncertainty in its absolute thermal structure.

These problems render any calculation of absolute strength rather pointless, and in the following section emphasis is placed on relative changes in lithospheric strength that accompany changes in the thermal structure of the lithosphere as a consequence of changes in the heat production parameters. While the precise magnitude of these relative strength changes will depend on the composition and thermal structure of the reference frame, they are likely to be far more robust than any absolute measure of strength, because most of the uncertainties alluded to above will cancel.

In previous sections it has been demonstrated that changes in heat source distribution, associated with tectonic processes affecting the central Australian crust, produced long-term changes in Moho temperatures of the order of $100°C$. The greatest changes are likely to be due to erosional denudation of deep crustal rocks in the cores of the orogens. Figure 9 shows how such changes in crustal thermal structure accompanying such denudation impact on the strength. In constructing Figure 9 we have used strength parameters appropriate to quartz-dominated upper crust (sedimentary succession), feldspar-dominated mid-lower crust, and olivine-dominated mantle (see Sandiford *et al.* 1988, for a more detailed account). For the parameters used in the calculation of Figure 9 ($q_c = 45\,\mathrm{mW\,m^{-2}}$, $k = 2.5\,\mathrm{W\,m^{-1}\,k^{-1}}$, $q_m = 25\,\mathrm{mW\,m^{-2}}$), 5 km of

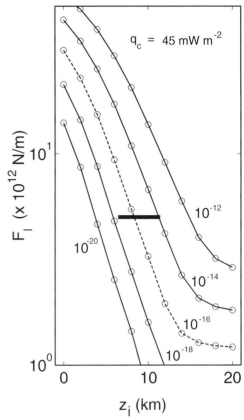

Fig. 9. Illustration of the rheological effects of progressive denudation of a radioactive basement using a 'Brace-Goetze' rheological model as discussed in the text, and the thermal parameter range used in Figure 5. The parameter z_i represents the locus of maximum heat production. In terms of the central Australian example, it corresponds to a depth several kilometres beneath the basement-cover unconformity. The calculated lithospheric strength (F_l) is shown for a range of effective strain rates ($= 10^{-12}$, 10^{-14}, 10^{-16}, 10^{-18} and $10^{-20}\,\mathrm{s^{-1}}$). The strength parameter (F_l) can be best viewed as the magnitude of the tectonic force driving deformation. At $F_l = 5 \times 10^{12}\,\mathrm{N\,m^{-1}}$ a decrease in the burial of the basement sequence by 5 km results in a corresponding decrease in strain rate of up to four orders of magnitude (solid bar), provided thermal equilibration takes place. Note that while the absolute values of the calculated strain rates at any given strength (or strengths at a given strain rate) are highly uncertain, rather more confidence is attached to estimates of the relative changes that accompany a given change in the thermal state.

denudation (i.e. c. $90°C$ Moho cooling; Fig. 8) leads to sufficient strengthening of the system to reduce effective strain rates by c. four orders of magnitude. These calculations highlight the extraordinary temperature sensitivity of the 'Brace-Goetze' lithosphere, particularly given that the rates of geologically-significant distributed tectonic deformation in the modern continents spans little more than about four orders of magnitude.

Recognizing that the h–q_c plane can be contoured for thermal structure (Fig. 8) allows it to be contoured for changes in strength parameters (Fig. 10), with the proviso that the tectonic processes that lead to alteration in the heat production parameters do not also lead to long-term

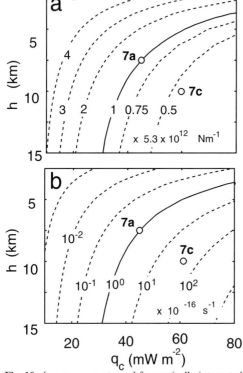

Fig. 10. h–q_c space contoured for vertically-integrated strength normalized against a lithosphere with $h = 75$ km and $q_c = 45$ mW m^{-2} (i.e., the notional central Australian lithosphere). Figure 10a shows variations in the normalized strength at a constant strain rate (10^{-16} s^{-1}). The estimated strength of the reference lithosphere is 5.3×10^{12} N m^{-1}. Figure 10b shows the strain rate that would apply to a lithosphere subject to a tectonic force of 5.3×10^{12} N m^{-1}. It should be noted that these estimates are dependent on the assumed compositional and rheological properties of the lithosphere. It is emphasized that these calculations are based on the assumption that there is no long-term change in Moho depth, as might be expected following extensional deformation and basin formation (see Sandiford 1999). Points 7a and 7c correspond to points **(a)** and **(b)** in Figure 7.

changes in the depth of the Moho (see below). In Figure 10, the mechanical response has been normalized against the reference frame, characterized by an initial configuration with $q_c = 45$ mW m^{-2} and $h = 7.5$ km, to emphasize relative changes in strength parameters. Figure 10a shows the change in strength at a specified strain rate. Note that the stretching – basin filling scenario illustrated in Figure 7a–c results in a two-fold variation in lithospheric strength. While strength envelope calculations are often presented as effective strength at a given strain rate (e.g. Fig. 10a), it is probably more useful to consider the rate of deformation that would apply as a result of an imposed force. Figure 10b illustrates the (normalized) deformation rates in response to an imposed tectonic force of 5.3×10^{12} N m^{-1}. The stretching-basin filling scenario illustrated in Figure 7a–c results in an effective change in rate of deformation of c. 1–2 orders of magnitude.

The calculations summarized in Figures 4–9 highlight an important and often overlooked aspect of continental rheology; namely, the rheological properties of the lithosphere are not time invariant, as is often implicitly assumed. Most importantly, the evolution of lithospheric strength is dependent on past tectonic activity. The calculations summarized here suggest that the amplitude of such time-dependent strength variations in continental interiors is large, and may be of similar magnitude to the spatial variations in strength arising from variations in the compositional make-up of the lithosphere. Our purpose in Figures 4–9 has been to highlight the role played by heat production distributions in modulating the long-term strength of the lithosphere. It must be noted, however, that Figure 10 only applies when Moho depth is held constant. As pointed out by Braun (1992) and Sandiford (1999), tectonic processes such as crustal extension leading to basin formation also typically lead to long-term changes in the depth of the Moho because they affect the density structure of the lithosphere. Changes in the depth of the Moho will also affect the strength of the lithosphere and therefore are also critical to the interplay between tectonism and long-term lithospheric strength.

2-D considerations

Our analysis of the thermal consequences of tectonic process that change heat source distribution has so far been 1-D. Such an analysis is only applicable if the processes operate on horizontal length-scales that are of similar order, or greater, than the thickness of domain over which conduction is occurring. In this case the thickness of the lithosphere defines the appropriate domain. In central Australia, the lithosphere is of the order of 200 km thick (Zeilhaus & van der Hilst 1996). While the broad depocentres associated with regional variations in isopach geometry within the Centralian Superbasin are of this order, the subsequent basement uplifts, and particularly the domains in which deep crustal rocks have been exhumed are rather narrower. For example, the width of the mafic granulite domains in the cores of both the Musgrave and Arunta Inliers is $c.\,30\text{--}50\,\text{km}$ (Fig. 1). In this section the role geometry plays in modulating the thermal response to the localized redistribution of a heat sources in 2-D is considered, focussing particularly on the long-term cooling associated with basement uplifts such as the Musgrave and Arunta Inliers.

To do this a variation of the heat production distribution used for the 1-D case examined is adopted to describe a spatially localized heat production anomaly:

$$H(x, z) = H_i \exp\left(\frac{-(z - z_i)^2}{hz_r^2}\right) * \exp\left(\frac{-x^2}{hx_r^2}\right) \quad (6)$$

where hz_r and hx_r, are the characteristic length scales of the heat production anomaly in the vertical and horizontal directions, respectively. Figure 11 shows the temperature variation (normalized against the solution for the 1-D approximation) directly beneath the centre of the heat production anomaly as a function of the parameter hx_r (the characteristic horizontal length scale). The calculations summarized in Figure 11 indicate that providing the characteristic length scale of the heat production anomaly is greater than about 50 km, then upper crustal temperatures will be greater than 50% of the 1-D approximation. Figure 11 also shows that for any given value of hx_r, the normalized temperature decreases with increasing depth. At Moho depths (i.e. 30–35 kms) it is $c.\,15\%$ lower than at 10 km depth. In view of the importance of Moho temperature in dictating the strength of the lithosphere, these calculations inform us about the horizontal scales in heat production distribution in the mid-upper crust required to produce significant mechanical effects. The across-strike dimensions greater than about 50 km are required in order for sub-basins, or subsequent uplifts, to seriously perturb lithospheric thermal structure at Moho levels. In central Australia, the sedimentary cover has been removed from the basement inliers over many

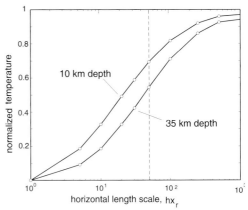

Fig. 11. In order to illustrate the effect of the horizontal length scale for the heat production anomaly, we show the temperatures at depths $z = 10$ and 35 km beneath the surface of an anomaly as a function of horizontal length-scale of the anomaly, hx_r (see text for discussion). The temperatures are normalized against the solution for the 1-D (infinite sheet) approximation, and thus span a range from 0 at $hx_r = 0$ (i.e., no anomalous heat production) to 1 at $hx_r = \infty$ (i.e. the 1-D solution).

hundreds of kilometres, while very deep levels of denudation in the cores of the inliers have characteristic across strike widths of $c.\,50\,\text{km}$. Consequently, the effective temperature changes at Moho levels are likely to be at least 50% of the estimate derived from the 1-D estimates (i.e., for central Australian heat production parameters about 9°C/km of erosion).

Rates of deformation and the mechanics of intraplate orogens

Previous sections have shown that tectonic processes, such as basin formation, deformation and erosion, have played a profound role in geochemically structuring the crust in the central Australian region. Moreover, the redistribution of heat production as a consequence of these processes has had profound long-term thermal consequences. Given our current understanding of the rheology of the lithosphere, these changes are likely to have profoundly affected its mechanical structure. The motivation for this analysis has been the unusual spatial and temporal patterns of deformation associated with intraplate deformation in this region. This distribution suggests that the strength parameters of this region have varied both in space and time in very significant ways. If this analysis is correct it raises the intriguing possibility that the changes in strength induced by this geochemical structuring have influenced the progression of ongoing deformation during the orogenic cycle. This section addresses this question using constraints on the duration, and rates, of deformation during the Alice Springs Orogeny. Particular interest is placed on relating this question to the origin, and long-term preservation, of the extraordinary gravity anomalies (Mathur 1976) that seem to be the hallmark of the central Australian intraplate orogenies. The Redbank Shear Zone is used as a prototype for the style of deformation associated with these orogenies.

The Redbank Shear Zone is a well-defined crustal-scale structure that can be traced on seismic profiles (Goleby *et al.* 1988, 1989) to depths of at least 40 km, where it clearly displaces the Moho by about 15 km. The crust is effectively ramped up along this $c.45°$ dipping segment of the Redbank Shear Zone such that it is now deeply denuded in the hanging wall of the structure (Fig. 13). It seems unlikely that such ramping could continue to much deeper levels than about 40 km, where displacement might be expected to be accommodated on shallowing detachments, or alternatively, partitioned into a corresponding downward directed flow in the deeper mantle (e.g. Braun & Shaw 1998; Fig. 13). The horizontal displacement of the Redbank Shear Zone and its associated splays such as the Ormiston Thrust during the Alice Springs Orogeny has recently been estimated at

20 km (Flöttmann & Hand 1999). While precise constraints on the duration of activity on the Redbank Shear Zone are not available, the maximum bound on the duration of the Alice Springs Orogen of 150 Ma provides a time-averaged lower bound on the displacement rate of $c. 0.125$ mm a^{-1}. An upper bound on the time-averaged displacement rate of $c. 0.4$ mm a^{-1} is provided by the evidence that sediment was shed from the hanging wall of the Redbank Shear Zone for at least 50 Ma, from $c. 390–340$ Ma (Shaw *et al.* 1991*a*).

While the history of displacement on the Redbank Shear Zone may well have been intermittent within these intervals, these bounds do provide an insight into the thermal response of the system to active deformation and denudation. In essence we are interested in whether the conductive response to changing heat source distributions overlap with the ongoing deformation. The qualitative thermal response of a deforming system is encapsulated in the thermal Peclet number, Pe_T, which provides a measure of the rate of advection of heat to conduction of heat (Fig. 12):

$$Pe_T = vl/\kappa \tag{7}$$

where v, is the characteristic velocity; l, is the appropriate length scale; and κ, is the thermal diffusivity. For $Pe_T > 10$, heat is carried with the deformation and so advection dominates the thermal evolution, whereas for $Pe_T < 1$ conduction dominates the evolution. Using the averaged displacement rates discussed above for the Redbank Shear Zone, where l is $c. 40$ km, κ $c. 10^{-6}$ m^2 s^{-1}, we estimate Pe_T to be $c. 0.1–0.4$, suggesting conductive heat transfer dominated the thermal evolution during deformation.

The notion that the Alice Springs Orogen evolved in a regime in which conduction dominated its ongoing thermal evolution is important. Plate margin orogens typically evolve at high Pe_T numbers because of the governing role played by subduction. The inference that the Alice Springs Orogen evolved at a low Pe_T raises the possibility that its mechanical evolution may have differed from the somewhat more familiar plate margin orogens. In particular, it raises the possibility that the cooling associated with exhumation of low-heat production deep crustal rocks during ongoing deformation has led to sufficient lithospheric strengthening to stall the deformation by providing a kind of 'thermal lock'. In order to constrain the thermal evolution during movement on the Redbank Shear Zone, a straightforward kinematic model has been used for deformation, as illustrated in Figure 13. The

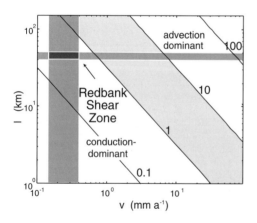

Fig. 12. Estimated values of mean displacement rates, v and appropriate length scales, l, for the movement on the Redbank Shear Zone during the Alice Springs Orogeny. Contours show values of thermal Peclet numbers, Pe_T, and indicate that the Alice Springs Orogeny operated in a regime where conduction was likely to dominate its thermal evolution. Note that typical plate margin orogenies evolve with $Pe_T \gg 10$.

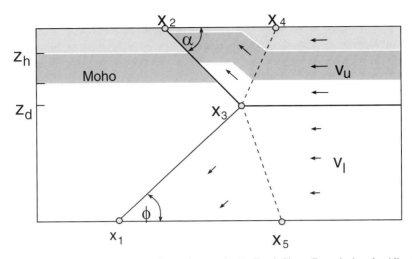

Fig. 13. Kinematic model for intraplate deformation on the Redbank Shear Zone during the Alice Springs Orogeny, based on the models of Beaumont *et al.* (1994). In this model the Redbank Shear Zone is assumed to sole at depth z_d slightly below the Moho. Displacement of the upper plate above the detachment occurs at rate v_u while a corresponding flow at rate v_l in the deeper mantle takes material down. See Figure 14 for details of thermal structure resulting from displacement at various values for v_u and v_l. Displacement causes the exhumation of an upper crustal heat producing layer, with erosion assumed to keep pace with deformation such that no substantial topography is generated during ongoing deformation.

results are shown in Figure 14. Two different scenarios have been explored, which apply to different behaviours in the deeper mantle lithosphere. In the first case it is assumed that mantle lithosphere shortens at the same velocity as the crust, inducing a singularity in the velocity field at x_3 (this follows closely from the models of Beaumont *et al.* 1994). In the second case it is assumed there is no deformation of mantle material at least in the vicinity of the crustal deformation. This implies a horizontal structure at depth z_d that detaches the crustal displacement from the deeper lithosphere. The model runs show a qualitatively variable behaviour depending on the rates of displacement, in accord with the general notions discussed above. Provided that displacement rates are $< c. 0.8 \, \text{mm a}^{-1}$ then conduction dominates the thermal evolution and cooling results from the ongoing deformation. At levels of 35–45 km this cooling amounts to $c. 10$–$35°C$ for the displacement rates appropriate to the Alice Springs Orogen (as illustrated by the shaded zone in Fig. 14), which amounts to 33–50% of the long-term cooling predicted for the denudation.

Discussion

Our prime purpose in this contribution has been to illustrate the role played by tectonic processes in modifying heat source distributions in the crust in intraplate settings. The recognition that such modification must lead to long-term changes in the thermal structure of the crust, provided that the processes operate at sufficient horizontal length-scales, naturally leads to the notion that past tectonic history will modulate the response to subsequent tectonic processes or, more simply, to the notion of 'tectonic feedback'. Such tectonic feedback appears to have been particularly effective in modulating the response to intraplate deformation in central Australia during the Phanerozoic, in part because the central Australian crust was already strongly differentiated and, at least in comparison with typical continental crust, relatively enriched in heat producing elements. However, it is believed that such tectonic feedback should be an essential feature of the evolution of continental interiors. This is clearly evident when the thermal consequences of tectonic processes are formulated in terms of changes in h and q_c, and illustrated on the h–q_c plane (Fig. 15d, f). For example, typical crust might be expected to have q_c of $30 \, \text{mW m}^{-2}$, with a characteristic length scale of 7–10 km. Figure 8 shows that for such a crust, a 10% strain coupled with a surface process that in the long-term restores the original crustal thickness, will lead to long-term temperature changes of $c. 20$–$30°C$ and that the repeated imposition of such strains over long

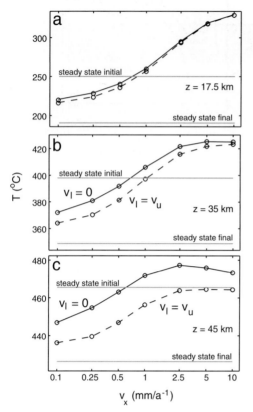

Fig. 14. Thermal consequences of displacement rate for kinematic model as shown in Figure 13. The temperatures shown correspond to positions above and below x_3 on Figure 13. The solid line shows the temperatures resulting after 15 km displacement (the inferred total Alice Springs displacement on the Redbank Shear Zone) at various rates of v_u with $v_l = 0$. The dashed line shows the equivalent temperatures with $v_l = v_u$. Steady state initial and final temperatures are indicated by the horizontal lines.

periods of geological time will lead to profound geochemical and thermal structuring of the crust (Fig. 15d, f). Even such subtle temperature changes are capable of producing dramatic changes in strength for a temperature-sensitive rheology, such as the 'Brace-Goetze' lithosphere (Fig. 9 & 10). Perhaps the most dramatic manifestation of this tectonic feedback is the extraordinary geophysical character of central Australia. Deformation during both the Petermann and Alice Springs Orogenies produced dramatic lateral variations in crustal density now manifest in the extraordinary gravity anomalies around the margins of the Amadeus Basin (Mathur 1976). The long-term preservation of these gravity anomalies implies significant lithospheric strength. The implication

of a strong lithosphere is at odds with the localization of regional-scale intraplate deformation. The calculations summarized in Figure 13 indicate that lithospheric cooling (and by inference strengthening) will accompany intraplate deformation provided mean displacement rates are less than *c.* 0.5 mm a^{-1}, and that erosion efficiently removes the topographic edifice. This potential for lithospheric-strengthening during ongoing deformation has important implications for the processes that lead to the cessation of deformation, and also to the way in which deformation shifts with time in the overall context of the orogen. If deformation were to cease as a result of ongoing lithospheric-strengthening, the lithosphere could potentially support a long-term isostatic imbalance by virtue of its acquired strength. Conceivably the preservation of the large gravity anomalies in central Australia reflects long-term strength increases arising as a consequence of the intraplate deformation. Furthermore, if slow rates of deformation lead to stripping of heat production over length-scales of ≥ 50 km, the distribution of relative lithospheric strength may change to the extent that deformation may shift to regions that have undergone less denudation. There is some evidence that regions in central Australia became 'stronger' as deformation and associated denudation proceeded. During the Alice Springs Orogeny, the major locus of deformation shifted away from the Redbank Shear Zone with time to similarly oriented structures 80–100 km into the interior of the orogen along the northern edge of the Ngalia Basin in the Early Carboniferous (Wells & Moss 1983). Conceivably this shift was controlled by an evolving thermal structure in the region around the Redbank Shear Zone that resulted in relative lithospheric strengthening despite the presence there of a crustal-scale discontinuity, favourably oriented to localize further deformation.

As a final discussion point, we consider the role that intraplate deformation, such as that observed in central Australia, plays in organizing the geochemical structure of the continental crust is considered. The chemical evolution of the continental crust is generally considered in terms of the primary crustal growth processes that arise from the depletion of the mantle in plate margin settings (e.g. O'Nions *et al.* 1979; McCulloch & Bennett 1994). While these models may account for the broad geochemical character of the continental crust, relatively little attention has been given to the subsequent geochemical structuring during post-crust forming processes. Geochemical structuring of the crust may occur in two main ways: (1) crustal differentiation

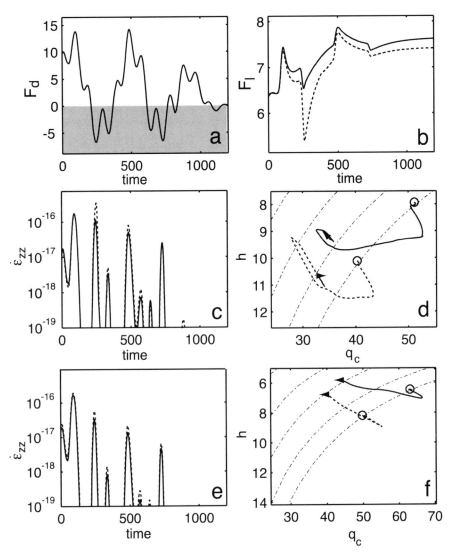

Fig. 15. The principle of "tectonic feedback" is illustrated with reference to numerical simulations in which a series of 1-D lithospheric columns (with temperature sensitive 'Brace-Goetze' rheology) are subject to a time-varying tectonic load (Fig. 15a, F_d in units of $10^{12}\,\mathrm{N\,m^{-1}}$). Units of time are 10^6 years. The load oscillates on a time scale of several 100 Ma from compression ($F_d > 0$) to extension ($F_d < 0$). The response is expressed in terms of vertical stretching rate (ε_{zz} in units of s^{-1}, Figures 14c & 14e) and is determined by the interaction between the tectonic loading function and the evolving thermal and compositional structure of the column. The heat production parameters (expressed in terms of h and q_c, Figures 14d & 14f) evolve as a consequence of the deformation as well as surface processes (erosion and sedimentation) that act to restore the long-term surface elevation (initial and final h-q_c configurations are indicated by the circle and arrow heads, respectively). The strength of the lithosphere (F_l in units of $10^{12}\,\mathrm{N\,m^{-1}}$, Fig. 15b) evolves as a function of its thermal and compositional structure, both due to transient and long-term changes. The figures show the evolution of a number of lithospheric columns that differ only in the way the heat sources are configured initially. Figure 15c & 14d show the results of simulations for two heat source distributions which initially approximate an inverse exponential dependence on depth, while Figure 15e & 14f show simulations for initial distributions in which heat production is confined to an upper crustal layer with thickness $\sim 2h$ of relatively uniform character. In all cases the heat production distributions are set so as to provide an identical Moho temperature so that initially the strength of the lithosphere is constant (Fig. 15b). In all simulations, the long-term effect of tectonic activity is to lower q_c (i.e. heat production is 'worked' out of the lithosphere) resulting in long-term increases in lithospheric strength. However, this is a strongly staged processes, with significant transient reductions in strength attendant with stretching events (i.e. at c. 250 and c. 750 Ma, Fig. 15b). Both Figure 15d & 14 illustrate a tendency for the long-term convergence of h-q_c parameters, suggesting that tectonic feedback provides may provide and effective agent for the long-term ordering of heat production in the continental crust.

associated with large-scale melt-transfer and (2) redistribution of material during deformation, sedimentation and erosion. The combination of these processes can dramatically affect the length scales associated with compositional variation in the crust. The formation of large mid-crustal granitic complexes of the type that characterize central Australia, effectively reduces the vertical length scale for the incompatible elements such as U, Th, K, Cs, Rb, Ba as well as SiO_2, by concentrating them at higher crustal levels. If these granites are subsequently exhumed, there is significant loss of the highly incompatible elements from the exhumed region. In an intraplate context, this material is deposited into intraplate basins, and may also be transported to continental margins in large drainage systems. This redistribution, which is a function of the tectonic history and hence thermal regime of the crust, fundamentally changes the chemical make up of the continental interior and, in so doing, its thermo-mechanical structure (Fig. 15). When compared to continental margins, there is very little likelihood of re-enriching a depleted continental interior via processes such as subduction. As such, intraplate deformation such as observed in central Australia, ultimately tends to drive existing heat production out of the crust (i.e., towards lower values of q_c – see Fig. 15). Indeed, such intraplate deformation may in fact be the prime reason that continental interiors have average heat production parameters of h c. 5–10 km and q_c c. 30–40 mW m^{-2}.

We thank Kerry Slater and the NTGS for supplying the airborne radiometric data used in the production of the heat production map of the Ormiston Gorge region, and Peter Haines for his contributions to our understanding of the duration of the Alice Springs Orogen. The organisers of the 'Orogenesis in the Outback' conference, particularly Jodie Miller, are thanked for their efforts in providing a wonderful and stimulating conference, during which the ideas discussed in this paper were presented. Ian Buick, Jean Braun and Bruce Goleby are thanked for their comments on the draft manuscript.

Appendix

This appendix outlines the reasoning that leads to the recognition that the length-scale in eqn. (4) is half the thickness of the heat producing layer, for the case where all crustal heat production is confined to an upper crustal layer of uniform heat production (Fig. 6a). For such a distribution, the temperature field in the layer is given by:

$$T_z = -\frac{q_r z}{k} + \frac{H_s z(h_r - z/2)}{k} \tag{A1}$$

where H_s, is the heat production in the heat producing layer; h_r, is the thickness of the heat producing layer; and k, is the thermal conductivity. In eqn. (A1) the first term on the right represents the component of the temperature field due to the heat flow from beneath the heat producing parts of the lithosphere (dashed line in Fig. 6b). The second term on the right represents the contribution due to heat sources in the crust (Fig. 6c). This is used to define the quantity, T_{q_c}, the temperature contribution at any due to heat production in the crust. Figure 6c shows that T_{q_c} reaches its maximum value (T'_{q_c}) at the base of the heat-producing layer. The appropriate expressions for T'_{q_c} for this model is therefore:

$$T'_{q_c} = \frac{H_s h_r^2}{2k} \tag{A2}$$

Equating eqn. (A2) with equation (3) and noting that q_c equals $H_s h_r$, yields $h = h_r/2$.

References

BEAUMONT, C., FULLSACK, P. & HAMILTON, J. 1994. Styles of crustal deformation caused by subduction of the underlying lithosphere. *Tectonophysics*, **232**, 119–132.

BRACE, W. F. & KOHLSTEDT, D. L. 1980. Limits on lithospheric strength imposed by laboratory experiments. *Journal of Geophysical Research*, **85**, 6248–6252.

BRADSHAW, J. D. & EVANS, P. R. 1988. Paleozoic tectonics, Amadeus Basin, central Australia. *In*: MOORE, D. B. (ed.) *Technical papers 1988 APEA conference*. Australian Petroleum Exploration Association, **28**, 267–282.

BRAUN, J. 1992. Postextensional mantle healing and episodic extension in the Canning Basin. *Journal of Geophysical Research*, **97**, 8977–8936.

BRAUN, J. & SHAW, R. D. 1998. Contrasting styles of lithospheric deformation along the northern margin of the Amadeus Basin, central Australia. *In*: BRAUN, J., DOOLEY, B., GOLEBY, B. R., VAN DER HILST, R. D. & KLOOTWIJK, C. (eds) *Structure and Evolution of the Australian Continent*. American Geophysical Union Geodynamic Series, **26**.

CAMACHO, A., VERNON, R. H. & FITZGERALD, J. D. 1995. Large volumes of anhydrous pseudotachylite in the Woodroffe Thrust, eastern Musgrave Ranges, Australia. *Journal of Structural Geology*, **17**, 371–383.

CAMACHO, A., COMPSTON, W., McCULLOCH, M. & McDOUGALL, I. 1997. Timing and exhumation of eclogite facies shear zones, Musgrave Inlier, Central Australia. *Journal of Metamorphic Geology*, **15**, 735–751.

CULL, J. P. 1982. An appraisal of Australian heat-flow data. *BMR Journal of Geology and Geophysics*, **7**, 11–21.

DUNLAP, W. J. & TEYSSIER, C. 1995. Palaeozoic deformation and isotopic disturbance in the southeastern Arunta Inlier, Australia. *Precambrian Research*, **71**, 229–250 1995.

DUNLAP, W. J., TEYSSIER, C., MCDOUGALL, I. & BALDWIN, S. 1995. Thermal and structural evolution of the intracratonic Arltunga nappe complex, central Australia. *Tectonics*, **14**, 1182–1204

EDGOOSE, C. J., CAMACHO, A., WAKELIN-KING, G. A. & SIMONS, B. A. 1993. Kulgera, Northern Territory. 1:250 000 Geological sheet and explanatory notes. Northern Territory Geological Survey, Darwin, Australia.

FLÖTTMANN, T. & HAND, M. 1999. Folded basement-cored tectonic wedges along the northern Amadeus Basin, Central Australia: constraints on intracontinental orogenic shortening. *Journal of Structural Geology*, **21**, 399–412.

FORMAN, D. J. 1971. The Arltunga Nappe Complex, MacDonnell Ranges, Northern Territory, Australia. *Journal of the Geological Society of Australia*, **18**, 173–182.

GOLEBY, B. R., WRIGHT, C., COLLINS, C. D. N. & KENNETT, B. L. N. 1988. Seismic reflection and refraction profiling across the Arunta Inlier and the Ngalia and Amadeus Basins. *Australian Journal of Earth Sciences*, **35**, 275–294.

GOLEBY, B. R., SHAW, R. D., WRIGHT, C., KENNETT, B. L. N. & LAMBECK, K. 1989. Geophysical evidence for 'thick-skinned' crustal deformation in central Australia. *Nature*, **337**(6205), 325–330.

HAND, M. & SANDIFORD, M. 1999. Intraplate deformation in central Australia, the link between subsidence and fault reactivation. *Tectonophysics*, **305**, 121–140.

HAINES., P. W., BAGAS, L., WYCHE, S., SIMONS, B. & MORRIS, D. G. 1991. *Barrow Creek*. 1:250 000 Geological sheet and explanatory notes. Northern Territory Geological Survey, Darwin, Australia.

HOSKINS, D. & LEMON, N. 1995. Tectonic development of the eastern Officer Basin. *Exploration Geophysics*, **26**, 395–402.

JONES, B. G. 1972. Upper Devonian to lower Carboniferous stratigraphy of the Pertnjara Group, Amadeus Basin, central Australia. *Journal of the Geological Society of Australia*, **19**, 229–249.

JONES, B. G. 1991. Fluvial and lacustrine facies in the Middle to Late Devonian Pertnjara Group, Amadeus Basin, Northern Territory, and their relationship to tectonic events and climate. *In*: KORSCH, R. J. & KENNARD, J. M. (eds) *Geological and geophysical studies of the Amadeus Basin, central Australia*. Bulletin of the Bureau of Mineral Resources, **236**, 333–348.

KORSCH, R. J., GOLEBY, B. R., LEVEN, J. H. & DRUMMOND, B. J. 1998. Crustal architecture of central Australia based on deep seismic reflection profiling. *Tectonophysics*, **288**, 57–69.

LAMBECK, K. & BURGESS, G. 1992. Deep crustal structure of the Musgrave Inlier, central Australia: results from teleseismic travel-time anomalies. *Australian Journal of Earth Sciences*, **39**, 1–19.

LINDSAY, J. F. & LEVEN, J. H. 1996. Evolution of a Neoproterozoic to Palaeozoic intracratonic setting, Officer Basin, South Australia. *Basin Research*, **8**, 403–424.

MABOKO, M. A. H., MCDOUGALL, I., ZEITLER, P. K. & WILLIAMS, I. S. 1992. Geochronological evidence for *c.* 530–550 Ma juxtaposition of two Proterozoic metamorphic terrains in the Musgrave Ranges, central Australia. *Australian Journal of Earth Sciences*, **39**, 457–471.

MATHUR, S. P. 1976. Relation of Bouguer anomalies to crustal structure in southwestern and central Australia. *BMR Journal of Australian Geology and Geophysics*, **1**, 277–286.

MAWBY, J., HAND., M. & FODEN, J. 1999. Sm–Nd evidence for high-grade Ordovician metamorphism in the Arunta Inlier, central Australia. *Journal of Metamorphic Geology*, **17**, 653–668.

MCLENNAN, S. M. & TAYLOR, S. R. 1996. Heat flow and chemical composition of continental crust. *Journal of Geology*, **104**, 369–377.

MCCULLOCH, M. T. & BENNETT, V. C. 1994. Progressive growth of the earth's continental crust and depleted mantle: geochemical constraints. *Geochimica et Cosmochimica Acta*, **58**, 4717–4738.

OFFE, L. A. 1978. *Mount Peake, Northern Territory*. 1:1:250 000 geological map series. Australian Bureau of Mineral Resources, Australia.

O'NIONS, R. K., EVENSEN, N. M. & HAMILTON, P. J. 1979. Geochemical modeling of mantle differentiation and crustal growth. *Journal of Geophysical Research*, **84**, 6091–6101.

PATERSON, M. S. 1987. Problems in the extrapolation of laboratory rheological data. *Tectonophysics*, **133**, 33–43.

SANDIFORD, M. 1999. Mechanics of basin inversion. *Tectonophysics*, **305**, 109–120.

SANDIFORD, M. & HAND, M. 1998a. Controls on the locus of Phanerozoic intraplate deformation in central Australia. *Earth and Planetary Science Letters*, **162**, 97–110.

SANDIFORD, M. & HAND, M. 1998b. Australian Proterozoic high-temperature metamorphism in the conductive limit. *In*: TRELOAR, P. & O'BRIEN, P. (eds) *What Controls Metamorphism*. Geological Society, London, Special Publications, **138**, 103–114.

SANDIFORD, M., PAUL, E. & FLÖTTMANN, T. 1988. Sedimentary thickness variations and deformation intensity during basin inversion in the Flinders Rangers, South Austrralia. *Journal of Structural Geology*, **30**, 1721–1731.

SCRIMGEOUR, I. & CLOSE, D. 1998. Regional high pressure metamorphism during intracratonic deformation: the Petermann Orogeny, central Australia. *Journal of Metamorphic Geology*, **17**, 557–572

SHAW, R. D. & BLACK, L. P. 1991. The history and tectonic implications of the Redbank Thrust Zone, central Australia, based on structural, metamorphic and Rb–Sr isotopic evidence. *Australian Journal of Earth Science*, **38**, 307–332.

SHAW, R. D., ETHERIDGE, M. A. & LAMBECK, K. 1991a. Development of the late Proterozoic to mid-Palaeozoic intracratonic Amadeus Basin in central Australia: A key to understanding tectonic forces in plate interiors. *Tectonics*, **10**, 688–721.

SHAW, R. D., KORSCH, R. J., WRIGHT, C. & GOLEBY, B. R. 1991b. Seismic interpretation and thrust tectonics of the Amadeus Basin, central Australia, along the BMR regional seismic line. *In*: KORSCH,

R. J. & KENNARD, J. M. (eds) *Geological and geophysical studies in the Amadeus Basin, central Australia.* Australian Bureau Mineral Resources Bulletin, **236**, 385–408.

SONDER, L. & ENGLAND, P. C. 1986. Vertical averages of rheology of the continental lithosphere; relation to thin sheet parameters. *Earth and Planetary Science Letters*, **77**, 81–90.

STEWART, A. J. 1995. Western extension of the Woodroffe Thrust, Musgrave Inlier, central Australia. *AGSO Journal of Geology and Geophysics*, **16**, 147–153.

TEYSSIER, C. 1985. A crustal thrust system in an intracratonic environment. *Journal of Structural Geology*, **7**, 689–700.

VAN DER HILST, R. D., KENNETT, B. L. N. & SHIBUTANI, T. 1998. Upper mantle structure beneath Australia from portable array deployments. *In*: BRAUN, J., DOOLEY, R., GOLEBY, B., VAN DER HILST, R. D. & KLOOTWIJK, C. (eds) *Structure and Evolution of the Australian Continent.* American Geophysical Union Geodynamic Series, **26**, 39–58.

WALTER, M. R. & GORTER, J. D. 1994. The Neoproterozoic Centralian Superbasin in Western Australia: the Savory and Officer Basins. *In*: PURCELL, P. G. & PURCELL, R. R. (eds) *The sedimentary basins of Western Australia.* Proceedings Petroleum Exploration Society Australia Symposium, Perth, 851–864.

WALTER, M. R., VEEVERS, J. J., CALVER, C. R. & GREY, K. 1995. Neoproterozoic stratigraphy of the Centralian Superbasin, Australia. *Precambrian Research*, **73**, 173–195.

WARREN, R. G. & HENSEN, B. J. 1989. The P–T evolution of the Proterozoic Arunta Inlier, central Australia, and implications for tectonic evolution. *In*: DALY, J. S., CLIFF, R. A. & YARDLEY, B. W. D. (eds) *Evolution of metamorphic belts.* Geological Society, London, Special Publications, **43**, 349–355.

WELLS, A. T. 1982. *Napperby, Northern Territory.* 1:1:250 000 geological map series. Australian Bureau of Mineral Resources (Canberra), Australia.

WELLS, A. T. & MOSS, F. J. 1983. The Ngalia Basin, Northern Territory; stratigraphy and structure. *Australian Bureau Mineral Resources Geology and Geophysics Bulletin*, **212**, 88.

WELLS, A. T., FORMAN, D. J., RANFORD, L. C. & COOK, P. J. 1970. Geology of the Amadeus Basin, central Australia. *Australian Bureau Mineral Research Geology and Geophysics Bulletin*, **100**.

YOUNG, D. N., FANNING, C. M., SHAW, R. D., EDGOOSE, C. J., BLAKE, D. H., PAGE, R. W. & CAMACHO, A. 1995. U–Pb zircon dating of tectonomagmatic events in the northern Arunta Inlier, central Australia. *Precambrian Research*, **71**, 45–68.

ZIELHUIS, A. & VAN DER HILST, R. D. 1996. Upper-mantle shear velocity beneath eastern Australia from inversion of waveforms from Skippy portable arrays. *Geophysical Journal International*, **127**, 1–16.

Long-term thermal consequences of tectonic activity at Mount Isa, Australia: Implications for polyphase tectonism in the Proterozoic

S. McLAREN* & M. SANDIFORD*

*Department of Geology & Geophysics, University of Adelaide, South Australia,
5005, Australia (e-mail: smclaren@unimelb.edu.au)*
** Present address: School of Earth Sciences, University of Melbourne,
Victoria 3010, Australia*

Abstract: Mount Isa is a Palaeo-Mesoproterozoic terrane in Northern Australia characterized by >300 Ma of episodic tectonic activity prior to effective cratonization. This tectonic activity has resulted in dramatic changes in the heat production distribution in the crust and must have been accompanied by long-term changes in thermal regimes. Primary differentiation of crust initially enriched in heat producing elements has been achieved by felsic magmatism over much of the 300 Ma history, often associated with extensional deformation. The flux of heat producing elements from lower to mid-upper crustal levels associated with this magmatism was sufficient to cause long-term lower crustal cooling of at least 200°C. The accumulation of the radiogenic intrusives (which comprise $c.\,23\%$ of surface outcrop and have heat production rates averaging $5.2\,\mu Wm^{-3}$) in the mid-upper crust resulted in a highly stratified heat production distribution. One consequence of this distribution is that small changes in the depth to this heat production, through processes such as deformation, erosion and the deposition of sediments, lead to significant changes in deep crustal temperatures (up to 100°C) and consequently lithospheric strength. These considerations suggest that the long-term evolution of the Mount Isa region partly reflects the progressive concentration of heat-producing elements in the upper crust leading to a long-term increase in lithospheric strength, and eventually to effective cratonization. The long-term cooling and strengthening trend was locally countered by the role of subsidence during basin formation which, through burial of heat producing elements in the existing crust and the accumulation of more heat production in insulating sediments, helped to localize subsequent contractional deformation.

Proterozoic crustal evolution is typically characterized by polyphase tectonism and tectonic reactivation (e.g. Dallmeyer & Keppie 1993; Kalsbeek 1995; O'Dea *et al.* 1997a; Betts *et al.* 1998). In many instances, semi-continuous tectonic activity seems to have occurred over many hundreds of millions of years prior to effective cratonization. Such extended periods of tectonic activity exceed that which can easily be ascribed to a single 'plate-tectonic cycle' and suggest that these terranes have either been at active plate margins for very extended periods, or that at least part of the record reflects tectonism in an intra-plate setting. Neither explanation is particularly satisfactory. In the former case we are confronted by questions concerning the very nature of active plate boundaries in and around Precambrian continents, while in the latter we recognize a distribution and style of deformation that, on face value at least, is very different from that evident in the continent interiors today.

Major deformation, both at plate boundaries and within continental interiors, requires that the tectonic stresses exceed yield strength through much of the lithosphere. Therefore, the distribution of active deformation of the lithosphere can be understood as a consequence of either stress amplification or strength reduction. In the modern Earth both factors seem to apply at plate boundaries where substantial stress amplification is related to processes such as subduction and continental collision, while significant thermal weakening occurs in response to subduction-related magmatism as well as crustal thickening and, possibly, shear heating. It is not surprising that much (but not all) of the deformation on the modern Earth is located along plate margin zones!

Conversely, the factors responsible for the distribution of intraplate deformation evident in the modern Earth are not so well understood. It is difficult to see how stresses can be greatly amplified in intraplate settings, although localized mantle down-welling associated with convective instability beneath continental interiors may provide an important source of tectonic stress amplification (e.g. Houseman *et al.* 1981;

From: MILLER, J. A., HOLDSWORTH, R. E., BUICK, I. S. & HAND, M. (eds) *Continental Reactivation and Reworking.* Geological Society, London, Special Publications, **184**, 219–236. 1-86239-080-0/01/$15.00 © The Geological Society of London 2001.

Beaumont *et al.* 1994). One problem here is that the magnitude and source of stresses that apply in any particular geological scenario are only poorly constrained. On the other hand, the mechanical properties of continental interiors are likely to vary significantly because of subtle variations in the thermal and compositional structure of the lithosphere (e.g. Neil & Houseman 1997). Because of the coupling between the thermal state and mechanical strength of the lithosphere, constraints on the way in which crustal thermal regimes have evolved through time may therefore provide particularly useful insights into the evolution of lithospheric strength, and the role it has played in contributing to tectonism.

The recognition that lithospheric strength and thermal regime are coupled may provide a solution to the old problem of long-lived tectonic activity in Precambrian terranes. Rather than attempting to reconcile whether such activity is strictly 'plate-tectonic' (in the sense that the modern Earth is understood, e.g. Myers *et al.* 1996) or 'intraplate' (e.g. Etheridge *et al.* 1987; Wyborn 1988), it may prove more rewarding to view it in terms of the way in which the history of tectonic activity has modified the thermal and mechananical properties of the lithosphere, and so influenced its subsequent tectonic evolution. The objective of this paper is to provide some understanding of how the evolution of thermal regime and, by inference, lithospheric strength may have influenced the tectonic development of the Palaeo-Mesoproterozoic Mount Isa Inlier (MII) in northern Australia. The Mount Isa region is particularly well suited to such an analysis because its protracted history of magmatism, sedimentation and deformation (both extensional and convergent) has been constrained at the outcrop scale by a number of detailed geochronological, structural and tectonic studies (e.g. Eriksson *et al.* 1993; Connors & Page 1995; O'Dea *et al.* 1997*a, b*; Betts *et al.* 1998; Page & Sweet 1998). The region records an *c.* 300 Ma polyphase tectonic history prior to effective cratonization, and the role of strength changes in this transition from persistent active tectonics to craton-like behaviour is of particular interest. Also, unusually good exposure allows for detailed characterization of the thermal properties of outcropping lithologies. Together with existing heat flow determinations, these data provide constraints on the present-day thermal structure of the Mount Isa crust and the way in which it has evolved through the various tectonic processes that have shaped it. By exploring the way in which the distribution of heat sources has evolved at the crustal scale as a function of tectonic activity, constraints can be placed on the long-

term thermal evolution of the Mount Isa crust. Thus, likely changes in its strength through time can be inferred.

This paper begins with a brief outline of the record of polyphase tectonism in the Mount Isa region. The way in which the distribution of heat sources evolved during this history is then described. The insights provided by these first-order observations suggest that the tectonic evolution of Mount Isa reflects, at least in part, long term crustal cooling and consequently strengthening, due to the progressive differentiation of heat producing elements within the crust.

A review of the tectono-stratigraphic history of the Mount Isa region

The Mount Isa Inlier is a metamorphic and igneous complex in NW Queensland (Fig. 1) consisting dominantly of Palaeo-Mesoproterozoic sedimentary sequences, felsic plutonic rocks and bimodal volcanics. The inlier records a complex

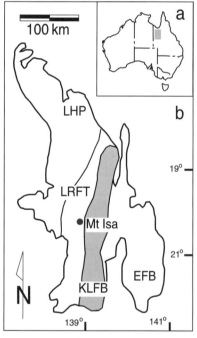

Fig. 1. (**a**) Location of the Mount Isa Inlier (MII) in north-western Queensland, Australia. (**b**) Structural domains of the Mount Isa area (after Blake & Stewart 1992). MI, Murphy Inlier; LHP, Lawn Hill Platform; LRFT, Leichhardt River Fault Trough; KLFB, Kalkadoon–Leichhardt Fold Belt; EFB, Eastern Fold Belt. Together the Leichhardt River Fault Trough and Lawn Hill Platform comprise the Western Fold Belt (WFB).

and protracted history of magmatic and orogenic activity extending throughout an interval of more than 300 Ma (Fig. 2). The Inlier can be divided into three main tectonic domains: the Western Fold Belt (WFB) (including the Lawn Hill Platform and Leichhardt River Fault Trough); the Kalkadoon–Leichhardt Fold Belt (KLFB); and the Eastern Fold Belt (EFB) (Fig. 1). In the following section the geological history of the Mount Isa region is reviewed through the description of those major events that have lead to the long-term geochemical and structural reorganization of the crust. Particular emphasis is placed on the description of the history of the Western Fold Belt where the most lengthy history of reactivation has been recorded, and where excellent exposure has permitted a number of detailed investigations (e.g. Blake 1987; Blake & Stewart 1992; Eriksson et al. 1993; Connors & Page 1995; O'Dea et al. 1997a, b; Betts et al. 1998; Page & Sweet 1998).

The earliest history of the Western and Kalkadoon–Leichhardt Fold Belts is associated with deposition and subsequent metamorphism of Palaeoproterozoic sedimentary packages (Yaringa Metamorphics, May Downs Gneiss and Tewinga Group) during the Barramundi Orogeny. This widespread orogenic event, occurring between 1840 and 1890 Ma (with peak metamorphism c. 1875 Ma; Page & Williams 1988; Wyborn 1988) is also recorded through other Australian Proterozoic terranes (e.g. Etheridge et al. 1987; Wyborn 1988). Metamorphism is typically high temperature and generally low pressure (<5 kbar) amphibolite to granulite facies. This event was associated with voluminous, extensive and dominantly felsic magmatism, characterized by emplacement of I-type granites (Wyborn 1988). At Mount Isa these include the Kalkadoon, Big Toby and Ewen Batholiths (e.g. Wyborn 1988). These granites share common Nd model ages of c. 2200 Ma (Wyborn 1998), suggesting a crustal prehistory of c. 300–400 Ma. Etheridge et al. (1987), Wyborn et al. (1988), and Wyborn (1998) have argued that they were derived from the melting of mafic rocks accreted to the base of the crust. Units associated with the Barramundi Orogeny form local basement throughout much of the MII and are best exposed in the deeply exhumed KLFB. Recent deep seismic results (e.g. Drummond et al. 1998) suggest Barramundi-aged rocks also underlie the WFB and EFB. The Barramundi Orogeny is inferred to have concluded around 1840 Ma (e.g. Page & Sun 1998) and from this time until c. 1790 Ma the terrane experienced a period of relative tectonic quiescence, particularly when compared to its later tectonic history.

Lying unconformably upon the metasediments and plutonic rocks of the Barramundi Orogeny are sequences of sediments and volcanics related to cycles of rifting and post-rift subsidence. These sequences form a series of 3 major stacked and unconformity bound basins in which the deposition of each package of sag related sediments is terminated by a further stage of rift-related extension (e.g. O'Dea et al. 1997b). Volcanics associated with each rift basin are characteristically bimodal, consisting predominantly of continental tholeiites and ignimbrites. Basin development accompanying rifting was best developed through the Western Fold Belt, although sedimentation associated with post-rift subsidence was continuous from the Lawn Hill Platform to the Eastern Fold Belt (e.g. Derrick et al. 1980).

Within the Western Fold Belt, rifting was strongly partitioned into the Leichhardt River Fault Trough (LRFT); a N–S trending fault bounded fold belt which represents the deformed axis of a major Proterozoic rift zone (O'Dea et al. 1997a, b). The two-stage Leichhardt Rift Event (LRE) (1790–1765 Ma, Page 1983; O'Dea et al. 1997b) follows the Barramundi Orogeny in this region. The first phase of LRE extension was associated with the deposition of a thick rift-sag sequence – the Bottletree Formation and Mount Guide Quartzite. The second phase of rifting resulted in the extrusion of voluminous continental tholeiites of the Eastern Creek Volcanics, and deposition of the sag-related Lena Quartzite. The Leichhardt Rift Event was terminated by a second major period of intermittent rifting and subsidence – the Myally Rift Event (MRE) (1765–1740 Ma) (Betts et al. 1998). Overlying sediments of the upper Myally Subgroup and Quilalar Formation are associated with post-rift subsidence and reflect a period of relative tectonic stability (Derrick et al. 1980).

A further period of major extension, the two-phase Mount Isa Rift Event (MIRE) (O'Dea et al. 1997a; Betts et al. 1998), was characterized by the reactivation of many earlier formed, mainly extensional faults (Betts et al. 1998). The first stage (c. 1708 ± 2 Ma, Page & Sweet 1998) was associated with extrusion of the felsic Fiery Creek Volcanics, and deposition of the Bigie Formation; the second resulted in the deposition of the rift-related Surprise Creek Formation and the thick (c. 12–14 km) sag-related shale and dolomite rich sequences of the Mount Isa Group (c. 1653–1595 Ma, Page et al. 1994; Page & Sweet 1998). The Mount Isa Group forms a thick (c. 12–14 km) package of sediments (Andrews 1998) across the Western Fold Belt and the Lawn Hill Platform (where it is termed the McNamara Group).

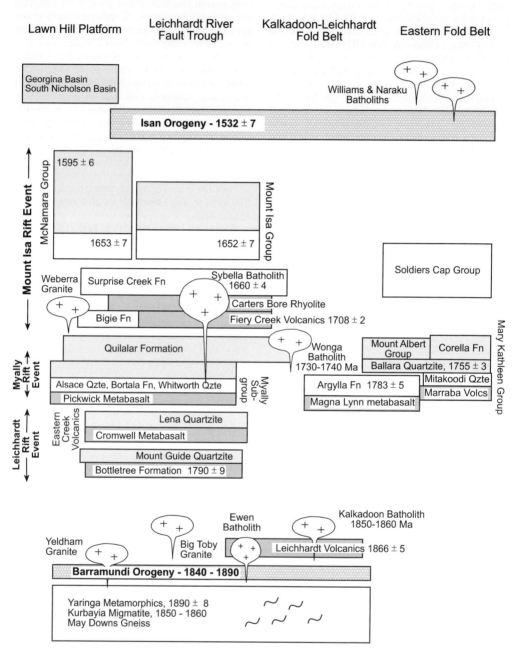

Fig. 2. Tectono-stratigraphic history of the Mount Isa Inlier. Major magmatic, and deformational events are shown for the Lawn Hill Platform, Western, Kalkadoon–Leichhardt and Eastern Fold Belts. Unshaded units denote rift-related sedimentary packages, light shading sag-related sediments, and dark shading volcanic sequences. Geochronological data from Connors & Page 1995 (Sybella Batholith and Isan Orogeny); Page & Sun 1998 (Eastern Fold Belt); Page & Sweet 1998 (Mount Isa and McNamara Group). All other geochronology from Wyborn *et al.* 1988 and Wyborn *et al.* 1998.

The Mount Isa Rift Event was also associated with the intrusion of a number of voluminous granitic batholiths (and comagmatic volcanics) in the LRFT, including the Sybella Batholith (c. 1655 Ma, Connors & Page 1995), Carters Bore Rhyolite (1678 ± 2 Ma, Page & Sweet 1998), and the Weberra Granite (c. 1698 Ma, Wyborn et al. 1988).

The Leichhardt Rift Event is manifest in the KLFB by extrusion of the Magna Lynn Metabasalt, and in both the KLFB and EFB by the sag-related Argylla Formation (c. 1783 ± 5 Ma, Page & Sun 1998). The Myally Rift Event in the EFB is represented by rift-related Marraba Volcanics and Mitakoodi Quartzite. Associated sag related sediments of the Mary Kathleen Group are present across both the EFB and KLFB. There are no post-Myally Rift Event sequences preserved in the KLFB. The Eastern Fold Belt records little evidence of the thick post-rift sedimentary sequences associated with the MIRE in the Western Fold Belt and Lawn Hill Platform. Stratigraphic studies however, imply the presence of a temporally equivalent pre-orogenic basin in this region (P. Southgate pers. comm., 1999).

The rift-related depositional record of the Western, Kalkadoon–Leichhardt and Eastern Fold Belts was terminated by basin inversion during the Isan Orogeny (c. 1590–1500 Ma, Blake 1987, O'Dea et al. 1997a, b; peak metamorphism at 1532 ± 7 Ma, Connors & Page 1995). In the WFB, compressional deformation was partitioned strongly into the LRFT, which contained the thickest rift-related sequences, however, it is also recorded throughout the KLFB and EFB. The latter suffered particularly intense deformation and associated high temperature–low pressure metamorphism (e.g. Loosveld & Etheridge 1990; Reinhardt 1992)

In the Western Fold Belt, compressional deformation during the Isan Orogeny terminated the record of Mesoproterozoic tectonic activity. In the Eastern Fold Belt tectonic activity terminated with the emplacement of the I-type Williams and Naraku Batholiths at c. 1510 Ma (Page & Sun 1998), the youngest granitic bodies in the MII. In contrast to the histories of both the WFB and EFB, and despite recording a similar early sedimentary history, the Lawn Hill Platform was relatively unaffected by the Isan Orogeny. Furthermore, it remained an active depocentre until at least the middle Cambrian. The Lawn Hill Platform is also characterized by a virtual absence of felsic magmatism.

The protracted, but spatially and temporally variable, tectonic history of the MII records a series of rift-related extensional events associated with the generation of voluminous felsic magmas, as well as convergent deformation. Granites throughout the terrane, from the Barramundi Igneous Association to the young granites of the EFB, show remarkable geochemical and isotopic homogeneity (e.g. Wyborn 1988, 1998). Consistency of Nd-isotopic data through time (Wyborn 1988) seems to reflect lithospheric recycling during major crustal tectonism rather than significant additions from the mantle. A lengthy period of crustal prehistory is further supported by the characteristic granite geochemical signature of Sr-depletion, implying derivation from crustal protoliths in which plagioclase was stable (Wyborn 1998).

The remainder of this paper investigates how crustal thermal regimes have evolved as a consequence of the major periods of crustal activity recorded at Mount Isa. In order to understand this it is useful to first consider the thermal regime of the modern crust, as reflected by modern heat flow and surface heat production data.

Table 1. *Global and Australian Proterozoic heat flow data*

Region	Average heat flow (m Wm^{-2})	Number of determinations
Mount Isa, Western Fold Belt	82	25
Blockade, Kalkadoon-Leichhardt Fold Belt	78	3
Dugald River, Eastern Fold Belt	72 ± 3	
Mount Dore, Eastern Fold Belt	98 ± 4	
Australian Proterozoic terranes	88 ± 22	26
Global Archaean*	41 ± 11	136
Global (early) Proterozoic*	51 ± 20	78
Global (late) Proterozoic*	54 ± 20	265
Global (early) Palaeozoic*	52 ± 16	88
Global (late) Palaeozoic*	61 ± 17	514
Global Mesozoic*	72 ± 27	85
Global Cenozoic*	71 ± 36	587

Note: Australian heat flow average taken from data in Cull 1982; Houseman et al. 1989; and Gallagher 1990. Mount Isa heat flow measurements from Hyndman & Sass 1966 and Cull & Denham 1979. Many individual heat flow values represent the average of a number of determinations at a single locality.
* Global heat flow averages taken from a compilation by Morgan 1984 and Chapman & Furlong 1977, with age taken as the time of the last tectonothermal event within a particular terrane.

The modern thermal regime in the Mount Isa Inlier

The MII is characterized by elevated levels of surface heat flow and heat production. Surface heat flow has been determined at four localities: Mount Isa in the Western Fold Belt; Blockade, $c.\,40$ km east of the city of Mount Isa in the Kalkadoon–Leichhardt Fold Belt; and Dugald River and Mount Dore in the Eastern Fold Belt. The mean heat flow at each of these locations is 82 and 78 m Wm^{-2} (Hyndman & Sass 1966) and 72 and 98 mW m^{-2} (Cull & Denham 1979), respectively. In contrast to these measurements, a single surface heat flow measurement from Camooweal, near the western limit of the Lawn Hill Platform, is much lower at $c.\,48$ m Wm^{-2} (Hyndman 1967). In terms of global averages (Table 1) the MII data are anomalous. However, they are not unusual in the context of Australian Proterozoic terranes where heat flow averages 88 m Wm^{-2} and is locally as high as 100 m Wm^{-2} (Cull 1982; Table 1). The relative contributions of mantle and crustal sources to these elevated high heat flow regimes can be inferred using several lines of evidence. In the first instance, seismic data indicate that the lithosphere beneath the MII is characterized by relatively high upper mantle velocities associated with an exceptionally thick lithosphere of $c.\,250$–300 km (Zielhaus & van der Hilst 1996). For such thick lithosphere, mantle heat flows of 10–20 m Wm^{-2} might be expected. For this to be consistent with the high surface heat flows, the crust must be enriched in heat production, contributing as much as 60–70 m Wm^{-2} of the observed surface heat flow.

Such crustal contributions to surface heat flow are exceptional when compared with other terranes of equivalent age (e.g. Chapman & Furlong 1977). Because there are only two regions from which heat flow measurements are available in the MII it is not possible to rigorously examine potential sources of error. However, one source of error which may be important is heat refraction associated with contrasting thermal conductivities. This is likely in the case of the Mount Isa

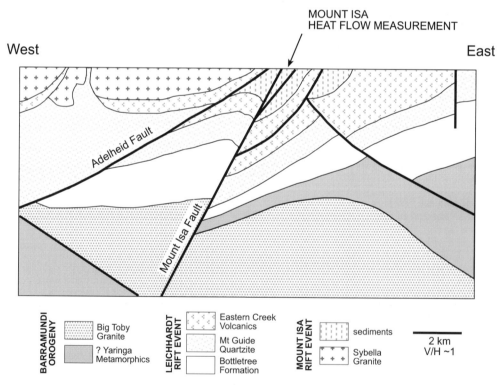

Fig. 3. Interpreted seismic section from the WFB, MII (compiled from Drummond *et al.* 1998 and Wyborn *et al.* 1996). In this region the Yaringa Metamorphics and Big Toby Granite comprise local Barramundi basement. Sediments form part of the Isa Superbasin. Note that the heat flow measurement at Mount Isa city ($c.\,82$ mWm^{-2}, Hyndman & Sass 1966) lies immediately to the east of the Adelheid and Mount Isa Faults, near the contact of the Sybella granite, Eastern Creek Volcanics and younger sedimentary cover.

measurements which lie close to the surface exposure of the contact between the Sybella Granite and the metasediments of the Isa Superbasin (Fig. 3). In theory, heat refraction due to conductivity contrasts can lead to substantial local amplification of heat flow density, possibly by as much as $10-20\,mWm^{-2}$ (e.g. England et al. 1980). In light of this possibility a more conservative estimate for the mean contribution of crustal heat sources to the observed heat flow of $50\,mWm^{-2}$ is used.

Calculated heat production values (Tables 2 and 3; Fig. 4) support the notion that the MII crust is exceptionally enriched in heat producing elements, and suggest that much of this enrichment is localized in granite rocks now near the present surface. Throughout the terrane, the average heat production of all Proterozoic granites (by area) is $5.2\,\mu Wm^{-3}$ (Table 2; Fig. 4). This is approximately two times the 'average' granite (Table 2). This value is all the more extraordinary given that the area of outcropping granite on which this estimate is based is c. $11\,000\,km^2$ constituting c. 23% of the total outcrop area of the inlier and, as noted earlier, Barramundi granites are known to underlie the Palaeoproterozoic supracrustal sequence. Examples of these high heat production granites include the Sybella Batholith in the LRFT (Fig. 4) where heat production averages c. $5\,\mu Wm^{-3}$, and in some individual

Table 2. *Geochemistry and heat production of selected granites of the Mount Isa Inlier*

Granite unit§	Age	# of samples	Area of outcrop (km²)	U* (ppm)	Th* (ppm)	K₂O* (wt%)	Th:U	Hp§ (µW m⁻³)	He§ (µW m⁻³)
Kalkadoon Batholith	1856 ± 10	5	3478	5	26	4.6	5.2	3.6	5.3
Big Toby Granite	1804 ± 15	4	29	5	19	3.6	3.8	3.0	4.4
Ewen Batholith	~1820	1	321	9	31	4.8	3.6	5.0	7.3
Wonga Batholith – main phase	1758 ± 8	30	335	8	46	5.2	5.7	5.9	8.1
Wonga Batholith – microgranite	1671 ± 8	11		11	66	5.5	6.0	8.2	10.9
Weberra Granite	1698 ± 24	5	47	5	34	7.8	6.8	4.5	6.5
Sybella Batholith – main phase	1655 ± 4	31	1231	8	35	5.1	4.3	5.1	7.0
Sybella Batholith – BQ phase		14		8	33	5.4	4.1	5.0	6.9
Sybella Batholith – microgranite		12		12	54	5.8	4.5	7.6	10.2
Naraku Batholith – main phase	1505 ± 5	19	395	10	57	3.5	5.7	7.1	8.9
Naraku Batholith – microgranite		3		19	58	4.9	3.1	9.6	12.6
Williams Batholith	1505 ± 5	44	1900	13	55	3.9	4.3	7.8	9.9
Terrane average (by area)†			7736	8	37	4.5		5.2	7.1
Average granite‖				4	15	3.5	3.8	2.5	

Note: Intrusive ages of granites are those determined by U–Pb zircon geochronology; SHRIMP data quoted where available. Kalkadoon and Ewen Batholith ages from Wyborn et al. 1998; Weberra and Big Toby Granite ages from Wyborn et al. 1998; Wonga Batholith ages from Page 1983, Pearson et al. 1992 and Wyborn et al. 1998; Sybella Batholith age from Connors & Page 1995; Williams and Naraku Batholith ages taken as that age of the Malakoff granite (from Page & Sun 1998).

* Average geochemical analyses from Wyborn et al. 1988; Wyborn 1988; and Wyborn et al. 1998.

† Terrane averages are weighted averages based on proportional outcrop area. Note that ungrouped granites constitute a further c. $3600\,km^2$ area of the inlier and are not included in this calculation.

‡ Area of outcrop includes all granites within that batholith, where applicable. Area of outcrop from Wyborn et al. 1998.

§ Hp, present heat production, calculated from present concentrations of U, Th, and K. He, heat production at the time of granite emplacement. As radioactive decay causes a reduction in the concentration of heat-producing elements through time, the value of the calculated heat production at the time of intrusion of each batholith is on average c. 25–30% higher than the present day value.

‖ Average granite geochemistry and heat productivity from Fowler 1990.

Table 3. *Geochemistry and heat production of major sedimentary units of the Mount Isa Inlier*

Sedimentary unit	# of samples	U (ppm)	Th (ppm)	K_2O (wt%)	H ($\mu W\,m^{-3}$)
Basin fill – sands and conglomerates[*]	288	2.3	8.3	2.0	1.39
Basin fill – siltstones and shales[*]	317	6.4	17.7	4.3	3.37
Basin fill – carbonates[*]	1176	1.5	3.0	1.5	0.75
Average sandstone[†]		0.6	1.8	0.9	0.37
Average siltstone and shale[†]		3.7	12	2.7	2.10
Average carbonate[†]		2.0	1.5	0.3	0.66

Note: H = present heat production, calculated from present concentrations of U Th, and K.
[*] Basin fill geochemistry from Australian Geological Survey Organization geochemical databases
[†] Average sedimentary heat productivity data from Haenel *et al.* 1988.

stocks is as high as $10\,\mu W\,m^{-3}$. Similarly, the late-stage Williams and Naraku Batholiths in the EFB have average heat production values of 7.8 and $7.1\,\mu W\,m^{-3}$, respectively (Table 2; Fig. 4). High values of heat production are not associated exclusively with granites and granite gneisses, with much of the sedimentary and volcanic basin fill also anomalously enriched in heat producing elements, when compared to 'average' values (Table 3).

These high values of surface heat production must be restricted to the upper crust in order to be consistent with the known range of surface heat flow, implying a strongly differentiated crust with regard to heat producing elements. For example, assuming that the average contribution of crustal heat producing elements is $50\,mW\,m^{-2}$ (as discussed above) then a layer of average MII granite some 9.6 km thick would account for the crustal contribution without any additional heat sources in the deeper crust. In the MII the crust is *c.* 40 km thick (e.g. Drummond *et al.* 1998) and if we allow that the mid to deep crust contributes on average *c.* $0.6\,\mu W\,m^{-3}$ (e.g. Fowler 1990), then the thickness of the average MII granite required to account for the surface heat flow is *c.* 6 km. This is consistent with available seismic evidence from the WFB (Fig. 3), where the Big Toby Granite is imaged as a unit at least 5 km thick.

The strongly differentiated distribution of heat production outlined above is clearly a consequence of the processes that have shaped the MII crust. In particular, it is recognized that the distribution of heat producing elements must have been very different prior to major periods of magmatism and has certainly been modified during deformation and crustal extension, and associated erosion and sedimentation. As the majority of heat producing elements are carried by granites that formed as a result of crustal recycling, granite generation and transport from the deep crust must have been particularly significant in redistributing the heat producing elements.

In order to understand the thermal consequences of this redistribution, the next section outlines a simple framework for the quantitative description of the distribution of crustal heat sources.

The distribution of crustal heat sources

Treating the lithosphere as a 1-D column, the way in which the crustal complement of heat producing elements is distributed can be described using just two independent parameters, termed h and q_c (see also Sandiford *et al.* 2001).

The parameter q_c represents the crustal contribution to measured surface heat flow (i.e. the depth integrated heat production). As outlined above, a mean value of q_c *c.* $50\,mW\,m^{-2}$ is estimated for the MII. It is noted that this is almost twice what other workers have suggested is typical of Proterozoic crust (Table 1; Chapman & Furlong 1977; Nyblade & Pollock 1993; McLennan & Taylor 1996). The parameter h describes the length scale over which this heat production is distributed. In the past the length scale of the distribution of heat sources within the crust has been defined with reference to some particular analytical form of that distribution. For example, for the 'classic' exponential distribution (e.g. Lachenbruch 1970) heat production (H_z) decays exponentially from a maximum value (H_s) at the surface, such that:

$$H_z = H_s * \exp\left(\frac{-z}{h_r}\right).$$

In this case, the length scale of the heat production distribution (h_r) is defined as that depth at which the heat production value is 1/e of the surface (maximum) value. In a similar way a characteristic length scale can also be defined for homogeneous or step-like heat production distributions. These largely idealized heat production distributions have in the past found much

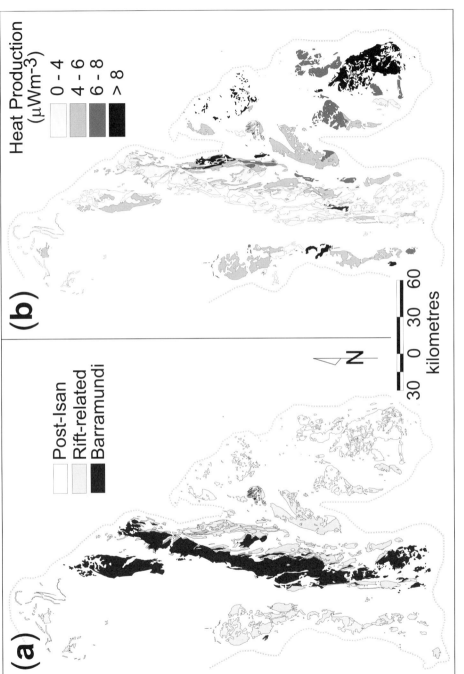

Fig. 4. Granites, and comagmatic volcanics, of the Mount Isa Inlier. (**a**) Granites are grouped according to age into three main subdivisions – Barramundi Granites (e.g. Wyborn *et al.* 1988); Rift-related granites (which include the Sybella granite and equivalents associated with the MIRE, and the Wonga Granite and equivalents associated with the Myally Rift Event) and the latest syn-post tectonic post-Isan Granites (including the Williams and Naraku Batholiths). (**b**) Granites are grouped by heat productivity. Heat production values are assigned to individual plutons from the average of a number of geochemical analyses from that pluton. Data and figure compiled from Wyborn *et al.* 1998. Note from Table 2 that 'average' granite has heat production c. $2.5 \mu Wm^{-3}$.

favour, largely because they permit the formulation of simple analytical expressions for the temperature field. However, since the crust has differentiated with respect to the incompatible trace elements, U and Th, it is difficult to see how any particular distribution of heat sources may have applied for any length of time, particularly during the early stages of crustal evolution. For this reason and in order to understand how the distribution of heat sources has changed through time, an alternative and more general description of the length-scale that makes no explicit assumption about the analytical form of the heat production distribution is required. Following

Sandiford *et al.* (2001) *h* the *effective depth* of the heat production distribution, can be formulated in terms of the additional temperature contribution, T'_{qc}, due to the presence of heat sources within the crust. T'_{qc} reaches a maximum value at the base of the heat producing layer, allowing *h* to be given by:

$$h = \frac{kT'_{qc}}{q_c}$$

where, k, is the thermal conductivity. In general terms a strongly differentiated crust (in which much of the crustal complement of heat producing elements resides in the uppermost crust) is

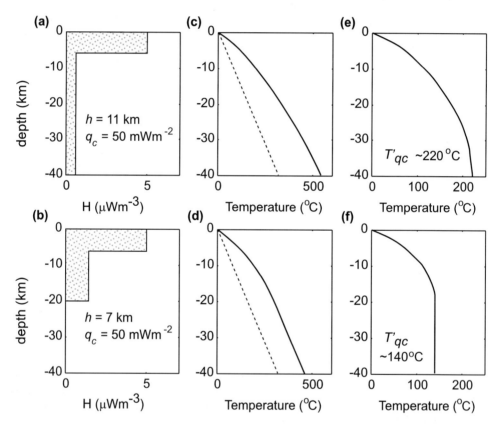

Fig. 5. Two alternative models for the distribution of heat sources in the modern crust in the MII. Both distributions give a total crustal contribution to surface heat flow q_c *c.* 50 m Wm^{-2} (see text for discussion). In both cases the upper layer represents the assumed thickness of average Mount Isa granite with *c.* 5 μWm^{-3} (see text), while the lower layers represent the largely depleted middle and lower crust. In (**a**) lower crustal heat production extends to the Moho with average productivity of 0.59 μWm^{-3}. For this distribution $h = 11$ km. In (**b**) the middle crust contributes 1.2 μWm^{-3} while the lowermost crust contains no significant heat sources. For this distribution $h = 7$ km. Both models are consistent with known surface heat flow and heat production. (**c**) and (**d**) show the geotherm resulting from the heat production described by Models (**a**) and (**b**) respectively. The dashed line is the geotherm that would apply in crust containing no heat producing elements (i.e. the mantle contribution). (**e**) and (**f**) show the additional temperature contribution, T'_{qc}, due to the crustal heat source distribution shown in (**a**) and (**b**) respectively. T'_{qc} reaches a maximum value at a point beneath which there are no additional heat sources. This value is *c.* 220°C for Model (**a**) and *c.* 140°C for Model (**b**).

characterized by low values of h; similarly undifferentiated crust will be characterized by high values of h. For the inferred total crustal contribution of heat sources at Mount Isa and the likely distribution of these heat sources (Fig. 5), the length scale of the heat production distribution, h in modern Mount Isa crust is estimated in the range $c.$ 7–11 km.

Processes which change either h or q_c, or both, will influence crustal thermal regimes. Therefore, the way the distribution of crustal heat sources changes with time, and during major tectono-orogenic processes can be 'mapped'. This parameterization is particularly useful in that it illustrates the thermal (and hence mechanical) consequences of the primary crustal processes (Sandiford et $al.$ 2001). The following section investigates the way in which each major period of crustal activity recorded in the MII and described above, impacts on the distribution of heat producing elements in the crust. Following Sandiford et $al.$ (2001) deformation and surface processes are linked through an isostatic response that, in the long-term, maintains crustal thickness. Thus, crustal thickening is coupled with erosion, while crustal extension is coupled with subsidence and basin formation.

In the following discussion, few assumptions about the starting configuration of the crustal system prior to tectonism are made (i.e. the initial values of the distribution parameters, h and q_c). One assumption made is that the heat producing elements were added to the crust from the lithospheric mantle relatively early in its history and have subsequently been redistributed during crustal scale processes. This assumption is supported at Mount Isa by both Nd and Sr-isotopic data from granites throughout the EFB, KLFB and WFB (Wyborn 1998). The fact that the modern crust is strongly differentiated in the heat producing elements and so characterized by low h infers that the crust has evolved with decreasing h through time.

Magmatic differentiation

Since much of the heat production in the MII is within granites, the generation, segregation and emplacement of these granites is central to the development of the inferred heat production distribution, with h $c.$ 7–11 km (Fig. 5). In this section some simple models for source region distributions are considered in order to quantify the long-term thermal effects of such magmatism. For the sake of simplicity modern day heat production values are used. The equivalent temperature changes in the Palaeoproterozoic to early Mesoproterozoic would be $c.$ 30% greater

than listed below. Moreover, the calculations below assume a modern crustal contribution of $c.$ 50 mWm^{-2} and so represent a minimum estimate of T'_{qc} as q_c may well have been much higher prior to the lengthy history of extensional deformation which occurred later in the tectonic history (see below for discussion).

As the MII granites are interpreted to represent lower crustal melting (e.g. Wyborn 1998), heat producing elements carried by these granites must have been concentrated at much deeper levels prior to granite generation. The process is started by exploring two simple scenarios that could explain the distribution of heat sources in the source region prior to granite generation. The first scenario considers a distribution in which the granitic melt was generated from a layer of equivalent thickness (and hence enrichment) in the lower crust yielding h $c.$ 30 km (Fig. 6a). In this scenario the change in the length scale required to generate the inferred present day distribution of heat producing elements is Δh $c.$ 19–23 km, implying a corresponding reduction in T'_{qc} of 380–460°C, assuming k equals 2.5 Wm^{-1}K^{-1}. However, it seems improbable that the lower crust was ever significantly more enriched in the heat producing elements than the upper crust, especially if it were largely mafic as suggested by Wyborn (1998), and a more reasonable scenario might have been an initially homogeneous distribution with h $c.$ 20 km (Fig. 6b). In this case a differentiation equivalent to Δh $c.$ 9–13 km is required to generate the inferred present day distribution of heat producing elements, which equates to a long-term reduction in T'_{qc} of 180–260°C (Fig. 7a, b).

These observations suggest that in the Mesoproterozoic the long-term cooling associated with the differentiation attributable to all magmatic processes in the MII is of the order of several hundred degrees. However, as noted previously this magmatism was the result of several distinct episodes, namely the Barramundi granites, the pre-Isan (Myally and Mount Isa Rift Events) granites and post-Isan granites (Fig. 4). This section attempts to constrain the relative contributions of each of the three main pulses of magmatism (Table 4).

The volumetrically most significant of these granite pulses was the Barramundi Igneous Association (e.g. the Kalkadoon, Big Toby and Ewen Batholiths). While these granites form only about 18% of the total surface exposure in the MII, seismic profiles show that Barramundi granites extend at depth beneath the WFB (Fig. 3), they are well exposed in the central KLFB where they comprise $c.$ 50% of surface outcrops, and

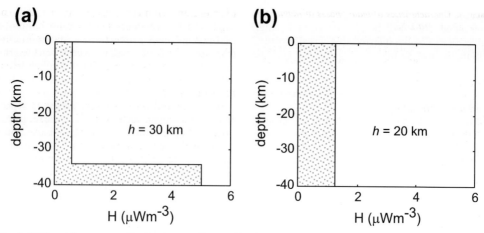

Fig. 6. Different heat source distributions as discussed in the text. Distributions (**a**) and (**b**) represent possible end-member starting configurations prior to crustal differentiation. h is the characteristic length scale of each distribution.

they most likely form basement to the younger sedimentary packages in the EFB (e.g. Drummond *et al.* 1998). Consequently the 'footprint' of these granites is assumed to be close to 100% of the area of the MII. Available geochemical data indicates an average heat production of $c.\,3.6\,\mu Wm^{-3}$ (Tables 2 and 4). Assuming that these Barramundi granites comprise $c.\,5\,km$ of the crust throughout the MII then they would account for $c.\,36\%$ of the total complement of

crustal heat production in the MII. The flux of heat producing elements associated with the generation of these granites equates with approximately a 7°C reduction in T'_{qc} for each kilometre of transport. For a vertical transport distance of 20 km, appropriate to transport from the deep to shallow crust, the reduction in T'_{qc} is $c.\,140°C$.

Younger granites in the MII (i.e. including the pre-Isan rift related granites and post-Isan granites), although often extraordinarily high heat

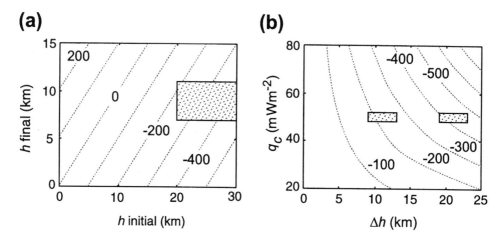

Fig. 7. Effects of changing heat source distribution on lower crustal temperature. (**a**) is contoured for the change in T'_{qc} (°C) for different relative changes in length scale, h and is constructed for q_c $c.\,50\,mWm^{-2}$ as appropriate to the Mount Isa Inlier. Thermal conductivity is assumed to be constant and temperature independent, k $c.\,2.5\,Wm^{-1}\,K^{-1}$. Negative $\Delta(T'_{qc})$ represents lower crustal cooling, such as would accompany magmatism (where $h_{initial} > h_{final}$). Positive $\Delta(T'_{qc})$ is equivalent to lower crustal heating resulting from, for example, the burial of a heat producing layer beneath a sedimentary basin (which has heat production such that $h_{initial} < h_{final}$). The shaded regions show the change in $\Delta(T'_{qc})$ for the inferred starting configurations (as shown in Fig. 6) and present-day configurations (as shown in Fig. 5).

Table 4. *Characteristics of main phases of magmatism in the Mount Isa Inlier*

Parameter	Barramundi granites	Rift-related granites	Post-Isan granites
Area (km^2)	7130	4310	1830
Area (%)	18	11	5
Thickness (km)	5	3	3
Heat Production (μWm^{-3})	3.6	5.1	7.7
HP (1) (%)	6.6	3.4	2.2
HP (2) (%)	19.4	6.7	4.6
Contribution to q_c (%)	36	30.6	46.2
ΔT (1)*	1.3	0.7	0.4
ΔT (2)*	7.2	6.1	9.2

Note: HP (1) is the percentage of total heat production in the MII crust based on the current area of outcrop and assumed thickness. As such HP (1) represents the minimum contribution of each pulse of granites to the total heat production in the MII. HP (2) is an estimate of the maximum percentage of total heat production in the MII based on the following estimates of the original 'foot-print' of each granite pulse: for the Post-Isan and Rift-related granites HP (2) is calculated for an area twice that which is outcropping; the Barramundi granites three times that area of outcrop (i.e. 54% of the terrane).
* ΔT (1) is the change in lower crustal temperature per kilometre of magma ascent when averaged over the entire inlier. ΔT (2) is the local change in lower crustal temperature per kilometre of magma ascent (i.e. that change in the immediate vicinity of each granite).

producing, are not as significant volumetrically as the older Barramundi granites. In particular, these granites are of limited areal extent comprising about 16% of the surface exposure in the MII (Table 4). While the current erosion surface only samples a proportion of all the granites generated at these times, it is unlikely that their total 'footprint' was vastly more extensive than represented in the current surface. In the following discussion the total 'footprint' for these granites is considered to be no more than two times their representation in the current surface. Seismic profiles (e.g. Drummond *et al.* 1998) and previous workers suggest that the Sybella Batholith is a thin sill-like body which intruded essentially parallel to stratigraphy, with a preserved thickness of *c.* 1.5–2 km in the plane of the seismic profile (Fig. 3). A number of observations suggest that the batholith was considerably thicker at the time of intrusion (McLaren *et al.* 1999), and here a typical thickness of *c.* 3 km is assumed. By area the pre-Isan rift-related granites account for about 11% of all exposure in the MII. An average heat production of 5.1 μWm^{-2}

(Tables 2 and 4) and an average thickness of 3 km implies that these granites account for 3.4–6.8% of the total heat production of the MII crust (the upper limit assuming a footprint twice the current surface exposure, Table 4). The generation of these granites would account for a reduction in T'_{qc} of 0.7–1.4°C per ascent kilometre, averaged over the area of entire inlier. Locally the significance would have been much greater. For example, assuming that the 'footprint' of the source area for individual granites was of the same dimension as the granite, then their segregation would lead to a reduction in T'_{qc} of 6.1°C per ascent kilometre, with a 20 km ascent path corresponding to a reduction in T'_{qc} of 122°C.

Similarly the post-Isan Orogeny granites such as the Williams and Naraku Batholiths are confined to the uppermost crust within the EFB, constituting about 5% of the total surface exposure of the MII. Seismic interpretation (McCready 1997) suggests that these late-orogenic granites are sill-shaped and *c.* 3–4 km in thickness. As such the granites host between 2.2 and 4.4% of the total MII crustal heat production (Table 4), and are estimated to have contributed to a regional reduction in T'_{qc} of 0.4–0.8°C per ascent kilometre. Locally their generation would have contributed to a reduction in T'_{qc} by as much as 9.2°C per ascent kilometre.

Extension and convergent deformation

Although the very enriched nature of granites in the MII implies that granite magmatism provided the principal means of crustal differentiation, both extension and convergent deformation are also likely to have significantly modified the distribution of crustal heat sources. Of particular importance is the recognition that crustal thermal regimes are sensitive to small changes in the distribution parameters that result from deformation and associated surface processes of erosion and deposition (Sandiford *et al.* 2001).

Crustal extension, when accompanied by basin formation and sedimentation, results in two fundamental changes to the distribution of heat sources within the crust. The relative magnitude of each is dependent on the amount and symmetry of basement extension, and the thickness and heat production of the basin-fill. In general terms, the effect of stretching during crustal extension is to attenuate the pre-existing heat production while at the same time moving it to deeper levels (via burial beneath the sedimentary cover deposited in response to the extension). It thereby results in an increase in *h* (Sandiford 1998; Sandiford *et al.* 2001). The change in q_c is dependent on the amount of stretching and the heat production of basin-fill relative to the attenuated basement. Figure 8 illustrates that for a basin-fill with average heat production of $2.5\,\mu Wm^{-3}$ as appropriate to the LRFT (McLaren *et al.* 1999; Table 3), crustal extension and basin formation will lead to an increase in both h and q_c (dashed lines in Fig. 8). For a stretching sufficient to produce a 7 km thick sedimentary pile (a minimum estimate of sediment thickness in the LRFT from AGSO unpublished data, P. Southgate pers. comm., 1999), the increases in h and q_c will lead to long-term increases in T'_{qc} in excess of 50°C.

Crustal thickening and erosion also change the distribution of heat sources within the crust. As crustal thickening normally results in erosion

of the uppermost crust, its effect on both h and q_c is dependant on the distribution of heat production prior to deformation. As shown by Sandiford *et al.* (2001), for a differentiated crust in which heat production is concentrated in the upper crust, crustal thickening followed by erosion will lead to a long-term reduction in both h and q_c. The solid lines in Figure 8 show the effect of deformation that homogeneously thickens the crust by 130% followed by erosion that returns crust to its present thickness of $c.$ 40 km (i.e. a deformation sufficient to generate average erosional levels of about 13 km, consistent with known geobarometry, e.g. Rubenach 1992). In all cases modelled in Figure 8 the reduction in q_c due to erosion leads to long-term reductions in T'_{qc} in excess of 100°C. Figure 8 shows that the

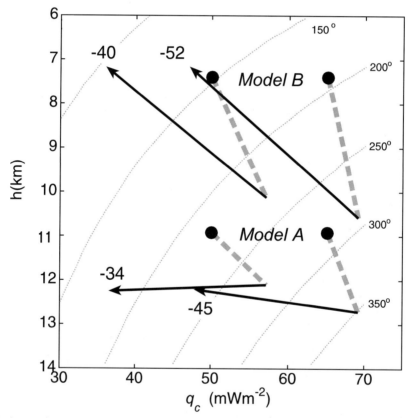

Fig. 8. h–q_c paths appropriate to the sequence of events associated with extension and basin formation (i.e., the Leichardt, Myally and Mount Isa Rift Events) and the crustal thickening and erosion during the Isan Orogeny. Rifting and basin formation sufficient to generate a 7 km thick basin following all long-term subsidence is indicated by the dashed lines. Crustal shortening followed by erosion sufficient to generate $c.$ 13 km of denudation is shown by the solid lines. Two different initial geometries of heat producing elements are considered. Model A corresponds to the distribution shown in Figure 5a. Model B corresponds to the distribution shown in Figure 5b. Initial values of h–q_c prior to deformation are indicated by the bullets. Deformation is considered to be homogeneous at the crustal scale. Contour intervals at 50°C show the values of T'_{qc} appropriate to a given set of h–q_c values (see Sandiford *et al.* 2001 for further discussion of h–q_c mappings).

successive imposition of crustal extension (with associated basin formation) followed by crustal thickening (with erosion) leads to long-term cooling, with the magnitude of this cooling dependant on the initial distribution of heat producing elements as well as the deformation parameters. For parameters believed to be appropriate to the evolution of the MII through the successive rifting events beginning with the Leichhardt event, through to the denudation following the Isan Orogeny, a long-term reduction of T'_{qc} of $c.\,35$–$50°C$ (Fig. 8) is estimated. Note also that in order to preserve a crust with a current complement of heat producing elements of $c.\,50\,mWm^{-2}$, it is likely that the heat production distribution in the crust prior to the various events was of the order of $c.\,65\,mWm^{-2}$.

Discussion

The geological record of the MII shows a protracted tectonic evolution spanning $>300\,Ma$. This history records episodic magmatism and deformation. Deposition of thick sedimentary sequences, and deep levels of erosional denudation, followed periods of extensional and convergent deformation, respectively. This paper attempted to show how this tectonic activity has shaped the distribution of heat producing elements inferred from analysis of present-day surface heat flow–heat production data. While it is not possible to rigorously con-strain the distribution of heat producing elements prior to the onset of tectonic activity, it is clear that the tectonic evolution has produced progressive differentiation of heat producing elements, leading to the long-term cooling of the deep crust. The principal mechanism for this differentiation appears to have been the generation and segregation of granites from the deep crust, although significant changes in the parameters that define the heat production distribution can be attributed to crustal deformation and the associated surficial responses of sedimentation and erosion. Assuming that the initial crust was essentially undifferentiated, as seems to be required in order to produce the enriched granites of the MII, then the crustal differentiation associated with granite magmatism was sufficient to cool the lower crust by at least $200°C$. Much of this differentiation is attributed to the volumetrically most abundant pulse of magmatism during the Barramundi Orogeny, however the subsequent periods of magmatism during the pre-Isan rifting events, and following the Isan Orogeny, are also significant, and locally may have contributed in excess of $100°C$ cooling of the deep crust. Redistribution of heat producing elements dur-

ing deformation associated with the Isan Orogeny, and subsequent erosional denudation is responsible for a further $c.\,100°C$ cooling of the deep crust. In contrast, rifting and basin-formation during the Leichhardt, Myally and Mount Isa rift events is estimated to have lead to a long-term lower crustal heating of the order of $50°C$ (Fig. 8).

The recognition that the tectonic record of the MII lead to the progressive cooling of the deep crust has profound implications for our understanding of its mechanical evolution. The strength of the continental lithosphere is known to be strongly temperature sensitive with Moho-temperature providing a useful proxy for the strength (Sonder & England 1986). Sandiford et al. (2001) have shown that a change in h–q_c parameters sufficient to generate $200°C$ reduction in Moho temperature, would equate with an eight-fold increase in the strength of the lithosphere or, alternatively, three orders of magnitude decrease in the rate of deformation that would apply to lithosphere subject to a tectonic load capable of localizing permanent deformation. It is believed that these considerations provide important new insights into the long-term tectonic record of the MII. In particular, they suggest that the tectonic record is one of progressive strengthening. The long-term cooling and strengthening trend was locally countered by the role of subsidence during basin formation which, through burial of heat producing elements in the existing crust and the accumulation of more heat production in insulating sediments, helped to localize subsequent contractional deformation.

Interpreting the tectonic history of the Mount Isa region by understanding the feedback between tectonic processes and crustal strength allows us to recognize some important controls on terrane reactivation and cratonization. In particular, it is recognized that within any terrane the mechanical properties, and hence propensity for reactivation, are a function of the abundance and distribution of heat producing elements within the crust. Furthermore, the redistribution of heat producing elements during ongoing tectonism is capable of radically altering the mechanical response to subsequent events. Since cratonization implies that the cumulative lithospheric strength exceeds any likely applied tectonic force, it is favoured by processes that progressively concentrate heat sources in the upper crust. These observations suggest that the protracted history of magmatism and deformation at Mount Isa resulted from long-term crustal weakening due to the presence of anomalous concentrations of heat producing elements deep in the crust. In turn both magmatism and deformation, which are very sensitive to lower crustal thermal regimes, lead to

the strengthening, and ultimately to the stabilization, of the crust through the progressive redistribution of heat producing elements to shallower
levels and the removal of heat production through
erosion. It is possible, therefore, that our model
represents a fundamentally different style of
tectonic evolution marking a time of transition
between a style of 'soft-plate' deformation (where
large portions of the crust were weak in a thermo-
mechanical sense) to the modern Palaeozoic
'rigid-plate' style (where deformation occurs
principally at plate boundaries due to plate-
plate collision).

L. Wyborn and I. Bastrakova are thanked for access to
and assistance with the Australian Geological Survey
Organization Granite GIS dataset, used in the generation of Figure 4. L. Wyborn is also thanked for
many helpful discussions regarding the evolution of
the Mount Isa region, and access to the ROCKCHEM
database used in the compilation of Tables 2 and 3 and
Figure 2. We thank M. Hand for contributing to our
knowledge of the evolution of the Mount Isa region,
and D. Scott and P. Southgate for many discussions regarding basin evolution. We are grateful to
P. Betts and N. Oliver for their detailed reviews of the
manuscript, and to I. Buick for his patience and
editorial assistance.

References

ANDREWS, S. J. 1998. Stratigraphy and depositional set-
    ting of the upper McNamara Group, Lawn Hills
    region, Northwest Queensland. In: WILLIAMS, P. J.
    (ed.) Metallogeny of the McArthur River-Mount
    Isa-Cloncurry minerals province. Economic Geol-
    ogy, 93, 1132–1152.
BETTS, P. G., LISTER, G. S. & O'DEA, M. G. 1998.
    Asymmetric extension of the Middle Proterozoic
    lithosphere, Mount Isa terrane, Queensland, Aus-
    tralia. Tectonophysics, 296, 293–316.
BEAUMONT, C., FULLSACK, P. & HAMILTON, J. 1994.
    Styles of crustal deformation in compressional
    orogens caused by subduction of the underlying
    lithosphere. In: CLOWES, R. M. & GREEN, A. G.
    (eds) Seismic reflection probing of the continents
    and their margins. Tectonophysics, 232, 119–132.
BLAKE, D. H. 1987. Geology of the Mount Isa Inlier and
    environs, Queensland and Northern Territory.
    Bureau of Mineral Resources, Geology and Geo-
    physics Bulletin, 225.
BLAKE, D. H. & STEWART, A. J. 1992. Stratigraphic and
    tectonic framework, Mount Isa Inlier. In: STEW-
    ART, A. J. & BLAKE, D. H. (eds) Detailed Studies of
    the Mount Isa Inlier. Australian Geological Survey
    Bulletin, 243, 1–13.
CHAPMAN, D. S. & FURLONG, K. P. 1977. Continental
    heat flow – age relationships. Eos, Transactions of
    the American Geophysical Union, 58, 1240.
CONNORS, K. A. & PAGE, R. W. 1995. Relationships be-
    tween magmatism and deformation in the western

Mount Isa Inlier, Australia. Precambrian Research,
    71, 131–153.
CULL, J. P. 1982. An appraisal of Australian heat-flow
    data. Journal of Australian Geology & Geophysics,
    7, 11–21.
CULL, J. P. & DENHAM, D. 1979. Regional variations in
    Australian heat flow. Journal of Australian Geol-
    ogy & Geophysics, 4, 1–13.
DALLMEYER, R. D. & KEPPIE, J. D. 1993. ^{40}Ar/^{39}Ar
    mineral ages from the southern Cape Breton High-
    lands and Creignish Hills, Cape Breton Island,
    Canada: evidence for a polyphase tectonothermal
    evolution. Journal of Geology, 101, 467–482.
DERRICK, G. M., WILSON, I. H. & SWEET, I. P. 1980.
    The Quilalar and Surprise Creek formations; new
    Proterozoic units from the Mount Isa Inlier; their
    regional sedimentology and application to region-
    al correlation. Journal of Australian Geology and
    Geophysics, 5, 215–223.
DRUMMOND, B. J., GOLEBY, B. R., GONCHAROV, A. G.,
    WYBORN, L. A. I., COLLINS, C. D. N. & MAC-
    CREADY, T. 1998. Crustal-scale structures in the
    Proterozoic Mount Isa inlier of north Australia:
    Their seismic response and influence on mineralisa-
    tion. Tectonophysics, 288, 43–56.
ENGLAND, P. C., OXBURGH, E. R. & RICHARDSON, S. W.
    1980. Heat refraction and heat production in and
    around granite plutons in North-east England.
    Geophysical Journal of the Royal Astronomical
    Society, 62, 439–455.
ERIKSSON, K. A., SIMPSON, E. L. & JACKSON, M. J.
    1993. Stratigraphic evolution of a Proterozoic syn-
    rift to post-rift basin: constraints on the nature of
    lithospheric extension in the Mount Isa Inlier,
    Australia. In: FROSTICK, L. E. & STEEL, R. J. (eds)
    Tectonic controls and signatures in sedimentary
    successions. International Association of Sedimen-
    tologists, Special Publications, 20, 203–221.
ETHERIDGE, M. A., RUTLAND, R. W. R. & WYBORN,
    L. A. I. 1987. Orogenesis and Tectonic process in
    the Early to Middle Proterozoic of Northern Aus-
    tralia. In: Kroner, A. (ed.) Precambrian Litho-
    spheric Evolution. American Geophysical Union,
    Geodynamic Series, 17, 131–147.
FOWLER, C. M. R. 1990. The solid earth: an introduc-
    tion to global geophysics. Cambridge University
    Press, New York.
GALLAGHER, K. 1990. Some strategies for estimating
    present day heat flow from exploration wells, with
    examples. Exploration Geophysics, 21, 145–159.
HAENEL, R., RYBACH, L. & STEGENA, L. (eds) 1988.
    Handbook of terrestrial heat-flow density determi-
    nation; with guidelines and recommendations of the
    International Heat Flow Commission. Kluwer
    Academic Publishers, Dordrecht, Netherlands.
HOUSEMAN, G. A., MCKENZIE, D. P. & MOLNAR, P.
    1981. Convective instability of a thickened bound-
    ary layer and its relevance for the thermal evolu-
    tion of continental convergent belts. Journal of
    Geophysical Research, 86, 6115–6132.
HOUSEMAN, G. A., CULL, J. P., MUIR, P. M. &
    PATERSON, H. L. 1989. Geothermal signatures and
    uranium ore deposits on the Stuart Shelf of South
    Australia. Geophysics, 54, 158–170.

HYNDMAN, R. D. 1967. Heat flow in Queensland and Northern Territory, Australia. *Journal of Geophysical Research*, **72**, 527–539.

HYNDMAN, R. D. & SASS, J. H. 1966. Geothermal measurements at Mount Isa, Queensland. *Journal of Geophysical Research*, **71**, 587–601.

KALSBEEK, F. 1995. Geochemistry, tectonic setting, and poly-orogenic history of Palaeoproterozoic basement rocks from the Caledonian fold belt of North-East Greenland. *Precambrian Research*, **72**, 301–315.

LACHENBRUCH, A. H. 1970. Crustal temperature and heat production: implications of the linear heat-flow relation. *Journal of Geophysical Research*, **75**, 3291–3300.

LOOSVELD, R. J. H. & ETHERIDGE, M. A. 1990. A model for low-pressure facies metamorphism during crustal thickening. *Journal of Metamorphic Geology*, **8**, 257–267.

MCCREADY, T. 1997. *Geologic interpretation of the Mount Isa Deep Seismic Transect*. PhD Thesis, Monash University, Australia.

MCLAREN, S., SANDIFORD, M. & HAND, M. 1999. High radiogenic heat-producing granites and metamorphism – An example from the Mount Isa Inlier, Australia. *Geology*, **27**, 679–682.

MCLENNAN, S. M. & TAYLOR, S. R. 1996. Heat flow and the chemical composition of continental crust. *Journal of Geology*, **104**, 369–377.

MORGAN, P. 1984. The thermal structure and thermal evolution of the continental lithosphere. *In*: POLLACK, H. N. & MURTHY, V. R. (eds) *Structure and evolution of the continental lithosphere*. Physics and Chemistry of the Earth, **15**, 107–185.

MYERS, J. S., SHAW, R. D. & TYLER, I. M. 1996. Tectonic evolution of Proterozoic Australia. *Tectonics*, **15**, 1431–1446.

NEIL, E. A. & HOUSEMAN, G. A. 1997. Geodynamics of the Tarim Basin and the Tian Shan in central Asia. *Tectonics*, **16**, 571–584.

NYBLADE, A. A. & POLLACK, H. N. 1993. A global analysis of heat flow from Precambrian terrains: Implications for the thermal structure of Archaean and Proterozoic lithosphere. *Journal of Geophysical Research*, **98**, 12207–12218.

O'DEA, M. G., LISTER, G. S., BETTS, P. G. & POUND, K. S. 1997*a*. A shortened intraplate rift system in the Proterozoic Mount Isa terrane, NW Queensland, Australia. *Tectonics*, **16**, 425–441.

O'DEA, M. G., LISTER, G. S., MACCREADY, T., BETTS, P. G., OLIVER, N. H. S., POUND, K. S., HUANG, W. & VALENTA, R. K. 1997*b*. Geodynamic evolution of the Proterozoic Mount Isa terrain. *In*: BURG, J. P. & FORD, M. (eds) *Orogeny through Time*. Geological Society, London, Special Publications, **121**, 99–122.

PAGE, R. W. 1983. Chronology of magmatism, skarn formation and uranium mineralization, Mary Kathleen, Queensland, Australia. *Economic Geology*, **78**, 838–853.

PAGE, R. W. & SUN, S-S. 1998. Aspects of geochronology and crustal evolution in the Eastern Fold Belt, Mount Isa Inlier. *Australian Journal of Earth Sciences*, **45**, 343–361.

PAGE, R. W. & SWEET, I. P. 1998. Geochronology of basin phases in the western Mount Isa Inlier, and correlation with the McArthur Basin. *Australian Journal of Earth Sciences*, **45**, 219–232.

PAGE, R. W. & WILLIAMS, I. S. 1988. Age of the Barramundi Orogeny in northern Australia by means of ion microprobe and conventional U–Pb zircon studies. *Precambrian Research*, **41**, 21–36.

PAGE, R. W., SUN, S-S. & CARR, G. 1994. Proterozoic sediment-hosted lead-zinc-silver deposits in northern Australia – U–Pb zircon and Pb isotopic studies. *Australian Journal of Earth Sciences*, **37**, 334–335.

PEARSON, P. J., HOLCOMBE, R. J. & PAGE, R. W. 1992. Syn-kinematic emplacement of the Middle Proterozoic Wonga Batholith into a mid-crustal extensional shear zone, Mount Isa Inlier, Queensland, Australia. *In*: STEWART, A. J. & BLAKE, D. H. (eds) *Detailed Studies of the Mount Isa Inlier*. Australian Geological Survey Bulletin, **243**, 289–238.

REINHARDT, J. 1992. Low-pressure, high-temperature metamorphism in a compressional tectonic setting: Mary Kathleen Fold Belt, northeastern Australia. *Geological Magazine*, **129**, 41–57.

RUBENACH, M. J. 1992. Proterozoic low-pressure/high-temperature metamorphism and anti-clockwise P-T-t paths for the Hazeldene area, Mount Isa Inlier, Queensland, Australia. *Journal of Metamorphic Geology*, **10**, 333–346.

SANDIFORD, M. 1998. Mechanics of Basin Inversion. *Tectonophysics*, **305**, 109–120.

SANDIFORD, M., HAND, M. & MCLAREN, S. N. 2001 Tectonic feedback, intraplate orogeny and the geochemical structure of the crust: a central Australian perspective. *In*: MILLER, J. A., HOLDSWORTH, R. E., BUICK, I. S. & HAND, M. (eds) *Continental Reactivation and Reworking*. Geological Society, London, Special Publications, **184**, 195–218.

SONDER, L. & ENGLAND, P. C. 1986. Vertical averages of rheology of the continental lithospheric; relation to thin sheet parameters. *Earth and Planetary Science Letters*, **77**, 81–90.

WYBORN, L. A. I. 1988. Petrology, geochemistry and origin of a major Australian 1880–1849 Ma felsic volcano-plutonic suite: A model for intra-continental felsic magma generation. *Precambrian Research*, **41**, 37–60.

WYBORN, L. A. I. 1998. Younger *ca* 1500 Ma granites of the Williams and Naraku Batholiths, Cloncurry district, eastern Mount Isa Inlier: geochemistry, origin, metallogenic significance and exploration indicators. *Australian Journal of Earth Sciences*, **45**, 397–411.

WYBORN, L. A. I., PAGE, R. W. & MCCULLOCH, M. T. 1988. Petrology, geochronology and isotope geochemistry of the post-1820 Ma granites of the Mount Isa Inlier; mechanisms for the generation of Proterozoic anorogenic granites. *Precambrian Research*, **41**, 509–541.

WYBORN, L. A. I., GOLEBY, B., DRUMMOND, B. & GALLAGHER, R. 1996. The Mount Isa geodynamic transect: a 2.5-dimensional metallogenic GIS analysis. *Australian Geological Survey Research Newsletter*, **24**, 10–12.

WYBORN, L. A. I., BUDD, A. R. & BASTRAKOVA, I. V. 1998. The metallogenic potential of Australian Proterozoic granites, Granite GIS (Partial Release) CD ROM, Australian Geological Survey Organization.

ZIELHUIS, A. & VAN DER HILST, R. D. 1996. Upper-mantle shear velocity beneath eastern Australia from inversion of waveforms from Skippy portable arrays. *Geophysical Journal International*, **127**, 1–16.

Tectonic evolution of the Reynolds–Anmatjira Ranges: a case study in terrain reworking from the Arunta Inlier, central Australia

M. HAND[1] & I. S. BUICK[2]

[1] Department of Geology and Geophysics, Adelaide University, South Australia.
5005 Australia (e-mail: martin.hand@adelaide.edu.au)
[2] Department of Earth Sciences, La Trobe University, Victoria 3083, Australia

Abstract: The Reynolds–Anmatjira Range region forms part of the Arunta Inlier in central Australia and has undergone four tectonothermal cycles that span an interval of c. 1450 Ma. The first two cycles were the Stafford Tectonic Event c. 1820 Ma, and the Strangways Orogeny c. 1770–1780 Ma, both of which were associated with regional low-pressure high-temperature metamorphism up to granulite grade that was coeval with the emplacement of voluminous sheet-like granites. The subsequent Chewings Orogeny occurred at around 1590–1570 Ma and was a long-lived event that produced regional low-pressure greenschist to granulite facies metamorphism without obvious associated magmatism. During the mid-Palaeozoic Alice Springs Orogeny (400–300 Ma), the terrain was dissected by a system of sub-greenschist to mid-amphibolite facies shear zones. In the Reynolds Range, the Proterozoic events produced a single regional foliation that is axial planar to simple large-scale folds. The composite regional Proterozoic foliation increases in grade smoothly from northwest to southeast, producing a pattern of isograds that is remarkably similar to those that formed during the mid-Palaeozoic Alice Springs Orogeny. Despite this simple pattern, the isograds reflect the superimposed metamorphic effects of four unrelated tectonothermal cycles. Without geochronological and stratigraphical information, the degree of terrain reworking in the Reynolds–Anmatjira Range region could have been largely obscured by the apparent simplicity of many of the structural and metamorphic relationships.

The evolution of the continents manifestly involves reworking of orogenic belts. This is particularly the case for Archaean and Proterozoic regions, which have had a substantial part of Earth history in which to accumulate the affects of multiple tectonic events. Reworking involves both mechanical and thermal processes that express themselves in the structural, metamorphic and magmatic character of metamorphic terrains. In many cases the legacy of an event or events exerts a fundamental control on the style and distribution of later events. This can be readily seen in regions where tectonism has been repeatedly focused into relatively narrow domains that flank more stable regions (e.g. Barton *et al.* 1994; D'Lemos *et al.* 1997).

In polytectonic terrains there is a danger of linking together structural and metamorphic features that formed during separate events, producing *apparent* histories that point toward tectonic processes that may not have occurred. However if the event history of complex terrains can be deciphered, it not only allows evaluation of the tectonic processes that shaped the evolution, but also provides insights into the potential

feedbacks (e.g. Sandiford *et al.* 2001) that may exist between the products of each stage of the evolution.

One good example of a terrain where there has been some confusion about the number and nature of events is the Reynolds–Anmatjira region in the Arunta Inlier in central Australia (Fig. 1). Despite being affected by at least four separate events, over an interval of c. 1450 Ma, the geological character of the Reynolds–Anmatjira region is relatively simple. The majority of the structural features associated with the multiple episodes of terrain reworking are approximately co-planar, and metamorphism is predominantly of low-pressure, high-temperature character, with relatively smooth regional metamorphic field gradients. The comparatively simple geological relationships in the Reynolds–Anmatjira region have led a number of workers to link various features of the separate tectonothermal events into single cycles rather than separate unrelated events (e.g. Clarke *et al.* 1990; Dirks & Wilson 1990; Clarke & Powell 1991; Collins & Vernon 1991; Dirks *et al.* 1991; Collins & Williams 1995). This paper describes the geological

From: MILLER, J. A., HOLDSWORTH, R. E., BUICK, I. S. & HAND, M. (eds) *Continental Reactivation and Reworking*. Geological Society, London, Special Publications, **184**, 237–260. 1-86239-080-0/01/$15.00 © The Geological Society of London 2001.

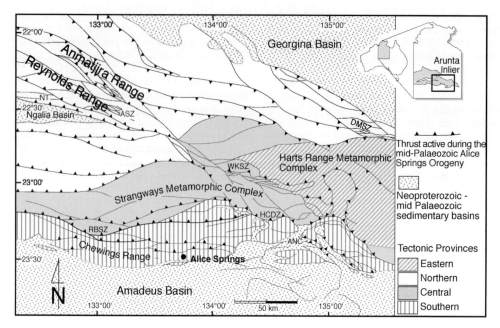

Fig. 1. Simplified map of the Arunta Inlier showing the location of the Anmatjira–Reynolds Ranges Region in the northern province of the inlier, together with the locations of the other tectonic provinces and major metamorphic complexes. The Amadeus, Georgina and Ngalia basins are structural remnants of a once continuous basin that cover the Arunta Block prior to the mid-Palaeozoic (400–300 Ma) Alice Springs Orogeny. Major shear zones associated with the Alice Springs Orogeny are: RBSZ, Redbank Shear Zone; ASZ, Aileron Shear Zone; WKSZ, Wallaby Knob Schist Zone; HCDZ, Harry Creek Deformed Zone; ISZ, Illogwa Shear Zone; DMSZ, Delny Mount Sainthill Fault Zone; NT, Napperby Thrust.

evolution of the Reynolds–Anmatjira Range region, and speculates on some of the long term consequences of the early tectonic history on later events.

Geological framework

The Reynolds–Anmatjira Ranges are located in the northern province of the Arunta Inlier, in central Australia. The Arunta Inlier is one of the largest exposed Proterozoic terrains in Australia, occupying an area of around 200 000 km^2. It forms a broadly E–W trending metamorphic complex bounded by a system of Neoproterozoic to mid-Palaeozoic intracratonic basins that form remnants of the Centralian Superbasin which once covered much of Proterozoic Australia. The Centralian Superbasin was broken up during two major intracratonic orogenies, the Late Neoproterozoic–early Cambrian Petermann Orogeny in southern Central Australia, and the Mid Palaeozoic Alice Springs Orogeny in central Australia. The Alice Springs Orogeny was responsible for the final exhumation of the Arunta Inlier.

The Arunta Inlier contains a polymetamorphic history that was shaped by two major intervals of tectonism. The first tectonic interval occurred during the Palaeo- to Mesoproterozoic (c. 1880–1560 Ma; Black et al. 1983; Black & Shaw 1992, 1995; Young et al. 1995; Collins & Shaw 1995; Williams et al. 1996; Rubatto et al. 1999), and was associated with multiple episodes of regional medium to high-temperature metamorphism and magmatism. In terms of affect on the overall geological character of the Inlier, the major tectonothermal event in the interval 1880–1560 Ma was the Strangways Orogeny (Collins & Shaw 1995), which produced regional granulite metamorphism throughout much of the central province of the Inlier (Fig. 1) over the interval 1780–1720 Ma (Collins & Shaw 1995). Deformation during the Strangways Orogeny produced regional scale west and southwest vergent structures in the Strangways Metamorphic Complex (Goscombe 1991; Collins & Sawyer 1996). The Strangways Orogeny was part of a protracted period of tectonism that is thought to reflect the development of magmatic arcs along the southern margin of the north Australian craton, and their subsequent accretion (Myers et al. 1996).

The other major event in the 1880–1560 Ma interval was the intracratonic Chewings Orogeny (*c.* 1600–1570 Ma) which produced a north-directed amphibolite-grade thrust system in the Chewings Range in the southern province (Teyssier *et al.* 1988; Collins & Shaw 1995; Fig. 1). Further north in the Strangways Metamorphic Complex in the central province, and in the Reynolds–Anmatjira region in the northern province, the Chewings Orogeny was associated with granulite facies metamorphism (Hand *et al.* 1995; Vry *et al.* 1996; Williams *et al.* 1996; Collins 2000). The Chewings Orogeny may reflect continental interior reworking during accretion of a strip of continental crust onto the southern margin of the north Australian Craton (Myers *et al.* 1996).

The second major phase of tectonism in the history of the Arunta Inlier occurred in the early to mid-Palaeozoic (*c.* 490 to 300 Ma), and was most pronounced in the Harts Range Metamorphic Complex which defines the eastern tectonic province (Fig. 1). This period began with N–S intraplate extension at around 490–480 Ma associated with granulite and upper amphibolite facies metamorphism (Hand *et al.* 1999*a*;

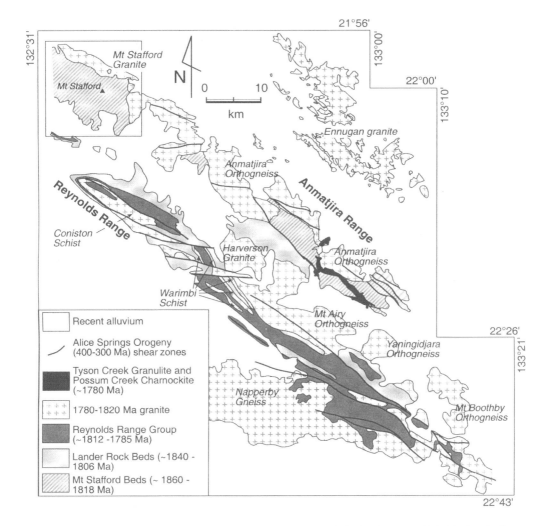

Fig. 2. Generalized geology of the Reynolds–Anmatjira region (modified after Stewart, 1981). The box in the vicinity of Mt Stafford in the northwestern Anmatjira Range shows the location of Figure 3. Magnetic data indicates that the bulk of the regions covered by recent alluvium are underlain by granite/granitic gneiss.

Mawby *et al.* 1999). This extensional event was linked to the development of a NW–SE trending marine corridor (the Larapintine Seaway) that stretched across Australia, connecting the Canning Basin in northwest Australia to the eastern margin. Subsequently N–S convergent deformation began at around 450 Ma (Mawby *et al.* 1999) heralding the onset of the intracratonic Alice Springs Orogeny. This was a protracted event terminating at around 300 Ma and developed synchronously with convergence along the eastern and northeastern margins of Australia.

The part of the Arunta Inlier in which the Reynolds and Anmatjira Ranges are located contains three principal sedimentary successions (Fig. 2) whose depositional ages are bracketed by detrital zircon U/Pb ages and the timing of granite emplacement. The oldest sedimentary sequence is the Lander Rock Beds, which are a succession of immature quartzose and relatively Mg-rich pelitic sediments that were deposited between 1806 and 1840 Ma (Vry *et al.* 1996). The Lander Rock Beds may be an equivalent of the Mt Stafford Beds, which are a comparatively Fe-rich sequence of psammitic and pelitic sediments deposited between 1860 and 1818 Ma (Collins & Williams 1995). These successions form basement to the Reynolds Range Group, which is a variably metamorphosed, shallow marine succession consisting of a basal conglomerate and quartz sandstones (Mt Thomas Quartzite), siltstones, arkoses and carbonates (Stewart *et al.* 1980; Dirks 1990) that were deposited between *c.* 1812 and 1785 Ma (Collins & Williams 1995; Williams *et al.* 1996). The (meta)-sedimentary successions have been intruded by several generations of sheet-like K-rich megacrystic bodies that in total occupy upwards of 60% of the terrain (Fig. 2). The important geochronological constraints on the emplacement ages of the granites in the region are shown in Table 1.

The metamorphic and structural history of the Reynolds–Anmatjira region has been the subject of considerable debate over the past ten years. Although there is a general consensus that the region has undergone several tectonothermal events, much debate has been concerned with: (1) the number of events (e.g. Collins & Vernon 1991; Hand *et al.* 1992; Vernon 1996); (2) their geographical extent and areal extent (e.g. Dirks & Wilson 1990; Collins *et al.* 1991; Collins & Vernon 1993; Hand *et al.* 1993; Buick *et al.* 1994); (3) the absolute timing of the events (e.g. Collins & Williams 1995; Williams *et al.* 1996; Vry *et al.* 1996); (4) conflicting correlations between the structural and metamorphic features in different regions (e.g Collins *et al.* 1991; Hand *et al.* 1992); (5) the P–T evolution of events (e.g. Clarke *et al.* 1990; Dirks *et al.* 1991; Hand *et al.* 1992; Vry & Cartwright 1994; Buick *et al.* 1998); and (6) the role of magmatism during metamorphism (e.g. Collins & Vernon 1991; Hand *et al.* 1992). As a result, there is no consis-tent terminology in place to describe the metamorphic and deformational history of the Reynolds–Anmatjira region. However, a growing geochronological database (Collins & Williams 1995; Hand *et al.* 1995; Vry *et al.* 1996; Williams *et al.* 1996; Buick *et al.* 1999; Rubatto *et al.* 1999) allows an increasingly coherent picture of the geological evolution of the Reynolds–Anmatjira Range region to emerge. The following sections briefly describe the important features of each event to have affected the Reynolds–Anmatjira Range region.

The Stafford tectonic event *c.* 1820 Ma

Metamorphism and deformation associated with this event was spatially associated with the *c.* 1818 Ma Mt Stafford and Harverson Granites (Figs 2 and 3). In the Mt Stafford region in the northwestern Anmatjira Range, metapelitic and metapsammitic sediments belonging to the Mt

Table 1. *Emplacement ages (zircon U–Pb) of granitic units in the Reynolds–Anmatjira region*

Granitic Unit	Emplacement age (Ma)	Intrusive Relationship
Mt Stafford Granite	1818 ± 15[1]	Intrudes Mt Stafford Beds
Harverson Granite	1818 ± 8[1]	Intrudes Lander Rock Beds
Yaningidjara Orthogneiss	1806 ± 6[2]	Intrudes Lander Rock Beds
Warimbi Schist	1785 ± 22[1]	Intrudes Reynolds Range Group
Napperby Gneiss	1788 ± 8[1]	Intrudes Reynolds Range Group
	1773 ± 19[3]	
Possum Creek Charnockite	1774 ± 6[1]	Intrudes Mt Stafford Beds
Tyson Creek Granulite	1767 ± 17[3]	Intrudes Mt Stafford Beds

[1] Collins & Williams (1995); [2] Vry *et al.* (1996); [3] Hand *et al.* (1995).

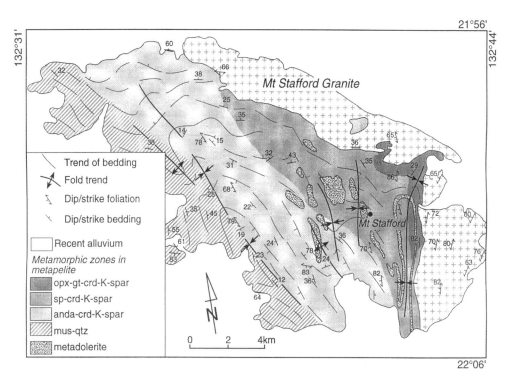

Fig. 3. Simplified geology of the Mt Stafford region in the northwestern Anmatjira Range (see Fig. 2 for location). The region contains open, northwest to north-trending folds that appear to be transected by high-temperature low-pressure isograds associated with the Mt Stafford Tectonic Event. The isograds show an increase in grade toward the east and northeast and are truncated by the 1818 Ma Mt Stafford Granite (Collins & Williams 1995). Figure modified from Stewart (1981) and Collins et al. (1991).

Stafford Beds and mafic sills underwent low-pressure metamorphism (*c.* 2.5 kbar) that was associated with a dramatic increase in temperature from *c.* 400°C to *c.* 800°C over a distance of around 4 km (Vernon *et al.* 1990; Collins *et al.* 1991; Fig. 3). The high-temperature metamorphism was not associated with pervasive deformation and the metamorphic character is best described as regional contact metamorphism (e.g. Clarke *et al.* 1990; Vernon *et al.* 1990; Collins & Vernon 1991).

The extreme vertical and horizontal metamorphic gradients, the low strain nature of the Mt Stafford terrain and the apparent absence of major detachment structures that would have facilitated fast tectonic unroofing (Stewart 1981; Clarke *et al.* 1990; Vernon *et al.* 1990; Collins & Vernon 1991; Greenfield *et al.* 1998) suggest that metamorphism occurred in magmatically driven system. However, the abundant metabasic sills in the area are themselves metamorphosed, and although the metamorphic grade appears to increase toward the 1818 ± 15 Ma Mt Stafford Granite (Fig. 3), the granite crosscuts the high-

temperature isograds (Collins & Williams 1995), suggesting that it postdates the peak metamorphism to some extent. Conceivably both the mafic rocks and the granites were causative factors in the metamorphism, and were emplaced throughout the high-temperature interval, with early magmatism dominated by mafic compositions, and late phases of the granite truncating isograds. If this is the case, then the metamorphism in the Mt Stafford region occurred at or before 1818 ± 15 Ma (Collins & Williams 1995). The high-temperature metamorphic isograds overprint map-scale, southeast to south-trending folds (Fig. 3), indicating that at least some of the folding in the northwestern Anmatjira Range occurred prior to the emplacement to the Mt Stafford Granite.

To the southeast of Mt Stafford, contact metamorphism of the Lander Rock Beds around the 1818 ± 8 Ma Harverson Granite reached lower grades than in the Mt Stafford region (Fig. 4), and typically involved the growth of andalusite and cordierite (Stewart *et al.* 1980; Dirks *et al.* 1991). The contact metamorphic blasts overprint

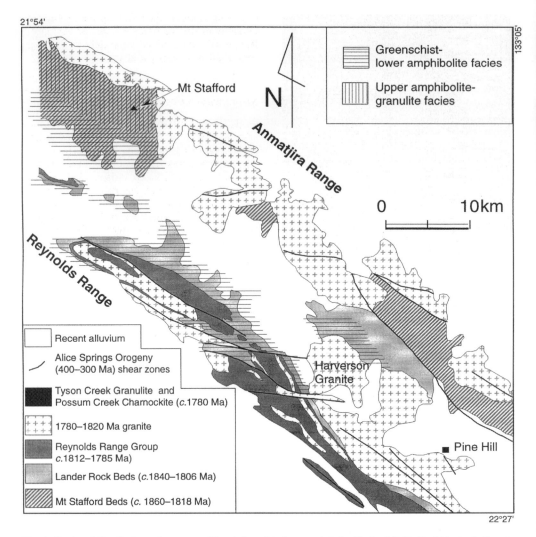

Fig. 4. Regional distribution of metamorphism inferred to be associated with the Mt Stafford Tectonic Event.

a biotite–quartz–muscovite foliation (Fig. 5a), indicating that regional deformation occurred in the Lander Rocks Beds in the northwestern Reynolds Range prior to 1818 ± 8 Ma. Elsewhere in the northwestern Reynolds Range, the Stafford Tectonic Event produced muscovite $\pm$ biotite-bearing greenschist assemblages (Dirks *et al.* 1991) that define a sub-vertical NW–SE trending foliation (Stewart 1981; Dirks & Wilson 1990). Fold structures in the Lander Rock Beds define a local unconformity with the overlying Reynolds Range Group that dies to the southeast, suggesting that the Stafford Tectonic Event may have

been localized in the northwestern part of the Reynolds–Anmatjira region.

The Strangways Orogeny *c.* 1780–1770 Ma

The Strangways Orogeny was the second major event to affect the rocks of the Reynolds–Anmatjira region. In the Arunta Inlier, the effects of the Strangways Orogeny are best seen outside the Reynolds–Anmatjira region, where they produced much of the metamorphic and structural character of the central province of

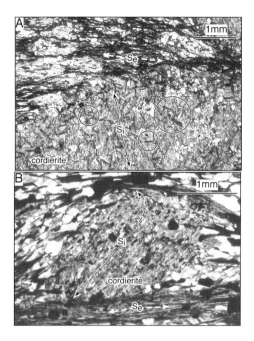

Fig. 5. (**a**) Contact metamorphic cordierite in the Lander Rock Beds immediately adjacent to the 1818 Ma Harverson Granite. The cordierite contains an internal foliation defined by biotite and muscovite that was overgrown by randomly oriented biotite (circled) prior to the growth of the cordierite. The presence of a foliation within the cordierite indicates that the Lander Rock Beds were deformed prior to 1818 Ma. (**b**) Contact metamorphic cordierite in the Reynolds Range Group surrounding the *c*. 1785 Ma Warimbi Granite. The cordierite contains an internal foliation defined by biotite and muscovite, indicating that the Reynolds Range Group was deformed prior to the emplacement of the Warimbi Granite. The external foliation (S_e) is axial planar to the regional upright southeastern-trending folds that run along the Reynolds Range.

the inlier during NE–SW directed shortening over the interval *c*. 1780–1720 Ma (Norman & Clarke 1990; Goscombe 1991; 1992; Black & Shaw 1992; Collins & Shaw 1995; Collins & Sawyer 1996; Möller *et al*. 1999). In the Reynolds–Anmatjira region, the most obvious manifestation of the Strangways Orogeny was the emplacement of the Possum Creek Charnockite and Anmatjira Orthogneiss and the granitic precursors to the Warimbi Schist and Napperby Gneiss (Fig. 2) at around 1785 Ma (Collins & Williams 1995; Hand *et al*. 1995; Collins & Sawyer 1996). The thermal affects associated with these granites produced regional high geothermal gradient metamorphism. In the

following sections, the affects of the Strangways Orogeny in the Reynolds Range and Anmatjira Range will be described separately as they vary somewhat in intensity and style.

Strangways Orogeny in the Reynolds Range

Evidence for metamorphism and deformation associated with the Strangways Orogeny is most easily seen in the Reynolds Range Group since this sequence was deposited after the Stafford tectonic event (Dirks & Wilson 1990; Collins & Williams 1995; Williams *et al*. 1996). Within the Reynolds Range Group, the northwestern part preserves the best evidence of Strangways Orogeny metamorphism due to the relatively low grade effects of later regional 1580 Ma metamorphism during the Chewings Orogeny (see below).

In the northwestern Reynolds Range, contact metamorphic assemblages formed around the granitic precursors to the Coniston and Warimbi Schists (Fig. 2), which intruded the Reynolds Range Group at around 1785 Ma (Collins & Williams 1995). The best developed contact aureoles are in the vicinity of the Warimbi Schist, where andalusite and cordierite-bearing assemblages occur in metapelite. For the bulk composition of the Reynolds Range Group metapelites, the stability of these assemblages suggests that maximum P–T conditions were around 550°C and 3.5 kbar (Xu *et al*. 1994; Mahar *et al*. 1997). This is supported by the relatively Ca-rich compositions of early scapolite porphyroblasts in anorthite-bearing marbles, which also suggest contact metamorphic temperatures of around 550°C in the vicinity of the granitic precursor of the Coniston Schist (Buick & Cartwright 1994). The contact metamorphic blasts surrounding the Warimbi Schist contain straight or gently curving internal foliations defined by muscovite–quartz ± biotite (Fig. 5b). The presence of curved inclusion trails suggests that the growth of the contact metamorphic assemblages occurred during deformation. In some instances inclusion trails within adjacent contact metamorphic blasts show systematic changes in orientation, defining gentle folds.

While the presence of inclusion trails within *c*. 1785 Ma contact metamorphic blasts indicates that the northwestern Reynolds Range Group was regionally deformed at or before 1785 Ma, the orientation of the deformational fabric is not clear. In many cases inclusion trails within the porphyroblasts, and also a foliation in the strain shadows of large blasts appears to be parallel to the external regional upright

SE-trending foliation. This may suggest that the *c.* 1785 Ma foliation was approximately upright and SE-trending, and that deformation occurred during NE–SW-directed shortening, as inferred in other parts of the Arunta Inlier during the Strangways Orogeny (Goscombe 1991; Collins & Sawyer 1996).

Although mid-amphibolite contact metamorphism was clearly associated with the Strangways Orogeny in the Reynolds Range, the regional metamorphic grade of the *c.* 1785 Ma fabrics is still not well established. Dirks *et al.* (1991) noted a systematic increase in the grade of inclusion assemblages within porphyroblasts along the length of the Reynolds Range, culminating in granulite-grade assemblages in the southeastern Reynolds Range. If these inclusion assemblages formed during the Strangways Orogeny, it implies that metamorphism in the southeastern Reynolds Range reached granulite grade at *c.* 1785 Ma. However, no Strangways Orogeny ages have been obtained from high-temperature chronometers (SHRIMP zircon and monazite U/Pb) from metasediments in the southeastern Reynolds Range (Vry *et al.* 1996; Williams *et al.* 1996; Rubatto *et al.* 1999), suggesting that metamorphic conditions were insufficient to promote growth of monazite and zircon, and as yet, there is no firm evidence for high-grade *c.* 1785 Ma metamorphism in the Reynolds Range.

Strangways Orogeny in the Anmatjira Range

In contrast to the Reynolds Range, which has a relatively simple structural and metamorphic character (e.g. Dirks & Wilson 1990; Dirks *et al.* 1991; Vry & Cartwright 1994; Buick *et al.* 1998), the structural and metamorphic character of the Anmatjira Range is superficially more complex, with several generations of granulite-grade fabric development and partial melting events (Clarke *et al.* 1990; Collins *et al.* 1991; Hand *et al.* 1992; Vernon 1996; Fig. 6).

The best evidence for a *c.* 1770–1780 Ma granulite event in the Anmatjira Range comes from the intrusive relationships between the 1774 ± 6 Ma Possum Creek Charnockite and mafic and felsic orthogneiss belonging to the Tyson Creek Granulite (Fig. 2) which intruded at 1767 ± 14 Ma (Hand *et al.* 1995). In low-strain regions, megacrystic charnockite cross-cuts a granulite grade foliation (Stewart *et al.* 1980; Collins & Williams 1995; Collins 2000), but also occurs as cm-scale migmatitic layers along the gneissic fabric, suggesting that the rocks were undergoing

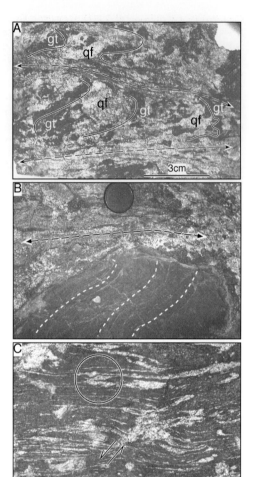

Fig. 6. Structural–metamorphic relationships in the southeastern Anmatjira Range. (**a**) Foliation defined by garnet–sillimanite–spinel–cordierite (solid line) overprinted by a cordierite–biotite–sillimanite-bearing fabric (dashed lined) that constitutes the principal form surface in the southeastern Anmatjira Range. (**b**) Gneissic foliation (white dashed lines) truncated at the margin of a mafic boudin by the regional gneissic layering (black dashed line). (**c**) Isoclinally folded (circled) and boudinaged leucosomes in felsic gneiss, overprinted by a younger episode of partial melting localized in a small shear band.

high-grade metamorphism and deformation at around 1774 ± 6 Ma. Although the Strangways Orogeny in the Anmatjira Range appears to have been associated with regional deformation, the regional-scale structural expression is unclear. Conceivably the regional gneissic fabric in the Anmatjira Range formed during the Strangways

Orogeny (e.g. Collins *et al.* 1991). However, since the 1774 ± 6 Ma Possum Creek Charnockite is itself intensely deformed and migmatized, it is entirely possible that the dominant structural fabric in the Anmatjira Range formed during a subsequent event to the Strangways Orogeny.

Metamorphic textures associated with *c.* 1770–1780 Ma tectonism in the Anmatjira Range are difficult to unequivocally identify. In metasediments, transposed and truncated foliations are defined by garnet–cordierite $\pm$ spinel $\pm$ biotite $\pm$ sillimanite $\pm$ quartz $\pm$ K-feldspar (Fig. 6a). In some assemblages garnet appears to have been partly replaced by cordierite (Fig. 7a) prior to the development of the regional foliation. In addition, garnet porphyroblasts (Figs. 7b) show evidence for several stages of growth that *may* reflect *c.* 1770–1780 Ma and 1580 Ma metamorphism. While these mineral textures do not prove a poly-granulite history, the complexity of the metamorphic textures in the southeastern Anmatjira Range (e.g. Clarke *et al.* 1990; Hand *et al.* 1992) contrasts markedly with the simplicity of the granulite assemblages in the southeastern Reynolds Range (Dirks *et al.* 1991; Hand *et al.* 1992; Buick *et al.* 1998), which appear to record only the metamorphic products of the 1580 Ma Chewings Orogeny (Vry *et al.* 1996; Williams *et al.* 1996; Buick *et al.* 1998; Rubatto *et al.* 1999).

The Chewings Orogeny (*c.* 1595–1560 Ma)

The Reynolds Range is dominated by a spectacular continuous NW–SE transition in metamorphic grade from greenschist to granulite (Fig. 8). The age of this transition has been the source of some uncertainty (e.g. Collins & Williams 1995; Williams *et al.* 1996). However, a growing geochronological database (Vry *et al.* 1996; Williams *et al.* 1996; Rubatto *et al.* 1999) suggests that it formed at around 1580 Ma during the Chewings Orogeny. Several lines of evidence support this notion. First, all the dated migmatized granitic orthogneisses in the terrain contain a population of relatively low Th/U zircons with ages around 1590–1570 Ma (Fig. 9). Zircons of this age do not occur in dated granites that are down-grade of the migmatite-in isograd (Collins & Williams 1995), suggesting that partial melting was a causative factor in the

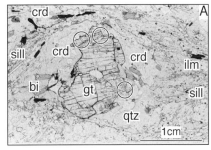

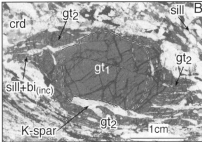

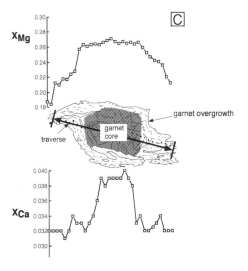

Fig. 7. Partially resorbed garnet surrounded by cordierite ($X_{Fe} = 0.41$) and enclosed by the regional foliation in the southeastern Anmatjira Range. The foliation is defined by sillimanite–biotite–ilmenite and cordierite ($X_{Fe} = 0.34$) and is deflected by the cordierite corona that partially encloses the garnet, suggesting replacement of garnet predated the formation of the foliation. (**a**) The circles contain sillimanite inclusions truncated by cordierite at the garnet margin. (**b**) Possible *c.* 1580 Ma garnet (gt_2) containing inclusion trails of biotite and sillimanite (sill + $bi_{(inc)}$) that are continuous with the external sillimanite–biotite–cordierite–K–spar–plagioclase-bearing regional foliation. The inclusion trails were deflected by an earlier generation of inclusion-free garnet (gt_1) (outlined by dashed line) prior to the growth of gt_2. (**c**) Compositional traverse through the garnet shown in (**b**). The core of the garnet (gt_1) is characterized by higher X_{Mg} and X_{Ca} than gt_2.

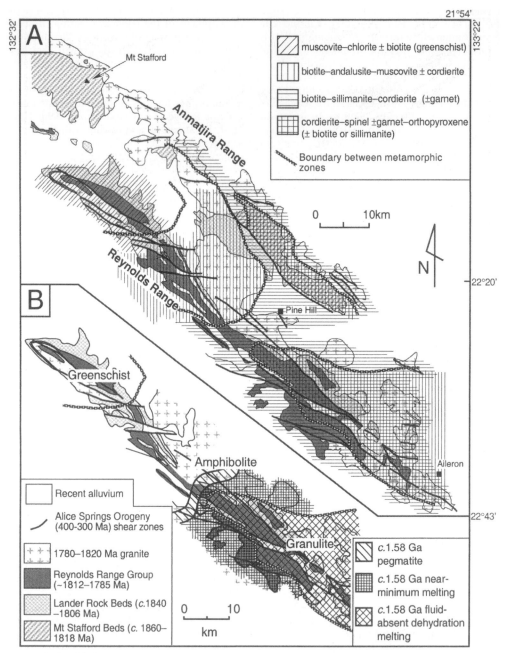

Fig. 8. (**a**) Simplified geological map of the Reynolds–Anmatjira Range region showing the metamorphic zones associated with the *c.* 1580 Ma Chewings Orogeny. (**b**) Metamorphic zones in the Reynolds Range defined by assemblages produced during partial melting. Assemblages in the near minimum melting zone include ilmenite–magnetite-bearing leucosomes in migmatized granite. In the granulite zone, leucosomes contain cordierite ± garnet ± orthopyroxene.

growth of zircon at around 1590–1570 Ma. Second, monazite and metamorphic zircon overgrowths in metasediments in the Reynolds Range have yielded ages only in the range 1595–

1570 Ma (Vry *et al.* 1996; Williams *et al.* 1996; Rubatto *et al.* 1999). Therefore, it is reasonable to suggest that the regional greenschist to granulite grade transition seen along the length of the

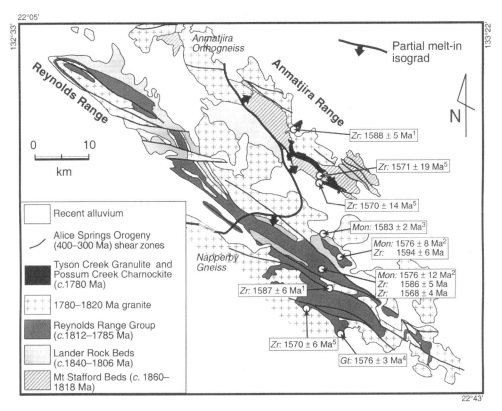

Fig. 9. Distribution of *c*. 1580 Ma U/Pb ages in the Reynolds–Anmatjira region from migmatised metasediments and granites. The heavy arrowed line is the migmatite-in isograd (arrows point in direction of increasing metamorphic grade). The abundance of 1580 Ma ages up-grade of the partial melt-in isograd and the apparent absence of 1580 Ma ages at lower grades implies that the regional increase in metamorphic grade to the southeast reflects 1580 Ma metamorphism. Data: [1] Collins & Williams (1995); [2] Williams *et al.* (1996); [3] Vry *et al.* (1996); [4] Buick *et al.* (1999); [5] Hand *et al.* (2000).

Reynolds Range, and a similar, although less well defined transition along the Anmatjira Range (Clarke *et al.* 1990; Collins & Vernon 1991) is related to *c*. 1580 Ma Chewings Orogeny.

Across the greenschist to granulite transition in the Reynolds Range, metapelitic rocks of the Reynolds Range Group are transformed from phyllites to andalusite ± cordierite-bearing schists and finally to migmatitic granulites (Dirks *et al.* 1991; Hand & Dirks 1992; Williams *et al.* 1996; Buick *et al.* 1998). In the mid-amphibolite grade part of the Reynolds Range, the andalusite–cordierite-bearing assemblages occur in contact aureoles around *c*. 1785 Ma granite sheets such as the Warimbi Schist, and therefore grew during the Strangways Orogeny. However, the contact blasts are texturally stable in the regional foliation, which can be traced continuously into the 1580 Ma granulites at the southeastern end of the range. The stability of *c*. 1785

contact metamorphic minerals during regional 1580 Ma deformation suggests at intermediate grades, the metamorphic conditions at around *c*. 1580 Ma were essentially identical to those attained during the earlier Strangways Orogeny.

The metamorphic transition in the Reynolds Range during the Chewings Orogeny corresponds to an increase in temperature from 400–500°C in the northwest to *c*. 750–800°C in the southeast (Dirks *et al.* 1991; Vry & Cartwright 1994; Buick *et al.* 1998). In metapelite the metamorphic field gradient is characterized by the following metamorphic zones (Fig. 8a); (1) muscovite–chlorite ± biotite, (2) texturally stable Strangways Orogeny andalusite and cordierite, (3) first appearance of sillimanite, and (4) stable co-existence of cordierite–spinel assemblages. At the highest grades, granulite grade metamorphism occurred at pressures of around 5 kbar (Vry & Cartwright 1994; Buick *et al.* 1998) and produced migmatitic cordierite–quartz–K-feldspar ± plagioclase ± biotite ± sillimanite ± spinel ± garnet-bearing assemblages.

The higher grade regions of the Reynolds Range can be further metamorphically subdivided by the leucosome assemblages that formed during partial melting (Fig. 8b). In the upper amphibolite regions immediately upgrade of the sillimanite-in isograd, volumetrically minor leucosomes are pegmatitic in character, with simple mineralogies that reflect water saturated melting (Buick et al. 1998). At slightly higher grades, the leucosomes contain ilmenite–magnetite intergrowths that formed via the breakdown of biotite (Hand & Dirks 1992). The highest grade

(granulite) leucosomes contain cordierite and/or garnet or orthopyroxene and formed during fluid-absent dehydration reactions that consumed biotite and sillimanite. The partial melting assemblages overprint the gneissic layering in the Reynolds Range (Fig. 10a), suggesting that to some extent, high-temperature metamorphism outlasted pervasive deformation.

In the highest-grade regions of the Anmatjira Range, peak metamorphism occurred at around 750–800°C and 5 kbar (Clarke et al. 1990; Hand et al. 1992). As in the Reynolds Range, the gneissic fabrics have been overprinted by leucosomes that formed during fluid-absent dehydration melting (Fig. 10b), suggesting once again

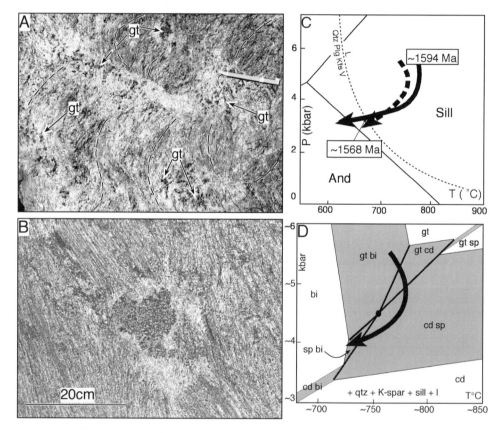

Fig. 10. (a) Garnet-bearing leucosomes overprinting the regional foliation in peraluminous granitic gneiss in the southeastern Reynolds Range. The leucosomes are interpreted to have formed via the generalized fluid-absent melting reaction: biotite + sillimanite + plagioclase + quartz = garnet + melt ± K-feldspar (Hand & Dirks 1992). (b) Cordierite–spinel-bearing migmatitic segregation overprinting the regional foliation in the southeastern Anmatjira Range. The leucosomes formed via fluid-absent partial melting of a sillimanite–biotite–garnet-bearing assemblage. (c) The P–T evolution of the 1580 Ma Chewings Orogeny in the southeastern Reynolds Range. Solid line from Buick et al. (1998), dashed line from Vry & Cartwright (1994). The age constraints are from Williams et al. (1996) and imply that the terrain remained above the solidus for c. 25 Ma. (d) Inferred P–T evolution of the Chewings Orogeny in the southeastern Anmatjira Range (adapted from Hand et al. 1992).

that high-temperature conditions outlasted pervasive deformation. Retrograde reaction textures and partial melting relationships in both the Reynolds and Anmatjira Ranges imply that both regions followed high-temperature clockwise P–T evolutions (Fig. 10c, d), suggesting they shared a common metamorphic history during the Chewings Orogeny.

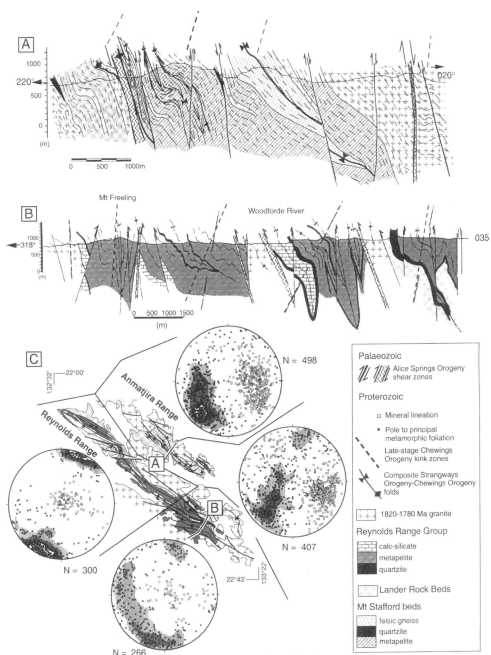

Fig. 11. Structural cross-sections through the southeastern Anmatjira Range (**a**), and the southeastern Reynolds Range (**b**, adapted from Dirks & Wilson 1990). (**c**) Orientation data for the major structural fabric in the Reynolds and Anmatjira Ranges. In both ranges the foliation becomes increasingly reoriented toward the southeast by folds that formed late in the Chewings Orogeny.

The mineral assemblages developed during the Chewings Orogeny are interpreted to define the regional structural fabrics (for an alternative interpretation see Collins *et al.* 1991). In the higher grade parts of the Reynolds Range *c.* 1595 Ma granulite and upper amphibolite-grade assemblages (Williams *et al.* 1996; Buick *et al.* 1998; Rubatto *et al.* 1999) are aligned parallel to the axial surface of the regional upright SE-trending isoclinal folds. The upright folds reflect a total of around 50% shortening (Dirks & Wilson 1990; Fig. 11), and can be traced along the length of the Reynolds Range to the lower grade northwestern Reynolds Range (Dirks & Wilson 1990), where the axial surface fabric overprints *c.* 1785 Ma contact metamorphic minerals that formed during the Strangways Orogeny. Within the *c.* 1785 Ma granites (the Warimbi and Coniston Schists), the intense subvertical foliation may be largely the consequence of the Chewings Orogeny. Although the Reynolds Range Group clearly underwent

deformation during the Strangways Orogeny at around 1785 Ma (see above), the Chewings Orogeny structures have a remarkably simple geometry (Fig. 11), and there is no evidence for regional inversions of stratigraphy that point to large-scale refolding (Stewart 1981; Dirks & Wilson 1990).

The NW–SE trending regional in folds in the Reynolds Range exhibit marked local plunge variations, and many of the macro-scale folds are doubly plunging (Stewart *et al.* 1980; Dirks & Wilson 1990), reflecting significant vertical extension. This is consistent with the presence of a generally steeply plunging mineral stretching lineation within the regional foliation (Fig. 11; Dirks & Wilson 1990). Although anticline hinge lines are commonly sheared out in high-strain zones, there is no consistent asymmetry to the folds and high strain zones show movement senses that change systematically across the folds (Dirks & Wilson 1990), suggesting that deformation was largely coaxial. In contrast in the

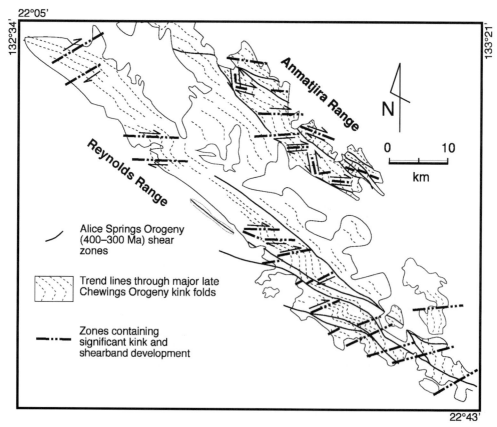

Fig. 12. Distribution of large-scale kink-style folds that formed late in the Chewings Orogeny. Data from the Reynolds Range is from Dirks & Wilson (1990).

Anmatjira Range, mineral lineations have some-what shallower easterly plunges than in the Rey-nolds Range (Fig. 11), and the fabric contains shear sense indicators suggesting sinistral and northeast over southwest flow.

In both the Reynolds and Anmatjira Ranges the regional fabric has been deformed on all scales by conjugate, steeply-dipping shear and crenulation bands, that in geometry resemble conjugate kink bands (Dirks & Wilson 1990; Hand & Dirks 1992). The dominant set in the conjugate pair trends approximately E–W

(Fig. 12) and plunges between 0 and 70° to the east. The subordinate set trends approximately N–S and plunges predominantly north. These folds increase in scale and complexity toward the southeast in both ranges, resulting in rota-tion of folds and foliations into shallow east or west-dipping to locally recumbent domains (Fig. 11). These reoriented domains form cor-ridors that may be up to 8 km wide in which the foliation has a dominantly northerly trend (Stewart 1981; Dirks & Wilson 1990). In the upper amphibolite and granulite regions, these

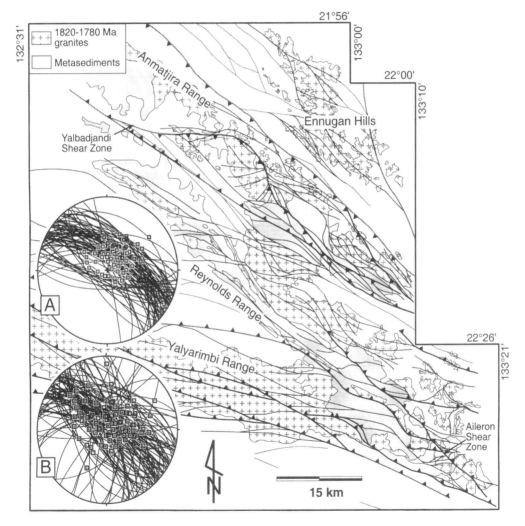

Fig. 13. Distribution of shear zones in the Reynolds–Anmatjira region that either formed or underwent reactivation during the mid-Palaeozoic Alice Springs Orogeny. The majority of the shear zones dip to the northeast. Hanging wall blocks are indicated by the barbs and are only shown for the major zones (100–200+ m wide), which are indicated by heavy lines. Inset (**a**) shows the orientation of the shear fabric and lineation for shear zones >50 m wide. Inset (**b**) shows shear fabric and lineation for minor shear zones (<50 m wide). Shear zones indicated by grey lines are inferred from magnetic data supplied by the Northern Territory Geological Survey.

conjugate structures are associated with extensive water saturated and fluid-absent dehydration melting (Hand & Dirks 1992). Zircons from leucosomes within the crenulation bands give ages around 1570 Ma (Hand et al. 1995; Williams et al. 1996), confirming that the structures formed during the Chewings Orogeny.

Palaeozoic shear zones: the Alice Springs Orogeny (c. 400–300 Ma)

The Proterozoic structures in the Reynolds–Anmatjira Range region are heavily dissected by southeast- and east-trending shear zones (Fig. 13) that form part of an Arunta-wide crustal to lithospheric scale shear zone system (Fig. 1) associated with the 400–300 Ma intracratonic Alice Springs Orogeny. In the Reynolds–Anmatjira Range region, micaceous

greenschist to lower amphibolite-grade shear zone assemblages give ^{40}Ar–^{39}Ar and Rb–Sr ages of around 330–320 Ma (Cartwright et al. 1999) confirming that they were active during the Alice Springs Orogeny. Several of the larger-scale structures such as the Aileron Shear Zone (Fig. 13), extend into the mid crust, where they sole out into low angle detachments (Goleby et al. 1989). The major shear zones are predominantly steeply north-dipping and generally contain steep northeast-plunging lineations (Fig. 13). The overall geometry of the shear zone system in the Reynolds–Anmatjira Ranges has been interpreted by Collins & Teyssier (1989) to have formed in a transpressional setting, with the northeast-plunging lineation representing a component of sinistral movement. Movement along the shear zones in part controls the distribution of metamorphic grade in the region, with juxtaposition of granulites against lower grade rocks along the margins of the southeastern Anmatjira

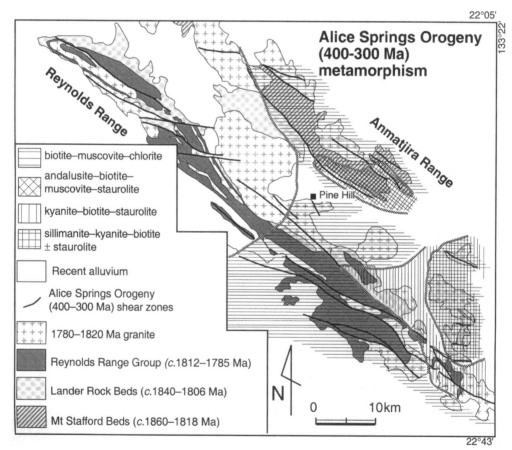

Fig. 14. Metamorphic zones defined by mid-Palaeozoic metapelitic shear zone assemblages in the Reynolds–Anmatjira Range region.

Range (Stewart 1981; Collins & Teyssier 1989), also in the southwestern Reynolds Range (Dirks *et al.* 1991).

Along the length of the Reynolds and Anmatjira Ranges, the metamorphic grade of the shear zones increases to the southeast (Fig. 14) in a pattern similar to that expressed by Chewings Orogeny assemblages (Fig. 8). In the southeastern Reynolds Range, the shear zones contain kyanite, staurolite and sillimanite-bearing assemblages in metapelite (Dirks *et al.* 1991), suggesting maximum P–T conditions around 5–5.5 kbar and 550–600°C. In the southeastern Anmatjira Range, the shear zones contain andalusite and staurolite-bearing assemblages in intermediate X_{Fe} metapelite, implying P–T conditions around 4 kbar and 580°C (e.g. Xu *et al.* 1994). In the northwest and central Reynolds and Anmatjira Ranges, the shear zones are associated with greenschist or lower-grade metamorphism (Dirks *et al.* 1991). The difference in pressure between the shear zones in the southeastern Anmatjira and Reynolds Ranges, suggests the regional field gradients shown in Figure 14 reflect increasing levels of denudation towards the southeast. Accompanying the increase in metamorphic grade is an increase in the number and width of the shear zones, to the extent that in the southeastern Reynolds and Anmatjira Ranges individual zones may be up to 300m wide.

Discussion

Summary of the tectonothermal evolution of the Reynolds–Anmatjira region

The Reynolds–Anmatjira Ranges region has experienced at least four major tectonothermal events after the deposition of the Lander Rock Beds and the Mt Stafford Beds at or before *c.* 1818 Ma. The first of these events (the Stafford Tectonic Event) occurred at around 1820 Ma and was associated with localized, extremely low-pressure granulite facies metamorphism, and more widespread greenschist to mid-amphibolite facies metamorphism spatially linked to *c.* 1820 Ma granites. In the Mt Stafford area in the northwestern Anmatjira Range the metamorphic assemblages overprint upright southeast to south-trending macro-scale folds, and in the northwestern Reynolds Range, the Mt Stafford Tectonic Event produced an upright southeast-trending fabric.

Following the deposition of the Reynolds Range Group sometime between *c.* 1812 and 1785 Ma, the region was deformed and metamorphosed during the Strangways Orogeny at around 1780–1770 Ma. This event was associated with the emplacement of voluminous sheet-like granites that produced low-to mid-amphibolite facies contact metamorphism. In Anmatjira Range regional low-pressure granulite facies metamorphism appears to have occurred synchronously with the emplacement of charnockite mafic rocks, and voluminous megacrystic granite. Deformation during the Strangways Orogeny may have occurred in a NE–SW oriented compressional setting.

Some 200 Ma later, at around 1595–1570 Ma, regional low-pressure high-temperature metamorphism occurred during NE–SW oriented compression (the Chewings Orogeny) that resulted in the development of the upright SE-trending folds. Regional metamorphic grade ranged from greenschist in the northwestern-most Reynolds Range to granulite in the southeastern Reynolds and Anmatjira Range. Retrograde reaction textures and geochronological data indicate the terrain followed a clockwise P–T evolution, with conditions remaining above the solidus for at least 25 Ma.

Some 1.2 Ga later during the 400–300 Ma intracratonic Alice Springs Orogeny, the region was dissected by SE- and E-trending shear zones that form part of an Arunta Inlier wide crustal and lithospheric-scale shear zone system. In the Reynolds Anmatjira Range, the shear zones exhibit an increase in metamorphic grade toward the southeast from sub-greenschist to mid amphibolite.

Structural and metamorphic superposition in the Reynolds–Anmatjira region

Despite the poly-tectonic evolution of the Reynolds–Anmatjira region, the overall structural and metamorphic character of the area is relatively simple. This apparent simplicity is particularly evident in the Reynolds Range Group metasediments, which preserve evidence for three of the four metamorphic cycles to have effected the northern Arunta Inlier. Within the Reynolds Range Group there is an apparently smooth increase in metamorphic grade within a simple structural package that is essentially characterized by upright SE-trending isoclinal folds that have been locally reoriented into flat-lying positions, and then cross-cut by steeply dipping NW-SE trending shear zones. In the underlying Lander Rock Beds, which have experienced four tectonothermal events, a similarly simple structural and metamorphic pattern is also evident. Aside from the presence of a local unconformity between the Reynolds Range Group and the underlying Lander Rock Beds in the north-western Reynolds Range, the structural–metamorphic relations in the region could easily be interpreted in terms of a simple tectonothermal evolution.

In the Reynolds Range, the relative intensity and style of metamorphism during the Mt Stafford Tectonic Event (1820 Ma), Strangways Orogeny (1780 Ma), Chewings Orogeny (1580 Ma) and Alice Springs Orogeny (400–300 Ma), highlights the subtle metamorphic record left by the superposition of four unrelated events. Textural evidence indicates that contact metamorphic andalusite and cordierite blasts developed around the c. 1818 Ma Harverson Granite and the c. 1785 Ma Warimbi Schist were stable during the development of a single regional SE-trending foliation which can be traced continuously into low-pressure granulite grade regions that only contain evidence for 1580 Ma metamorphism (Vry et al. 1996; Williams et al. 1996; Rubatto et al. 1999). The regional increase in metamorphic grade related to the 1580 Ma Chewings Orogeny obliterates the Strangways Orogeny-aged assemblages in the southeastern Reynolds Range, and is virtually identical to the regional pattern of 400–300 Ma metamorphism. The consistency of the metamorphic field gradients, coupled with, (1) apparant NE–SW convergent deformation for each of the structural stages, (2) the growth of low-pressure high-temperature assemblages around granitic sills, and (3) the presence of relatively low pressure retrograde shear zones points superficially to a single, magmatically driven tectonothermal cycle characterized by an anticlockwise P–T loop (e.g. Sandiford & Powell 1991; Sandiford et al. 1991). However, the geochronological constraints indicate that the metamorphic field gradients are in part composite features that formed during the superposition of separate events. Without the geochronological framework, much of the terrain reworking would be obscured by the apparent simplicity of the metamorphic and structural geology.

Within the time scales of orogenic cycles, typically tens of millions of years, a number of regional deformational events may produce structures that are distinct both in terms of geometry and, kinematic and metamorphic character. While the duration of individual deformational events is relatively poorly constrained, there is a general agreement that events last in the order of 1–5 Ma (e.g. Pfiffner & Ramsay 1982). One consequence of this view is that individual geometries are commonly thought to reflect a single phase of deformation within a tectonic cycle. However, several studies (e.g. Tobisch & Fiske 1982; Tavarnelli & Holdsworth 1999) have shown that essentially, similar, sub-parallel structures can form during deformational events that are significantly separated in time. Such a case appears to also exist in the Reynolds–Anmatjira region. During each of the four events to have affected the region, the regional-scale structures appear to have approximately the same strike, and in the case of the Lander Rock Beds in the Reynolds Range, shortening during high-temperature low-pressure metamorphism associated with the 1820 Ma Stafford Tectonic Event, 1780 Ma Strangways Orogeny and the 1580 Ma Chewings Orogeny appears to have been approximately co-planar.

Throughout the Arunta Inlier, structures associated with the 1780–1730 Ma Strangways Orogeny reflect approximately NE–SW shortening (Norman & Clarke 1990; Goscombe, 1991; Collins & Shaw 1995; Young et al. 1995; Collins & Sawyer 1996). At around 1600–1580 Ma, deformation in the Chewings Range in the southern Arunta Inlier occurred in response to N–S shortening (Teyssier et al. 1988). In the Reynolds Range region, deformation at c. 1580 Ma appears to have been primarily in response to NE–SW shortening. During the mid-Palaeozoic Alice Springs Orogeny, the Reynolds Range was deformed within an Arunta-wide system of moderately to steeply dipping E–W and NW–SE-trending shear zones were active during NNE–SSW shortening (Collins & Teyssier 1989; Shaw & Black 1991). In many parts of the Arunta Inlier, including the southeastern Reynolds and Anmatjira Ranges, these shear zones truncate Proterozoic structures at high angles (Stewart 1981; Shaw & Langworthy 1984; Shaw et al. 1984; Collins & Teyssier 1989; Dirks & Wilson 1990), indicating that their geometries were not necessarily controlled by the enclosing structural fabrics. The formation of co-planer and near-coplanar structures during unrelated events may reflect the controlling influence of a pervasive lithospheric fabric, or result from fortuitously oriented far field stresses. It is not clear to what extent the orientation of superimposed structures in the Reynolds–Anmatjira region is controlled by the initial lithospheric structure, or reflects prevailing far field stresses at different times in the evolution of the Arunta Inlier. However, the structures that formed at different times during the evolution of the Reynolds–Anmatjira region show geometric similarities to features elsewhere in the Arunta Inlier, suggesting the orientation of structures may have been largely controlled by regionally operating processes.

In the Anmatjira Range, the metamorphic grade of the Strangways Orogeny was apparently higher than in the Reynolds Range and the $c.$ 1770–1780 Ma assemblages and fabrics may not have been obliterated during the Chewings Orogeny, despite granulite grade conditions at around 1580 Ma. As a consequence, the metamorphic character of the Anmatjira Range is very different to that of the Reynolds Range. Hand et al. (1992) suggested that the isobaric cooling paths inferred from metamorphic textures in the Anmatjira Range (e.g. Clarke et al. 1990; Collins & Vernon 1991; Collins et al. 1991), reflected the interference effects caused by the superposition of two unrelated events, and therefore the cooling paths were apparent, rather than real. The contrast in metamorphic complexity between the Reynolds and Anmatjira Ranges highlights the variable record left by the terrain reworking in the Reynolds–Anmatjira region, and demonstrates how the nature of an earlier event may control the expression of later reworking.

Thermal controls on polymetamorphism in the Reynolds–Anmatjira region

There appears little doubt that metamorphism during the $c.$ 1820 Ma Stafford Tectonic Event and the $c.$ 1780 Ma Strangways Orogeny was strongly controlled by magmatic processes. Although the heat source for extremely low-pressure granulite facies metamorphism in the Mt Stafford region has not be unequivocally identified, the extreme T/P conditions ($\geq 750°C$, 3 kbar) and the large lateral temperature gradients appear to require a localized high-temperature heat source (Vernon et al. 1990; Greenfield et al. 1998). Similarly, magmatic heating is indicated by the presence of andalusite and cordierite porphyroblasts immediately surrounding the 1818 ± 8 Ma Harverson Granite.

During the Strangways Orogeny, the terrain was invaded by voluminous sheet-like granites such as the Napperby Gneiss (1778 ± 8 Ma, Collins & Williams 1995) and probably the Anmatjira Orthogneiss. In addition, in the southeastern Anmatjira Range, the Possum Creek Charnockite (1774 ± 6 Ma, Collins & Williams 1995) and the mafic dominated Tyson Creek granulite (1767 ± 14 Ma, Hand et al. 1995) indicate high-temperature magmatism.

At present, there are no age constraints on the time interval over which magmatism took place at $c.$ 1770 Ma. In the Anmatjira Range, the 1767 ± 14 Ma Tyson Creek Granulite contains a solid-state foliation that has been cross cut by the 1774 ± 6 Ma Possum Creek Charnockite.

This indicates that magmatism at around 1770 Ma occurred over a sufficiently long interval for early intrusive phases to accumulate deformation. Other intrusive relationships indicate that the Possum Creek Charnockite was intruded by granites, which were subsequently foliated prior to the emplacement of sills up to 8 km long and 1.5 km thick (Stewart et al. 1980). Thus in the Anmatjira Range at least, there appears to have been ongoing deformation, metamorphism and magmatism at around 1770 Ma. If the inferred metamorphism at $c.$ 1770–1780 Ma is driven by advective processes, it is likely to have been relatively short-lived and spatially restricted to the vicinity of the highest temperature magmatic bodies. This would logically explain why the Reynolds Range, which does not contain high temperature intrusives such as charnockite and mafic rocks, does not appear to have undergone high-T metamorphism at around 1770–1780 Ma.

While the $c.$ 1820 Ma and 1770–1780 Ma metamorphism in the Reynolds–Anmatjira region can be plausibly linked to magmatism, the regional low-pressure high-temperature metamorphism that affected both the southeastern Reynolds and Anmatjira Ranges during the Chewings Orogeny at around 1590–1570 Ma is more problematic. As yet, no $c.$ 1590–1570 Ma igneous bodies have been identified in either the Reynolds or the Anmatjira Range (Collins & Williams 1995; Hand et al. 1995; Vry et al. 1996; Buick et al. 1998). In addition, there is no metamorphic evidence that the terrain underwent rapid denudation from a deep crustal location during the Chewings Orogeny (e.g. Clarke et al. 1990; Clarke & Powell 1991; Dirks et al. 1991; Hand et al. 1992; Buick et al. 1994, 1998; Vry & Cartwright 1994), apparently ruling out fast unroofing as the cause for regional low-pressure granulite facies metamorphism.

Assuming that rapid denudation was not the causative mechanism for high-temperature metamorphism at around 1580 Ma, the absence of syn-metamorphic igneous bodies poses a major problem in applying conventional models to account for the $c.$ 1590–1570 Ma granulite facies metamorphism in the Reynolds–Anmatjira region. Another problem that is difficult to explain is the apparent duration of high-temperature conditions during the Chewings Orogeny in the southeastern Reynolds Range, where Williams et al. (1996) showed that conditions remained above the solidus for 26 ± 7 Ma. A duration of this magnitude is far in excess of that normally associated with advectively driven high-T metamorphism (e.g. Hanson & Barton 1989; De Yoreo et al. 1991), and could only be generated by numerous intrusive events. This makes the absence of syn-metamorphic magmatism even more problematic.

In a series of recent papers Sandiford & Hand (1998), Sandiford *et al.* (1998), McLaren *et al.* (1999) and Hand *et al.* (1999*b*) have shown that low-pressure metamorphism up to granulite grade can occur without advective heat sources in terrains that have abnormally high levels of crustal heat production. In the Reynolds–Anmatjira region, the voluminous 1820–1770 Ma granites contain relatively high levels of crustal heat production, with regional-scale bodies such as the Anmatjira Orthogneiss and the Napperby Gneiss averaging around $9 \mu Wm^{-3}$ (Fig. 15). To put this into perspective, a 4 km thick sheet of Napperby Gneiss generates around 50% more heat flow than the global average for Proterozoic crust in its entirety (Taylor & McLennan 1985). Although the thickness of the large granitic

bodies in the Reynolds Range is not known with any certainty, they must be in the order of several kilometres thick to be preserved in both the hangingwalls and footwalls of major Alice-Springs-aged shear zones.

Preliminary calculations based on the heat production content of the granites in the Reynolds–Anmatjira region (Hand *et al.* 1999*b*), indicate that temperatures up to 750°C at 18 km depth are attainable for a realistic range of thermal parameters without melting a refractory lower crust. Importantly, vertical thermal gradients above the high-heat production granitic layer are up to 40°C per km (see also McLaren *et al.* 1999), indicating that minor differential denudation will produce significant lateral temperature gradients. This could explain the increase in temperatures from around 500–550°C at 3.5 kbar in the northwestern Reynolds Range to *c.* 750°C at 5 kbar in the southeastern part of the range

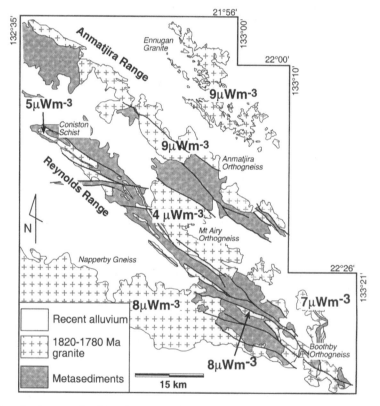

Fig. 15. Heat production rates for the major outcropping granitic bodies in the Reynolds and Anmatjira Range region. The granitic bodies occur in both the hanging walls and footwalls of large-scale Alice Springs Orogeny thrusts (Collins & Teyssier 1989), indicating they are at least several kilometers thick. Gravity and magnetic data suggest that much of the sub-crop is also composed of granitic gneiss, meaning that the average regional heat production is probably $>5 \mu Wm^{-3}$. U–Th–K data from; Hazel *et al.* (1997); Hand *et al.* (1999*c*).

(Dirks *et al.* 1991; Buick *et al.* 1998) during the Chewings Orogeny. Another aspect of the calculations is that the temperatures are strongly dependent of the depth of burial of the heat production, with the consequence that mid-crustal temperatures will remain high while the heat production is deeply buried. This introduces the possibility of relatively long-lived high-temperature low-pressure metamorphic events, and could offer a plausible explanation for the apparent *c.* 25 Ma duration of 650C + mid-crustal temperatures in the southeastern Reynolds Range (Williams *et al.* 1996).

If elevated levels of crustal heat production played a role in regional low-pressure high-temperature metamorphism during the Chewings Orogeny, it highlights an intriguing similarity between isograds that formed during the Chewings Orogeny (Fig. 8), and those associated with mid-Palaeozoic shear zones associated with the Alice Springs Orogeny (Fig. 14). In both cases, metamorphic grade increases toward the southeast with the isograd pattern during the Alice Springs Orogeny suggestive of a similarly distributed, but less intense, heat source as that which generated the *c.* 1590–1570 Ma metamorphism. Although speculative at this stage, it is suggested that the pattern of isograds during the Alice Springs Orogeny was also strongly controlled by the crustal heat production in the terrain. Between 1580 and 350 Ma, the crustal heat production in the Reynolds–Anmatjira Ranges would have declined by around 15%. However, by global standards the heat production rates are still extremely high for regional-scale bodies, and are likely to have made a major contribution to the mid-Palaeozoic thermal regime.

The possibility that regional low-pressure high-temperature metamorphism at around 1590–1570 Ma, and medium-grade metamorphism during the mid-Palaeozoic Alice Springs Orogeny at around 400–300 Ma was strongly controlled by heat production contained in 1820–1780 Ma granites points to a long-term interdependency between the separate events which influences the style of terrain reworking. During the 1820 Ma Stafford Tectonic Event and the 1780–1770 Ma Strangways Orogeny, the emplacement of voluminous granites in the mid-crust radically changed the distribution of heat producing elements in the crust. The immediate thermal consequences of the magmatism were to produce relatively localized, advectively driven metamorphism at *c.* 1820 Ma and 1780–1770 Ma. However, the magmatism also enriched the mid to upper crust in heat production, and unless deep levels of denudation strip the heat production away, or extreme crustal extension attenuates the granitic layer, the heat production concentration will remain relatively intact in the mid to upper crust. Burial of this granitic layer during the lead up to the Chewings Orogeny may have resulted in conductively driven high-temperature, mid-crustal metamorphism (Sandiford & Hand 1998; Sandiford *et al.* 1998; McLaren *et al.* 1999; Hand *et al.* 1999b) at around 1590–1570 Ma. Given the preservation of *c.* 3–5 kbar Proterozoic assemblages in the Reynolds–Anmatjira region, deep levels of denudation did not occur during, or following the Chewings Orogeny. In this case, a significant component of the high heat production in the terrain would have remained, with the possibility that the character of later events such as the Alice Springs Orogeny-aged metamorphism would also have been strongly shaped by the heat production contained in the 1820–1770 Ma granites. Viewed from this perspective, magmatism during the Stafford Tectonic Event and the Strangways Orogeny exerted a major control on the style and distribution of subsequent tectonothermal events in the Reynolds–Anmatjira region. Given that the terrain still contains significantly elevated levels of crustal heat production, the possibility obviously exists that future episodes of terrain reworking will show a similar thermal pattern to the events that have already affected the terrain. Conceivably, reworking in intracratonic settings such as the Reynolds–Anmatjira Ranges, which undergo limited episodes of denudation during a number of events, is more sensitively influenced by the past tectonic activity compared to continental margins which often undergo dramatic modification during tectonic activity.

Discussions with Julie Vry, Bill Collins and Ian Cartwright are gratefully acknowledged. Comprehensive and careful reviews by Tim Wawrzyniec and an anonymous reviewer were greatly appreciated. MH and ISB acknowledge Australian Research Council Australian Research and Senior Research Fellowships respectively. This research was supported Australian Research Council small grants to MH and ISB.

References

BARTON, JR, J. M., HOLZER, L., KAMBER, B., DOIG, R., KRAMERS, J. D. & NYFELER, D. 1994. Discrete metamorphic events in the Limpopo Belt, southern Africa: implications for the application of P–T paths in complex metamorphic terranes. *Geology*, **22**, 1035–1038.

BLACK, L. P. & SHAW, R. D. 1992. U/Pb zircon chronolgy of prograde Proterozoic events in the central and southern provinces of Arunta Block, central Australia. *Australian Journal of Earth Sciences*, **39**, 153–171.

BLACK, L. P. & SHAW, R. D. 1995. An assessment, based on U/Pb zircon data, of Rb–Sr dating in the Arunta Inlier, central Australia. *Precambrian Research*, **71**, 3–16.

BLACK, L. P., SHAW, R. D. & STEWART, A. J. 1983. Rb–Sr geochronology of Proterozoic events in the Arunta Inlier, central Australia. *BMR Journal of Geology and Geophysics*, **8**, 129–138.

BUICK, I. S. & CARTWRIGHT, I. 1994. The significance of early scapolite in greenschist facies marbles from the Reynolds Range Group, central Australia. *Journal of Geological Society of London*, **151**, 803–812.

BUICK, I. S., CARTWRIGHT, I. & HARLEY, S. L. 1998. The retrograde P–T–t path for low-pressure granulites from the Reynolds Range, central Australia: petrological constraints and implications for low-P/high-T metamorphism. *Journal of Metamorphic Geology*, **16**, 511–529.

BUICK, I. S., FREI, R. & CARTWRIGHT, I. 1999. The timing of high-temperature retrogression in the Reynolds range, central Australia: constraints from single mineral Pb–Pb dating. *Contributions to Mineralogy and Petrology*, **135**, 244–254.

BUICK, I. S., CARTWRIGHT, I., HAND, M. & POWELL, R. 1994. Evidence for pre-regional metamorphic fluid infiltration of the lower Calcsilicate Unit, Reynolds Range Group (central Australia). *Journal of Metamorphic Geology*, **12**, 789–810.

CARTWRIGHT, I., BUICK, I. S., FOSTER, D. A. & LAMBERT, D. D. 1999. Alice Springs age shear zones from the southeastern Reynolds Range, central Australia. *Australian Journal of Earth Sciences*, **46**, 355–363.

CLARKE, G. L. & POWELL, R. 1991. Proterozoic granulite facies metamorphism in the southeastern Reynolds Range, central Australia: Geological context, P–T path and overprinting relationships. *Journal of Metamorphic Geology*, **9**, 267–281.

CLARKE, G. L., COLLINS, W. J. & VERNON, R. H. 1990. Successive overprinting granulite facies metamorphic events in the Anmatjira Range, central Australia. *Journal of Metamorphic Geology*, **8**, 65–88.

COLLINS, W. J. 2000. *Granite magma segregation and transfer during compressional deformation in the deep crust?* Proterozoic Arunta Inlier, Central Australia Conference Excursion Guide, 15th Australian Geological Congress.

COLLINS, W. J. & SAWYER, E. 1996. Pervasive granitoid magma transfer through the lower-middle crust during non-coaxial compressional deformation. *Journal of Metamorphic Geology*, **14**, 565–579.

COLLINS, W. J. & SHAW, R. D. 1995. Geochronological constraints on orogenic events in the Arunta Inlier, a review. *Precambrian Research*, **71**, 315–346.

COLLINS, W. J. & TEYSSIER, C. 1989. Crustal scale ductile fault systems in the Arunta Inlier, central Australia. *Tectonophysics*, **158**, 49–66.

COLLINS, W. J. & VERNON, R. H. 1991. Orogeny associated with anticlockwise P–T–t paths: Evidence from low-P, high-T metamorphic terranes in the Arunta Inlier, central Australia. *Geology*, **19**, 835–838.

COLLINS, W. J. & VERNON, R. H. 1993. How well established is isobaric cooling in Proterozoic orogenic belts? An example from the Arunta Inlier, central Australia – Comment. *Geology*, **21**, 953–954.

COLLINS, W. J. & WILLIAMS, I. S. 1995. SHRIMP ionprobe dating of short-lived Proterozoic tectonic cycles in the northern Arunta Inlier, central Australia. *Precambrian Research*, **71**, 69–90.

COLLINS, W. J., VERNON, R. H. & CLARKE, G. L. 1991. Discrete Proterozoic structural terranes associated with low-P, high-T metamorphism, Anmatjira Range, Arunta Inlier, central Australia: tectonic implications. *Journal of Structural Geology*, **13**, 1157–1171.

D'LEMOS, R. S., SCHOFIELD, D. I., HOLDSWORTH, R. E. & KING, T. R. 1997. Deep crustal and local rheological controls on the siting and reactivation of fault and shear zones, northeastern Newfoundland. *Journal of the Geological Society, London*, **154**, 117–121.

DE YOREO, J. J., LUX, D. R. & GUIDOTTI, C. V. 1991. Thermal modelling in low-pressure/high-temperature metamorphic belts. *Tectonophysics*, **188**, 209–238.

DIRKS, P. H. G. M. 1990. Inter-tidal and sub-tidal sedimentation during an Early Proterozoic marine transgression, Reynolds Range Group, Arunta Block, central Australia. *Australian Journal of Earth Sciences*, **37**, 409–422.

DIRKS, P. H. G. M. & WILSON, C. J. L. 1990. The geological evolution of the Reynolds Range, Central Australia: Evidence for three distinct structural/metamorphic cycles. *Journal of Structural Geology*, **12**, 651–665.

DIRKS, P. H. G. M., HAND, M. & POWELL, R. 1991. The P–T-deformation path for a mid-Proterozoic, low-pressure terrain: the Reynolds Range, central Australia. *Journal of Metamorphic Geology*, **9**, 641–661.

GOLEBY, B. R., SHAW, R. D., WRIGHT, C., KENNETT, B. L. N. & LAMBECK, K. 1989. Geophysical evidence for 'thick-skinned' crustal deformation in central Australia. *Nature*, **337**, 6205, 325–330.

GOSCOMBE, B. 1991. Intense non-coaxial shear and the development of mega-scale sheath folds in the Arunta Block, central Australia. *Journal of Structural Geology*, **13**, 299–318.

GOSCOMBE, B. 1992. High-grade reworking of central Australia granulites, part 2: metamorphic evolution. *Journal of Petrology*, **33**, 917–962.

GREENFIELD, J. E., CLARKE, G. L. & WHITE, R. W. 1998. A sequence of partial melting reactions at Mt Stafford, central Australia. *Journal of Metamorphic Geology*, **16**, 363–378.

HAND, M. & DIRKS, P. H. G. M. 1992. The influence of deformation on the formation of axial planar leucosomes and the segregation of small melt bodies within the migmatitic Napperby Gneiss, Central Australia. *Journal of Structural Geology*, **14**, 591–604.

HAND, M., FANNING, C. M., SANDIFORD, M. 1995. Low-P High metamorphism and the role of high-heat producing granites in the northern Arunta Inlier. *Australian Geological Society, Abstracts*, **40**, 60–61.

HAND, M., SANDIFORD, M. & WYBORN, L. 1999*b*. Some thermal consequences of high heat production in the Australian Proterozoic. *Australian Geological Survey Organisation Newsletter*, **30**, 20–22.

HAND, M. & DIRKS, P. H. G. M., POWELL, R. & BUICK, I. S. 1992. How well established is isobaric cooling in Proterozoic terranes: an example from central Australia. *Geology*, **20**, 649–652.

HAND, M. & DIRKS, P. H. G. M., POWELL, R. & BUICK, I. S. 1993. Reply to comment on How well established is isobaric cooling in Proterozoic terranes: an example from central Australia. *Geology*, **21**, 893.

HAND, M., MAWBY, J., KINNY, P. & FODEN, J. 1999*a*. SHRIMP evidence for multiple high-T Palaeozoic events in the Arunta Inlier, central Australia. *Journal of the Geological Society, London*, **156**, 715–730.

HAND, M., SLATER, K., MCLAREN, S. & SANDIFORD, M. 1999*c*. Heat production rates in the Australian Proterozoic. *Australian Geological Society, Abstracts*, **54**, 35.

HANSON, R. B. & BARTON, M. D. 1989. Thermal development of low-pressure metamorphic belts: results from two-dimensional numerical models. *Journal of Geophysical Research*, **94**, 10 363–10 377.

HAZEL, M., BUDD, A. R., KILGOUR, B. & WYBORN, L. A. I. 1997. *Rockchem database Release 3*. Australian Geological Survey Organisation (AGSO) Record, **1997/60**.

MAHAR, E. M., BAKER, J. M., POWELL, R., HOLLAND, T. J. B. & HOWELL, N. 1997. The effect of Mn on mineral stability in metapelites. *Journal of Metamorphic Geology*, **15**, 223–238.

MAWBY, J., HAND, M. & FODEN, J. 1999. Sm/Nd evidence for high-grade Ordovician metamorphism in the Arunta Block, central Australia. *Journal of Metamorphic Geology*, **17**, 653–668.

MCLAREN, S., SANDIFORD, M. & HAND, M. 1999. High radiogenic heat-producing granites and metamorphism – An example from the western Mount Isa inlier, Australia. *Geology*, **27**, 679–682.

MÖLLER, A., ARMSTRONG, R. A., BALLÈVRE, M., HENSEN, B. J. & MEZGER, K. 1999. Crustal growth, metamorphism and deformation in the Strangways Metamorphic Complex: a summary of recent U/Pb and Sm/Nd geochronology. *Australian Geological Society, Abstracts*. **54**, 69.

MYERS, J. S., SHAW, R. D. & TYLER, I. M. 1996. The tectonic evolution of Proterozoic Australia. *Tectonics*, **15**, 1431–1446.

NORMAN, A. R. & CLARKE, G. L. 1990. A barometric response to late compression in the Strangways Metamorphic Complex, Arunta Block, Central Australia. *Journal of Structural Geology*, **12**, 667–684.

PFIFFNER, O. A. & RAMSAY, J. G. 1982. Constraints on geological strain rates: Arguments from finite strain states of naturally deformed rocks. *Journal of Geophysical Research*, **87**, 311–321.

RUBATTO, D., WILLIAMS, I. S. & BUICK, I. S. 1999. Zircon and monazite record of Proterozoic metamorphism in the Reynolds Range, central Australia. *Australian Geological Society, Abstracts*, **54**, 90.

SANDIFORD, M. & HAND, M. 1998. Australian Proterozoic high-temperature metamorphism in the conductive limit. *In*: TRELOAR, P. & O'BRIEN, P. (eds) *What Controls Metamorphism*. Geological Society, London, Special Publications, **138**, 103–114.

SANDIFORD, M. & POWELL, R. 1991. Some remarks on high-temperature-low-pressure metamorphism in convergent orogens. *Journal of Metamorphic Geology*, **9**, 333–340.

SANDIFORD, M., HAND, M. & MCLAREN, S. 1998. High geothermal gradient metamorphism during thermal subsidence. *Earth and Planetary Science Letters*, **163**, 149–165.

SANDIFORD, M., HAND, M. & MCLAREN, S. 2000. Tectonic feedback, intraplate orogeny and the geochemical structure of the crust: a central Australian perspective. *In*: MILLER, J. A., HOLDSWORTH, R. E., BUICK, I. S. & HAND, M. (eds) *Continental Reactivation and Reworking*. Geological Society, London, Special Publications, **184**, 195–218.

SANDIFORD, M., MARTIN, N., ZHOU, S. & FRASER, G. 1991. Mechanical consequences of granite emplacement during high-T, low-P metamorphism and the origin of 'anticlockwise' P–T paths. *Earth and Planetary Science Letters*, **107**, 164–172.

SHAW, R. D. & BLACK, L. P. 1991. The history and tectonic implications of the Redbank Thrust Zone, central Australia, based on structural, metamorphic and Rb–Sr isotopic evidence. *Australian Journal of Earth Sciences*, **38**, 307–332.

SHAW, R. D. & LANGWORTHY, A. P. 1984. *Strangways Range Region, Northern Territory, 1:100 000 geological map series*. Australian Bureau of Mineral Resources, Geology and Geophysics, Canberra, Australia.

SHAW, R. D., STEWART, A. J. & BLACK, L. P. 1984. The Arunta Inlier, a complex ensialic mobile belt in central Australia, Part 2 Tectonic history. *Australian Journal of Earth Sciences*, **31**, 457–484.

STEWART, A. J. 1981. *Reynolds Range region. Northern Territory, 1:100 000 geological map series*. Australian Bureau of Mineral Resources, Geology and Geophysics, Canberra, Australia.

STEWART, A. J., OFFE, L. A., GLIKSON, A. J., WARREN, R. G. & BLACK, L. P. 1980. *Geology of the northern Arunta Block, Northern Territory*. Australian Bureau of Mineral Resources, Geology and Geophysics Record, **1980/83**.

TAVARNELLI, E. & HOLDSWORTH, R. E. 1999. How long do structures take to form in transpression zones? A cautionary tale from California. *Geology*, **27**, 1063–1066.

TAYLOR, S. R. & MCLENNAN, S. M. 1985. *The Continental Crust: Its Composition and Evolution*. Blackwell, London.

TEYSSIER, C., AMRI, C. & HOBBS, B. E. 1988. South Arunta Block: the internal zones of a Proterozoic overthrust in central Australia. *Precambrian Research*, **40/41**, 157–173.

TOBISCH, O. T. & FISKE, R. S. 1982. Repeated parallel deformation in part of the eastern Sierra Nevada, California and its implications for dating structural events. *Journal of Structural Geology*, **4**, 177–195.

VERNON, R. H. 1996. Problems with inferring P–T–t paths in low-P granulite facies rocks. *Journal of Metamorphic Geology*, **14**, 143–153.

VERNON, R. H., CLARKE, G. L. & COLLINS, W. J. 1990. Mid-crustal granulite facies metamorphism: low-pressure metamorphism and melting, Mount Stafford, central Australia. *In*: ASHWORTH, J. R. & BROWN, M. (eds) *High temperature Metamorphism and Crustal Anatexis*. Special Publication of the Mineralogical Society, **2**, 272–319.

VRY, J. K. & CARTWRIGHT, I. 1994. Sapphirine-kornerupine rocks from the Reynolds Range, central Australia: constraints on the uplift history of a Proterozoic low-pressure terrain. *Contributions to Mineralogy and Petrology*, **116**, 78–91.

VRY, J. K. , COMPSTON, W. & CARTWRIGHT, I. 1996. SHRIMP II dating of zircons and monazites: reassessing the timing of high-grade metamorphism and fluid flow in the Reynolds Range, northern Arunta Block, Australia. *Journal of Metamorphic Geology*, **14**, 566–587.

WILLIAMS, I. S., BUICK, I. S. & CARTWRIGHT, I. 1996. An extended episode of early Mesoproterozoic metamorphic fluid flow in the Reynolds Range, central Australia. *Journal of Metamorphic Geology*, **14**, 29–47.

XU, G., WILL, T. M. & POWELL, R. 1994. A calculated petrogenetic grid for the system $K_2O–FeO–MgO–Al_2O_3–SiO_2–H_2O$, with particular reference to contact-metamorphosed pelites. *Journal of Metamorphic Geology*, **12**, 99–119.

YOUNG, D. N., FANNING, C. M., SHAW, R. D., EDGOOSE, C. J., BLAKE, D. H., PAGE, R. W. & CAMACHO, A. 1995. U/Pb zircon dating of tectonomagmatic events in the northern Arunta Inlier, central Australia. *Precambrian Research*, **71**, 45–68.

High-grade reworking of Proterozoic granulites during Ordovician intraplate transpression, eastern Arunta Inlier, central Australia

IAN SCRIMGEOUR[1,2] & JOHANN G. RAITH[1]

[1] Institute of Geological Sciences, University of Leoben, Peter-Tunner Straße 5, A-8700 Leoben, Austria

[2] Northern Territory Geological Survey, P.O. Box 2655, Alice Springs, N.T., 0871, Australia
(e-mail: ian.scrimgeour@nt.gov.au)

Abstract: In the Huckitta region of the eastern Arunta Inlier, central Australia, two terrains with distinct metamorphic histories are separated by a zone of sinistral strike-slip mylonitic deformation and reworking, the Entire Point Shear Zone (EPSZ). To the south of the EPSZ, in the Harts Range Group, Ordovician (c. 470 Ma) intraplate granulite facies metamorphism (c. 800°C, 8–10 kbar) was followed by decompression to c. 7 kbar. In contrast, the Kanandra Granulite, to the north of the EPSZ, is characterized by Palaeoproterozoic high-grade metamorphism at 770–850°C and 5–7 kbar, followed by inferred near-isobaric cooling. Juxtaposition of these terrains along the EPSZ occurred at upper amphibolite facies conditions (700°C, 7 kbar), and resulted in extensive reworking of the Kanandra Granulite. Monazite growth within EPSZ mylonites is dated at 445 ± 5 Ma, whilst a garnet amphibolite gives a Sm–Nd isochron age of 434 ± 6. The timing of this deformation is broadly coincident with the inferred onset of south-vergent compressional deformation in the Harts Range region to the south. This suggests that juxtaposition of the Ordovician granulite terrain with the surrounding Proterozoic terrains occurred during intraplate sinistral transpression in the late Ordovician. Further reworking of the Kanandra Granulite occurred at mid-amphibolite to greenschist facies conditions, during north-vergent mylonitic deformation that exhumed the Ordovician high-grade terrain during the 400–300 Ma Alice Springs Orogeny. Although this zone of Palaeozoic reworking is <5 km wide, it forms the northern margin of Palaeozoic high-grade intraplate deformation and represents a major tectonic boundary in central Australia.

Deciphering the metamorphic and structural evolution of a reworked or polymetamorphic high-grade terrain is a challenge that requires integrated structural, petrological and chronological studies (e.g. Hand *et al.* 1994; Vernon 1996; Bartlett *et al.* 1998). A good example of a multiply reworked high-grade terrain is the Arunta Inlier in central Australia, which has undergone a complex stratigraphical, structural and metamorphic history extending from the Palaeoproterozoic to the Palaeozoic (Shaw *et al.* 1984*a*; Collins & Shaw 1995). This tectonic evolution can be broadly divided into a period of multiple high-grade events during the Palaeo- to Mesoproterozoic (c. 1870–1560 Ma; e.g. Black & Shaw 1992; Collins & Shaw 1995; Vry *et al.* 1996) and a period of intraplate tectonism during the Palaeozoic (500–300 Ma), which is largely concentrated in the southeastern Arunta Inlier (e.g. Dunlap & Teyssier 1995; Hand *et al.* 1999*a*). Due to the polymetamorphic nature of this terrain, recognition and identification of the separate overprinting events has proved difficult, and has been

the source of considerable debate (e.g. Hand *et al.* 1992; Collins & Vernon 1993; Vernon 1996), although recent studies accompanied by precise geochronology have led to an improved understanding of the expression and distribution of the various Proterozoic thermal events (Vry *et al.* 1996; Williams *et al.* 1996; Möller *et al.* 1999). Until recently, the prevailing model for the evolution of the Arunta Inlier was that all high-grade (upper amphibolite to granulite facies) metamorphism occurred during these Proterozoic events, with subsequent exhumation of high-grade rocks occurring during the 400–300 Ma intraplate Alice Springs Orogeny, associated with mid-amphibolite to greenschist facies metamorphism. However, recent studies in the Harts Range region of the eastern Arunta Inlier (Mawby *et al.* 1998, 1999; Miller *et al.* 1998; Hand *et al.* 1999*a*) have identified a previously unrecognized high-grade event during the Ordovician (480–460 Ma), suggesting that intraplate tectonics in central Australia is more complex than was previously believed, and may have

From: MILLER, J. A., HOLDSWORTH, R. E., BUICK, I. S. & HAND, M. (eds) *Continental Reactivation and Reworking.* Geological Society, London, Special Publications, **184**, 261–287. 1-86239-080-0/01/$15.00 © The Geological Society of London 2001.

contributed to high-grade reworking of Proterozoic granulites.

Palaeozoic intraplate deformation and reworking in central Australia

High-grade reworking of metamorphic terrains is most commonly attributed to the incorporation of previously metamorphosed crust into zones of deformation occurring at or near active plate boundaries, often related to the breakup and collision of continental fragments (e.g. Groenewald et al. 1995; Tucker et al. 1999). In comparison, examples of high-grade reworking which can be shown to have occurred within the interior of a continental plate are relatively rare. An exception is in central Australia, where a well-preserved record of intraplate basin formation and orogenesis extends from c. 900–300 Ma (e.g. Shaw et al. 1991; Sandiford & Hand 1998). This intraplate activity includes two major orogenic events; the 580–520 Ma Petermann Orogeny (Camacho et al. 1997; Scrimgeour & Close 1999) and the 400–300 Ma Alice Springs Orogeny (Shaw et al. 1992; Dunlap & Teyssier 1995; Cartwright et al. 1999). These intraplate orogenic events resulted in thick-skinned reworking of previously metamorphosed Palaeo- to Mesoproterozoic terrains

and the exhumation and exposure of these basement terrains from beneath a broad intracratonic basin (the Centralian Superbasin, Walter et al. 1995). The Petermann Orogeny resulted in garnet granulite to sub-eclogite facies reworking of Proterozoic granulites of the Musgrave Block, with exhumation of rocks from >40 km depth (Camacho et al. 1997; Scrimgeour & Close 1999). In comparison, the Devonian to Carboniferous Alice Springs Orogeny resulted in less intense mid-amphibolite to greenschist facies metamorphism in the Arunta Inlier (Cartwright et al. 1999; Hand et al. 1999a) and, until recently, all upper amphibolite to granulite facies metamorphism in the Arunta Inlier was considered to be Proterozoic in age.

Recent metamorphic and geochronological studies in the eastern Arunta Inlier (Mawby et al. 1998, 1999; Miller et al. 1998; Hand et al. 1999a) have led to a radical re-interpretation of the evolution of the region, with the recognition of a third, previously unrecognized high-grade intraplate event at c. 480–460 Ma in the Harts Range region (Fig. 1). In comparison to the Petermann and Alice Springs Orogenies, the Ordovician high-grade metamorphism in the southeastern Arunta Inlier did not result in exhumation of the basement terrain, but instead occurred during a deepening of the overlying

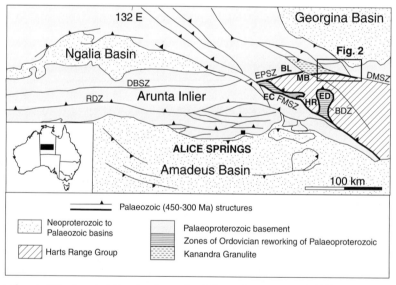

Fig. 1. Map of part of the Arunta Inlier showing major Palaeozoic structures. Bold lines represent the major structures which bound the Harts Range Group (EPSZ, FMSZ, BDZ). Zones of documented Ordovician reworking of Proterozoic granulite are from Mawby et al. (1999) and Hand et al. (1999a,b). Abbreviations: HR, Harts Range; MB, Mallee Bore; EC, Edwards Creek; ED, Entia Dome; BL, Bleechmore Granulite; DBSZ, Desert Bore Shear Zone; RDZ, Redbank Deformed Zone; FMSZ, Florence–Muller Shear Zone; EPSZ, Entire Point Shear Zone; DMSZ, Delny–Mt Sainthill Shear Zone.

intracratonic basin (Shaw *et al.* 1991). This led Hand *et al.* (1999*a*) and Mawby *et al.* (1999) to suggest that the 480–460 Ma high-grade metamorphism and associated shallowly dipping fabrics were associated with extensional deformation. Within metasediments and mafic rocks of the Harts Range Group (Shaw *et al.* 1984*a*), Ordovician metamorphism occurred at upper amphibolite to granulite facies (8–12 kb, $\geq$800°C) followed by decompression to *c.* 7 kbar (Miller *et al.* 1997; Mawby *et al.* 1999). The timing of deposition of the protoliths to the Harts Range Group is still unclear, but it preserves no evidence for Palaeoproterozoic metamorphism or magmatism (Mawby *et al.* 1999). The Harts Range Group is structurally underlain by the Palaeoproterozoic Entia Gneiss Complex, which outcrops in the Entia Dome (Fig. 1). The Entia Gneiss Complex was metamorphosed to granulite facies at *c.* 1730 Ma (Cooper *et al.* 1988) and was then strongly reworked at *c.* 700°C and 8–9 kbar during the Palaeozoic (*c.* 480–470 Ma; Mawby *et al.* 1999). Juxtaposition of the Harts Range Group over the Entia Gneiss Complex occurred along shallowly dipping detachments, and is interpreted to have occurred during compressional deformation at *c.* 450 Ma (D3 of Mawby *et al.* 1999).

The southwestern margin of the Ordovician high-grade terrain in the Harts Range region is defined by major south-directed mid- to upper-amphibolite facies mylonites (Florence–Muller Shear Zone, Fig. 1; Ding & James 1985; James & Ding 1988; Hand *et al.* 1999*b*) which separate it from Proterozoic granulites of the Strangways Metamorphic Complex (Stewart *et al.* 1980). These mylonites are interpreted to be coeval with the *c.* 450 Ma D3 event in the Harts Range (Mawby *et al.* 1999). The Strangways Metamorphic Complex was metamorphosed during the Strangways Event between 1780 and 1710 Ma (Lafrance *et al.* 1995; Möller *et al.* 1999). Rocks belonging to the Strangways Metamorphic Complex are reworked in the Florence–Muller Shear Zone along the contact with the Harts Range Group (James & Ding 1988; Mawby *et al.* 1999), but elsewhere they show no evidence of high-grade metamorphism associated with the 480–460 Ma event. Therefore, the Florence–Muller Shear Zone represents a major structural boundary that juxtaposes these terrains. In comparison, the northern margin of the Ordovician high-grade terrain is poorly understood, due to a relative lack of outcrop to the north of the Harts Range. Documented in this paper is the evolution of the Huckitta region, north of the Harts Range, where 3–5 km of complexly reworked Proterozoic granulites separate Ordovician gran-

ulites of the Harts Range Group from a granite-dominated terrain, which underlies unmetamorphosed Neoproterozoic sediments. This zone of reworking represents the northern margin of the Palaeozoic metamorphic overprint in the eastern Arunta Inlier, and is of fundamental importance for understanding the tectonics of intraplate deformation in central Australia.

Regional setting of the Huckitta region

The Huckitta region is situated near the northern margin of the eastern Arunta Inlier (Fig. 1), immediately south of the Neoproterozoic to Palaeozoic Georgina Basin, and can be broadly divided into three main tectonic elements, separated by major E–W trending shear zone systems (Fig. 2). The southernmost of these three elements comprises upper amphibolite to granulite facies metasediments belonging the Irindina Supracrustal Assemblage of the Harts Range Group (Shaw *et al.* 1984*b*). Dominant lithologies include migmatitic metapelite, metabasite, garnet–biotite gneiss and subordinate calc-silicate rock, marble and quartzite. The lithological associations and metamorphic grade are very similar to rocks of the Irindina Supracrustal Assemblage elsewhere in the eastern Arunta Inlier, which underwent peak metamorphism (>800°C, 8–12 kbar) at 480–460 Ma (e.g. Mallee Bore, Miller *et al.* 1997, 1998; Harts Range, Hand *et al.* 1999*a*; Mawby *et al.* 1999).

To the north of the Harts Range Group is the Kanandra Granulite, which belongs to the Palaeoproterozoic Strangways Metamorphic Complex (Shaw & Warren 1975). It comprises felsic and mafic granulites with garnet-bearing pelitic and semi-pelitic migmatites and rare calc-silicate rock, intruded by deformed granites. Limited studies suggest that it has undergone medium-pressure granulite facies metamorphism (Warren *et al.* 1987; Warren & Hensen 1989), and has a Rb–Sr whole rock age of 1790 ± 35 Ma (Black *et al.* 1983). It forms part of a *c.* 170 km long E–W trending belt of intermittently outcropping pelitic and mafic granulites including the Bleechmore Granulite to the west (Fig. 1; Shaw & Warren 1975). The Huckitta region occurs near the eastern end of this belt of granulites, where the Kanandra Granulite is strongly reworked in amphibolite to greenschist facies mylonite zones.

The third major geological element in the Huckitta region, located to the north of the Kanandra Granulite, is the Jinka province, which contains Palaeoproterozoic low-pressure granulite to upper amphibolite facies metasediments

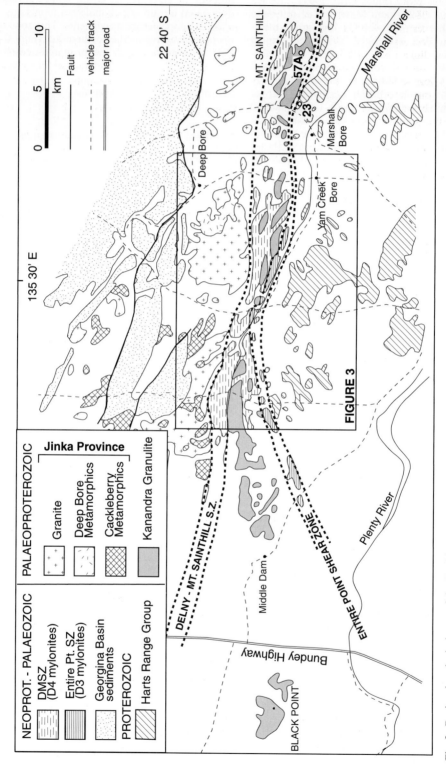

Fig. 2. Regional geological map of the Huckitta region of the eastern Arunta Inlier, showing the two major shear zone systems (Entire Point Shear Zone and Delny-Mt Sainthill Shear Zone). The location is given in Figure 1. Numbered localities refer to samples mentioned in the text, which occur outside of the area shown in Figure 3.

(Deep Bore and Cackleberry Metamorphics; Shaw *et al.* 1984*b*), with extensive syn- to post-tectonic granites that intruded at 1780–1710 Ma (Zhao & Bennett 1995). Palaeoproterozoic rocks of the Jinka province are unconformably overlain by unmetamorphosed, gently folded Neoproterozoic sediments of the intracratonic Georgina Basin, which forms part of the Centralian Superbasin (Walter *et al.* 1995).

Two major shear zone systems separate the three tectonic elements in this region: the Entire Point Shear Zone, which separates the Harts Range Group from the Kanandra Granulite, and the Delny–Mt Sainthill Shear Zone, which separates the Kanandra Granulite from the Jinka province (Fig. 2). The Entire Point Shear Zone (EPSZ; formerly the Entire Point Fault; Shaw *et al.* 1984*b*; Collins & Teyssier 1989; Figs 1 and 2), trends E–NE, dipping steeply to the south, and merges with the E–SE striking Delny–Mt Sainthill Shear Zone in the Huckitta region. The metamorphic grade of this zone is largely upper amphibolite facies (Shaw *et al.* 1984*b*; Kelsey 1998), with a largely strike-slip (sinistral) sense of movement (Collins & Teyssier 1989). The timing of this movement on the EPSZ is not well constrained, but postdates peak metamorphism in the Harts Range Group (Shaw *et al.* 1984*b*).

The Delny–Mt Sainthill Shear Zone (DMSZ; formerly the Delny–Mt Sainthill Fault Zone), is a major E–SE striking structure separating the Kanandra Granulite from the Jinka province and Georgina Basin, and is >150 km in length (Warren 1978). A substantial gravity gradient is evident across the DMSZ (Warren 1978), implying that it is a major crustal-scale feature. The DMSZ comprises anastomosing steeply south-dipping mid-amphibolite to greenschist facies mylonite zones (Shaw *et al.* 1984*b*), and is locally up to 3 km wide. The timing of movement on the DMSZ is not well constrained. Warren (1978) considered it to be largely Proterozoic in age with a north-up sense of movement, but more recent studies (e.g. Collins & Teyssier 1989; Kelsey 1998) have concluded that significant movement on the DMSZ occurred associated with the mid-Palaeozoic Alice Springs Orogeny, with a south-up, reverse sense of movement.

Field relations

In the Huckitta region (Fig. 3), the Kanandra Granulite comprises pelitic granulite, mafic granulite, garnet-biotite migmatite, garnet leucogranite and biotite granite, with rare calc-silicates and small ultramafic bodies. Preserved granulite facies assemblages occur within large relics (0.01 to 1.5 km in width) between extensive anastomosing mylonites related to the DMSZ and EPSZ. West of the Huckitta region, in the vicinity of Black Point (Fig. 2) there are more widespread but poorly outcropping granulites that occur between the mylonites of the EPSZ to the south and the DMSZ to the north. The structural and metamorphic evolution of the Kanandra Granulite can be subdivided into four main phases (D1-4/M1-4), on the basis of structural style, metamorphic grade and over-printing relationships. However, each of these phases is more complex than a single fabric-forming event, particularly where the rocks are inferred to have undergone progressive mylonitic deformation. The distribution of lithologies and zones of reworking within the Kanandra Granulite are shown in Figure 3 along with stereonets summarizing the structural data relevant to the various events.

Granulite facies gneisses (D1-2)

The granulite facies gneisses comprise meta-sedimentary migmatites interlayered with mafic granulites, and have been intruded by garnet leucogranites, and post-D1 biotite granites. The migmatitic gneisses can be subdivided into two main groups:

(1) A relatively homogeneous garnet-biotite ± sillimanite migmatite in which leucosomes contain coarse garnets. Sillimanite occurs locally within these gneisses but is typically restricted to biotite-rich selvages around leucosomes.

(2) More compositionally heterogeneous migmatites, dominated by garnet–sillimanite–biotite ± cordierite ± spinel metapelites with garnet-bearing leucosomes, with psammitic layers and rare calc-silicate horizons.

In addition to the migmatites, two-pyroxene mafic granulites with orthopyroxene-bearing leucosomes occur as layers and boudins up to 200 m wide within the migmatites.

Pelitic and semi-pelitic migmatites preserve an S1 fabric, which is defined by parallel leucosomes and locally by coarse oriented biotite and sillimanite or elongate aggregates of garnet, spinel and biotite. An early generation of leucosomes is locally isoclinally folded and truncated by axial planar leucosomes. Leucosomes locally accumulate to form bodies of garnet leucogranite, up to tens of metres in width, which truncate the fabric.

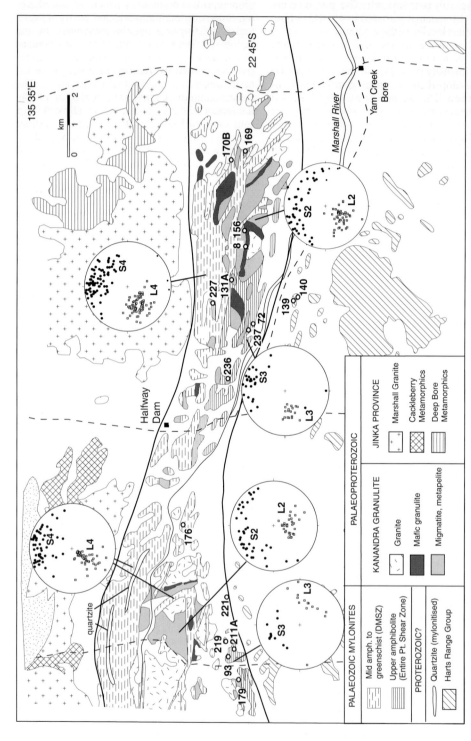

Fig. 3. Map of the primary region of study, with Kanandra Granulite reworked by anastomosing D3 and D4 shear zones. Bold lines represent the inferred northern and southern margins of the Kanandra Granulite. Stereonets summarize the structural data for D2–D4 in the Kanandra Granulite, for structural domains to the east and west of Halfway Dam. Numbered localities refer to sample numbers referred to in the text and tables (prefix ISHU98.).

S0, defined by psammitic layers within metapelites, is typically parallel to S1. The granulites are intruded by extensive post-D1, pre-D2 biotite granites that locally contain garnet.

The S1 fabric is overprinted by, and locally transposed into, a planar high strain fabric of variable intensity (S2), which is characterized by a well-developed south-plunging quartz stretching lineation (L2). In metapelites this lineation is also defined by sillimanite, which forms part of a biotite-sillimanite fabric that envelops M1 garnet. In zones of high D2 strain, most leucosomes within metapelites are strongly deformed and transposed into S2, although rare late melts are axial planar to F2 folds, or truncate S2 at low angles. Mafic granulites have recrystallized in S2 to assemblages that contain two pyroxenes and hornblende. The S2 fabric is folded openly along south plunging fold axes, and generally dips steeply to the southeast to southwest, with lineations plunging towards 170–220°. In some localities, S2 is strongly mylonitic, but can generally be distinguished from subsequent mylonites by granulite facies assemblages in mafic rocks and localized occurrences of small leucosomes truncating the fabric in metapelites.

D3 – Upper amphibolite facies mylonites (Entire Point Shear Zone)

The Kanandra Granulite has been reworked within the EPSZ by mylonites (D3) that formed at upper amphibolite facies conditions. These mylonites extend up to 2 km into the Kanandra Granulite, where they form an anastomosing network of shear zones 0.1–100 m in width. Pelites have recrystallized in D3 mylonites to biotite–sillimanite–garnet bearing assemblages, whilst mafic lithologies have recrystallized to hornblende amphibolites that locally contain garnet. Sillimanite is often very fine-grained and elongated in the lineation, whilst garnet forms larger porphyroclasts, and less abundant, small subhedral grains. Calc-silicate layers within D3 mylonites contain the assemblage diopside–quartz–calcite–scapolite–titanite ± anorthite.

West of Halfway Dam (Fig. 3), the dominant mylonitic fabric (S3b) trends in a west-south-westerly orientation. L3b mineral lineations, defined by sillimanite in metapelites and quartz elongation in felsic rocks, range from near horizontal to plunging moderately to the east or E–SE, on a plane which typically dips steeply towards 150–160°. Kinematic indicators consistently suggest a sinistral sense of strike-slip movement. East of Halfway Dam, the orientation of

S3b changes to become steeply south-southwest dipping, with a moderately west to W–SW plunging lineation and kinematic indicators suggesting sinistral-reverse oblique movement. In this region the northern margin of D3b mylonitization is unclear, as the D3b mylonites are extensively reworked by subsequent lower grade D4 mylonites.

Less commonly, mafic rocks preserve a locally developed, steeply south-dipping fabric with a south to southeast plunging mineral lineation, which contains garnet–hornblende–plagioclase–quartz assemblages. This fabric is best preserved within large boudins in the EPSZ, and is designated S3a. It clearly postdates S1-2 granulite facies fabrics, but is locally strongly reworked by S3b mylonites, which recrystallize the rock to finer-grained garnet-hornblende bearing assemblages. Geothermobarometry also suggests that S3a and S3b formed at different P–T conditions (see below).

D4 – Mid-amphibolite to greenschist facies mylonites (DMSZ)

The Kanandra Granulite and D3 high-grade mylonites have been substantially reworked in lower grade amphibolite to greenschist facies mylonite zones (D4). The metamorphic grade in these mylonite zones decreases from mid amphibolite facies (biotite–muscovite ± sillimanite, staurolite or garnet in pelitic rocks) in the south through to greenschist facies (chlorite–muscovite) in the north. In the northern parts of the DMSZ, relict gneisses in low strain zones have undergone extensive hydrous retrogression to chlorite and muscovite. The mylonitic fabrics typically dip steeply south to southwest, with mineral lineations that plunge variably towards the southwest. These mylonites have abundant kinematic indicators, particularly S–C fabrics, which show a consistent southwest over northeast (reverse) sense of movement. These mylonites form the Delny–Mt Sainthill Shear Zone (DMSZ), and the intensity of D4 deformation increases towards the north, where they form a zone of pervasive mylonitization and retrogression c. 1 km in width. Zones of mylonitized quartzite of an uncertain origin are intercalated with strongly mylonitized and retrogressed granulite and granite within the northern DMSZ west of Halfway Dam (Fig. 3), and form quartz–muscovite schists and phyllites. The northern margin of the DMSZ is defined by a >50 m wide zone of poor outcrop with abundant vein quartz, bounded to the north by granites of the Jinka province.

Harts Range Group

Scattered outcrops of metasediments that occur south of the EPSZ were mapped as the Irindina Supracrustal Assemblage of the Harts Range Group by Shaw et al. (1984b). The following description refers only to Harts Range Group outcrops that occur within 5 km of the Kanandra Granulite in the Huckitta region. For the purpose of this description, fabrics in the Harts Range Group have been designated S1h-3h, to distinguish them from S1-3 in the Kanandra Granulite.

The Harts Range Group in the Huckitta region comprises migmatitic gneisses, dominated by biotite $\pm$ garnet bearing felsic and semi-pelitic gneiss, with less abundant metabasite, garnet–hornblende–biotite gneiss, metapelite and calc-silicate rock. The sequence is dominated by a planar fabric (S2h), which has largely transposed an earlier migmatitic layering (S1h). Boudins of low S2h strain are locally preserved in which metabasites contain coarse garnet- and clinopyr-oxene-bearing leucosomes. Garnet-bearing leucosomes also occur within metapelites, although these are typically transposed into S2h. The S2h fabric typically dips moderately to the northwest, with shallowly southwest plunging mineral elongation lineations (L2h). S2h is locally folded by moderately to shallowly west-plunging tight folds that are inclined to the north and also fold L2h. Mafic lithologies have recrystallized in S2h to hornblende amphibolites with rare garnet, whilst pelitic lithologies have recrystallized to biotite-sillimanite-garnet bearing assemblages.

A steeply south-dipping mylonitic fabric (S3h) is strongly developed in the Harts Range Group within 1 km of its contact with the Kanandra Granulite. Metapelites have a biotite-sillimanite garnet fabric that envelops porphyroclastic garnet, quartz and feldspar, with a shallowly west-plunging sillimanite lineation. A near-continuous exposure from the Harts Range Group into the Kanandra Granulite is exposed 2–3 km south of Mt. Sainthill (Fig. 2). In this region the S3h fabric intensifies to the north into a c. 250 m wide

mylonite comprising the EPSZ, separating the Harts Range Group from the Kanandra Granulite. This fabric continues to the north, where it becomes S3b in the Kanandra Granulite.

Petrology of the Kanandra Granulite

The following petrological description of the Kanandra Granulite refers only to assemblages that contain information pertinent to the pressure–temperature evolution of the terrain. Tables summarizing the petrology of samples of metapelitic Kanandra Granulite examined during this study are available upon request from the first author.

Cordierite-absent pelitic assemblages (D1-2/M1-2)

In most pelitic migmatites in the Kanandra Granulite, the peak metamorphic (M1) assemblage comprises garnet, biotite, sillimanite, ilmenite, quartz, K-feldspar and plagioclase. M1 garnet is typically elongated in S1, and contains inclusions of biotite, sillimanite, quartz and ilmenite, with or without spinel. Within the matrix, S1 is defined by the alignment of coarse sillimanite and biotite. Sillimanite locally contains inclusions of spinel. This fabric generally deflects around garnet, although sillimanite and biotite inclusion trails within garnet locally continue directly into the fabric in the matrix. Within leucosomes, M1 garnet is typically equant and poikilitic, with numerous quartz inclusions.

The M1 assemblage has been overprinted by a biotite-sillimanite S2 fabric that varies in intensity between different samples. In rocks in which S2 is weakly developed, fine-grained M2 biotite and sillimanite occur along feldspar grain boundaries and also overgrow coarser M1 sillimanite and biotite. M1 sillimanite locally contains narrow fractures perpendicular to S2, which are infilled with spinel–biotite–ilmenite $\pm$ corundum (Fig. 4a). These spinel-bearing inter-growths only occur in sillimanite-rich zones and

---

Fig. 4. Photomicrographs from the Kanandra Granulite in the Huckitta region. (**a**) Sillimanite-rich zone within metapelite (ISHU98.44B), in which M1 sillimanite is fractured perpendicular to the L2 lineation and fractures are infilled with very fine-grained spinel, biotite and ilmenite. Width of field of view is 0.7 mm. (**b**) Cordierite-bearing granulite facies metapelite (sample ISHU98.131A), showing an M1 garnet–cordierite–sillimanite assemblage overprinted by fine grained intergrown M2 biotite and sillimanite. Euhedral M2 garnet (Grt2) overgrows M1 garnet (Grt1) and contains numerous small sillimanite inclusions (not visible). Width of field of view is 2mm. (**c**) M3b pelitic mylonite (ISHU98.227) showing euhedral garnet in the mylonitic fabric, which is defined by biotite and fine grained sillimanite. Colourless minerals are quartz, K-feldspar and plagioclase. Width of field of view is 2 mm. (**d**) Recrystallized intermediate to mafic rock from the EPSZ (ISHU98.211A) showing a relatively coarse grained garnet–hornblende–plagioclase–quartz M3a assemblage recrystallized in to a finer grained garnet–hornblende–plagioclase–quartz assemblage in the S3b mylonitic fabric. Width of field of view is 4 mm.

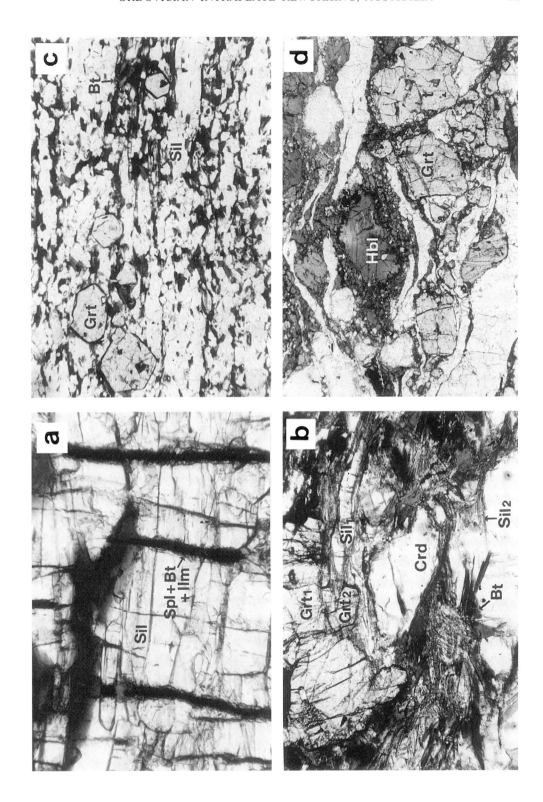

are never in contact with quartz. Fractures in M1 garnet contain M2 biotite with minor sillimanite ± spinel. Where S2 deformation is more intense, a biotite-sillimanite ± ilmenite fabric envelops and overgrows M1 garnet, sillimanite and biotite, and M1 garnet is significantly fractured and corroded and replaced by M2 biotite ± sillimanite. Spinel is typically absent from more highly strained rocks. Within biotite-rich rocks, smaller subhedral garnets with fine-grained sillimanite inclusions occur in the S2 fabric and locally overgrow M1 garnet.

Cordierite-bearing pelitic assemblages (D1-2/M1-2)

The peak (M1) assemblage in cordierite-bearing rocks contains garnet porphyroblasts with inclusions of sillimanite, biotite, quartz, rutile and ilmenite, and coarse sillimanite which locally contains inclusions of hercynitic spinel, ilmenite and rutile, and which variably defines the S1 fabric. In addition, the M1 assemblage also contains coarse granoblastic cordierite, quartz, K-feldspar and plagioclase. The M1 garnet–cordierite–sillimanite assemblage is overprinted by a relatively fine-grained M2 biotite–sillimanite assemblage associated with the mylonitic S2 fabric. M1 garnet is locally fractured perpendicular to S2, and these fractures are infilled by biotite. Cordierite, quartz and feldspar have recrystallized at grain boundaries associated with the growth of M2 biotite and sillimanite (Fig. 4b). M1 garnet and sillimanite are enveloped and corroded by fine-grained intergrown biotite and M2 sillimanite, with less abundant ilmenite and rutile. M1 garnet locally has overgrowths of symplectitic garnet-quartz, whilst elsewhere it is overgrown by euhedral garnet with inclusions of fibrous sillimanite (Fig. 4b). This secondary garnet growth locally occurs at contacts between M1 garnet and M2 biotite-sillimanite, whilst small euhedral garnets with fine-grained sillimanite inclusions also occur in S2.

D3/M3 pelitic mylonites

Within metapelites, the mineral assemblage associated with S3b is garnet–biotite–sillimanite–K-feldspar–quartz–plagioclase–ilmenite ± muscovite. Fine-grained biotite and sillimanite define the S3b fabric and L3b lineation. Porphyroclasts of M1 garnet, sillimanite, quartz and feldspar are commonly present, and in general the porphyroclastic garnet is rounded (but not corroded),

whilst M1 sillimanite is typically partially corroded at contacts with biotite. In many D3b mylonites, small euhedral garnets occur in the S3b biotite–sillimanite fabric (Fig. 4c). M3b garnets contain rare inclusions of quartz, biotite and sillimanite. In less common bulk compositions, muscovite also occurs in S3b, particularly in the vicinity of K-feldspar porphyroclasts.

D4/M4 pelitic mylonites

The mineral assemblage in most S4 pelitic mylonites is dominated by biotite–muscovite–quartz–plagioclase–ilmenite, with or without sillimanite, garnet or staurolite. Less commonly, muscovite-absent garnet–biotite or sillimanite–biotite mylonites occur. Garnet and sillimanite are never stable together in S4. Sillimanite-bearing mylonites typically contain coarse porphyroclastic sillimanite as well as finer grained sillimanite, which defines the S4 fabric with biotite ± muscovite. Garnet-bearing mylonites contain small euhedral garnets that occur in the fabric with biotite ± muscovite, and less abundant porphyroclastic garnet and locally preserved corroded porphyroclastic sillimanite. Staurolite-bearing mylonites contain an assemblage of staurolite, muscovite, biotite, quartz, and ilmenite, defining a strong mylonitic fabric with locally well-developed S–C fabrics. Further north within the Delny–Mt Sainthill Shear Zone, D4 pelitic mylonites comprise muscovite–biotite–chlorite–quartz assemblages, and less deformed granulite relics have chlorite–muscovite pseudomorphs after garnet.

Mafic lithologies

The peak (M1) assemblage in mafic granulites comprises granoblastic orthopyroxene, clinopyroxene, hornblende and plagioclase, with minor ilmenite ± biotite ± quartz. Where mafic rocks have recrystallized in the S2 fabric, the assemblage remains the same, but with a higher modal proportion of hornblende relative to pyroxene. In the S3a fabric, mafic rocks have recrystallized to a garnet–hornblende–plagioclase–quartz–ilmenite ± rutile assemblage. In general the texture is granoblastic, although hornblende shows a variable degree of preferred orientation in the S3a fabric. Garnet often has abundant inclusions of hornblende, plagioclase, ilmenite and quartz. The coarse S3a assemblage is often partially recrystallized to a finer-grained garnet–hornblende–plagioclase–ilmenite–quartz assemblage associated with mylonitic S3b deformation (Fig. 4d). However, some S3b mylonites contain garnet-absent hornblende amphibolites. S4 mylonites contain equigranular hornblende–plagioclase–ilmenite $\pm$ quartz assemblages, grading into lower grade epidote-bearing assemblages in the northern DMSZ.

Mineral chemistry

Mineral analyses were performed on an ARL–SEMQ electron microprobe at the University of Leoben (20 kV, 15 nA), using wavelength-dispersive spectrometry. Mineral recalculations and estimation of Fe^{3+} content were made using the unpublished program aX (96v2.2) by T. Holland. Representative analyses are available upon request from the first author.

M1-2 metapelites

Garnet in M1 metapelites is typically almandine-pyrope, with low grossular and low to negligible spessartine contents. Compositional variations consistently occur between cordierite-absent pelites ($Alm_{70-78}Py_{19-25}Grs_{2-4}Sps_{0-2}$) and cordierite-bearing pelites ($Alm_{56-61}Py_{35-37}Grs_{3-5}Sps_{0-1}$). M2 garnets are generally similar in composition to M1 garnets, although where M1 and M2 garnets occur in the same rock, M2 garnets generally have higher almandine and grossular, and lower pyrope contents. *Biotite* has c. 3.0–4.5 wt% TiO_2 in all bulk compositions, with an X_{Fe} ($Fe/(Fe+Mg)$) that varies from 0.45–0.48 for cordierite-absent pelites, to 0.33–0.35 for cordierite-bearing pelites. *Cordierite* has an X_{Fe} of 0.14–0.17, with negligible compositional zoning. *Spinel* in biotite-bearing breakdown reactions after sillimanite has a variable X_{Fe} of 0.72–0.86, with 2.0–2.5 wt% ZnO and negligible Cr. *Plagioclase* is typically andesine or oligoclase (An_{22-35}); M2 plagioclase has a similar compositional range to M1 plagioclase but where both occur in the same sample, M2 plagioclase typically has a slightly higher Na content.

M3-4 pelitic mylonites

Garnets in M3 pelitic mylonites are almandine-rich ($Alm_{64-73}Py_{9-17}Sps_{4-13}Grs_{5-8}$), and show minor zoning, with a rimward increase in Ca and Mn, and a corresponding decrease in Mg. Porphyroclastic garnets show more complex zoning due to the retention of M1-2 compositions in the cores, and were not used for thermobarometry. *Biotite* in M3 mylonites typically has

an X_{Fe} of 0.44–0.55, and contains 2.4–3.0 wt% TiO_2. *Plagioclase* is largely oligoclase (An_{25-30}) and less commonly andesine (An_{30-38}), and shows no consistent compositional zoning.

The one analysed M4 mylonite is relatively Fe-rich, with almandine-rich *garnet* ($Alm_{80-83}Py_{10-12}Sps_{4-5}Grs_{2-3}$) containing no significant zoning other than retrograde Fe–Mg exchange with adjacent biotite. *Biotite* has a X_{Fe} of 0.56–0.59 and 2.1–2.4% TiO_2, whilst *plagioclase* has oligoclase compositions (An_{27-30}).

Mafic lithologies

Minerals recrystallized within the granoblastic M3a assemblage in mafic rocks are chemically distinct from minerals in the later, mylonitic M3b assemblage. *Garnets* in M3a assemblages are almandine-rich with a significant grossular component ($Alm_{61-71}Py_{5-11}Sps_{2-5}Grs_{15-30}$). They are typically unzoned, although locally there is a slight increase in Ca towards the rim. M3b garnets have broadly similar compositions ($Alm_{60-67}Py_{5-9}Sps_{1-6}Grs_{18-25}$), but where M3a and M3b garnets occur in the same rock, M3b garnets consistently have higher Ca and lower Mg contents. *Plagioclase* is typically bytownite (An_{74-91}) in M3a assemblages, whilst in comparison, plagioclase in M3b assemblages is significantly more sodic (An_{41-60}). In rocks that contain both M3a and M3b plagioclase, intermediate compositions exist which are likely to reflect partial re-equilibration of M3a plagioclase during M3b. *Hornblende* typically has tschermakitic to ferro-tschermakitic hornblende compositions, with a X_{Fe} of 0.48–0.69. M3a hornblendes typically have a higher X_{Fe} than M3b hornblendes in the same rock.

P–T conditions of metamorphism

Methods

Pressure–temperature estimates of metamorphism were derived using a combination of conventional exchange thermometers and net transfer equilibria, as well as internally consistent datasets, and are summarized in Table 1. To calculate the pressures and temperatures of peak metamorphism in the Kanandra Granulite, mineral analyses were used from the core of minerals inferred to be in textural equilibrium. For mylonitic assemblages, samples were chosen, wherever possible, in which all minerals including garnet had recrystallized in the mylonitic fabric, in which case core mineral analyses were used.

Table 1. *Summary of conventional P–T estimates and Thermocalc results for selected samples from the Kanandra Granulite*

M1 / M2 samples

	Grt–Bt HS82 T (@6 kbar)	GASP HC85 P (@800°C)	Grt–Bi–Pl–Qtz H90 Mg	Fe	Thermocalc PH88, HP98 avP aH2O=0.35	avT
M1						
57A	696–865 / 782±65; 7	6.0–6.6 / 6.3±0.2; 7	5.5–6.3 / 5.9±0.3; 7	6.2–7.1 / 6.6±0.4; 7	6.0±2.0	789±52
131A	749–824 / 801±35; 4	5.5–6.1 / 5.8±0.2; 4	5.7–6.0 / 5.9±0.2; 4	5.5–6.0 / 5.8±0.2; 4	6.1±0.8	800±62
169	750–839 / 804±33; 7	5.6–6.2 / 5.8±0.2; 7	5.7–5.9 / 5.8±0.1; 7	5.8–6.2 / 5.9±0.2; 7	5.7±1.6	767±49
236	810–866 / 829±32; 3	5.6–6.1 / 5.8±0.2; 4	5.0–5.4 / 5.2±0.2; 3	4.8–5.4 / 5.1±0.4; 3	5.9±1.6	817±49
average (2σ error)†					*5.9±1.5*	*793±53*
M2		P (@750°C)				
131A	740–792 / 762±18; 6	5.7–6.5 / 6.2±0.4; 6	6.0–6.5 / 6.2±0.2; 4	5.9–6.5 / 6.2±0.2; 4	5.4±0.7	753±40
179	692–778 / 736±35; 4	4.9–5.3 / 5.1±0.2; 4	4.7–5.1 / 4.9±0.2; 4	5.0–5.3 / 5.2±0.1; 4	4.9±0.5	760±51
average (2σ error)†						

M3a / M3b samples

	Grt–Hbl GP82 T (@6 kbar)	Grt–Hbl–Pl–Qtz‡ KS90 P — Mg	Fe	Thermocalc PH88, HP98 avP aH2O=0.35	avT
M3a		*P (@650°C)*			
8	625–659 / 632±12; 5	4.5–5.1 / 4.7±0.3; 4	5.6–6.0 / 5.7±0.2; 4	4.6±1.3	661±105
156	619–649 / 641±15; 4	4.1–4.6 / 4.3±0.2; 4	5.3–6.0 / 5.5±0.3; 4	5.4±2.6	632±121
211A	610–645 / 630±19; 3	4.4–5.3 / 4.9±0.5; 3	5.6–7.0 / 6.2±0.7; 3	4.3±1.0	735±102
221				5.3±2.3	672±107
average				*4.9±1.8*	*673±109*
M3b		*P (@700°C)*			
211A	580–701 / 647±54; 4	5.2–7.6 / 6.5±0.9; 4	7.2–8.2 / 7.6±0.4; 4	–	–
219	622–666 / 641±19; 4	5.2–7.9 / 7.0±1.3; 4	6.1–8.6 / 7.8±1.2; 4	–	–

M3b / M4 samples

	Grt–Bt HS82 T (@7 kbar)	Grt–Phe GH82	GASP HC85 P (@700°C)	Grt–Bt–Pl–Qtz H90 Mg	Fe	Grt–Pl–Ms–Bt HC95 Mg	Fe	Grt–Pl–Ms–Qtz H90 Mg	Fe	Grt–Pl–Ms–Sil HC95	Grt–Ms–Bt–Sil HC95	Thermocalc PH88, HP98 avP aH2O=0.35	avT
M3b				*P (@650°C)*				*P (@650°C)*					
72	640–728 / 695±37; 5	672–703 / 685±15; 5	6.6–7.1 / 6.9±0.2; 6	6.6–6.9 / 6.7±0.1; 4	6.6–7.0 / 6.8±0.2; 4	6.6–7.0 / 6.8±0.1; 5		7.1±0.1; 3	7.0–7.1	6.5–7.0 / 6.8±0.2; 5	6.6–7.4 / 7.1±0.4; 5	6.8±1.3	686±43
93A	619–725 / 653±40; 6		6.1–6.6 / 6.4±0.2; 6	5.8–6.4	6.2–6.8							6.3±1.3	666±45
227	674–695 / 681±10; 4		6.4–7.2 / 6.7±0.3; 5	6.3–6.8	6.5–7.2							7.2±1.1	709±45
237	640–763 / 689±42; 7	702–786 / 728±32; 5	6.6–7.6 / 7.2±0.4; 5	6.5–7.5 / 6.9±0.5; 4	6.8–7.3 / 7.1±0.4; 4	6.7–7.4 / 7.1±0.3; 4	6.7–7.4	6.0–7.2 / 6.4±0.2; 3	6.7–7.8 / 6.4±0.5; 5	6.7–8.2 / 7.7±0.6; 5	6.6–6.8 / 6.7±0.1; 4	7.2±1.3	699±30
23*	722–780 / 743±25; 2		5.9–6.9 / 6.4±0.5; 4	5.9–6.3	6.2–6.3							6.3±1.4	698±43
average (2σ error)†			6.6±0.5; 4	6.5±0.1; 3	6.3±0.1; 3							*6.8±1.1*	*692±37*
M4			*P (@5 kbar)*									*aH2O=1.0*	
170B	607–660 / 634±19; 6	608–656 / 636±14; 6	3.7–4.4 / 4.1±0.3; 6	3.7–4.9 / 4.4±0.5; 6		4.2–6.0 / 5.0±0.6; 6		3.3–5.3 / 4.1±0.7; 6	4.4–6.0 / 5.0±0.7; 6			5.1±1.3	653±34

Errors on averaged conventional estimates are standard deviation from the mean, value after the semi-colon is the number of estimates from which the mean is derived.
† 2σ error is calculated by dividing the average 1σ error by the square root of the number of analyses and dividing this number by two. Errors for individual samples are 1σ.
* Denotes sample from Harts Range Group.
‡ Composition of M3a garnets are outside of the range for which the Kohn & Spear (1990) barometer is calibrated.
Abbreviations: HS82, Hodges & Spear (1982); HC85, Hodges & Crowley (1985); H90, Hoisch (1990); GP, Graham & Powell (1982); KS90, Kohn & Spear (1990); PH88, Powell & Holland (1988); HP98, Holland & Powell (1998); GH82, Green & Hellman (1982). Mineral abbreviations after Kretz (1983).

Conventional thermobarometry was performed using the unpublished program Thermobarometry 2.1 (GTB2.1) by Kohn & Spear (1990). For each thermometer and barometer, calculations were made from between three and seven independent groups of minerals from each sample, and these were then statistically averaged and are presented in Table 1. For biotite–sillimanite–garnet ± muscovite bearing metapelites, the Hodges & Spear (1982) calibration of the biotite Fe–Mg exchange thermometer was used in conjunction with the pelite barometers of Hodges & Crowley (1985) due to the internal consistency between these calibrations. Average P–T calculations were made using the program Thermocalc (Powell & Holland 1988; v.2.5), using the method of Powell & Holland (1994) and the internally consistent dataset of Holland & Powell (1998). Mineral end-member activities were calculated from microprobe data using the unpublished program aX (96v2.2) by T. Holland. Water activities were chosen on the basis of the mineral assemblages present, and by comparison with fluid-independent barometers and thermometers (see discussion below). Average P–T results from Thermocalc are shown in Table 1 along with combined averages of the P–T estimates for each metamorphic event, with a calculated 2σ error.

Kanandra Granulite – M1 stage

A sample of cordierite-bearing metapelite (ISHU98.131A) and three samples of garnet–biotite–sillimanite bearing pelites (ISHU98.57A, 168 & 236) were selected for thermobarometric calculations of M1 in the Kanandra Granulite, and the results are summarized in Table 1. Garnet–biotite thermometry on core analyses using the calibration of Hodges & Spear (1982) gives temperatures of 696–866°C with averaged temperature estimates for each sample ranging between 782 and 829°C. Pressure estimates using conventional barometers are typically between 5.2 and 6.6 kbar for a reference temperature of 800°C (Table 1). Average P–T conditions using Thermocalc (Powell & Holland 1988; Holland & Powell 1998) yield relatively consistent results between samples (Table 1) which give a combined average P–T estimate of 793 ± 53°C and 5.9 ± 1.5 kbar (2σ error) for an aH$_2$O of 0.35. Due to the dependence of these estimates on biotite- and cordierite-bearing equilibria, the selected water activity had a significant influence on the results. For example, an aH$_2$O of 0.5 typically resulted in estimates approximately 50°C

and 0.5 kbar higher, whilst an aH$_2$O of 0.1 resulted in unrealistic temperatures of <700°C.

Additional constraints on the peak M1 conditions of the Kanandra Granulite are given by the presence of extensive garnet-bearing melts produced in metapelites during M1, which are interpreted to have resulted from the biotite dehydration reaction:

biotite + quartz + plagioclase + sillimanite

= garnet + K-feldspar + melt

Experimental studies by Le Breton & Thompson (1988) showed that this melting reaction is likely to commence at c. 760–770°C for pressures of 6 kbar, thus giving a lower constraint on peak metamorphic temperatures. The degree of melting and the formation of locally derived bodies of garnet leucogranite suggest that temperatures may have significantly exceeded this minimum value. In summary, the results of thermobarometry suggest that P–T conditions during M1 in the Kanandra Granulite were 770–850°C and 5–7 kbar.

Kanandra Granulite – M2 stage

Thermobarometry relating to the development of the S2 fabric in the Kanandra Granulite was undertaken on two samples that contained the M2 assemblage garnet–biotite–sillimanite–plagioclase–quartz–K-feldspar (ISHU98.131A, 179). Garnet–biotite thermometry using the Hodges & Spear (1982) calibration gives temperatures of 692–792°C. The GASP barometer of Hodges & Crowley (1985) and Grt–Bt–Pl–Qtz barometer of Hoisch (1990) give estimates of 5.7–6. kbar for a reference temperature of 750°C for sample 131A, and give estimates c. 1 kbar lower for 179 (Table 1). Thermocalc estimates for an aH$_2$O of 0.35 suggest P–T conditions of 753 ± 40°C and 5.4 ± 0.7 kbar for 131A, and 760 ± 51°C and 4.9 ± 0.5 kbar for 179 (1σ errors). In summary, temperatures during M2 are likely to have been c. 50°C lower than during M1, at similar or slightly lower pressures.

Harts Range Group (M1h)

Pressure–temperature estimates of peak metamorphism in the Harts Range Group were obtained from a metapelite (sample 140) containing the assemblage garnet (Alm$_{67}$Py$_{19}$Grs$_9$Sps$_6$), biotite (X$_{Fe}$ = 0.50), sillimanite, plagioclase (An$_{46}$), quartz and K-feldspar, and a metabasite (139) containing garnet (Alm$_{53}$Py$_{10}$Grs$_{32}$Sps$_5$), hornblende (X$_{Fe}$ = 0.52), clinopyroxene (X$_{fe}$ = 0.46), plagioclase (An_{40}) and quartz. For sample 140, temperature estimates were 798–832°C at 9 kbar using the garnet–biotite thermometer of Hodges & Spear (1982), whilst the GASP barometer of Hodges & Crowley (1985) gives pressures of 9.2–9.5 kbar (at 800°C). The Grt–Bt–Pl–Qtz barometer of Hoisch (1990) gave pressures of 9.4 kbar and 9.3 kbar for the Mg and Fe end-members, respectively. For sample 139, the Grt–Cpx–Pl–Qtz barometer of Newton & Perkins (1982) gives a pressure of 9.0 kbar, whilst the Grt–Hbl–Pl–Qtz barometer of Kohn & Spear (1990) gives pressures of 9.0 and 11.2 kbar for the Fe and Mg end-members, respectively. Average P–T estimates using Thermocalc give conditions of 772 ± 85°C and 8.4 ± 1.4 kbar for sample 139, and 793 ± 57 and 8.5 ± 1.4 kbar for sample 140 (1σ, $aH_2O = 0.8$). These estimates are consistent with experimental studies on garnet-producing melting reactions for pelitic and mafic bulk compositions similar to those in the Harts Range Group (e.g. Le Breton & Thompson 1988; Wyllie & Wolf 1993; Wolf & Wyllie 1994), and also with P–T estimates from elsewhere in the Harts Range Group in the Harts Range, 60–120 km to the SW of the Huckitta region (c. 800°C, 10.5 kbar; Mawby et al. 1999) and Mallee Bore region, 80 km to the west (800–875°C, 8–12 kbar; Miller et al. 1997; Fig. 1).

Entire Point Shear Zone (M3)

Three metapelitic samples from the EPSZ yield relatively consistent pressure-temperature estimates for the assemblage garnet–biotite–sillimanite–quartz–plagioclase–K-feldspar ± muscovite. The Hodges & Spear (1982) garnet–biotite thermometer gives temperatures of 640–763°C for a pressure of 7 kbar, averaging between 680 and 700°C, whilst the garnet–phengite barometer of Green & Hellman (1982) gives temperatures ranging from 672 to 786°C. For temperatures of 700°C, the GASP barometer of Hodges & Crowley (1985) gives pressure estimates between 6.4 and 7.6 kbar, and five other pelite barometers give results within the same range (Table 1). One muscovite-absent sample (ISHU98.93A) gives slightly lower temperatures of 619–725°C, and pressure estimates of 6.1–6.6 kbar. A combination of average P–T estimates using Thermocalc for all four samples, plus one sample of a D3b mylonite from the Harts Range Group (ISHU98.23) yields an estimate of 692 ± 37°C and 6.8 ± 1.1 kbar (2σ errors) for a water activity of 0.35 (see discussion below). Together, these data suggest that P–T conditions during mylonitic deformation on the Entire Point Shear

Zone were 670–720°C and 6.5–7.3 kbar, i.e. at pressures c. 1–2 kbar higher than during granulite metamorphism.

Garnet-bearing recrystallized mafic rocks yielded P–T estimates that are less well constrained than results from the metapelites. However, the existence of two generations of garnet, hornblende, plagioclase and quartz in many of these rocks can provide constraints on the P–T trajectory during M3b, as M3a assemblages clearly predate the mylonitic M3b assemblage (e.g. Fig. 4d). For three samples containing the granoblastic M3a assemblage, application of the garnet–hornblende thermometer of Graham & Powell (1984) gives estimates in the range 620–659°C, whilst the garnet–hornblende–plagioclase–quartz tschermakite barometer of Kohn & Spear (1990) gives estimates of between 5.6 and 7.0 for the Fe end-member, and 4.1–5.3 kbar for the Mg end-member. Application of these thermobarometers to samples containing the overprinting M3b mylonitic garnet–hornblende–plagioclase assemblage (ISHU98.211A, 219) yields temperatures of 580–701°C, and pressures of 6.1–8.6 kbar for the Fe end-member and 5.2–7.9 kbar for the Mg end-member. Although these estimates are relatively imprecise, they suggest that the mylonitic M3b assemblage formed at higher pressures (and probably slightly higher temperatures) than the earlier coarse grained M3a assemblage. In particular, application of identical thermobarometers to the M3a and M3b assemblages within the one rock (211A) suggest a pressure increase of approximately 1–2 kbar from M3a to M3b.

Estimation of water activity in the Entire Point Shear Zone

Numerous lines of evidence suggest that aH_2O during D3b mylonitization was <1.0. Fluid independent barometers and thermometers consistently suggest P–T conditions of 670–730° and 6.4–7.6 kbar. Comparison of these estimates with average P–T estimates for varying aH_2O using Thermocalc suggest an aH_2O in the range 0.3–0.5 during D3 mylonitization. Higher water activities give temperatures of 750–850°C, which would preclude the stability of muscovite and thus appear unrealistic, whilst an aH_2O of <0.3 results in P–T estimates which are in the kyanite stability field. Additional evidence for an aH_2O of <1.0 during D_3 is the abundance of calcite and scapolite, and absence of garnet or wollastonite, in M3 calc-silicate rocks at c. 700°C, suggesting an aCO_2 of >c. 0.25 (e.g. Fitzsimons & Harley 1994). Furthermore, comparison with experimentally derived melting curves for granitic bulk compositions (e.g. Bohlen *et al.* 1982) suggests that the lack of migmatization of felsic lithologies deformed at *c*.700°C during D3 is consistent with an $aH_2O \leq 0.5$.

Delny–Mt Sainthill Shear Zone (M4)

Due to a scarcity of garnet-bearing assemblages in M4 mylonites, quantitative P–T estimates were only performed on one sample (ISHU98.170B), which contained the assemblage garnet–biotite–muscovite–quartz–plagioclase, with rare ilmenite and K-feldspar. The garnet–biotite thermometer of Hodges & Spear (1982) and garnet–phengite thermometer of Green & Hellman (1982) both give temperature estimates which range between 607 and 660°C, whilst various pelite barometers of Hodges & Crowley (1985) and Hoisch (1990) give pressure estimates between 3.3 and 6.0 kbar, but which are most typically in the range 4.0–5.5 kbar (Table 1). Average P–T estimates on this assemblage using Thermocalc are $653 \pm 34°C$ and 5.1 ± 1.3 kbar for an aH_2O of 1.0, and $631 \pm 33°C$ and 4.9 ± 1.2 kbar for an aH_2O of 0.8. Temperatures of 600–650°C at *c*. 5 kbar are consistent with the stability of the assemblages staurolite–muscovite–biotite–quartz (ISHU98.176) and sillimanite–biotite–muscovite–quartz (ISHU98.174) in differing bulk compositions in the KFMASH system (e.g. Xu *et al.* 1994). These estimates reflect maximum conditions (i.e. mid-amphibolite facies) for mylonites in the DMSZ, but chlorite- and epidote-bearing assemblages and quartz–muscovite phyllites in the northern DMSZ reflect a decrease in grade to lower amphibolite and greenschist facies conditions to the north.

Geochronology

To interpret the regional significance of the high grade mylonitic reworking of the Kanandra Granulite, it is crucial to constrain the timing of deformation on the EPSZ (D3). Monazite is a particularly useful mineral in dating upper amphibolite facies metamorphism, as it grows readily at these grades, and is likely to have an effective Pb-diffusion temperature of $\geq 725°C$ (e.g. Copeland *et al.* 1988; Lanzarotti & Hanson 1996). Therefore, SHRIMP U–Pb dating of new monazite growth within S3b pelitic mylonites is likely to be the most reliable method of dating D3b. In addition, a Sm–Nd mineral isochron was obtained from a garnet–hornblende assemblage to attempt to constrain the timing of S3a.

SHRIMP U–Pb monazite

The sample selected for monazite dating was a S3b pelitic mylonite reworking the Kanandra Granulite (sample ISHU98.237; Fig. 3). The mylonitic fabric is defined by fine-grained biotite, sillimanite, quartz, K-feldspar and plagioclase, with minor ilmenite and muscovite. Most garnet is porphyroclastic, but smaller subhedral garnets locally occur in the fabric. In thin section, monazite occurs as small rounded grains, 5–100 μm in diameter, often aligned in the fabric. The lineation on the mylonitic fabric in this locality plunges moderately towards 265°. Thermobarometry on the mylonitic assemblage yielded estimates of *c*. 700°C and 7 kbar (Table 1).

Monazite was separated at Research School of Earth Sciences (RSES), Australian National University. The monazite concentrate was handpicked and mounted in epoxy together with the RSES SHRIMP monazite standard WB.T.329 from the Thompson Mine, Manitoba, Canada (Williams *et al.* 1996). Prior to analysis, monazite grains were photographed in transmitted and reflected light and examined on a SEM using backscatter electron imaging (BSE). Analyses were carried out using the SHRIMP II; analytical procedures follow those of Ireland & Gibson (1998). U/Pb ratios in the grains were normalized to a value of 0.3152 (1766 Ma) for the Thompson Mine standard. Uncertainties in the individual isotopic ratios and ages in Table 2 (and in the error ellipses in the plotted data) are reported at the 1σ level, but uncertainties on any age calculations are reported as 95% confidence limits. All statistical evaluations have been done using the software Isoplot/Ex version 2.0 (Ludwig 1999).

The sample yielded clear, light yellow monazite of good quality. The separated monazites were of variable size and form, but were generally quite rounded, and BSE imaging showed some zonation. Inclusions are present, but not abundant, and in one case a whole zircon grain is enclosed in a monazite crystal. Sixteen analyses were done on eleven different monazites (Table 2). The data are plotted on a conventional Wetherill U–Pb concordia diagram (Fig. 5) and it is clear that within the measured uncertainties all analyses conform to a single concordant age population. Pb–Pb, Th–Pb and U–Pb ages can be calculated for this data set and all show that the monazites are *c*. 445 Ma old and that there are no older (inherited) grains or parts of grains. The calculated ages are listed in Table 3. Although there is some variation in the three ages calculated, they do agree within the 95% confidence limits cited. Nevertheless, it is deemed

Table 2. *Summary of SHRIMP U–Th–Pb monazite results for sample ISHU98.237*

Grain. spot	U (ppm)	Th (ppm)	Th/U	Pb* (ppm)	$\frac{204Pb}{206Pb}$	f_{206} %‡†	Radiogenic Ratios $\frac{208Pb}{232Th}$	±1σ	$\frac{206Pb}{238U}$	±1σ	$\frac{207Pb}{235U}$	±1σ	$\frac{207Pb}{206Pb}$	±1σ	Ages (in Ma) $\frac{208Pb}{232Th}$	±1σ	$\frac{206Pb}{238U}$	±1σ	$\frac{207Pb}{235U}$	±1σ	$\frac{207Pb}{206Pb}$	±1σ	Conc.§ %
1.1	745	5107	6.86	149	0.000099	0.18	0.0224	0.0005	0.0738	0.0015	0.570	0.015	0.0561	0.0007	448	10	459	9	458	9	454	28	101
1.2	999	6285	6.29	186	0.000088	0.16	0.0222	0.0006	0.0717	0.0016	0.545	0.016	0.0552	0.0009	443	11	446	10	442	10	420	35	106
2.1	935	3772	4.04	130	0.000043	0.08	0.0212	0.0005	0.0709	0.0017	0.552	0.015	0.0564	0.0006	424	11	441	10	446	10	470	23	94
2.2	765	6396	8.36	167	0.000175	0.32	0.0214	0.0005	0.0699	0.0015	0.529	0.016	0.0549	0.0010	429	10	435	9	431	10	410	39	106
2.3	1231	7299	5.93	216	0.000131	0.24	0.0217	0.0005	0.0708	0.0015	0.539	0.016	0.0552	0.0009	433	10	441	9	438	10	420	38	105
3.1	537	6139	11.44	150	0.000068	0.12	0.0218	0.0006	0.0702	0.0017	0.543	0.016	0.0561	0.0009	435	11	437	10	440	11	456	36	96
4.1	872	6575	7.54	180	0.000095	0.17	0.0218	0.0005	0.0707	0.0015	0.546	0.017	0.0560	0.0011	436	10	440	9	442	11	452	42	97
5.1	1159	6873	5.93	205	0.000071	0.13	0.0218	0.0005	0.0714	0.0016	0.545	0.017	0.0554	0.0011	435	10	444	9	442	11	430	43	104
6.1	1213	7222	5.96	213	0.000259	0.47	0.0216	0.0005	0.0700	0.0016	0.530	0.021	0.0550	0.0016	433	10	436	9	432	14	411	67	106
7.1	1088	4868	4.47	165	0.000053	0.10	0.0220	0.0005	0.0728	0.0016	0.566	0.015	0.0564	0.0007	439	10	453	9	455	10	467	29	97
8.1	968	9004	9.30	238	0.000068	0.12	0.0224	0.0005	0.0723	0.0016	0.559	0.015	0.0561	0.0008	447	10	450	9	451	10	457	31	99
8.2	671	6833	10.19	170	0.000202	0.37	0.0215	0.0006	0.0696	0.0016	0.520	0.018	0.0542	0.0012	430	11	433	10	425	12	379	51	114
9.1	871	7058	8.10	187	0.000193	0.35	0.0215	0.0005	0.0708	0.0016	0.535	0.023	0.0548	0.0019	429	10	441	9	435	15	405	79	109
9.2	2053	11680	5.69	362	0.000010	0.02	0.0223	0.0005	0.0727	0.0015	0.562	0.013	0.0561	0.0004	447	10	452	9	453	8	455	16	99
10.1	729	6636	9.10	176	0.000084	0.15	0.0222	0.0005	0.0726	0.0016	0.558	0.016	0.0558	0.0009	444	11	452	10	450	11	443	35	102
11.1	1323	6043	4.57	207	0.000159	0.29	0.0226	0.0005	0.0739	0.0015	0.567	0.016	0.0556	0.0009	451	10	460	9	456	10	436	37	105

† f_{206} denotes the percentage of ^{206}Pb that is common Pb.
‡ Correction for common Pb made using the measured $^{204}Pb/^{206}Pb$ ratio.
§ 100% denotes a concordant analysis.

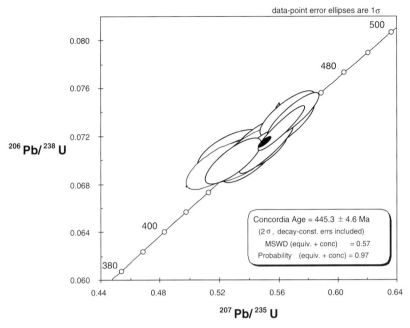

Fig. 5 U–Pb concordia plot of SHRIMP monazite data for sample ISHU98.237 from the Entire Point Shear Zone. The calculated Concordia Age is shown graphically by the black ellipse. The concordia curve is calibrated in Ma.

Table 3. *Summary of results for SHRIMP U–Th–Pb analyses, sample ISHU98.237*

Method	Analyses	MSWD	Probablity	Age (Ma)*
$^{238}U/^{206}Pb$	All data	0.75	0.73	445 ± 5
$^{208}Pb/^{232}Th$	All data	0.63	0.86	438 ± 5
$^{207}Pb/^{206}Pb$	All data	0.44	0.97	447 ± 16
Concordia Age	All data	0.59†	0.97†	445 ± 5

* 95% confidence limits, except for the Concordia Age which is 2σ.
† Calculated for concordance and equivalence.

most appropriate to quote the 'Concordia Age' (Ludwig 1998) as calculated using Isoplot/Ex (Ludwig 1999). Taking all data points into account the Concordia Age for these monazites is 445 ± 5 Ma (2σ error, including decay constant error; MSWD calculated for concordance and equivalence = 0.57, $P = 0.97$). This age is interpreted to represent timing of crystallization of monazite within the EPSZ.

Sm–Nd mineral isochron

Sample ISHU98.156 is a mafic rock from the Kanandra Granulite, which contains the M3a assemblage garnet–hornblende–plagioclase–quartz with less abundant ilmenite and apatite. It occurs within a large region of Kanandra Granulite that

escaped intense S3b mylonitization (Fig. 3), and contains a weakly defined S3a fabric with a S–SE plunging mineral lineation. Whilst the S3a fabric is locally overprinted by subsequent S3b mylonites, the sample selected for dating has no overprinting fabric. The texture is granoblastic with a grainsize typically between 0.3 and 1.5 mm. Garnet contains numerous inclusions of hornblende, quartz and plagioclase, whilst hornblende locally has inclusions of quartz. Sm–Nd analysis was performed on Finnigan MAT 261 and 262 mass spectrometers at the University of Adelaide. Sample preparation and analytical techniques follow those of Mawby *et al.* (1999). Typical internal precision on the La Jolla Nd-standard during the course of the study was 0.511835 ± 20 (2σ). The Sm and Nd blanks contribution was 0.3 ng. Isochron calculations follow Ludwig

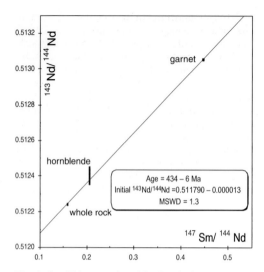

Fig. 6. Sm–Nd garnet–hornblende–whole rock isochron for sample ISHU98.156, Kanandra Granulite.

(1999). The results of Sm–Nd isotopic analysis of sample ISHU98.156 are presented in Figure 6 and Table 4 and give an 3-point garnet–hornblende–WR mineral isochron age of 434 ± 6 Ma.

Interpretation of geochronological data

Given the SHRIMP U–Pb age of 445 ± 5 Ma for monazite growth during M3b, the Sm–Nd age of 434 ± 6 Ma for the M3a assemblage appears inconsistent with the field and petrological evidence suggesting that M3a predates M3b. A problem in the interpretation of Sm–Nd mineral isochrons is the uncertainty relating to the closure temperature of the Sm–Nd in garnet, which may be as low as 600–650°C (e.g. Burton *et al.* 1995) or as high as 750–800°C, depending on factors such as grain size and cooling rate (Becker 1997). An additional source of uncertainty in comparing U–Pb monazite and Sm–Nd garnet ages in upper amphibolite facies rocks is the relative timing of monazite growth, which is controlled by monazite-producing mineral

reactions that do not necessarily coincide with the peak of metamorphism (e.g. Lanzarotti & Hanson 1996). In spite of these uncertainties, it is proposed that the most likely interpretation of the geochronological data is that the monazite age reflects the timing of D3b mylonitization, and the Sm–Nd age reflects cooling through the closure temperature following peak conditions at 445 Ma. This interpretation is supported by the probable slow cooling following M3b (see discussion below), and the relatively fine grain size of garnet in sample 156 (0.5–1 mm), which both lead to a lowering in the closure temperature of Nd in garnet (Becker 1997).

Discussion

P–T evolution of the Kanandra Granulite, and regional correlations

The Kanandra Granulite in the Huckitta region is a multiply reworked metamorphic terrain. A summary of the evolution of the Kanandra Granulite, and correlations with other regions in the eastern Arunta Inlier are given in Figure 7 and Table 5. Peak M1 metamorphism in the Kanandra Granulite involved fluid-absent partial melting of metapelites at 770–850°C and 5–7 kbar. Reaction textures related to development of the S2 fabric are all consistent with P–T estimates that suggest that S2 developed at lower temperatures (700–750°C) but at similar pressures to M1. Whilst it is possible that M1 and M2 represent two unrelated metamorphic events, it is suggested that they form a continuum, with M2 reflecting near-isobaric cooling from peak conditions (Fig. 7). A similar evolution has been recorded elsewhere in the Strangways Metamorphic Complex, including the Edwards Creek region of the Strangways Range, where a peak M1 assemblage (800°C, 6–8 kbar) is overprinted by a planar S2 fabric with associated cooling textures (Warren 1983; Ballèvre *et al.* 1997). SHRIMP U–Pb dating of zircons in leucosomes related to these two events by Möller *et al.* (1999), derived ages of 1727 ± 2 Ma for M1, and 1717 ± 2 Ma for M2. A similar U–Pb age of

Table 4. *Summary of Sm–Nd analytical data for sample ISHU98.156*

Sample	Sm ppm	Nd ppm	$^{147}Sm/^{144}Nd$	$^{143}Nd/^{144}Nd$	$\pm 2\sigma$
Garnet	0.77	1.05	0.444	0.513056	±000007
Hornblende	4.74	13.92	0.206	0.512403	±000050
Whole rock	3.07	11.64	0.160	0.512241	±000008

† 0.3% error is used for $^{147}Sm/^{144}Nd$.
$^{143}Nd/^{144}Nd$ ratios are normalized to $^{146}Nd/^{144}Nd = 0.7129$.

1730 ± 1 Ma for U-rich zircons within amphibolite in the Entia Gneiss (Cooper et al. 1988) suggests that an event of regional significance occurred at c. 1730 Ma. Due to the similar geological evolution of the Kanandra Granulite and the Edwards Creek region of the Strangways Range we tentatively correlate M1 and M2 between these geographically separate areas; hence it is suggested that M1 and M2 in the Kanandra Granulite occurred during the c. 1730–1715 Ma event proposed by Möller et al. (1999) (Table 5).

Palaeozoic reworking of the Kanandra Granulite commenced with upper amphibolite facies mylonitization at c. 700°C and 7 kbar associated

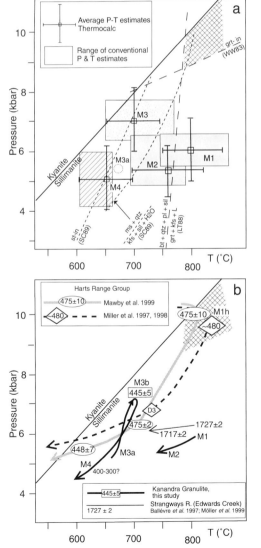

with the juxtaposition of the Kanandra Granulite and Harts Range Group along the EPSZ. Textural and metamorphic evidence in recrystallized mafic rocks suggests that the Kanandra Granulite underwent an up-pressure path prior to peak M3 metamorphism, with the overprinting of a granoblastic M3a assemblage by a finer grained mylonitic M3b assemblage which formed at higher pressures. This inferred early up-pressure history leads to an unusual P–T path for the Kanandra Granulite during the Ordovician, with an increase in pressure after the terrain was already at elevated temperatures (Fig. 7). Thermal modelling by Kelsey (1998) suggested that the Kanandra Granulite may have had an elevated ambient thermal regime immediately prior to Palaeozoic deformation, due to the presence of high heat-producing granites in the basement and an insulating sedimentary cover. This ambient thermal perturbation may also have contributed to the localization of strain in an intraplate setting in this region (e.g. Sandiford & Hand 1998).

SHRIMP U–Pb dating of monazite from the S3 fabric in the EPSZ suggests that peak M3 metamorphism occurred at 445 ± 5 Ma. A Sm–Nd age of 434 ± 6 Ma for garnet-bearing mafic

Fig. 7. (**a**) Summary of P–T constraints for different stages of metamorphism in the Kanandra Granulite. Error bars represent 2σ uncertainties on combined average P–T estimates using Thermocalc, except for M4, which shows a 1σ uncertainty for a single P–T estimate. Boxed fields represent the range of pressures and temperatures calculated using conventional geothermobarometry. The dashed circle represents the approximate P–T estimates for the M3a assemblage (errors are ± c. 100°C and ±1.5–2 kbar). Short dashed lines represent the calculated approximate upper stabilty limits of staurolite and muscovite in the pelitic KFMASH system (SC89 – Spear & Cheney 1989). Long dashed lines represent the experimentally determined position of melting reactions for bulk compositions similar to those in the Harts Range Group: the steep line represents dehydration melting of pelite to produce garnet + melt (LB88 – Le Breton & Thompson 1988) and the shallow line represents dehydration melting of amphibolite to produce a garnet-bearing melt (WW93 – Wyllie & Wolf 1993). These melting reactions provide approximate minimum P and T constraints on peak metamorphism in the Harts Range Group in the Huckitta region (cross-hatched area). (**b**) Proposed P–T paths for the two events in the Kanandra Granulite, with age constraints, in comparison to P–T paths and age constraints for Strangways Complex granulites at Edwards Creek (Möller et al. 1999) and the Harts Range Group at Mallee Bore (Miller et al. 1997, 1998) and the Harts Range (Mawby et al. 1999). 'D3' on the Mallee Bore P–T path represents the P–T conditions of D3 mylonitization at Mallee Bore (Miller et al. 1997). Discussion in text.

Table 5. *Summary of the evolution of Harts Range Group and Strangways Complex rocks from various localities within the eastern Arunta Inlier, with correlations*

Age	Harts Range Group			Strangways Complex and equivalents		
	Harts Range[2,3]	Mallee Bore[4,5]	Huckitta region*	Kanandra Granulite*	Entia Gneiss[2]	Edwards Creek[1,8,9]
1730 Ma[1,6] Strangways Event 1715 Ma[1]				Granulite facies metamorphism (M1) (5–7 kbars, 770–850°C) Granulite facies metamorphism (M2) 5–6 kbars, 700–750°C	Granulite facies metamorphism	Granulite facies metamorphism (M_1) 6–8 kbars, c. 800°C Granulite facies metamorphism (M_2) 6–8 bars, <750°C
Neoproterozoic?[4]	Deposition of Harts Range Group					??
480–460 Ma[2,3,4] Ordovician high grade event	High P–T metamorphism (800°C, 9–10 kbar), extensional deformation, decompression to 7 kbar	High P–T metamorphism (800°C, 8–12 kbar)	High P–T metamorphism (M1h) (>770°C, >8–9 kbar) decompression to 7 kbar	??	Kyanite-grade amphibolite facies metamorphism (700°C, 8–9 kbar)	??
450–440 Ma*[1,2] Juxtaposition of terrains, onset of compressional deformation	Compressional S-directed mylonitic deformation (M3) (650C, 5–6 kbar)	Mylonitic deformation, Upper amphibolite facies (c. 700°C, 7 kbar), Sinistral strike-slip movement	Mylonitic deformation (M3h), Upper amphibolite facies (700°C, 7 kbar), Sinistral strike-slip movement (Entire Point S.Z.)	Mylonitic deformation (M3b), Upper amphibolite facies (700°C, 7 kbar), Sinistral strike-slip movement (Entire Point S.Z.)	Compressional S-directed mylonitic deformation (650C, 5–6 kbar)	New zircon growth within amphibolite facies shear zones
400–300 Ma[1,3,7] Alice Springs Orogeny	Amphibolite (600°C, 6 kbar) to greenschist facies south-directed shear zones	Minor hydrous retrogression	Minor chlorite–muscovite retrogression, pegmatite intrusion	North-directed steeply south-dipping shear zones (M4) Amphibolite (600–650°C, 5–6 kbar) to greenschist facies (Delny-Mt S. S.Z.)	Amphibolite (600°C, 6 kbar) to greenschist facies south-directed shear zones	Amphibolite (600°C, 6 kbar) to greenschist facies south-directed shear zones

References *This study; [1] Möller *et al.* (1999); [2] Mawby *et al.* (1999); [3] Hand *et al.* (1999a); [4] Miller *et al.* (1999); [5] Miller *et al.* (1998); [6] Cooper *et al.* (1988); [7] Dunlap & Teyssier (1995); [8] Warren & Hensen (1989); [9] Ballèvre *et al.* (1997).

rocks provides further evidence for high-T metamorphism in the Kanandra Granulite at this time. This postdates the peak metamorphism in the Harts Range Group, which has been dated at 480–460 Ma (Miller *et al.* 1998; Mawby *et al.* 1999). There is no evidence for high-T deformation of the Kanandra Granulite during peak metamorphism of the Harts Range Group. Instead, evidence from this study suggests that prograde heating and burial of the Kanandra Granulite occurred at *c.* 450–440 Ma, following or during retrograde decompression of the Harts Range Group to 6–7 kbar (Fig. 7; Hand *et al.* 1999*a*; Mawby *et al.* 1999).

No precise constraints currently exist on the timing of mid-amphibolite to greenschist facies reworking of the Kanandra Granulite in the DMSZ. However, evidence from elsewhere in the eastern Arunta Inlier (Dunlap & Teyssier 1995; Hand *et al.* 1999*a*) strongly suggests that this deformation and metamorphism is associated with the 400–300 Ma Alice Springs Orogeny. If this is the case, then the presence of sillimanite within some DMSZ mylonites implies that either the temperatures remained elevated (>600°C) for ≥50 Ma after upper amphibolite facies metamorphism at 445 Ma, or there was a later thermal pulse associated with the Alice Springs Orogeny. Modelling of zonation profiles in porphyroclastic garnets from the EPSZ by Kelsey (1998) suggested that temperatures were elevated above 600°C for >20 Ma during Palaeozoic metamorphism, implying that the Kanandra Granulite may have experienced a prolonged high-T history during the Palaeozoic. Therefore, it is possible that the estimates of 600–650°C and 5 kbar for M4 (DMSZ) mylonites reflect renewed deformation following relatively slow cooling accompanying exhumation of the Kanandra Granulite, rather than a second thermal event. Studies elsewhere in the eastern Arunta Inlier have suggested that deformation and exhumation related to the Alice Springs Orogeny may have been episodic over a period of *c.* 100 Ma (e.g. Dunlap & Teyssier 1995; Hand *et al.* 1999*a*), and therefore movement on the DMSZ may also be complex, and probably involved more than a single episode of deformation. Precise geochronology on DMSZ mylonites is required to address these questions.

Long-lived mid-crustal residence or a second cycle of burial and exhumation?

A common problem in the interpretation of near-isobarically cooled medium pressure high-grade terrains is identifying the mechanism by which the terrain was exhumed to the surface. A common interpretation is that these terrains often reside at depth for a prolonged period, before being exhumed during an unrelated episode of tectonic reworking (e.g. Harley 1989). Given the apparent cooling-dominated history following M1, and the similarities in pressure estimates between M1 and M3, it is tempting to suggest that the Kanandra Granulite remained at depth following Palaeoproterozoic metamorphism, before being reworked and exhumed during the Palaeozoic. However, in assessing this notion, it is worth considering the evolution of the Bleechmore Granulite, which forms the western continuation of the Kanandra Granulite 90 km W of the Huckitta region (Fig. 1). The Bleechmore Granulite has a very similar metamorphic evolution to the Kanandra Granulite, with peak granulite facies metamorphism at *c.* 6 kbar (Warren & Hensen 1989) overprinted by apparent cooling textures such as biotite–sillimanite–garnet fabrics overprinting cordierite-bearing assemblages (I. Scrimgeour & M. Hand unpub. data). The Bleechmore Granulite is unconformably overlain by a sequence of lower to mid-amphibolite facies metasediments, the Mendip Metamorphics (Shaw & Warren 1975; Scrimgeour & Hand unpub. data). The age of the Mendip Metamorphics is not known, but the existence of this sequence implies that the Bleechmore Granulite was exhumed and reburied between the Palaeoproterozoic and the Palaeozoic, which is the latest metamorphic episode in the eastern Arunta Inlier (Collins & Shaw 1995). This suggests that the Palaeozoic event may have involved a second cycle of burial, deformation and prograde metamorphism, rather than reworking of granulites which had remained at depth since the Palaeoproterozoic. This is consistent with evidence that the Harts Range Group, which is believed to have been deposited after the Palaeoproterozoic and possibly as late as the Neoproterozoic (Miller *et al.* 1998), was buried to >30 km depth by the mid-Ordovician (Mawby *et al.* 1999).

The preserved record in the Kanandra Granulite of near-isobaric cooling following granulite facies metamorphism, then reactivation and exhumation from a similar crustal depth, is similar to other terrains worldwide where prolonged mid-crustal residence of high-grade rocks has been postulated (e.g. Harley 1985; Warren & Stewart 1988). The fact that the Bleechmore Granulite was exposed to the surface and then reburied between the Palaeoproterozoic and Palaeozoic, without any record of such decompression and reburial being preserved within the granulites, suggests that a similar scenario may apply to the Kanandra Granulite. This highlights the fact that caution must be used before inferring a history of prolonged crustal residence of granulites purely on the basis of later reworking and exhumation from a similar crustal depth.

The 450–440 Ma event in the eastern Arunta Inlier

The juxtaposition of the Harts Range Group and Kanandra Granulite along the EPSZ at 445 ± 5 Ma forms part of a more widespread, newly recognized 450–440 Ma event in eastern Arunta Inlier (Table 5). In the Harts Range region, the metasediments of the Harts Range Group are separated from Palaeoproterozoic granulites of the Entia Gneiss Complex and Strangways Metamorphic Complex by shallowly dipping south-vergent detachments (the Bruna Detachment Zone and Florence-Muller Shear Zone; Ding & James 1985; Collins & Teyssier 1989; Hand et al. 1999b). These south-directed thrusts (D3 of Mawby et al. 1999) formed at upper amphibolite facies conditions, postdating the peak of metamorphism in the Harts Range Group, and have Sm–Nd isochron ages of 448 ± 7 Ma and 445 ± 51 Ma (Mawby et al. 1999). Mawby et al. (1999) interpreted the D3 event in the Harts Range as reflecting a change from extensional to compressional deformation in the eastern Arunta Inlier, due to apparent south-directed reverse kinematics in D3 shear zones (James et al. 1989) as well as a change in the distribution of sedimentation in the surrounding basins at c. 450 Ma (e.g. Shaw et al. 1991). P–T conditions during D3 in the Harts Range were 600–650°C and 5–6 kbar (Mawby et al. 1999) suggesting that the EPSZ mylonites now exposed in the Huckitta region formed at deeper crustal levels than D3 mylonites in the Harts Range.

Further evidence for an event at 450–440 Ma is provided by Möller et al. (1999) who derived an age of 443 ± 6 Ma from zircon overgrowths in a staurolite-bearing shear zone in the Edwards Creek region in the Strangways Complex. Although no 450–440 Ma ages have yet emerged from the Mallee Bore area, Miller et al. (1997) document localized mylonitization at 680–730°C and 5.8–7.7 kbar which overprints the peak metamorphic assemblage in the Harts Range Group (c. 480 Ma; Miller et al. 1998), and has a sinistral strike-slip sense of movement (Hand et al. 1994b). As the Mallee Bore area occurs within 10–15 km of the inferred western continuation of the EPSZ, these mylonites are tentatively correlated with other 450–440 Ma mylonites in

the eastern Arunta Inlier. The distribution of 450–440 Ma ages for mylonites in the eastern Arunta Inlier implies that it was an event of regional significance, that led to the structural juxtaposition of the Harts Range Group and Entia Gneiss Complex against Proterozoic granulites that had seen no previous high-grade Palaeozoic deformation. The evidence of new zircon growth in the Strangways Range at c. 445 Ma (Möller et al. 1999) suggests that reworking of the Strangways Metamorphic Complex occurred at the same time (i.e. 450–440 Ma) both north and south of the Harts Range region. The timing of the apparent onset of intraplate compressional deformation in the Arunta Inlier at 450–440 Ma (Mawby et al. 1999) is 40–50 Ma earlier than the traditionally accepted onset of the Alice Springs Orogeny (c. 400 Ma; Dunlap & Teyssier 1995).

Tectonic implications for Palaeozoic intraplate compression in central Australia

The existence of steeply-dipping Palaeozoic structures in the Huckitta region has implications for the tectonics of the eastern Arunta Inlier during intraplate deformation. The style of deformation related to the EPSZ differs markedly from D3 shear zones in the Harts Range, in terms of both orientation and kinematics. Whilst the D3 shear zones in the Harts Range are shallowly-dipping south-directed thrusts, the EPSZ is steeply south-dipping, with shallowly-plunging lineations and a sinistral strike-slip sense of movement (Fig. 8). This implies that compressional deformation associated with the 450–440 Ma event was more complex than simple south-directed compression, and involved significant partitioning of strain, with dominantly dip-slip movement in the south and strike-slip movement in the north. The existence of a steeply-dipping strike-slip zone (typically within a reverse hinterland-directed component) located behind a thrust-dominated wedge is a feature of numerous orogenic belts worldwide, particularly along the Alpine–Himalayan system (e.g. the Periadriatic/Insubric Line in the Alps, Schmid et al. 1989; the Altyn Tagh Fault, Tibet, Tapponnier et al. 1986; the Karakoram Fault, Ladakh, Searle et al. 1998).

Subsequent deformation on the DMSZ, which involved oblique reverse deformation and juxtaposed upper amphibolite facies Palaeozoic rocks in the south against sub-greenschist facies Neoproterozoic sediments to the north (Fig. 8), can best be described as a steep 'retro-shear' (e.g. Willett et al. 1993; Beaumont et al. 1996). This

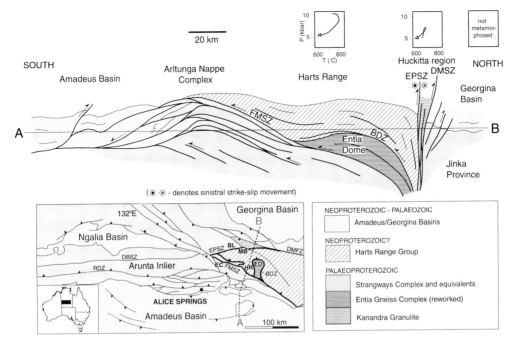

Fig. 8. Schematic N–S cross-section across the eastern Arunta Inlier (adapted and modified after Dunlap & Teyssier 1995). Abbreviations as in Figure 1.

suggests that although most of the exposed late Ordovician to Carboniferous structures in the Arunta Inlier are south-vergent (Dunlap *et al.* 1995), the orogen can be considered, as a whole, to be doubly-vergent (Collins & Teyssier 1989), with significant exhumation along steep structures at the back of the south-vergent thrust stack (Fig. 8). Therefore, in spite of the intraplate setting of Palaeozoic compression in central Australia, the large-scale geometry of the orogen has affinities with the collisional stage of Alpine-type orogens (e.g. Dewey *et al.* 1986; Beaumont *et al.* 1996).

Conclusions

The Kanandra Granulite, in the Huckitta region of the eastern Arunta Inlier, has a complex high-grade history extending from the Palaeoproterozoic to the Palaeozoic. Peak metamorphism of the Kanandra Granulite during the Palaeoproterozoic ($c.\,800°C$, 5–7 kbar) was followed by inferred near-isobaric cooling. Reworking of these granulites at $c.\,700°C$ and 7 kbar occurred during sinistral strike-slip to transpressional deformation along the steeply south-dipping Entire Point Shear Zone (EPSZ) at $c.\,440$ Ma. South of the EPSZ is the Harts Range Group, which was

metamorphosed at $c.\,800°C$ and 8–10 kbar during inferred extensional deformation at 480–460 Ma. The EPSZ juxtaposed the Harts Range Group and Kanandra Granulite, and is related to a major 450–440 Ma event that is inferred to represent the onset of Palaeozoic compressional intraplate deformation in the Arunta Inlier (Mawby *et al.* 1999). The significant strike-slip component of movement on the EPSZ implies that deformation at 450–440 Ma was, at least locally, transpressional in nature. The Kanandra Granulite records an up-pressure evolution prior to peak metamorphism in the EPSZ, whilst the Harts Range Group underwent decompression, implying a component of south-up reverse movement on the EPSZ. The rare exposure of a metamorphosed cover sequence overlying the Bleechmore Granulite to the west suggests that the Palaeoproterozoic granulites may have been exhumed and reburied prior to reworking, rather than undergoing prolonged residence in the mid-crust.

Further reworking of the Kanandra Granulite at mid-amphibolite to greenschist facies was associated with steeply south-dipping reverse mylonites of the Delny–Mt Sainthill Shear Zone (DMSZ), which juxtaposed the granulites adjacent to unmetamorphosed Neoproterozoic sediments. This second phase of reworking is interpreted to have occurred during the 400–300 Ma Alice Springs Orogeny. The total width of the zone of reworked Kanandra Granulite, which incorporates the EPSZ and DMSZ is 3–5 km. This zone of reworking separates the Ordovician high-grade terrain from the Neoproterozoic sediments to the north, and therefore represents a structural zone of profound importance in the intraplate evolution of central Australia. It is interpreted to reflect long-lived strike-slip and north-directed reverse movement on steeply-dipping structures in the hinterland of a southward propagating intraplate thrust stack. This north-vergent and strike-slip deformation has received relatively little attention in previous studies of intraplate deformation in central Australia, but resulted in upper amphibolite facies reworking of Proterozoic granulites and significantly contributed to the exhumation of high-grade rocks within an intraplate setting.

The authors wish to acknowledge the contribution of M. Hand and J. Mawby towards the development of ideas presented in this paper, and thank P. Haines and D. Kelsey for introducing them to the area. SHRIMP U–Pb monazite dating was undertaken by R. Armstrong and Sm–Nd analyses were done by J. Mawby. Logistical support from the Northern Territory Geological Survey is gratefully acknowledged. The manuscript benefited from constructive reviews by S. Harley, B. Hensen and I. Buick. This work forms part of the FWF (Austrian Science Foundation) project P11879-TEC, and the financial support from the FWF is acknowledged.

References

BALLÈVRE, M., HENSEN, B. J. & REYNARD, B. 1997. Orthopyroxene-andalusite symplectites replacing cordierite in granulites from the Strangways Range (Arunta block, central Australia): A new twist to the pressure–temperature history. *Geology*, 25, 215–218.

BARTLETT, J. M., DOUGHERTY-PAGE, J. S., HARRIS, N. B. W., HAWKESWORTH, C. J. & SANTOSH, M. 1998. The application of single zircon evaporation and model Nd ages to the interpretation of polymetamorphic terrains: an example from the Proterozoic mobile belt of south India. *Contributions to Mineralogy and Petrology*, 131, 181–195.

BEAUMONT, C., ELLIS, S., HAMILTON, J. & FULLSACK, P. 1996. Mechanical model for subduction-collision tectonics of Alpine-type compressional orogens. *Geology*, 24, 675–678.

BECKER, H. 1997. Sm–Nd garnet ages and cooling history of high-temperature garnet peridotite massifs and high-pressure granulites from Lower Austria. *Contributions to Mineralogy and Petrology*, 127, 224–236.

BLACK, L. P. & SHAW, R. D. 1992. U–Pb zircon chronology of prograde Proterozoic events in the central and southern provinces of the Arunta Block, central Australia. *Australian Journal of Earth Sciences*, 39, 153–171.

BLACK, L. P., SHAW, R. D. & STEWART, A. J. 1983. Rb–Sr geochronology of Proterozoic events in the Arunta Inlier, central Australia. *B.M.R. Journal of Geology and Geophysics*, 8, 129–138.

BOHLEN, S. R., BOETTCHER, A. L. & WALL, V. J. 1982. The system albite–H_2O–CO_2: a model for melting and activities of water at high pressures. *American Mineralogist*, 68, 1049–1058.

BURTON, K. W., KOHN, M. J., COHEN, A. S. & O'NIONS, R. K. 1995. The relative diffusivities of Pb, Nd, Sr and O in garnet. *Earth and Planetary Science Letters*, 133, 199–211.

CAMACHO, A., COMPSTON, W., McCULLOCH, M. & McDOUGALL, I. 1997. Timing and exhumation of eclogite facies shear zones, Musgrave Block, central Australia. *Journal of Metamorphic Geology*, 15, 735–751.

CARTWRIGHT, I., BUICK, I. S., FOSTER, D. A. & LAMBERT, D. D. 1999. Alice Springs age shear zones from the southeastern Reynolds Range, central Australia. *Australian Journal of Earth Sciences*, 46, 355–363.

COLLINS, W. J. & SHAW, R. D. 1995. Geochronological constraints on orogenic events in the Arunta Inlier: a review. *Precambrian Research*, 71, 69–89.

COLLINS, W. J. & TEYSSIER, C. 1989. Crustal scale ductile fault systems in the Arunta Inlier, central Australia. *Tectonophysics*, 158, 49–66.

COLLINS, W. J. & VERNON, R. H. 1993. How well established is isobaric cooling in Proterozoic orogenic belts? An example from the Arunta Inlier, central Australia – Comment. *Geology*, 21, 953–954.

COOPER, J. A., MORTIMER, G. E. & JAMES, P. R. 1988. Rate of Arunta Inlier evolution at the eastern margin of the Entia Dome, central Australia. *Precambrian Research*, 40/41, 217–231.

COPELAND, P., PARRISH, R. R. & HARRISON, T. M. 1988. Identification of inherited radiogenic Pb in monazite and its implications for U–Pb systematics. *Nature*, 333, 760–763.

DEWEY, J. F., HEMPTON, M. R., KIDD, W. S. F., SAROGLU, F. & SENGÖR, A. M. C. 1986. Shortening of continental lithosphere: the neotectonics of Eastern Anatolia – a young collision zone. *In*: COWARD, M. P. & RIES, A. C. (eds) *Collisional Tectonics*. Geological Society, London, Special Publications, 19, 3–36.

DING, P. & JAMES, P. R. 1985. Structural evolution of the Harts Range area and its implication for the development of the Arunta Block, central Australia. *Precambrian Research*, 27, 251–276.

DUNLAP, W. J. & TEYSSIER, C. 1995. Palaeozoic deformation and isotopic disturbance in the southeastern Arunta Block, central Australia. *Precambrian Research*, 71, 229–250.

DUNLAP, W. J., TEYSSIER, C., McDOUGALL, I. & BALDWIN, S. 1995. Thermal and structural evolution of the intracratonic Arltunga nappe complex, central Australia. *Tectonics*, 14, 1182–1204.

FITZSIMONS, I. C. W. & HARLEY, S. L. 1994. Garnet coronas in scapolite-wollastonite calc-silicates from East Antarctica: the application and limitation of activity corrected grids. *Journal of Metamorphic Geology*, **12**, 761–777.

GRAHAM, C. M. & POWELL, R. 1984. A garnet-hornblende geothermometer: calibration, testing, and application to the Pelona Schist, Southern California. *Journal of Metamorphic Geology*, **2**, 13–21.

GROENEWALD, P. B., MOYES, A. B., GRANTHAM, G. H. & KRYNAUW, J. R. 1995. East Antarctic crustal evolution: geological constraints and modelling in western Dronning Maud Land. *Precambrian Research*, **75**, 231–250.

GREEN, T. H. & HELLMAN, P. L. 1982. Fe–Mg partitioning between coexisting garnet and phengite at high pressure, and comments on a garnet-phengite geothermometer. *Lithos*, **15**, 253–266.

HAND, M., DIRKS, P. G., H. M., POWELL, R. & BUICK, I. S. 1992. How well established is isobaric cooling in Proterozoic orogenic belts? An example from the Arunta Inlier, central Australia. *Geology*, **20**, 649–652.

HAND, M., MAWBY, J., KINNY, P. & FODEN, J. 1999*a*. U–Pb ages from the Harts Range, central Australia: evidence for early-Ordovician extension and constraints on Carboniferous metamorphism. *Journal of the Geological Society, London*, **156**, 715–730.

HAND, M., MAWBY, J., MILLER, J. A., BALLÈVRE, M., HENSEN, B. J., MÖLLER, A. & BUICK, I. S. 1999*b*. Tectonothermal evolution of the Harts and Strangways Range region, eastern Arunta Inlier, central Australia. *Specialist Group in Geochemistry, Mineralogy and Petrology Field Guide*, **4**, Geological Society of Australia.

HAND, M., SCRIMGEOUR, I., POWELL, R., STÜWE, K. & WILSON, C. J. L. 1994. Metapelitic granulites from Jetty Peninsula, east Antarctica: formation during a single event or by polymetamorphism? *Journal of Metamorphic Geology*, **12**, 557–573.

HARLEY, S. L. 1985. Paragenetic and mineral-chemical relationships in orthoamphibole-bearing gneisses from Enderby Land, east Antarctica: a record of Proterozoic uplift. *Journal of Metamorphic Geology*, **3**, 179–200.

HARLEY, S. L. 1989. The origins of granulites: a metamorphic perspective. *Geological Magazine*, **126**, 215–247.

HODGES, K. V. & CROWLEY, P. D. 1985. Error estimation and empirical geothermobarometry for pelitic systems. *American Mineralogist*, **70**, 702–709.

HODGES, K. V. & SPEAR, F. S. 1982. Geothermometry, geobarometry and the Al_2SiO_5 triple point at Mt. Moosilauke, New Hampshire. *American Mineralogist*, **67**, 1118–1134.

HOISCH, T. D. 1990. Empirical calibration of six geobarometers for the mineral assemblage quartz + muscovite + biotite + plagioclase + garnet. *Contributions to Mineralogy and Petrology*, **104**, 225–234.

HOLLAND, T. J. B. & POWELL, R. 1998. An internally consistent thermodynamic dataset for phases of petrological interest. *Journal of Metamorphic Geology*, **16**, 309–343.

IRELAND, T. R. & GIBSON, G. M. 1998. SHRIMP monazite and zircon geochronology of high-grade metamorphism in New Zealand: *Journal of Metamorphic Geology*, **16**, 149–167.

JAMES, P. R. & DING, P. 1988. 'Caterpillar tectonics' in the Harts Range area: a kinship between two sequential extension-collision orogenic belts within the Arunta Inlier of central Australia. *Precambrian Research*, **40/41**, 199–216.

JAMES, P. R., MACDONALD, P. & PARKER, M. 1989. Strain and displacement in the Harts Range detachment zone: a structural study of the Bruna Gneiss from the western margin of the Entia Dome, central Australia. *Tectonophysics*, **158**, 23–48.

KELSEY, D. 1998. Controls on the localisation of Palaeozoic deformation in the northeastern Arunta Inlier. Unpublished BSc (Hons) Thesis, University of Adelaide.

KOHN, M. J. & SPEAR, F. S. 1990. Two new barometers for garnet amphibolites with applications to eastern Vermont. *American Mineralogist*, **75**, 89–96.

KRETZ, R. 1983. Symbols for rock forming minerals. *American Mineralogist*, **68**, 277–279.

LAFRANCE, B., CLARKE, G. L., COLLINS, W. J. & WILLIAMS, I. S. 1995. The emplacement of the Wuluma Granite: melt generation and migration along steeply dipping extensional fractures at the close of the late Strangways orogenic event, Arunta Block, central Australia. *Precambrian Research*, **72**, 43–67.

LANZAROTTI, A. & HANSON, G. N. 1996. Geochronology and geochemistry of multiple generations of monazite from the Wepawaug Schist, Connecticut, USA: implications for monazite stability in metamorphic rocks. *Contributions to Mineralogy and Petrology*, **125**, 332–340.

LE BRETON, N. & THOMPSON, A. B. 1988. Fluid-absent (dehydration) melting of biotite in metapelites in the early stages of crustal anatexis. *Contributions to Mineralogy and Petrology*, **99**, 226–237.

LUDWIG, K. R. 1998. On the treatment of concordant uranium-lead ages. *Geochimica et Cosmochimica Acta*, **62**, 665–676.

LUDWIG, K. R. 1999. Isoplot/Ex version 2.00: A Geochronological Toolkit for Microsoft Excel. *Berkeley Geochronology Center Special Publication*, **1a**, 46pp.

MAWBY, J., HAND, M. & FODEN, J. 1999. Sm–Nd evidence for Ordovician granulite facies metamorphism in an intraplate setting in the Arunta Inlier, central Australia. *Journal of Metamorphic Geology*, **17**, 653–668.

MAWBY, J., HAND, M., FODEN, J. & KINNY, P. 1998. Ordovician granulites in the eastern Arunta Inlier: a new twist in the Palaeozoic history of central Australia. *Geological Society of Australia, Abstracts*, **49**, 296.

MILLER, J. A., CARTWRIGHT, I. & BUICK, I. 1997. Granulite facies metamorphism in the Mallee Bore area, northern Harts Range: implications for the thermal evolution of the eastern Arunta Inlier, central Australia. *Journal of Metamorphic Geology*, **15**, 613–629.

MILLER, J. A., BUICK, I. S., WILLIAMS, I. S. & CARTWRIGHT, I. 1998. Re-evaluating the metamorphic and tectonic history of the eastern Arunta Block, central Australia. *Geological Society of Australia, Abstracts*, **49**, 316.

MÖLLER, A., ARMSTRONG, R., HENSEN, B. J. & WILLIAMS, I. S. 1999. Dating metamorphic events and deformation: SHRIMP U–Pb zircon examples from the Strangways Metamorphic Complex, Arunta Inlier, Australia. *Journal of Conference Abstracts*, **4**, 711.

NEWTON, R. C. & PERKINS, D. III. 1982. Thermodynamic calibration of geobarometers based on the assemblages garnet–plagioclase–orthopyroxene (clinopyroxene)–quartz. *American Mineralogist*, **67**, 203–222.

POWELL, R. & HOLLAND, T. J. B. 1988. An internally consistent thermodynamic dataset with uncertainties and correlations: 3. Applications to geobarometry, worked examples and a computer program. *Journal of Metamorphic Geology*, **6**, 173–204.

POWELL, R. & HOLLAND, T. J. B. 1994. Optimal geothermometry and geobarometry. *American Mineralogist*, **79**, 120–133.

SANDIFORD, M. & HAND, M. 1998. Controls on the locus of Phanerozoic intraplate deformation in central Australia. *Earth and Planetary Science Letters*, **162**, 97–110.

SCRIMGEOUR, I. & CLOSE, D. 1999. Regional high pressure metamorphism during intracratonic deformation: the Petermann Orogeny, central Australia. *Journal of Metamorphic Geology*, **17**, 557–572.

SCHMID, S. M., AEBLI, H. R., HELLER, F. & ZINGG, A. 1989. The role of the Periadriatic Line in the tectonic evolution of the Alps. *In*: COWARD, M. P., DIETRICH, D. & PARK, R. G. (eds) *Alpine Tectonics*. Geological Society, London, Special Publications, **45**, 153–171.

SEARLE, M. P., WEINBERG, R. F. & DUNLAP, W. J. 1998. Transpressional tectonics along the Karakoram Fault Zone, northern Ladakh: constraints on Tibetan extrusion. *In*: HOLDSWORTH, R. E., STRACHAN, R. A. & DEWEY, J. D. (eds) *Continental Transpressional and Transtensional Tectonics*. Geological Society, London, Special Publications, **135**, 307–326.

SHAW, R. D. & WARREN, R. G. 1975. Alcoota, Northern Territory. *1:250000 geological sheet and explanatory notes, Bureau of Mineral Resources, Canberra, Australia*.

SHAW, R. D., ETHERIDGE, M. A. & LAMBECK, K. 1991. Development of the late Proterozoic to mid-Palaeozoic intracratonic Amadeus Basin in central Australia: A key to understanding tectonic forces in plate interiors. *Tectonics*, **10**, 688–721.

SHAW, R. D., STEWART, A. J. & BLACK, L. P. 1984a. The Arunta Inlier: A complex ensialic mobile belt in central Australia. Part 2: Tectonic history. *Australian Journal of Earth Sciences*, **31**, 457–484.

SHAW, R. D., WARREN, R. G., OFFE, L. A., FREEMAN, M. J. & HORSFALL, C. L. 1984b. *Geology of*

Arunta Block in the southern part of the Huckitta 1:250000 sheet area – Preliminary data 1980 survey. Bureau of Mineral Resources, Geology and Geophysics, Record **1984/3**.

SHAW, R. D., ZEITLER, P. K., McDOUGALL, I. & TINGATE, P. 1992. The Palaeozoic history of an unusual intracratonic thrust belt in central Australia based on ^{40}Ar–^{39}Ar, K–Ar and fission track dating. *Journal of the Geological Society, London*, **149**, 937–954.

STEWART, A. J., SHAW, R. D., OFFE, L. A., LANGWORTHY, A. P., WARREN, R. G., ALLEN, A. R. & CLARK, D. B. 1980. Stratigraphic definitions of named units in the Arunta Block, Northern Territory. *Bureau of Mineral Resources, Australia, Report*, **221**.

TAPPONNIER, P., PELTZER, G. & ARMIJO, R. 1986. On the mechanics of the collision between India and Asia. *In*: COWARD, M. P. & RIES, A. C. (eds) *Collisional Tectonics*. Geological Society, London, Special Publications, **19**, 115–157.

TUCKER, R. D., ASHWAL, L. D., HANDKE, M. J., HAMILTON, M. A., LEGRANGE, M. & RAMBELOSON, R. A. 1999. U–Pb geochronology and isotope geochemistry of the Archaean and Proterozoic rocks of north-central Madagascar. *Journal of Geology*, **107**, 135–153.

VERNON, R. H. 1996. Problems with inferring P–T–t paths in low-P granulite facies rocks. *Journal of Metamorphic Geology*, **14**, 143–153.

VRY, J., COMPSTON, W. & CARTWRIGHT, I. 1996. SHRIMP II dating of zircons and monazites: reassessing the timing of high-grade metamorphism and fluid flow in the Reynolds Range, northern Arunta Block, Australia. *Journal of Metamorphic Geology*, **14**, 335–350.

WALTER, M. R., VEEVERS, J. J., CALVER, C. R. & GREY, K. 1995. Neoproterozoic stratigraphy of the Centralian Superbasin, Australia. *Precambrian Research*, **73**, 173–195.

WARREN, R. G. 1978. Delny-Mount Sainthill Fault System, eastern Arunta Block, central Australia, *B.M.R. Journal of Australian Geology and Geophysics*, **3**, 76–79.

WARREN, R. G. 1983. Metamorphic and tectonic evolution of granulites from the Arunta Block, central Australia. *Nature*, **305**, 300–303.

WARREN, R. G. & HENSEN, B. J. 1989. The P–T evolution of the Proterozoic Arunta Block, central Australia, and implications for tectonic evolution. *In*: DALY, J. S., CLIFF, R. A. & YARDLEY, B. W. D. (eds) *Evolution of Metamorphic Belts*. Geological Society, London, Special Publications, **43**, 349–355.

WARREN, R. G. & STEWART, A. J. 1988. Isobaric cooling of Proterozoic high-temperature metamorphites in the northern Arunta Block, central Australia: implications for tectonic evolution. *Precambrian Research*, **40/41**, 175–198.

WARREN, R. G., HENSEN, B. J. & RYBURN, R. J. 1987. Wollastonite and scapolite in Precambrian calc-silicate granulites from Australia and Antarctica. *Journal of Metamorphic Geology*, **5**, 213–223.

WILLETT, S., BEAUMONT, C. & FULLSACK, P. 1993. A mechanical model for the tectonics of doubly-vergent compressional orogens. *Geology*, **21**, 371–374.

WILLIAMS, I. S., BUICK, I. S. & CARTWRIGHT, I. 1996. An extended episode of early Mesoproterozoic metamorphic fluid flow in the Reynolds Range, central Australia. *Journal of Metamorphic Geology*, **14**, 29–48.

WOLF, M. B. & WYLLIE, P. J. 1994. Dehydration-melting of amphibolite at 10 kbar; the effects of temperature and time. *Contributions to Mineralogy and Petrology*, **115**, 369–383.

WYLLIE, P. J. & WOLF, M. B. 1993. Amphibolite dehydration-melting: sorting out the solidus. *In*: PRIT-
CHARD, H. M., Alabaster, T., HARRIS, N. B. W. & NEARY, C. R. (eds) *Magmatic Processes and Plate Tectonics*. Geological Society, London, Special Publications, **76**, 405–416.

XU, G., WILL, T. M. & POWELL, R. 1994. A calculated petrogenetic grid for the system K_2O–FeO–MgO–Al_2O_3–SiO_2–H_2O, with particular reference to contact metamorphosed pelites. *Journal of Metamorphic Geology*, **12**, 99–119.

ZHAO, J.-X. & BENNETT, V. C. 1995. SHRIMP U–Pb zircon geochronology of granites in the Arunta Inlier, central Australia: implications for Proterozoic crustal evolution. *Precambrian Research*, **71**, 17–43.

The response of mineral chronometers to metamorphism and deformation in orogenic belts

RANDALL R. PARRISH

Department of Geology, University of Leicester & NERC Isotope Geoscience Laboratories, British Geological Survey, Keyworth, Nottingham NG12 5GG, UK
(e-mail: r.parrish@nigl.nerc.ac.uk)

Abstract: Mineral chronometers, especially accessory minerals using the U–Pb decay system, can reveal important information regarding the environmental conditions and duration of metamorphic–deformation events during the re-working of older rocks. Minerals such as zircon can newly grow during amphibolite facies or granulite facies events, providing direct ages of metamorphism. Pre-existing minerals like monazite, allanite, and titanite can preserve a component of their original age in spite of upper amphibolite facies re-working and very thorough recrystallization of the rock fabric during mylonite development. The degree of Pb loss can be used to deduce, at least semi-quantitatively, the temperature and duration of the subsequent event. In well-studied examples, the relative retentivity of Pb is highly predictable, and this helps place strong constraints on relative closure temperatures, even when laboratory experimental data are lacking or inconclusive. A number of examples are presented from a wide variety of geological environments to illustrate the response of U–Pb isotope systematics within accessory minerals to superimposed deformation, metamorphism and/or mineral growth.

The *response* of mineral isotope systematics to metamorphism and deformation refers to how pre-existing minerals are affected by new environmental conditions, and to how this response can be used to infer quantitative aspects such as temperature, timing and duration of a subsequent event. This paper reviews this topic as it pertains to the reactivation and reworking of older terrains within orogenic belts.

Although various isotopic systems have been applied to the cooling history and dating of rocks, this paper emphasizes the U–Pb scheme with accessory minerals because of their widespread application to Precambrian rocks. U–Pb geochronology has developed in the past 20 years into one of the most important geological tools and has revolutionized the interpretation of Precambrian orogens and their temporal evolution. One of the most important legacies of U–Pb dating using both conventional isotope dilution thermal ionization mass spectrometry (TIMS) and secondary ionization mass spectrometry (SIMS), especially SHRIMP; Compston *et al.* 1984; Stern 1999) has been the extraction of primary crystallization ages of igneous protoliths from complexly deformed and metamorphosed lithologies. The importance of this geological information is plain to see. For example, these type of studies have revealed the age of the world's oldest rocks

in the Acasta gneisses of the western Slave Province (Bowring & Williams 1999) that would otherwise have been indeterminate with any degree of precision. The ability of zircon to withstand the battering of multiple high grade metamorphic events and retain the signature of its original crystallization is well known. This is perhaps its most valuable aspect to earth scientists.

By contrast, the application of U–Pb dating using minerals other than zircon was slower to enter the mainstream. These alternative minerals can reveal the thermal and deformational history of complex orogenic belts. Minerals that either recrystallize or lose some or all of their radiogenic Pb during metamorphic–deformational events are the most important minerals to be employed in reconstructing the chronology of the thermal–metamorphic history.

The relevance of the U–Pb system in relation to other dating systems (i.e. K–Ar, Rb–Sr) lies in three important facts. First, the accessory minerals datable by the U–Pb method have quite variable Pb retentivities, allowing them to be used over a wide range of conditions. Second, many of these minerals have significant enrichment of uranium coupled with little or no incorporation of initial lead, leading to a more robust interpretation, and the dating can be undertaken with one mineral using single grain or intra-grain analysis.

From: MILLER, J. A., HOLDSWORTH, R. E., BUICK, I. S. & HAND, M. (eds) *Continental Reactivation and Reworking.* Geological Society, London, Special Publications, **184**, 289–301. 1-86239-080-0/01/$15.00 © The Geological Society of London 2001.

Third, the coupled decay scheme of two uranium parent isotopes decaying to two different radiogenic lead isotopes potentially allows no loss of precision across the entire spectrum of geological time with about one million year resolution. A drawback however, is the difficulty in relating the growth of accessory minerals to the development of specific textures or mineral reactions with P–T significance.

Many of the accessory minerals used in U–Pb geochronology, such as zircon, titanite, monazite, xenotime, rutile, apatite, and allanite are involved in metamorphic reactions with other minerals including garnet, clinopyroxene, amphibole, plagioclase, etc. Although the pressure–temperature–fluid conditions of such reactions are at present very poorly understood, there is great potential to coordinate the isotopic and P–T data into a robust P–T–time path which could improve significantly the interpretation of calculated ages. This in turn would quantifiably improve knowledge of the response of pre-existing datable minerals to subsequent tectonic events.

This paper will introduce the main elements that govern how these mineral chronometers respond to subsequent events, present several illustrative examples, and discuss the metamorphic petrology of accessory minerals that have undergone recrystallization during reworking, relating dating to mineral-forming reactions.

Processes affecting the isotopic response in mineral chronometers

Mineral cooling and closure temperature

The literature on the theoretical, experimental, and empirical aspects of closure temperature in minerals is well-developed, having its origins in the work of Armstrong (1966). In that study of orogenic belts, widespread re-setting of K–Ar ages by subsequent re-heating and metamorphism was clearly documented. The theory of radiogenic isotope retention in cooling rocks was elegantly quantified by Dodson (1973), whose name is synonymous with closure temperature. This concept is now widely applied to reveal aspects of the thermal history of cooling orogens. Provided that minerals cooled from temperatures where there was no retention of radiogenic daughter isotope to temperatures where complete retention prevailed, the approach is generally valid. Taking advantage of the variable closure temperatures of a variety of minerals using several analytical methods, it is possible to determine the cooling history of a terrain over >600°C. Figure 1 is an example from British Columbia of a

relatively complete thermal history using this approach (Spear & Parrish 1996), using dating schemes discussed further in this paper.

Relative retentivities of radiogenic daughter isotopes in minerals

Many studies, both experimental and empirical, have helped to elucidate the approximate closure temperatures of minerals used in thermochronology. Table 1 summarizes this body of knowledge. The closure temperature of some minerals is entirely within the brittle field of deformation (i.e. fission track dates in apatite), whereas that of other minerals are within middle to upper amphibolite facies conditions where metamorphic recrystallization leading to ductile flow is common (i.e. titanite U–Pb). The value and relevance of mineral chronometers to revealing conditions of orogenic reworking is therefore quite variable.

The isotopic response of minerals to heating

In reworked terrains, incomplete loss of radiogenic isotopes (i.e. partial re-setting) during a heating event is common, and in these circumstances, the theory is less well developed with few well-studied examples. Examples are the presence of complex argon release spectra in hornblende, mica or feldspar (McDougall & Harrison 1988), complex Rb–Sr mineral isochrons, discordance in U–Pb dates, and inherited xenocrystic zircon in anatectic magma.

In an influential early study by Hart (1964), the pattern of age re-setting in Precambrian gneisses in Colorado as a function of distance from the Tertiary Eldora stock was well documented. Hornblende K–Ar dates were fully reset only within a few metres of the contact whereas biotite K–Ar dates reflected the age of intrusion up to 100 m away from the granite–gneiss boundary. This study was one of the first to establish the relative retentivities of radiogenic Sr and Ar in minerals. From this type of study it is possible to calculate the *fractional loss* of radiogenic daughter isotope that occurred during heating. Using equations of diffusion (i.e. Carslaw & Jaeger 1959), mineral diffusivities, duration of heating and effective grain size, the maximum temperature of heating can be deduced, in theory. Alternatively, if the temperature was well-constrained, the maximum duration of a heating event could be determined, which in turn could have important tectonic consequences.

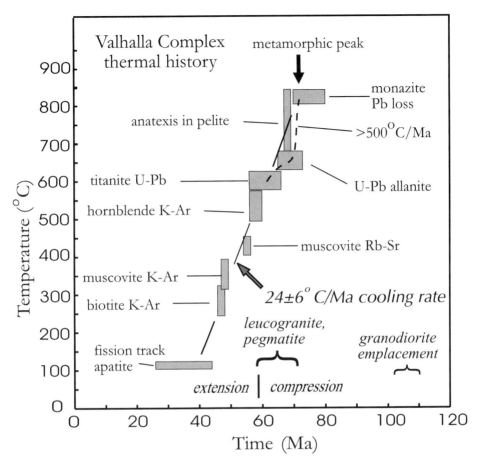

Fig. 1. Valhalla core complex thermal history, from Spear & Parrish (1996). The mineral chronometers used allow the reconstruction of the full cooling history of this high-grade terrain, with an average rate of cooling of about 25°C Ma⁻¹. Higher rates of short-lived cooling may have occurred immediately after reaching peak temperatures, possibly the result of thrust juxtaposition at deep crustal levels, as suggested by the dotted line. Fission track ages vary with altitude with a total range of nearly 20 Ma. Note that the temperature used for the Rb–Sr muscovite age of *c.* 55 Ma is not the Rb–Sr muscovite closure temperature from Table 1 because the dated mica grew at lower greenschist facies metamorphic conditions (about 400°C), well below the nominal 500°C closure temperature; this muscovite is therefore not a *cooling* age.

Table 1. *Mineral isotopic closure temperatures*

Mineral	Closure temperature	Reference
U–Th–Pb zircon	>900°C	Lee *et al.* (1997)
U–Th–Pb monazite	>750°C	Spear & Parrish (1996)
U–Pb xenotime	>650°C(?)	Heaman & Parrish (1991)
U–Th–Pb allanite	650°C	Heaman & Parrish (1991)
U–Th–Pb titanite	600°C	Heaman & Parrish (1991); Cherniak (1993)
U–Pb rutile	400°C	Mezger *et al.* (1989)
K–Ar, ^{39}Ar–^{40}Ar hornblende	530 ± 40°C	Harrison (1981); MacDougall & Harrison (1988)
Rb–Sr muscovite	500°C	Purdy & Jäger (1976), Jäger *et al.* (1967)
K–Ar, ^{39}Ar–^{40}Ar muscovite	350°C	Purdy & Jäger (1976)
K–Ar, ^{39}Ar–^{40}Ar biotite	280 ± 40°C	Harrison *et al.* (1985)
Fission Track apatite	100–120°C	Naeser (1979)

An integrated example of basement reheating

In a study from the southeastern Canadian Cordillera, Crowley (1999), Crowley & Parrish (1999), and Parrish (1995) studied the response of mineral chronometers in c. 1.9–2.1 Ga protolith gneisses to re-heating during a 50–60 Ma event involving intra-crustal thrusting. In this area high grade (sillimanite–K-feldspar–anatexis) hanging wall terrain (the Selkirk Allochthon) was emplaced within a thick ductile shear zone (the Monashee decollement), onto the footwall consisting of Precambrian granitic/gneissic rocks (the Monashee Complex). The heating of the footwall was variable, with deformational and metamorphic effects decreasing downwards. Significant reworking of the footwall is present within the shear zone up to 1.0 km beneath the thrust, with widespread recrystallization of quartz, feldspar, mica and amphibole. Beneath the zone of re-working, Precambrian structures can be recognized (Fig. 2).

1.86 Ga titanite, monazite, and zircon are present within amphibolite and leucogranite of the footwall. Zircon is the least well-behaved of the minerals in terms of U–Pb systematics, due to a combination of Pb loss events. However, the U–Pb systematics of titanite and monazite reveal a linear array of analyses connecting c. 1.86 Ga with c. 50–60 Ma (summarized in Fig. 3 from a much larger data set). There appears to be a correlation between U–Pb discordance and proximity to the hanging wall, with discordance varying widely from 6% to 96%. There is also limited evidence that discordance may partly be a function of grain size, but this is not simple because of partial recrystallization. Notwithstanding the complexities of recrystallization manifested by a coronitic zoning of apatite + thorite on original monazite, the maximum fractional loss of radiogenic Pb from monazite and titanite from lower structural levels during the 60 Ma event can be broadly estimated from these analyses to be <30% and 65%, respectively (Fig. 3).

For the sake of argument, if one accepts that diffusion coefficients for these minerals can be estimated, even with inconsistent experimental values (Heaman & Parrish 1991; Cherniak 1993; Suzuki et al. 1994; Smith & Giletti 1997), a family of maximum temperature and heating duration combinations can produce the documented fractional loss, for the grain sizes of minerals that were analysed. Using maximum temperatures derived from petrological arguments (i.e. 600–700°C), the duration of heating can be deduced. This analysis was done as part of the Parrish (1995) and Crowley & Parrish (1999) studies; these heating duration estimates were much shorter than

intuitively expected, generally a maximum of two million years. Because the shear zone has displacement exceeding 100 km, relatively rapid rates of emplacement are implied. Although this analysis has limitations due to uncertain input parameters, it indicates the potential of the approach in providing a quantitative estimate of heating rates, that in turn provide constraints on the rates of movement and strain in shear zones. This information was extracted mainly from mineral chronometers in the reworked older gneisses. By contrast, mineral dates on all hanging wall minerals other than zircon are Cretaceous or younger and these provide no particular insight into the conditions prevailing in the reworked footwall rocks. This example establishes how the thermal history, including inferences about the duration of events, can be deduced by the response of minerals to orogenic re-heating during reworking.

Armouring of minerals: isotopic effects

The diffusion of radiogenic isotopes from mineral to matrix occurs when the mineral's diffusivity is sufficiently high at a given temperature, provided the matrix can incorporate the diffusing element. This process becomes limited when a mineral is enclosed within another mineral of limited diffusivity, for example when armoured by garnet. Recent papers by Zhu et al. (1997) and Zhu & O'Nions (1999) documented monazite crystals within garnet, in high grade pelite of the late Archaean Lewisian basement of NW Scotland. In these samples, monazite ages on armoured inclusions are generally older than those of the matrix, by up to 200 Ma or more. This example illustrates how a protracted metamorphic history may be extracted from a single thin section. The metamorphic reaction(s) that caused primary growth of metamorphic monazite, and the associated pressure–temperature–fluid environment, are however, less clear.

Consequences of metamorphism and recrystallization to isotope systematics

When older rocks are reworked by a subsequent event, there is an imposition of new environmental conditions. These comprise different pressure and temperature, and the introduction of a fluid phase into an otherwise relatively dry rock. Variable strain is also imposed, a condition favouring neoblast growth and variable recrystallization of pre-existing mineral phases. The new environmental conditions can result in prograde

Fig. 2. Photographs of Early Proterozoic structures re-worked by a Tertiary structural–metamorphic event in British Columbia. (**a**) Folded gneissosity developed in 2.1 Ga orthogneiss and amphibolite, cut by foliated 2.05 Ga amphibolite dyke, and then deformed during subsequent events at 1.85 Ga and 50–60 Ma; hammer is 30 cm in length. (**b**) Folded and sheared ortho- and paragneisses of 1.86–2.2 Ga age that were subsequently deformed 50–60 Ma ago. A variety of structures can be discerned, the latest of which are about 50 Ma old. The person standing is 1.6 m in height.

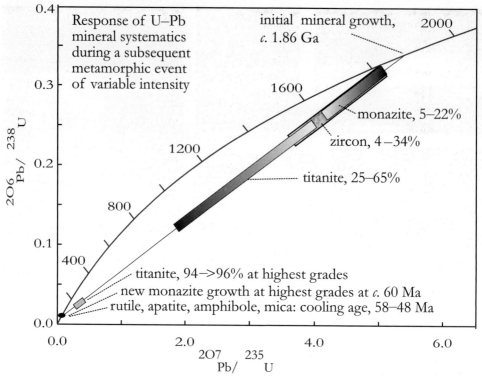

Fig. 3. U–Pb concordia showing a summary of more than 50 analyses of zircon, monazite and titanite from rocks of the Monashee Complex of southeastern British Columbia, drawing on the work of Parrish (1995), Crowley (1999) and Crowley & Parrish (1999). The rocks dated consist of Early Proterozoic gneiss with a strong *c*. 1.85 Ga orogenic signature that were involved in a subsequent orogenic event under amphibolite facies conditions *c*. 50–60 Ma ago. During the younger event the Monashee Complex footwall was overthrust by a thick allochthon (the Selkirk allochthon) undergoing uppermost amphibolite facies metamorphism, and this juxtaposition preserved inverted metamorphic isograds in the sheared footwall. Accessory minerals from Neoproterozoic gneisses of the footwall were disturbed by Pb loss and new mineral growth during the 50–60 Ma event, and the effect of this event was more intense at higher structural levels. In the figure, the field of analyses from the deeper structural levels are shown as the three larger patterned boxes with the least discordance. Note that titanite is more discordant than monazite. The highest grade of metamorphism in the footwall is just beneath the decollement and the analyses from these samples show much greater discordance and substantial new mineral growth. The discordance of minerals is likely the result of radiogenic Pb loss and/or new crystal growth onto an existing mineral. Studies of this type clearly demonstrate that the relative retentivity of Pb for monazite is higher than titanite.

metamorphism of the older rocks, for example heating mafic lithologies to eclogite or granulite facies. Also encountered in orogenic belts is the retrogressive effect of reworking of granitic rocks with a pre-existing high temperature igneous mineralogy. Fabrics associated with the younger event are often penetrative and characterized by mineralogy consistent with the superimposed conditions.

Prograde reworking: eclogite and granulite development

High pressure and high temperature metamorphism of reworked basement is a common feature of

collisional orogens when subduction of continental crust occurs. Metamorphic zircon is a common feature of certain lithologies, mainly mafic rocks, and it provides an excellent means to date the metamorphic event (Davidson & van Breemen 1988; Figs 4a, 5a, c).

The Norwegian Caledonides contain a world class example of this type where footwall gneisses of Mesoproterozoic age were overthrust by Caledonian nappes during continental collision (Gee 1978). A regional eclogite terrane was developed, manifest by the preservation of eclogite in appropriate mafic lithologies, often occurring only as boudins (Wain 1997). These older mafic rocks, of presumed pre-Caledonian amphibolite or lower metamorphic facies, were deformed

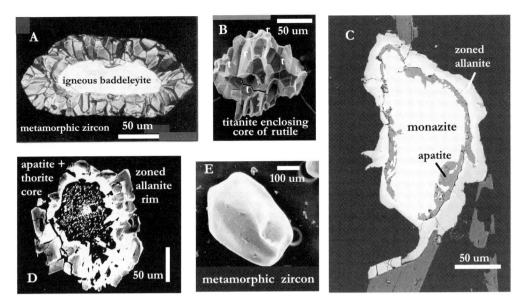

Fig. 4. Scanning electron micrographs of reaction textures involving accessory minerals. (**a**) back scattered electron microscope image of metamorphic zircon replacing baddeleyite (ZrO_2), in coronitic reaction textures within gabbroic rocks of the metamorphosed Sudbury dykes of Ontario (Davidson & van Breemen 1988); (**b**) secondary electron image of rutile replaced by titanite 'buds' in the retrogressed eclogite from Glenelg, Scotland; note the similar morphology of these replacement grains to those of figure (**a**); (**c**) backscattered electron image of apatite and zoned allanite–epidote corona surrounding a core of monazite, evidence of monazite breakdown in schist in the Hunza region of northern Pakistan; (**d**) backscattered electron image of monazite breakdown reaction to produce a core of apatite + thorite (all monazite appears to have reacted, though some small relicts may be preserved in the bright central region), surrounded by REE-zoned allanite–epidote, also from Hunza, Pakistan; (**e**) secondary electron image of metamorphic zircon with round aspect, growing in retrogressed zone within eclogite, from near Glenelg, Scotland.

into discontinuous sheets and boudins during shearing, with significant changes to mineralogy and texture. In adjacent 1.6 Ga granitic gneisses, shear fabrics were superimposed but mineralogical evidence for high pressure metamorphism is less evident (Dewey *et al.* 1993).

In a study of the titanite occurring in the 1.6 Ga protoliths of the Norwegian Caledonides, Tucker *et al.* (1987) showed that a younger generation of titanite was present, in addition to older crystals, with core-rim zoning being observed. The U–Pb systematics of titanite from various samples preserved a striking linear array on the concordia diagram connecting the original titanite age (1.66 Ga) with the 395 Ma age of the Caledonian metamorphic event. Titanite U–Pb systematics indicate 6–100% discordance along this linear array. Although it is not possible to fully differentiate the effects of Pb loss and new growth on the isotope systematics, the age and relatively short duration of the Caledonian event was accurately determined. The tectonic events, including continental subduction, eclogite facies metamorphism, and rapid tectonic denudation

and cooling, occurred rapidly (Krabbendam & Dewey 1998).

In a study of variably retrogressed eclogite preserved in the Lewisian Glenelg inlier of NW Scotland, Sanders *et al.* (1984) determined a 1.0–1.1 Ga Sm–Nd age of garnet–clinopyroxene pairs, suggesting 'Grenvillian' eclogite development within host gneisses that were inferred to be >2.5 Ga. Subsequent detailed study of one eclogite (T. S. Brewer & others, pers. comm., 1999) revealed that metamorphic zircon occurred in these samples (Fig. 4e), usually as round to irregular grains with very low uranium contents of <10 ppm. The rare occurrence of these grains in thin section suggests that they preferentially occur within zones of hydration and retrogression. Rutile in equilibrium with the eclogite facies assemblage is partially replaced by titanite (Fig. 4b) produced during retrogression. Titanite also occurs as matrix grains. An analysis of the U–Pb systematics of zircon, titanite and rutile reveals a complex picture of isotopic disequilibrium. The titanite is similar in age, within error, to the metamorphic zircon at very close to 1.00 Ga. Rutile,

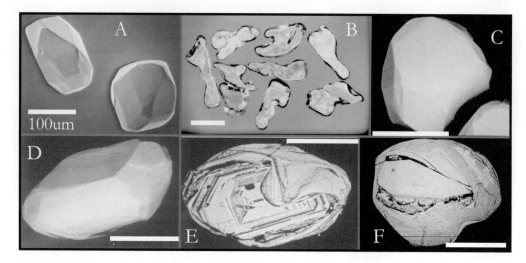

Fig. 5. Microscopic images of external and internal morphologies of zircon from re-worked gniessic rocks. Scale bar on all images are 100 μm. (**a**) secondary electron image of 55 Ma metamorphic zircon growing in the solid state in a metamorphized 740 Ma nepheline syenite gneiss from the Monashee Complex of British Columbia; (**b**) transmitted light image of highly irregular shapes of zircon from a recrystallized mylonite developed from a 100 Ma hornblende–biotite grantoic protolith of the Valhalla Complex of southern British Columbia. The irregularity of the grains is predominantly the result of strong resorption of igneous grains, and comparison of some of these shapes to that of figure (e) indicates strong similarities; (**c**) secondary electron image of 50 Ma anhedral metamorphic zircon grown in the solid state in a hornblende–plagioclase amphibolite of probable Late Palaeozoic age, with multiple planar growth faces that do not clearly correspond to crystallographic faces; southern British Columbia; (**d**) secondary electron image of moderately resorbed subhedral zircon with significant metamorphically-induced rounding of crystal faces, Valhalla Complex of southern British Columbia; (**e**) backscattered electron image of etched internal surface of highly resorbed Early Proterozoic zircon from *c.* 1.95 Ga granulite facies Thelon tectonic zone of the northwest Canadian Shield. The >2.0 Ga core is overgrown by new zircon of *c.* 1.95 Ga age; note the overall round shape masking the internal complexity and evidence of strong resorption; (**f**) backscattered electron image of composite granulite facies zircon from mylonitic straight gneiss of the Thelon zone, northwest Canadian Shield; the resorbed >2.0 Ga core is overgrown by a complex region of wedge-shaped growth zones, most of which are in turn truncated and overgrown by new zones, resembling 'cross-stratification; this may have been produced by repeated dissolution and precipitation of zircon rotating in the rock fabric during shearing, exposing sides of the crystal to differential chemical and strain gradients during rotation.

on the other hand, is *c.* 430 Ma. The rutile age can be explained by its low Pb retentivity, with final closure of its Pb mobility in Caledonian time.

These examples show that in reworked older rocks, zircon and titanite can provide information on the age of subsequent metamorphic events during both eclogite facies and later retrogression. Careful study of the fabrics associated with these mineralogical changes may allow inferences about the absolute age of the fabric development to complement geological or regional arguments.

Retrogression of granulites during reworking

In the Ungava peninsula, northern Quebec, the Late Archaean crust of the Superior Province was overridden by the metasedimentary and metavolcanic allochthons of the Cape Smith Belt

(St-Onge *et al.* 1992). The Archaean rocks beneath the basal decollement preserve evidence of both granulite and amphibolite facies metamorphism, whereas the allochthons are of greenschist to lower amphibolite facies. The zone of footwall deformation (Lucas & St-Onge 1992) is tens to hundreds of metres thick, and is associated with hydration and metamorphic retrogression. Within granulites, titanite is absent, rutile being the stable Ti-bearing phase. Within the hornblende-bearing retrogressed zones, original orthopyroxene and rutile are rare to absent, and titanite is stable, in part overgrowing ilmenite. The retrogressive reaction(s) were probably catalyzed by the presence of water. In a study of the isotopic response of these samples, Scott & St-Onge (1995) dated titanite from numerous samples of retrogressed granulite within this zone. The ages were nearly concordant, ranging from

1814 to 1789 Ma, and these dates were interpreted to reflect the age of titanite growth, fabric development, retrogression, and shear strain. In this case petrographic evidence of new titanite growth is very clear. A factor considered by the authors was the extent to which the titanite ages were cooling ages, since the temperatures of retrogressive metamorphism ($670°C$) are similar to the titanite closure temperature estimate (Table 1). If the titanite dates were in fact cooling ages, then they place only a minimum constraint on the ages of the retrogression and shearing. This situation illustrates an interpretation ambiguity that shall remain until better diffusion information is obtained.

Zircon within these rocks is virtually unaffected by the Proterozoic retrogression event, and this can provide no useful information on its age. Because of the relatively low metamorphic grade, no pegmatite or other intrusive rocks are associated with the shearing, rendering the dating of the shear zone problematic. Dates using the K–Ar and ^{40}Ar–^{39}Ar methods on hornblende and mica are <1750 Ma and they shed little light on the age of basement reworking. Titanite growth, a product of mineral reactions during shearing and associated retrogression, provides instead the most important timing constraint of shear zone development.

Reworking of rocks under amphibolite facies conditions

Hornblende–biotite bearing granitic rocks are a common granitoid type in continental crust. They are commonly involved in subsequent re-working events. If subjected to middle to upper amphibolite facies conditions during shearing and recrystallization, the mineralogy is unlikely to change substantially, though there may be penetrative fabric development. In the re-heating and recrystallization of these types of rocks, Fe–Mg mica and hornblende will be completely reset, losing all radiogenic argon (and Sr) and will yield cooling ages if dated. Accessory minerals such as allanite, titanite, zircon and apatite may be present, and for the most part will remain stable mineralogically during middle to upper amphibolite facies re-working. Although recrystallization of these accessory minerals does occur and is sometimes visible as core/rim overgrowth relationships (Tucker et al. 1987), titanite and allanite appear to have greater internal strength than most rock forming minerals. They often resist recrystallization during penetrative fabric development, although a common feature is a partial resorption of sharp edges and surfaces of euhedral pheno-

crysts to produce a rounded subhedral appearance, as illustrated by zircon in Figure 5d and e. Sometimes, resorption can be extreme, producing highly irregular crystal shapes (Fig. 5b).

In a study of ultramylonites within the Brevard zone of the Appalachians, Wayne & Sinha (1988, 1992) showed that zircon was little-affected by intense fracturing and by the thorough recrystallization and grain size reduction in the rock's matrix. These studies show that little information can be obtained on the nature of the subsequent deformational event from zircon; instead, zircon clearly preserves the protolith age.

The survival of original igneous grains in spite of major recrystallization can be proven by the U–Pb systematics in a study of the reworking of 100 Ma and 360 Ma granodiorite protoliths during an upper amphibolite facies event c. 60 Ma ago (Fig. 6; Heaman & Parrish 1991; Spear & Parrish 1996). In that study, allanite was found to be 75–90% discordant, but not fully reset. Titanite from the same samples was much more discordant, approximately 90–99%, but still not totally reset. Aside from demonstrating survival of original grains, these data clearly show that allanite has a stronger retention of Pb than titanite. These discordant arrays can be used to infer both initial age of protolith (along with zircon), age of the subsequent event, and semiquantitative information on the temperature and duration of the later event. These minerals are thus a powerful bridge between the information obtained from U–Pb zircon on the one hand, and K–Ar or Rb–Sr mica ages on the other.

Metamorphic and isotopic effects during lower temperature reworking

During lower amphibolite and greenschist facies metamorphic conditions, granitic lithologies usually contain a mineralogy substantially modified from the protolith. For example, hornblende may react to form biotite, epidote, and/or chlorite, and relatively calcic plagioclase may become sausseritized with the formation of albite, epidote and calcite. With more leucocratic lithologies (mica-bearing granitoid rocks) affected by lower greenschist conditions, biotite reacts to form chlorite, or muscovite if metasomatism is significant. Because the lower temperature stability of hornblende virtually coincides with its closure temperature ($c. 500°C$), argon dating of hornblende will rarely, if ever, provide an unambiguous age of mineral growth, excepting rapidly cooled and unaltered volcanic rocks. Argon dating of mica has the same problem.

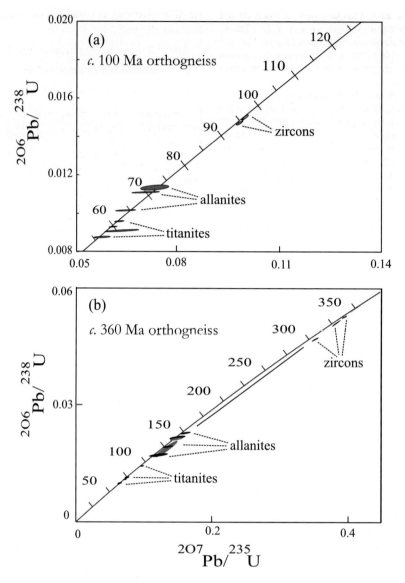

Fig. 6. U–Pb concordia diagram of zircon, allanite and titanite from granitic rocks reworked by subsequent upper amphibolite facies heating and recrystallisation. These samples are from the metamorphic core zone of southeast British Columbia, including the Selkirk Allochthon and Valhalla Complex. **(a)** the 100 Ma Mulvey gneiss of the Valhalla Complex (Spear & Parrish 1995) with clear separation of zircon, allanite, and titanite analyses illustrating variable Pb loss; **(b)** similar pattern of discordance within Devonian orthogneiss of the Selkirk Allochthon (see Heaman & Parrish 1991). Both samples were thoroughly recrystallized at *c.* 60 Ma during metamorphism and deformation. Although experiencing some resorption and new crystal growth, the accessory minerals preserve some of the isotopic signature of the protolith, indicating that Pb loss was not complete. This proves that allanite and titanite did not undergo the thorough recrystallization experienced by the matrix minerals.

Useful isotopic information on the age of crystal growth can be extracted from Rb–Sr dating of medium to coarse-grained muscovite. In an insightful study, Piasecki & van Breemen (1983) recognized correctly that Rb–Sr dating of large books of muscovite, even when subjected to an amphibolite facies Caledonian event, preserved evidence of their crystallization more than 700 Ma ago. This can be explained by the positive correlation of grain size with closure temperature (Dodson 1973); very large muscovite books several centimetres in diameter may retain their radiogenic Sr to upper amphibolite facies temperatures.

The studies of Meffan-Main & Cliff (1997) and Freeman et al. (1997) in the Swiss Alps applied Rb–Sr dating to fabric-forming minerals in the re-worked Variscan granitoid basement nappes of the central Alps. They used micro-sampling techniques to date new mica, calcite, and feldspar that grew during the Alpine event. This work has allowed robust direct dating of Alpine deformation and has also provided an explanation of why bulk dating of mica and feldspar from similar samples have ambiguous ages that are difficult to interpret. Isotopic disequilibrium amongst the major phases, including older feldspar megacrysts, is widespread in this greenschist grade environment.

In an application to the dating of shear zone fabrics, Parrish et al. (1988) studied porphyroblastic muscovite growing in a sheared biotite granite during lower greenschist facies conditions. In this sample, biotite disappeared from the protolith upon significant ductile shearing and recrystallization at this low grade. Metasomatism accompanied by major fluid fluxing induced the growth of muscovite at the expense of biotite, with only minor formation of chlorite. Muscovite 'fish' that formed concurrently with C–S fabric development, yielded Rb–Sr ages calculated with feldspar that accurately reflected the age of shearing. Because the closure temperature for Sr diffusion in muscovite is about 500°C, the mica grew more than 100°C below its closure temperature, a condition ensuring full radiogenic Sr retention and a robust age determination; these mica ages are shown in the thermal history of Figure 1.

Allanite–monazite equilibria in retrogressed rocks

There are several good examples of reaction textures involving allanite, monazite, apatite, and/or thorite from samples exposed to either a subsequent metamorphic event or a retrogressive metamorphism following peak temperature. Finger et al (1998), Fraser (2000), and Simpson et al. (2000) describe granitoid and/or pelitic samples that contain monazite surrounded by a complex reaction rim involving apatite, allanite, and/or thorite. This is probably a retrogressive reaction involving the addition of water, possibly represented by the equation:

$$\text{monazite} + \text{calcic plagioclase} + \text{water}$$
$$= \text{allanite} + \text{apatite} + \text{thorite}$$
$$+ \text{Ca-poorer plagioclase}$$

In these studies, these striking textures (Fig. 3c, d) have failed to progress to completion, probably limited by diffusion and possibly the availability of water. Their importance lies in the potential to date the retrogressive reaction by *in situ* isotopic techniques, and hypothetically, to determine P–T conditions of such reactions. Because little is known of the phase equilibria of the CaO–REE–phosphate–ThO–H$_2$O system in rocks, quantitative P–T–fluid conditions cannot be determined at present.

Summary

Notwithstanding the fact that most dating studies focus on determining protolith ages, the age of peak metamorphism, and/or the age of deformation events using magmatic rocks, there is much to be learned about the full temperature–time path and the response of rocks to reworking by the study of accessory minerals. Minerals other than zircon, namely allanite, monazite, titanite, and rutile often provide key information pertaining to the full thermal and metamorphic history of samples. This is possible because of their differential retention of radiogenic Pb, their structural ability to withstand intense recrystallization of the rock's matrix, and the nature of the U–Th–Pb isotope systematics. The understanding of the thermochemistry of mineral reactions involving accessory phases is in its infancy, but fascinating textures involving these minerals indicate that they are dynamically responding to changing environmental conditions during subsequent events. Careful study of complex samples can reveal both protolith crystallization ages, protolith cooling ages, ages of the subsequent metamorphic–deformational event, and semi-quantitative information about the temperature and duration of subsequent events. The number of careful studies in this field is small at present, but is increasing steadily. The ability of accessory minerals and in some cases other rock forming minerals (micas) to record complexities of the chronology within a rock's reacting matrix ensures that fruitful research in this field will continue for years to come.

The author thanks T. Brewer, J. Crowley, J. Fraser, R. Simpson, and O. van Breemen for illustrations used in the text, and R. E. Holdsworth and J. A. Miller for the invitation to contribute to this volume. This work has been supported by NERC grant GR3/13006, NERC funding to the NERC Isotope Geosciences Laboratory, and Leicester University.

References

ARMSTRONG, R. L. 1966. K–Ar dating of plutonic and volcanic rocks in orogenic belts. *In*: SCHAEFFER, O. A. & ZAHRINGER, J. (eds) *Potassium–Argon Dating.* Springer-Verlag, New York, 117–133.

BOWRING, S. A. & WILLIAMS, I. S. 1999. Priscoan (4.00 Ga) orthogneisses from northwestern Canada. *Contributions to Mineralogy and Petrology,* **134**, 3–16.

CARSLAW, H. S. & JAEGER, J. C. 1959. *Conduction of Heat in Solids,* 2nd edition. Clarendon, Oxford.

CHERNIAK, D. J. 1993. Lead diffusion in titanite and preliminary results on the effects of radiation damage on Pb transport. *Chemical Geology,* **110**, 177–194.

COMPSTON, W., WILLIAMS, I. & MEYER, C. 1984. U–Pb geochronology of zircons from lunar breccia 73217 using a sensitive high mass-resolution ion microprobe. *Journal of Geophysical Research,* **89B**, 525–534.

CROWLEY, J. L. 1999. U–Pb geochronologic constraints on Paleoproterozoic tectonism in the Monashee complex, Canadian Cordillera: elucidating an overprinted geologic history. *Geological Society of America Bulletin,* **111**, 560–577.

CROWLEY, J. L. & PARRISH, R. R. 1999. U–Pb isotopic constraints on diachronous metamorphism in the northern Monashee complex, southern Canadian Cordillera. *Journal of Metamorphic Geology,* **17**, 483–502.

DAVIDSON, A. & VAN BREEMEN, O. 1988. Baddeleyite–zircon relationships in coronitic metagabbro, Grenville Province, Ontario: implications for geochronology. *Contributions to Mineralogy and Petrology,* **100**, 291–299.

DEWEY, J. F., RYAN, P. D. & ANDERSEN, T. B. 1993. Orogenic uplift and collapse, crustal thickness, fabrics and metamorphic phase changes: the role of eclogites. *In*: ALABASTER, H. M., HARRIS, N. B. W. & NEARY, C. R. (eds) *Magmatic Processes and Plate Tectonics.* Geological Society, London, Special Publications, **76**, 325–343.

DODSON, M. H. 1973. Closure temperatures in cooling geochronological and petrological systems. *Contributions to Mineralogy and Petrology,* **40**, 259–274.

FINGER, F., BROSKA, I., ROBERTS, M. P. & SCHERMAIER, A. 1998. Replacement of primary monazite by apatite-allanite-epidote coronas in an amphibolite facies granite gneiss from the eastern Alps. *American Mineralogist,* **83**, 248–258.

FRASER, J. E. 2000. *The structural and metamorphic evolution of the deep crust in the Hunza Karakoram, north Pakistan.* DPhil Thesis, Oxford University.

FREEMAN, S. R., INGER, S., BUTLER, R. W. H. & CLIFF, R. A. 1997. Dating deformation using Rb–Sr in white micas: greenschist facies deformation ages for the Enterlor Shear Zone, Italian Alps. *Tectonics,* **16**, 57–76.

GEE, D. G. 1978. Nappe displacement in the Scandinavian Caledonides. *Tectonophysics,* **47**, 393–419.

HARRISON, T. M. 1981. Diffusion of ^{40}Ar in hornblende. *Contributions to Mineralogy and Petrology,* **78**, 324–331.

HARRISON, T. M., DUNCAN, I. & McDOUGALL, I. 1985. Diffusion of ^{40}Ar in biotite: Temperature, pressure and compositional effects. *Geochimica Cosmochimica Acta,* **49**, 2461–2468.

HART, S. R. 1964. The petrology and isotopic-mineral age relations of a contact zone in the Front Range, Colorado. *Journal of Geology,* **72**, 493–525.

HEAMAN, L. & PARRISH, R. R. 1991. U–Pb geochronology of accessory minerals. *In*: HEAMAN, L. & LUDDEN, J. N. (eds) *Applications of Radiogenic Isotope Systems to Problems in Geology.* Mineralogical Association of Canada, Short Course Handbook, **19**, 59–102.

JÄGER, E., NIGGLI, E. & WENK, E. 1967. Rb–Sr Altersbestimmungen an Glimmern der Zentralalpen. *Beiträge zur Geologischen Karte der Schweiz,* **134**.

KRABBENDAM, M. & DEWEY, J. F. 1998. Exhumation of UHP rocks by transtension in the Western Gneiss Region, Scandinavian Caledonides. *In*: HOLDSWORTH, R. E., STRACHAN, R. A. & DEWEY, J. F. (eds) *Continental Transpressional and Transtensional Tectonics.* Geological Society, London, Special Publications, **135**, 159–181.

LEE, J. K. W., WILLIAMS, I. S. & ELLIS, D. J. 1997. Pb, U and Th diffusion in natural zircon. *Nature,* **390**, 159–162.

LUCAS, S. & ST-ONGE, M. 1992. Terrane accretion in the internal zone of the Ungava orogen, northern Quebec. Part 2: Structural and metamorphic history. *Canadian Journal of Earth Sciences,* **29**, 765–782.

McDOUGALL, I. & HARRISON, T. M. 1988. *Geochronology and Thermochronology by the $^{40}Ar/^{39}Ar$ method.* Oxford Monographs on Geology and Geophysics **9**.

MEFFAN-MAIN, S. & CLIFF, R. A. 1997. Reliable dating of metamorphic fabrics in Alpine basement rocks using Rb–Sr microsampling. *Terra Nova,* **9**, Abstract supplement 1, 486–487.

MEZGER, K., HANSON, G. N. & BOHLEN, S. R. 1989. High-precision U–Pb ages of metamorphic rutile: application to the cooling history of high-grade terranes. *Earth and Planetary Science Letters,* **96**, 106–118.

NAESER, C. W. 1979. Fission track dating and geological annealing of fission tracks. *In*: JAGER, E. & HUNZIKER, J. C. (eds) *Lectures in Isotope Geology.* Springer-Verlag, New York, 154–169.

PARRISH, R. R. 1995. Thermal and tectonic evolution of the southeastern Canadian Cordillera, *Canadian Journal of Earth Sciences,* **32**, 1618–1642.

PARRISH, R. R., CARR, S. C. & PARKINSON, D. 1988. Eocene extensional tectonics and geochronology of the southern Omineca belt, British Columbia and Washington. *Tectonics,* **7**, 181–212.

PIASECKI, M. A. J. & VAN BREEMEN, O. 1983. Field and isotopic evidence for a *c.* 750 Ma tectonothermal event in Moine rocks in the Central Highland region of the Scottish Caledonides. *Transactions of the Royal Society of Edinburgh: Earth Sciences*, **73**, 119–134.

PURDY, J. W. & JÄGER, E. 1976. *K–Ar ages on rock-forming minerals from the Central Alps*. Institute of Geology and Mineralogy, University of Padova, Memoirs, **30**, 1–31.

SANDERS, I. S., VAN CALSTEREN, P. W. C. & HAWKESWORTH, C. J. 1984. A Grenville Sm–Nd age for the Glenelg eclogite in north-west Scotland. *Nature*, **312**, 439–440.

SCOTT, D. J & ST-ONGE, M. 1995. Constraints on Pb closure temperature in titanite based on rocks from the Ungava orogen, Canada: Implications for U–Pb geochronology and P–T–t path determinations. *Geology*, **23**, 1123–1126.

SIMPSON, R. L., PARRISH, R. R., SEARLE, M. P. & WATERS, D. J. 2000. Two episodes of monazite crystallisation during prograde metamorphism and crustal melting in the Everest region of the Nepalese Himalaya. *Geology*, **28**, 403–406.

SMITH, H. A. & GILETTI, B. J. 1997. Lead diffusion in monazite. *Geochimica et Cosmochimica Acta*, **61**, 1047–1055.

SPEAR, F. & PARRISH, R. R. 1996. P–T–t evolution of the Valhalla Complex, British Columbia, Canada. *Journal of Petrology*, **37**, 733–765.

STERN, R. A. 1999. *In situ* analysis of radiogenic isotopes with emphasis on ion microprobe techniques and applications. *In*: LAMBERT, D. D. & RUIZ, J. (eds) *Application of radiogenic isotopes to ore deposit research and exploration*. Reviews in Economic Geology, **12**, 173–199.

ST-ONGE, M. R., LUCAS, S. B. & PARRISH, R. R. 1992. Terrane accretion in the internal zone of the Ungava orogen, northern Quebec. Part 1: Tectonostratigraphic assemblages and their tectonic implications. *Canadian Journal of Earth Sciences*, **29**, 746–764.

SUZUKI, K., ADACHI, M. & KAJIZUKA, I. 1994. Electron microprobe observations of Pb diffusion in metamorphosed detrital monazites. *Earth and Planetary Science Letters*, **81**, 203–211.

TUCKER, R. D., RÅHEIM, A., KROGH, T. E. & CORFU, F. 1987. Uranium–lead zircon and titanite ages from the northern portion of the Western Gneiss Region, south-central Norway. *Earth and Planetary Science Letters*, **81**, 203–211.

WAIN, A. 1997. New coesite-eclogite occurrences in western Norway: the nature of an ultrahigh pressure province in the Western Gneiss Region. *Geology*, **25**, 927–930.

WAYNE, D. M. & SINHA, A. K. 1988. Physical and chemical response of zircons to deformation. *Contributions to Mineralogy and Petrology*, **98**, 109–121.

WAYNE, D. M. & SINHA, A. K. 1992. Stability of zircon U–Pb systematics in a greenschist grade mylonite: An example from the Rockfish Valley fault zone, central Virginia, USA. *Journal of Geology*, **100**, 593–603.

ZHU, X. K. & O'NIONS, R. K. 1999. Zonation of monazite in metamorphic rocks and its implications for high temperature thermochronology: a case study from the Lewisian terrain. *Earth and Planetary Science Letters*, **171**, 209–220.

ZHU, X. K., O'NIONS, R. K., BELSHAW, N. S. & GIBB, A. J. 1997. Lewisian crustal history from *in situ* SIMS mineral chronometry and related metamorphic textures. *Chemical Geology*, **136**, 205–218.

Polyphase deformation and metamorphism at the Kalahari Craton – Mozambique Belt boundary

A. D. S. T. MANHICA[1], G. H. GRANTHAM[2], R. A. ARMSTRONG[3],
P. G. GUISE[4] & F. J. KRUGER[5]

[1] Anglo American Corporation, Johannesburg, South Africa
[2] Council for Geoscience, P/Bag X112, Pretoria, 0001, South Africa
(e-mail: grantham@geoscience.org.za)
[3] PRISE, Australian National University, Canberra, Australia
[4] University of Leeds, Leeds, UK
[5] Hugh Allsopp Laboratory, University of the Witwatersrand, Private Bag 3, P.O. Wits, 2050

Abstract: The rocks of the Kalahari Craton in central western Mozambique have crystallization ages of between c. 2300 and 3400 Ma and comprise dominantly granite–greenstones, peraluminous two-mica granites, subordinate younger mafic and granitic intrusions of uncertain age and cover sedimentary rocks. The rocks of the Mozambique Belt comprise c. 1100 Ma intrusive granitoids as well as mafic intrusives and supracrustal migmatite gneisses of uncertain age. The boundary zone between and including these two crustal provinces is characterized by a strong N–S penetrative planar and migmatitic fabric. Sparse kinematic indicators suggest a sinistral sense of displacement along this shear zone. The metamorphic gradient increases from west to east from low grade on the Kalahari Craton to high-grade in the east, characterized by two generations of anatectic migmatization. $^{40}Ar/^{39}Ar$ thermochronology on mica suggests that the Kalahari Craton lithologies have experienced heating above at least c. 300°C during the c. 1100 Ma Grenville age orogeny and again at c. 530 Ma during the Pan-African Orogeny, possibly related to the collisional amalgamation of East and West Gondwana. The Mozambique Belt lithologies record a c. 550 Ma thermal overprint with the lithologies in the vicinity of the N–S shear zone recording thermal reactivation at c. 470 Ma. Comparisons of the new data with that from western Dronning Maud Land, which was adjacent to the study area prior to Gondwana fragmentation, yield many similarities.

The Mozambique Belt, which has been shown by Pinna *et al.* (1993) to have c. 1100 Ma crystallization ages with c. 550 Ma overprints, forms part of a high grade metamorphic belt, which, on reconstruction of Gondwana, stretched at least from Namaqualand through Natal in South Africa, through the Falkland Islands, Haag Nunatacks (West Antarctica), through the Maud Province of western Dronning Maud Land into southern Mozambique and beyond (Grantham *et al.* 1988; Groenewald *et al.* 1991; Jacobs *et al.* 1993; Grantham *et al.* 1997) into East Africa (Fig. 1). This metamorphic belt is considered to have formed by the collisional accretion of mostly juvenile crust onto the southern and eastern margins of the Kalahari Craton during the amalgamation of the Rodinia Supercontinent at c. 1000–1200 Ma (Jacobs *et al.* 1993; Grantham *et al.* 1997; Wareham *et al.* 1998). The subsequent fragmentation of Rodinia and reassembly of those fragments in different configurations resulted in the amalgamation of East and West Gondwana resulting in the Pan African overprint on the c. Grenvillian-age lithologies. Uncertainty regarding the exact location of the suture between East and West Gondwana exists. Implicit in this uncertainty is that if the suture, which defines the plate margins of East and West Gondwana, is not recognized in the study area, it is probably located further east or the Pan African overprint in this sector of the Mozambique Belt could possibly represent intraplate reworking. The former possibility has been proposed by Jacobs *et al.* (1999). This study was formulated to examine and determine the nature of the boundary between the Kalahari Craton and the Mozambique Belt and also to determine whether the suture between East and West Gondwana is located in this area (Fig. 1). Pinna *et al.* (1993) described the geology of northern Mozambique but no study of the evolution of this geological boundary has previously been published. Previous work in the study area (Phaup 1937; Oberholzer 1964; Obretenov 1977)

From: MILLER, J. A., HOLDSWORTH, R. E., BUICK, I. S. & HAND, M. (eds) *Continental Reactivation and Reworking.* Geological Society, London, Special Publications, **184**, 303–322. 1-86239-080-0/01/$15.00 © The Geological Society of London 2001.

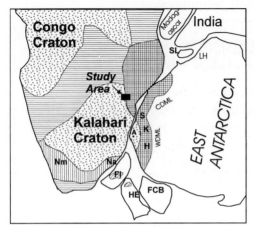

[dotted pattern] Archaean Cratons

[vertical lines pattern] Grenvillian-age Mobile Belt

[horizontal lines pattern] Pan African Mobile Belt

[cross-hatch pattern] Grenvillian Belt overprinted by Pan African

Fig. 1. Locality map showing the study area astride the boundary between the Zimbabwe segment of the Kalahari Craton and the Mozambique Mobile Belt, and the location of the study area in a broader Gondwana reconstruction context. A, Anndagstoppane; S, Sverdrupfjella; K, Kirwanveggen; H, Heimefrontfjella; Nm, Namaqualand; Na, Natal; FI, Falkland Islands; HE, Haag Elsworth–Whitmore block. WDML, Western Dronning Maud Land; CDML, Central Dronning Maud Land; FCB, Filchner Crustal Block. SL, Sri Lanka; LH, Lutzholm Bay.

concentrated on the Archaean-age Mutare–Manica greenstone belt because of the gold-bearing lithologies exposed therein.

Lithological units

The distribution of the various lithologies is shown in Figure 2. The lithologies can be subdivided into those that underlie the Kalahari Craton in the west and the Mozambique Belt in the east with the boundary being defined by the eastern limit of the Messica Granite Gneiss (Fig. 2).

Kalahari Craton lithologies

The lithologies which underlie the Kalahari Craton include the Vumba Granite Gneiss, the Mutare–Manica Greenstone Belt, the Messica

Granite Gneiss, the Frontier Formation meta-sediments, the Tchinhadzandze Granodiorite Gneiss and mafic intrusions (Fig. 2).

The Vumba Granitic Gneiss is typical of the tonalite–trondjhemite component of most Archaean granite–greenstone terrains. Mineralogically, the gneisses contain biotite and hornblende varieties with quartz (25%), plagioclase (40%) and K-feldspar (23%). Biotite and hornblende are partially altered to chlorite whereas plagioclase is locally sausseritized. Accessory phases include titanite, apatite and zircon. At least two phases of pegmatitic veining are recognized.

The Mutare–Manica Greenstone Belt contains serpentinites, actinolite schists, meta-conglomerates, metapelites and quartz sericite schists. Subordinate metapelitic units locally contain andalusite grains with post-tectonic textural morphologies. The age of the andalusite is however uncertain.

The Messica Granite Gneiss is a leucocratic granite characterized by biotite and subordinate muscovite and locally, garnet. In some areas the rocks are equigranular whereas at most localities the rocks are inequigranular with porphyritic K-feldspar grains up to 2 cm in length. These phenocrysts are particularly evident on weathered surfaces. In the west, the rocks appear undeformed but progressing eastwards, become increasingly strongly foliated.

The Frontier Formation is dominated by quartzites with subordinate metapelitic schists and forms N–S oriented synformal outliers. Poor exposure precludes direct observation of the contact with the underlying Messica Granite Gneiss. No xenoliths of quartzite are seen in the granite suggesting the contact is not intrusive but is either an unconformable sedimentary contact or a tectonic contact. One such quartzite outlier, which overlies the Mutare–Manica Greenstone sequence, contains both sillimanite and kyanite suggesting that it has been tectonically emplaced over the andalusite-bearing greenstone belt sequence.

The Tchinhadzandze Granodiorite Gneiss (Fig. 2) forms rounded inselbergs. The rocks are granodioritic in composition and contain biotite and hornblende as the main ferromagnesian phase. The distribution of the intrusions is restricted to the cratonic areas although the age of the gneiss is uncertain. The rocks contain planar fabrics which have widely varying, unsystematic orientations on outcrop scale (Manhica 1998) with the implication that the rocks are essentially undeformed (see stereonets later) and were possibly emplaced post-tectonically and consequently, are possibly late Pan-African in age.

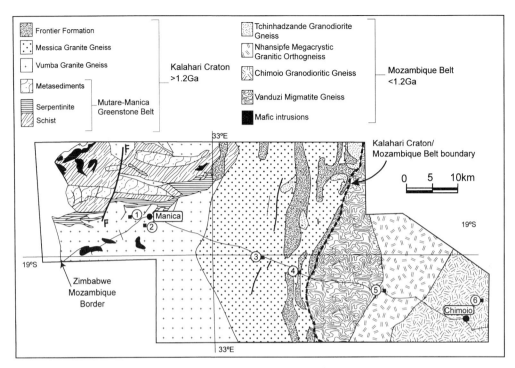

Fig. 2. Geological map of the study area. The sample localities of the $^{40}Ar/^{39}Ar$, U/Pb and Rb/Sr determinations are shown and correlated in Table 1.

Mafic dyke intrusions are found in both the Kalahari Craton and the Mozambique Belt sectors. They are mostly oriented $c.$ N–S and are sub-vertical. The grade of metamorphism within the dykes increases from west to east. Those from the western portion of the study area are fine-grained and still preserve igneous textures although plagioclase is partially sausseritized whereas pyroxene is partially replaced by chlorite and actinolite. Progressing toward the craton margin, approximately in the centre of the area mapped, the rocks still preserve a relict porphyritic igneous texture but contain an assemblage of garnet + hornblende + plagioclase + clinopyroxene, typical of upper amphibolite facies mafic rocks.

Mozambique metamorphic province lithologies

Progressing eastward from the craton margin, the first lithological unit exposed is a heterogenous migmatitic supracrustal sequence, the Vandusi Migmatite Gneiss (Fig. 2), which contains layered, dominantly felsic, rocks with subordinate mafic rocks and semipelitic gneisses. This unit is particularly poorly exposed. Two phases of

migmatization are recognized (Fig. 3a, b). The earlier migmatitic layering is developed parallel to the gneissic layering and is therefore considered to have developed during D_1/M_1 (Fig. 3a, b). The younger generation is clearly discordant and is seen as near vertical $c.$ N–S oriented lenses that are typically 2–3 cm wide and >10 cm long (Fig. 3a, b).

East of these migmatites, the Nhansipfe Megacrystic Granitic Gneiss is exposed (Fig. 2) in isolated inselbergs. The rock is mesocratic and is characterized by large (up to 2 cm) feldspar phenocrysts which are ovoid in low strain zones but form porphyroclastic augen in foliated varieties. The ferromagnesian mineralogy includes biotite, hornblende and garnet with accessory opaques. Two phases of migmatization are recorded in the Nhansipfe Granite Gneiss.

The Chimoio Granodioritic Gneiss underlies the eastern limits of the study area (Fig. 2). The Chimoio Granodiorite Gneiss is also characterized by two phases of migmatization (Fig. 3c). Mafic dyke intrusions are sparse in the Mozambique Belt sector but the few exposed examples are strongly deformed along their margins and contain a single generation of N–S oriented lenticular leucosomes (Fig. 3d). The development of only the N–S oriented lenticular leucosomes

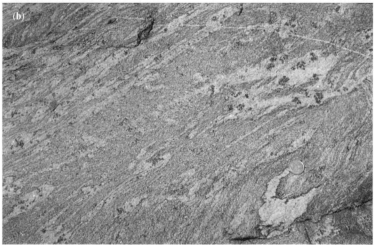

Fig. 3. (**a, b**) Two phases of migmatite development in the Vandusi Migmatite Gneiss Complex. (**a**) Shows early layer parallel folded migmatite banding cut by younger discordant layering whereas (**b**) shows a folded leucosome with melanosome as well as undeformed planar leucosomes with garnet patches, typical of vapour absent melting. (**c**) Two phases of migmatization in the Chimoio Granodiorite Gneiss. (**d**) Deformed mafic dyke in the Nhansipfe Megacrystic Granite Gneiss showing a single generation of lenticular leucosome development.

in the mafic dykes suggests that the mafic dykes were emplaced after the first period of migmatization but prior to the strong N–S fabric forming event.

Structural domains

Four structural domains are recognized (Fig. 4). The Kalahari Craton domain (Domain 1, Fig. 4), which includes the Vumba Granite Gneiss and the Mutare–Manica Greenstone Belt, is characterized by c. E–W- to slightly ENE–WSW striking, N–S dipping, planar deformational fabrics (stereonet in Domain 1, Fig. 4) as well as primary layering in the greenstone belt sequence. The planar fabrics in the Mutare–Manica Greenstone belt have an E–W strike in the west but swing around toward ENE–WSW, progressing eastwards. This change in strike may suggest a sinistral sense of shear along the craton margin. The western exposures of the Messica Granite Gneisses adjacent to the Vumba Granite Gneisses do not contain the same E–W fabric in

Fig. 3. (*continued*)

those gneisses and appear undeformed. This suggests that the planar fabrics in the Vumba Granite Gneiss and the Mutare–Manica Greenstone belt predated the emplacement of the Messica Granite Gneiss.

The Craton Boundary domain (Domain 2, Fig. 4), which includes the eastern portion of the Messica Granite Gneiss and the synformal Frontier Formation outliers, is characterized by near-vertical N–S oriented planar fabrics (stereonet in Domain 2, Fig. 4). As the eastern margin of the Messica Granite Gneiss is approached, the intensity of the planar fabric increases. No lineations are recognized however.

The Mozambique Belt domain (Domain 3, Fig. 4) is characterized by variably oriented fabrics with a dominating, strong, N–S oriented component similar to domain 2 (stereonet in Domain 3, Fig. 4). Foliations of this domain define relatively shallow dipping early S_1? fabrics which are overprinted by later, near vertical, S_2 shear zones. Lineations and fold axes are sparse in the steeply dipping zones but generally plunge shallowly up to 45° to the N or S suggesting a significant strike-slip component. Shear sense indicators in the Nhansipfe Granitic Orthogneiss are sparse with the few found suggesting sinistral displacements.

Structural domain 4 (Fig. 4) is restricted to the exposure of Tchinadzandze Granodiorite Gneiss in the north of the study area. The strikes of the planar fabrics in this rock unit vary from N–S to E–W and dip toward the NE and SW sectors at generally >45°. The orientations of these planar fabrics vary unsystematically on outcrop scale.

Sample descriptions and geochronological analytical methods

New data presented below were derived from SHRIMP U/Pb analysis of two zircon samples, conventional whole rock Rb/Sr isotope analysis of six samples and $^{40}Ar/^{39}Ar$ step heating analysis of six mica samples (5 biotite, 1 muscovite) selected across the Kalahari Craton–Mozambique Metamorphic Province Boundary. The sample localities are shown in Figure 2 with Table 1 describing which methods were applied to which samples from the various localities. The zircons were analysed at ANU, Canberra, Australia; the whole rock powders for Rb/Sr at the Hugh Allsopp Laboratory, University of the Witwatersrand, Johannesburg, South Africa and the mica's for $^{40}Ar/^{39}Ar$ at the University of Leeds, Leeds, UK.

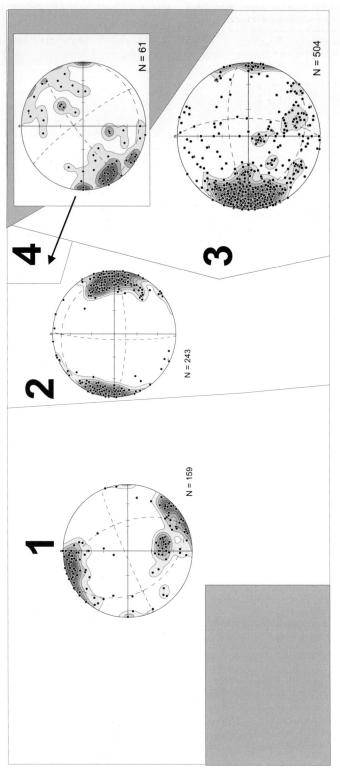

Fig. 4. Map showing the four structural domains recognized in the study area with the stereonets for each domain.

Table 1. *Table correlating the sample localities shown in Figure 2 with the geochronological data and methods shown in Figures 5, 7 and 8*

Sample locality	$^{40}Ar/^{39}Ar$	Rb/Sr (whole/rock)	U/Pb SHRIMP zircon
1	Fig. 8		
2	Fig. 8		
3	Fig. 8	Fig. 5	
4	Fig. 8		
5	Fig. 8		Fig. 7a
6	Fig. 8		Fig. 7b

Petrography of samples

Vumba Granite Gneiss. The Vumba Granite Gneiss typically has a granitic hypidiomorphic granular texture except in samples in which a strong foliation is developed. The Vumba Granite Gneiss consists of K-feldspar, plagioclase, quartz, biotite and hornblende and accessory minerals epidote, allanite, titanite, apatite, chlorite and zircon. Plagioclase exceeds K-feldspar by *c.* 1.5:1 to 2:1 indicating that these rocks are granodioritic to tonalitic in composition. K-feldspar is medium to coarse-grained and is essentially anhedral, tartan twinned microcline and is locally sericitized. Medium to coarse-grained plagioclase is anhedral, locally sausseritized and shows Carlsbad, polysynthetic and albite twins. Quartz is represented by two types namely, strained, anhedral, medium to coarse grains and fine grained strain-free grains commonly exhibiting a fine-grained recrystallized granoblastic texture. Biotite occurs as isolated grains or grain aggregate whose shapes vary from irregular to straight subhedral grains. It is locally poikilitic and partially altered to chlorite. Hornblende occurs as euhedral to subhedral coarse grains and exhibits a preferred orientation that defines the foliation. Epidote is commonly associated with plagioclase from which it is a reaction product. Allanite occurs as isolated hexagonal crystals commonly exhibiting pleochroic haloes or as fine inclusions in plagioclase. Titanite is included in plagioclase and hornblende. Apatite forms euhedral hexagonal crystals included in plagioclase and amphibole. Chlorite occurs as alteration products of biotite. Zircon is included in plagioclase.

Messica Granite Gneiss. The Messica Granite Gneisses are commonly medium-grained equigranular to porphyritic in which feldspars constitute the phenocrysts >5 mm in grain size. The fabric varies from randomly oriented feldspar phenocrysts to one in which strong preferred orientations of deformed quartz and feldspar porphyroclasts are seen. Biotite laths wrap around the feldspars in the deformed samples defining a planar fabric. Myrmekite is common. The mineralogy of Messica Granite Gneiss includes K-feldspar, plagioclase, quartz, biotite, muscovite, zircon, titanite, apatite and opaque minerals. K-feldspar, commonly microcline, is characterized by cross-hatch twinning and is locally perthitic while plagioclase exhibits Carlsbad and albite twins. Both feldspars are commonly poikilitic, enclosing finer quartz and feldspar grains, and are anhedral. Biotite occurs as isolated grains or grain aggregates with subhedral grain shapes. Some chloritized biotite grains exhibit rutile needle intergrowths. Muscovite occurs as small grains associated with biotite and as inclusions in feldspar. These relationships are interpreted to represent an alteration product resulting from retrogressive reactions. Zircon occurs as isolated prismatic grains. Titanite occurs as short prismatic irregular anhedral isolated grains and grain aggregates. Apatite commonly occurs as a fine hexagonal euhedral crystals. Opaque minerals occur as euhedral to anhedral grains, commonly associated with biotite and titanite and, in some cases titanite forms embayments to the opaque minerals.

Frontier Formation. The near mono-mineralic quartzitic samples of the Frontier Formation are characterized by a coarse granoblastic texture. The pelitic samples are coarse, schistose and contain quartz, biotite, muscovite, sillimanite and garnet. Quartz forms granoblastic lenses separated by the mica and sillimanite which define a strong planar fabric which wraps around pre-tectonic poikiloblastic garnet.

The Nhansipfe Granite Gneiss. The mineralogy of these rocks consists of alkali and plagioclase feldspar, quartz, biotite and hornblende with or without garnet. Accessory minerals include titanite, zircon and opaque minerals. The rocks are generally inequigranular and characterized by very coarse-grained porphyroclastic feldspar (>2 cm). They commonly have planar fabric defined by biotite and/or hornblende. Perthitic

K-feldspar and antiperthitic plagioclase occur as coarse xenomorphic grains exhibiting cross-hatched, Carlsbad and Carlsbad, albite and polysynthetic twinning respectively. Plagioclase grains are commonly zoned and contain inclusions of finer feldspar and quartz grains. Some are cracked and commonly contain fine grains of sericite and epidote. The coarse feldspars are deformed and elongated along foliation planes and locally form mosaics of non-zoned grains with granoblastic grain boundaries. Quartz is anhedral and commonly contains inclusions of feldspar and shows undulatory extinction. Hornblende forms anhedral grains with inclusions of quartz, zircon, titanite and opaque minerals. Some grains show preferred orientation and are deformed and oriented along foliation planes. Biotite occurs as aggregates of flakes or isolated laths and is generally spatially associated with hornblende. In some samples, biotite is the only ferromagnesian phase but when hornblende is present they have similar proportions. Biotite laths exhibit preferred orientation along foliation planes. Garnet is developed locally along contacts between plagioclase and hornblende. Coarse garnet grains appear mantled by hornblende. Titanite is common in samples rich in ferromagnesian minerals and is spatially associated with them. Coarse grains of titanite generally contain inclusions of felsic minerals. Zircon occurs as isolated prismatic grains as inclusions in both felsic and mafic minerals. Apatite is rare but occurs as hexagonal grains usually included in felsic minerals.

The Chimoio Granodiorite Gneiss. The mineralogical assemblage of Chimoio Gneiss consists of plagioclase, K-feldspar quartz, hornblende, biotite and accessory quantities of opaque minerals, apatite, allanite, muscovite and zircon. The dominant texture is granitic inequigranular medium to coarse grained and, locally, myrmekite is developed. The samples are poor in quartz and contain considerable hornblende and biotite up to 17%. Plagioclase is locally sausseritized and sericitized. Quartz is anhedral, partially recrystallized and shows undulatory extinction. Green to green/brown hornblende is poikiloblastic with inclusions of feldspar, quartz, apatite, zircon and allanite. Biotite is generally spatially associated with hornblende and contains inclusions of quartz and feldspar. Muscovite is rare and generally pseudomorphically replaces biotite and also forms an alteration product of plagioclase (as sericite). Allanite occurs as brownish fine grains associated with or as inclusions in the ferromagnesian and opaque minerals. Zircon is short prismatic generally enclosed by feldspar

and hornblende. Apatite occurs as fine hexagonal grains generally included in feldspar, hornblende and quartz.

Methods

U/Pb SHRIMP zircon geochronology. The zircon grains were mounted in epoxy together with the zircon standard AS3 (Duluth Complex gabbroic anorthosite; Paces & Miller 1989) and the RSES standard SLI3. The grains were then sectioned approximately in half, polished and photographed. All zircons were then examined by cathodoluminescence imaging. The SHRIMP data were reduced in a manner similar to that described by Compston *et al.* (1992) and Williams & Claesson (1987). U/Pb in the unknowns were normalized to a $^{206}*Pb/^{238}U$ value of 0.1859 (equivalent to an age of 1099.1 Ma) for AS3. The U and Th concentrations were determined relative to those measured in the SL 13 standard. Common Pb corrections were made using the measured $^{204}Pb/^{206}Pb$ ratios and the appropriate Pb compositions from the Cumming & Richards (1975) model. Uncertainties in the isotopic ratios and ages in the data tables (and in the error bars in the plotted data) are reported at the 1σ level, but unless otherwise stated in the text the final weighted mean ages are reported as 95% confidence limits. All age calculations and statistical assessments of the final data have been carried out with the software Isoplot/Ex of Ludwig (1999).

Rb/Sr whole-rock geochronology. Rb/Sr isotopic analysis at the Hugh Allsopp Laboratory, University of Witwatersrand uses conventional TIMS methods. Analytical errors were 0.53% on Rb/Sr and 0.0092% on $^{87}Sr/^{86}Sr$. Uncertainties in the isotopic ratios in the data tables are reported at the 1σ level, but unless otherwise stated in the text the final ages are reported as 95% confidence limits. The data were reduced using 'GEODATE for Windows (1998)' a Windows implementation of the recommendations published by Harmer & Eglington (1990).

$^{40}Ar/^{39}Ar$ *thermochronology.* Samples were irradiated at the Risø Reactor, Roskilde, Denmark. The fast neutron dose was monitored by Leeds biotite standard Tinto, 409.2 Ma (Rex & Guise 1986), HB3gr (Turner *et al.* 1971) and MMHb-1 (Alexander *et al.* 1978), ages used for the last two were as given in Roddick (1983). Flux variation over the package length was of the order of 3%. Interference correction factors were: (40/39)K, 0.048; (36/39)Ca, 0.38; and (37/39)Ca, 1492.

Isotopic analyses were performed with a modified MS10 mass spectrometer. The measured atmospheric $^{40}Ar/^{36}Ar$ (typically 289.2 ± 0.2 for these analyses) and the sensitivity (typically 1.4×10^{-7} Vcm^{-3} STP) change with filament life.

The J-value uncertainty is included in the errors for the total gas ages but the individual step ages have analytical errors only. All errors are quoted at the 1σ level unless otherwise stated. The $^{40}Ar/^{39}Ar$ ratio, age, and errors for each gas fraction were calculated using formulae similar to those given by Dalrymple & Lanphere (1971). Errors in these ratios were evaluated by numerical differentiation of the equation used to determine the isotope ratios and quadratically propagating the errors in the measured ratios. Ages were calculated using the constants recommended by Steiger & Jäger (1977). Data for isotope correlation plots are reduced using 'Isoplot' Ludwig (1990). Best fit lines and intercepts were calculated using Yorkfit 1 and errors taken from the 95% confidence level values.

Age of major lithological units

The ages of the lithological units will be described as progressing from west to east. The age of the Vumba Granitic Gneiss and the Mutare–Manica Greenstone belt were established by Manuel (1992) as being between $c.2800$ and $c.3400$ Ma. A suite of samples from granitoids exposed north of the Mutare–Manica Greenstone Belt yielded a date of 3385 ± 255 Ma (Rb/Sr, whole rock, Manuel 1992) whereas samples from granitoids south of the Mutare–Manica Greenstone Belt yielded a date of 2527 ± 632 Ma (Rb/Sr, whole rock Manuel 1992). Metavolcanic horizons within the Mutare–Manica Greenstone Belt have yielded a date of 2801 ± 42 Ma (Rb/Sr, whole rock, Manuel 1992).

Rb/Sr data for the Messica granite Gneiss are tabulated (Table 2) and shown in the conventional manner in Figure 5. The data define an errorchron of 2618 ± 472 (Ro = 0.7054, MSWD = 28.91). Exclusion of one sample

Table 2. *Rb/Sr isotopic data for the Messica Granitic Gneiss*

Sample	Rb	Sr	$^{87}Rb/^{86}Sr$	$1\sigma\%$	$^{87}Sr/^{86}Sr$	Precision	$1\sigma\%$
1	132.50	185.00	2.0873	0.53	0.7824	0.00007	0.0092
2	126.80	203.40	1.8155	0.53	0.7748	0.00011	0.0092
3	134.60	181.90	2.1574	0.53	0.7863	0.00019	0.0092
4	132.50	192.80	2.0024	0.53	0.7797	0.00008	0.0092
5	132.10	229.00	1.6790	0.53	0.7691	0.00017	0.0092
6	129.10	189.10	1.9900	0.53	0.7842	0.00008	0.0092

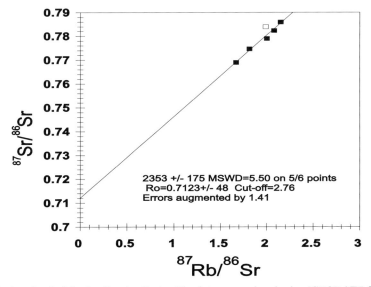

Fig. 5. Rb/Sr data for the Messica Granite Gneiss. The data were reduced using 'GEODATE for Windows (1998)' a Windows implementation of the recommendations published by Harmer & Eglington (1990). Errors are 95% confidence limits and have been augmented by Sqrt(MSWD/Cut-off).

yields an errorchron of 2353 ± 175 Ma (Ro =
0.7123, MSWD = 5.50 on 5/6 points). These
data confirm the age of the Messica Granite
Gneiss as being of late Archaean–early Proterozoic age and consequently forms part of the pre
c. 1100 Ma Kalahari Craton. This lithogical unit
is similar to the Chilimanzi Granite Suite of
Zimbabwe (P. Dirks, pers comm., 1998) which
has yielded a conventional TIMS U/Pb zircon
age of 2601 ± 14 Ma (Jelsma *et al.* 1996).

The age of the Frontier Formation is uncertain. Vail (1965) using the K–Ar method,
reported an age of 465 ± 20 Ma in muscovite
from the Chicamba quartzite (Frontier Formation, within the study area). This age is interpreted as a cooling age and is similar to that
reported below for muscovite (sample locality 4).
The Frontier Formation is intruded by the
Tchinadzandze Granodiorite Gneiss and is
underlain by rocks >*c*. 2300 Ma including the
Messica Granite Gneiss and the Vumba Granitic
Gneiss. Its age therefore is <*c*. 2300 Ma and
>*c*. 465 Ma.

The Tchinadzandze Granodiorite Gneiss in-
trudes the Frontier Formation and the Messica
Granite Gneiss. It contains a relatively weak
variably oriented planar fabric, which contrasts
strongly with the strong planar fabric developed
in the Frontier Formation and Messica Granite
Gneiss in the vicinity of the relatively small
intrusions (Fig. 4). This contrast may suggest
that the Tchinadzandze Gneiss was emplaced
after the strong development of the N–S fabric
developed in its host rocks. This is supported by
the absence of a granoblastic texture in these
rocks indicating they have not experienced textural recrystallization during metamorphism.

The age of the Vanduzi Migmatite Gneiss is
uncertain. Two phases of migmatization are
evident in these gneisses and consequently they
appear to be at least as old as the Nhansipfe and
Chimoio Granitic Gneisses.

The zircons from the Nhansipfe Gneiss
(Sample no. NHF) are clear and up to $500 \mu m$
in length. SEM cathodoluminescence imaging
reveals strong compositional zoning and the
occasional rounded core (Fig. 6a). The zircons
tend to have rounded terminations and irregular prism faces suggesting a degree of meta-
morphic(?) resorption and rounding. Acicular
inclusions and irregular cavities are common.
The U–Th–Pb results of the SHRIMP analyses
for this sample are reported in Table 3 and are
plotted on a Wetherill U–Pb concordia diagram
in Figure 7a. Most analyses were sited in the
zoned (magmatic) sections of the zircons and
these data all plot on or within error of the
concordia. A weighted mean $^{207}Pb/^{206}Pb$ age

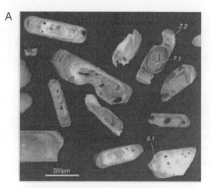

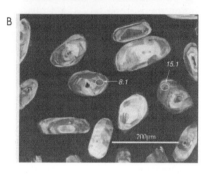

Fig. 6. (**a**) Cathodoluminescence images of
representative zircons from sample NHF. The elliptical
spot represents the area of the grain analysed by
SHRIMP, with numbering corresponding to the
analyses as listed in Table 1. Analysis 7.1 is of an
Archaean core. (**b**) Cathodoluminescence images of
representative zircons from sample CVGN.

of 1112 ± 18 Ma (MSWD = 0.52; probability =
0.89) can be calculated for these data, which is
interpreted as the crystallization age of these
gneisses. The lack of better precision in the
$^{207}Pb/^{206}Pb$ age calculation probably reflects
some real scatter in the data, possibly as a
consequence of intermediate Pb-loss. Analyses
10.1 and 13.1 clearly fall outside this group and
were not included in the above age calculation
(note that in view of the unacceptably poor
precision on analysis 10.1 this data point was
not plotted in Figure 7a). A single analysis of a
core identified by cathodoluminescence gave an
Archaean age (2685 ± 11 Ma).

The Chimoio Granodiorite Gneiss (sample
no. CVGN) yielded numerous clear and multi-
faceted zircons between 100 and $200 \mu m$ in
length (Fig. 6b). All are strongly zoned (with
both sector and concentric compositional zoning)
and are typical of igneous zircons. Pyramidal
terminations have slight metamorphic rounding.
This sample was analysed in three sessions as the
initial analyses showed some unusual reverse

Table 3. *Summary of SHRIMP U–Pb zircon results for sample NHF*

Grain spot	U (ppm)	Th (ppm)	Th/U	Pb* (ppm)	$^{204}Pb/^{206}Pb$	f_{206} %	Radiogenic Ratios						Ages (in Ma)						Conc. %
							$^{206}Pb/^{238}U$	±	$^{207}Pb/^{235}U$	±	$^{207}Pb/^{206}Pb$	±	$^{206}Pb/^{238}U$	±	$^{207}Pb/^{235}U$	±	$^{207}Pb/^{206}Pb$	±	
1.1	223	144	0.65	48	0.00001	0.017	0.1970	0.0055	2.091	0.064	0.0770	0.0008	1159	30	1146	21	1121	20	103
2.1	154	86	0.55	33	0.00001	0.017	0.1974	0.0057	2.091	0.067	0.0768	0.0009	1161	30	1146	22	1116	23	104
3.1	98	61	0.62	20	0.00001	0.017	0.1920	0.0057	2.013	0.068	0.0761	0.0010	1132	31	1120	23	1096	26	103
4.1	81	82	1.00	19	0.00003	0.053	0.1929	0.0068	2.049	0.089	0.0770	0.0017	1137	37	1132	30	1122	43	101
5.1	175	92	0.52	35	0.00009	0.148	0.1885	0.0053	1.959	0.064	0.0754	0.0010	1113	29	1102	22	1079	27	103
6.1	204	79	0.39	40	0.00003	0.057	0.1890	0.0054	2.014	0.067	0.0773	0.0011	1116	29	1120	23	1129	28	99
7.1	187	86	0.46	111	0.00001	0.014	0.5246	0.0156	13.274	0.416	0.1835	0.0012	2719	66	2699	30	2685	11	101
7.2	149	91	0.61	30	0.00008	0.140	0.1868	0.0057	1.968	0.080	0.0764	0.0018	1104	31	1105	28	1105	47	100
9.1	72	53	0.73	15	0.00001	0.017	0.1893	0.0061	2.039	0.077	0.0781	0.0013	1118	33	1129	26	1149	32	97
10.1	75	45	0.60	13	0.00133	2.253	0.1712	0.0065	1.519	0.220	0.0643	0.0087	1019	36	938	93	752	314	136
11.1	207	135	0.66	42	0.00006	0.102	0.1834	0.0056	1.932	0.070	0.0764	0.0012	1086	31	1092	25	1106	33	98
12.1	75	52	0.69	15	0.00011	0.179	0.1796	0.0059	1.859	0.082	0.0751	0.0019	1065	32	1067	30	1070	52	100
13.1	188	108	0.57	38	0.00033	0.553	0.1928	0.0060	1.893	0.098	0.0712	0.0027	1137	32	1079	35	963	80	118
14.1	108	69	0.64	21	0.00041	0.696	0.1763	0.0061	1.815	0.097	0.0747	0.0027	1047	34	1051	36	1060	75	99
15.1	90	56	0.63	19	0.00012	0.208	0.1878	0.0068	2.058	0.131	0.0795	0.0038	1109	37	1135	44	1184	98	94

Notes: 1. Uncertainties given at the one σ level.
2. f_{206} % denotes the percentage of 2 Pb that is common Pb.
3. Correction for common Pb made using the measured $^{204}Pb/^{206}Pb$ ratio.
4. For % Concentration, 100% denotes a concordant analysis.

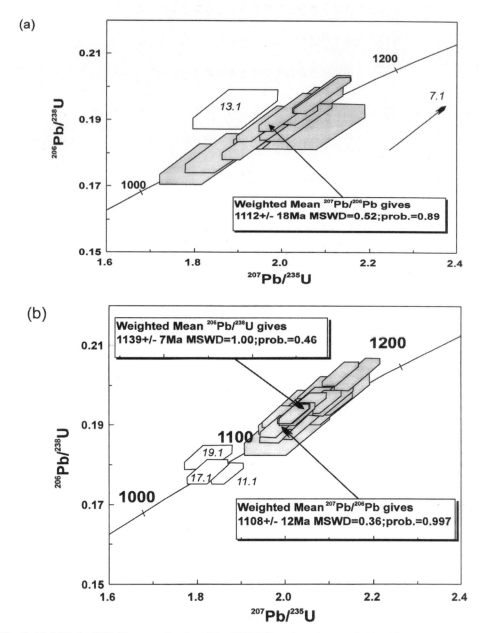

Fig. 7. (a) A Wetherill U–Pb concordia plot of the SHRIMP analyses for sample NHF from the Nhansipfhe Megacrystic Granite Gneiss showing the analyses of the magmatic population. The age calculation excludes analyses 13.1 (shown here as an unfilled error box) and 10. 1 (not plotted). (b) U–Pb concordia plot of SHRIMP analyses for sample CVGN from the Chimoio Granodiorite Gneiss. The discordant analyses represented by the unfilled error ellipses are not included in age calculation shown.

discordance. This feature has been confirmed by the subsequent analyses. The data for all 25 analyses are presented in Table 4 and are plotted on a conventional Wetherill U–Pb concordia (Fig. 7b). The preferred interpretation of the data is that the weighted mean $^{207}Pb/^{206}Pb$ age of 1108 ± 12 Ma (calculated from the 22 'concordant' analyses; MSWD $= 0.36$; probability $= 0.997$) represents the best estimate of the crystallization age of this rock. Note, however,

Table 4. *Summary of SHRIMP U–Pb zircon results for sample CVGN*

Grain spot	U (ppm)	Th (ppm)	Th/U	Pb* (ppm)	$^{204}Pb/^{206}Pb$	f_{206} %	$^{206}Pb/^{238}U$	±	$^{207}Pb/^{235}U$	±	$^{207}Pb/^{206}Pb$	±	$^{206}Pb/^{238}U$	±	$^{207}Pb/^{235}U$	±	$^{207}Pb/^{206}Pb$	±	Conc. %
1.1	128	114	0.89	29	0.00023	0.396	0.1965	0.0065	2.098	0.092	0.0774	0.0019	1156	35	1148	31	1133	51	102
1.2	81	58	0.71	17	0.000131	0.22	0.1893	0.0031	2.011	0.052	0.0770	0.0014	1118	17	1119	18	1122	37	100
2.1	109	41	0.38	21	0.00028	0.468	0.1884	0.0061	2.003	0.097	0.0771	0.0025	1113	33	1117	33	1125	66	99
2.2	174	165	0.95	40	0.000147	0.25	0.1951	0.0029	2.051	0.042	0.0763	0.0009	1149	16	1133	14	1102	24	104
3.1	97	87	0.90	22	0.00001	0.017	0.1954	0.0031	2.097	0.055	0.0778	0.0015	1151	17	1148	18	1143	38	101
4.1	75	41	0.55	16	0.00001	0.017	0.1960	0.0033	2.067	0.052	0.0765	0.0013	1154	18	1138	17	1107	33	104
5.1	65	57	0.88	15	0.00012	0.202	0.1984	0.0034	2.088	0.060	0.0763	0.0016	1167	22	1145	20	1104	42	106
6.1	87	76	0.88	20	0.00001	0.017	0.2026	0.0040	2.162	0.051	0.0774	0.0009	1190	17	1169	17	1131	22	105
6.2	111	40	0.36	23	0.00001	0.017	0.2026	0.0031	2.124	0.041	0.0761	0.0008	1189	17	1157	14	1097	21	108
7.1	220	207	0.94	49	0.00010	0.161	0.1912	0.0024	2.025	0.037	0.0768	0.0009	1128	13	1124	13	1117	24	101
8.1	162	150	0.93	36	0.00010	0.175	0.1927	0.0026	2.019	0.037	0.0760	0.0008	1136	14	1122	12	1095	21	104
9.1	100	75	0.75	21	0.00020	0.341	0.1919	0.0028	1.991	0.045	0.0753	0.0012	1132	15	1112	15	1075	31	105
10.1	137	129	0.94	31	0.00006	0.093	0.1946	0.0031	2.043	0.044	0.0761	0.0009	1147	17	1130	15	1098	25	104
11.1	153	100	0.65	31	0.00001	0.017	0.1777	0.0027	1.867	0.037	0.0762	0.0009	1054	15	1070	13	1101	22	96
12.1	209	194	0.93	47	0.00013	0.212	0.1926	0.0024	2.019	0.034	0.0760	0.0007	1136	13	1122	11	1096	20	104
13.1	171	121	0.71	37	0.00001	0.017	0.1957	0.0029	2.084	0.043	0.0773	0.0010	1152	16	1144	14	1128	25	102
14.1	51	30	0.58	11	0.00015	0.254	0.1943	0.0042	2.040	0.075	0.0761	0.0020	1145	23	1129	25	1099	55	104
15.1	143	118	0.83	31	0.000036	0.06	0.1926	0.0028	2.027	0.041	0.0763	0.0009	1135	15	1125	14	1104	24	103
16.1	133	116	0.87	29	0.000104	0.18	0.1895	0.0028	1.982	0.041	0.0759	0.0010	1119	15	1109	14	1092	25	102
17.1	93	83	0.89	20	0.000129	0.22	0.1781	0.0030	1.822	0.046	0.0742	0.0013	1057	17	1053	17	1046	34	101
18.1	87	71	0.82	19	0.000080	0.13	0.1961	0.0039	2.101	0.053	0.0777	0.0010	1155	21	1149	17	1139	26	101
19.1	77	54	0.69	16	0.000329	0.56	0.1816	0.0030	1.824	0.054	0.0729	0.0017	1076	16	1054	20	1010	47	107
20.1	66	67	1.01	15	0.000157	0.27	0.1933	0.0031	2.034	0.056	0.0763	0.0015	1139	17	1127	19	1103	41	103
21.1	178	259	1.45	44	0.000136	0.23	0.1882	0.0028	1.985	0.044	0.0765	0.0012	1112	15	1110	15	1107	30	100
22.1	85	70	0.83	18	0.000288	0.49	0.1911	0.0035	2.007	0.069	0.0762	0.0021	1127	19	1118	24	1100	55	103

Notes: 1. Uncertainties given at the one σ level.
2. f_{206} % denotes the percentage of ^{206}Pb that is common Pb.
3. Correction for common Pb made using the measured $^{204}Pb/^{206}Pb$ ratio.
4. For % Concentration, 100% denotes a concordant analysis.

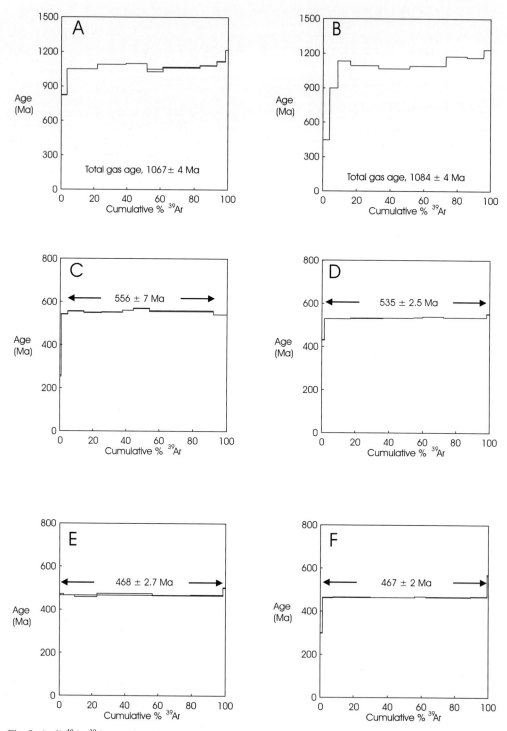

Fig. 8. (**a–f**) ^{40}Ar/^{39}Ar step heating profiles for six mica samples taken across the study area.

that the effect of the reverse discordance noted above results in the calculation of a higher apparent $^{206}Pb/^{238}U$ age of 1139 ± 7 Ma (MSWD = 1.00; probability 0.46). For both age calculations the three discordant points 11.1, 17.1 and 19.1 were excluded.

$^{40}Ar/^{39}Ar$ thermochronology

The data for the samples are shown in Figure 8 and are tabulated in Table 5. The mica in all samples is coarse with grain sizes >1mm being handpicked. Three samples (sample positions 1–3 in Fig. 2) were taken from structural domain 1 (the Kalahari Craton) and include two biotite samples from the Vumba Granite Gneiss near Manica (Fig. 8a, b; Fig. 2), and one biotite sample from the Messica Granite Gneiss (Fig. 8d) nearly midway between Manica and Chimoio (Fig. 2). The Messica Granite Gneiss at this locality shows no planar fabric and appears undeformed. The two biotite samples from the Vumba Granite Gneiss do not define reliable plateau ages although the total gas ages are similar and yield ages of 1084 ± 4 Ma and 1067 ± 4 Ma progressing from west to east. The errors reflected here define the precision of analysis of the total gas collected and do not define a statistical calculation utilizing the individual steps with associated errors. These data are preliminarily interpreted to reflect c. Grenvillian-age thermal overprints of the Archaean gneisses of the Kalahari Craton, because they are similar to the SHRIMP zircon ages from the Nhansipfe Megacrystic Granite Gneiss and the Chimoio Granodiorite Gneiss intrusions exposed c. 30–40 km to the east. Rb/Sr analyses currently being conducted at the Council for Geoscience of South Africa, using mineral–whole rock pairs are aimed either at confirming this c. Grenvillian-age overprint or to see whether the $^{40}Ar/^{39}Ar$ data are the result of 'excess argon' implying no meaningful age with no correlation to any geological event.

The biotite from the weakly deformed Messica Granite Gneiss yields a well constrained date of 535 ± 2.5 Ma (Fig. 8d) with a well defined plateau. This date is interpreted as representing a resetting cooling age for the biotite in the c. 2300 Ma Messica Granite. Consequently, the Kalahari Craton appears to record thermal reactivation or thermal overprinting during both the c. Grenvillian-age orogenic accretion of the Mozambique Belt and the Pan African Orogeny.

One biotite sample from the Chimoio Granodioritic Gneiss was taken from the extreme eastern limit of the study area, east of Chimoio

(Fig. 2), well within the Mozambique Belt in structural domain 3. This sample provided a well defined plateau date of 556 ± 7 Ma (Fig. 8c), similar to that for biotite from the Messica Granite Gneiss. This date is similarly interpreted as representing a reset cooling age for the biotite in the c. 1100 Ma Chimoio Granodiorite Gneiss.

Two samples were taken from the vicinity of the Kalahari Craton boundary between structural domain 1 and 2, and comprise muscovite from schists of the Frontier Formation and biotite from western exposures of the Nhansipfe Megacrystic Granitic Gneiss (Fig. 8e, f) (See Fig. 2 for sample localities). These samples therefore appear to be related to those areas characterized by the strong N–S migmatitic planar fabric and are located in a zone boundered by the two c. 530–560 Ma ages of Figures 8c and d. The ages of the muscovite from the Frontier Formation and the biotite from the Nhansipfe Megacrystic Orthogneiss are 468 ± 2.7 Ma and 467 ± 2 Ma respectively. These ages are therefore considerably younger than those from the bounding areas of c. 530 and 560 Ma to the west and east respectively. While it is recognized that the data does not reflect the age of N–S oriented migmatization in this zone, with the migmatites requiring temperatures >c. 700°C and the blocking temperatures for biotite and muscovite in the $^{40}Ar/^{39}Ar$ system being c. 295–410°C (Blanckenburg et al. 1989) and c. 350°C, respectively (Hames & Bowring 1994), it does suggest that the area characterized by the strong N–S fabric experienced thermal reactivation in a localized zone in the vicinity of the craton margin, with cooling through c. 350°C at 467 ± 2 Ma. It is uncertain whether the N–S migmatization is of Pan African or of Grenvillian age. The prejudice of the authors is that it is of Pan African age.

Discussion and interpretation

The data presented suggests that the eastern margin of the Kalahari Craton experienced thermal reactivation during the accretion of the c. 1100 Ma Mozambique Belt. This accretion has been interpreted to have involved a convergent margin setting in which calc-alkaline rocks, typical of island-arcs, collided with the Kalahari Craton along its eastern and southern margins (Jacobs et al. 1993; Grantham et al. 1995; Wareham et al. 1998) during the orogeny related to the amalgamation of Rodinia. The terminal phase of this orogeny involved the emplacement of megacrystic granitoid plutons, which have chemical compositions typical of A-type granites and locally (Natal, Namaqualand and to a

Table 5 $^{40}Ar/^{39}Ar$ data for six mica samples from the study area

Temp °C	$^{39}Ar_K$	$^{37}Ar_{Ca}$	$^{38}Ar_{Cl}$	Ca	$*^{40}Ar$	%Atm ^{40}Ar	Age	Error	$\%^{39}Ar_K$
		Vol. $\times 10^{-9}$ cm^3		K	$\%^{39}Ar_K$		Ma		
1 Messica Granite Gneiss									
595	4.7	0.17	0.42	0.072	41.38	17.2	431.0	0.6	1.8
700	41.7	0.26	0.93	0.012	52.41	1.4	530.3	0.1	15.6
730	50.8	0.08	1.10	0.003	52.75	0.3	533.3	0.1	19.0
795	47.9	0.16	1.04	0.007	52.87	0.3	534.3	0.1	17.9
860	15.1	0.06	0.33	0.008	53.06	0.4	536.0	0.2	5.6
960	33.4	0.18	0.74	0.011	53.63	0.4	540.9	0.1	12.5
1025	41.0	0.19	0.90	0.009	53.14	0.3	536.7	0.2	15.4
1115	28.2	0.20	0.62	0.014	53.12	0.6	536.5	0.2	10.6
1300	4.3	0.20	0.09	0.090	55.05	8.3	553.3	0.5	1.6
2 Vumba Granite Gneiss									
600	4.3	0.69	0.27	0.32	43.11	16.3	446.9	0.6	4.0
680	5.5	0.31	0.23	0.11	99.12	5.3	899.3	0.4	5.0
720	8.2	0.19	0.33	0.05	134.01	1.7	1132.8	0.2	7.5
765	18.1	0.25	0.69	0.03	128.06	0.4	1095.1	0.3	16.5
860	20.1	0.63	0.82	0.06	123.67	0.5	1066.8	0.2	18.3
975	24.1	4.16	1.15	0.34	127.44	0.6	1091.2	0.2	22.0
1035	13.7	4.62	0.67	0.67	140.65	0.8	1174.1	0.2	12.6
1120	11.1	5.53	0.44	0.99	139.18	1.0	1165.0	0.3	10.2
1300	4.3	5.37	0.17	2.49	150.48	1.3	1233.4	1.0	3.9
3 Frontier Formation Quartzite									
600	1.1	0.00	0.111	0.00	54.32	24.4	547.0	6.4	0.5
690	4.4	0.00	0.059	0.00	45.69	6.8	470.4	2.9	1.9
735	15.0	0.00	0.194	0.00	45.44	2.8	468.1	0.3	6.7
770	30.6	0.04	0.388	0.00	44.94	1.7	463.6	3.6	13.6
840	75.4	0.00	0.937	0.00	45.83	0.9	471.7	4.1	33.6
950	33.1	0.07	0.426	0.00	45.23	1.2	466.2	0.8	14.8
1000	18.6	0.01	0.242	0.00	45.35	1.4	467.3	0.6	8.3
1120	43.3	0.07	0.553	0.00	45.35	0.8	467.4	1.5	19.3
1300	3.0	0.00	0.035	0.00	49.27	3.5	502.6	1.3	1.3
4 Chimoio Granite Gneiss									
590	2.7	0.10	0.23	0.069	23.49	31.1	256.7	2.7	1.0
675	10.1	0.08	0.51	0.015	54.03	5.7	543.8	0.7	3.7
730	26.5	0.17	1.28	0.013	55.55	0.7	556.9	0.5	9.8
770	29.2	0.10	1.40	0.007	54.86	0.4	551.0	0.6	10.7
890	34.5	0.15	1.66	0.009	55.12	0.3	553.2	0.3	12.7
960	18.6	0.25	0.92	0.027	56.14	0.6	562.0	0.5	6.8
1010	25.2	0.65	1.27	0.051	57.15	0.5	570.7	0.6	9.3
1110	102.4	3.27	5.05	0.063	55.60	0.2	557.3	1.7	37.7
1300	22.5	1.27	1.07	0.112	53.85	0.7	542.2	0.3	8.3
5 Vumba Granite Gneiss									
590	8.9	0.45	0.56	0.10	88.90	6.2	823.8	0.5	3.5
685	46.3	0.81	1.86	0.04	121.39	0.9	1050.6	0.3	18.5
725	43.1	0.37	1.71	0.02	127.56	0.3	1090.7	0.3	17.2
780	32.1	0.44	1.26	0.03	128.67	0.4	1097.8	0.3	12.8
875	23.9	2.53	0.90	0.21	119.68	0.7	1039.4	10.8	9.6
975	54.0	3.75	2.10	0.14	123.31	0.6	1063.2	4.1	21.6
1030	25.8	2.15	1.01	0.17	126.05	0.5	1081.0	1.2	10.3
1120	12.7	6.99	0.50	1.10	131.80	0.7	1117.7	0.3	5.1
1300	3.7	3.55	0.13	1.93	148.19	3.9	1218.4	1.0	1.5

(*continued*)

Table 5 (*continued*)

Temp °C	$^{39}Ar_K$	$^{37}Ar_{Ca}$	$^{38}Ar_{Cl}$	Ca	$*^{40}Ar$	%Atm	Age	Error	$\%^{39}Ar_K$
		Vol. $\times 10^{-9}$ cm^3		$\dfrac{}{K}$	$\%^{39}Ar_K$	^{40}Ar	Ma		

^{6} Nhansipfe Granite Gneiss									
580	3.8	0.22	0.40	0.115	27.86	25.9	300.3	1.5	1.5
650	16.1	0.21	1.59	0.025	45.23	5.1	465.0	1.6	6.2
720	56.4	0.16	5.65	0.005	45.48	0.6	467.2	1.1	21.8
760	35.0	0.10	3.50	0.006	45.35	0.3	466.1	0.2	13.5
805	21.4	0.08	2.15	0.008	45.34	0.3	466.0	0.2	8.3
855	11.8	0.03	1.18	0.004	45.34	0.4	465.9	0.3	4.6
930	17.6	0.13	1.76	0.015	45.76	0.5	469.7	0.3	6.8
1020	69.2	0.27	6.91	0.008	45.42	0.3	466.7	0.1	26.8
1150	25.6	0.50	2.57	0.039	45.63	0.2	468.6	0.1	9.9
1320	1.6	0.39	0.16	0.501	56.93	0.2	568.0	3.0	0.6

Errors are 1s. $*^{40}Ar°$ volume of Radiogenic ^{40}Ar, gas volumes corrected to STP.
[1] Biotite from Messica Granite Gneiss (Pd Gr/P), run 2400 weight = 0.07244 g, J value = 0.00652 ± 0.5%. Total gas age 534 ± 2.5 Ma (weight %K = 8.1, $*^{40}Ar = 1946 \times 10^{-7}$ cm^3 g^{-1}).
[2] Biotite from Vumba Granite Gneiss (Vgr – 3b), run 2401 weight = 0.0589 g, J value = 0.00652 ± 0.5%. Total gas age 1084 ± 4 Ma (weight %K = 4.1, $*^{40}Ar = 2347 \times 10^{-7}$ cm^3 g^{-1}).
[3] Muscovite from Frontier Formation Quartzite (Ch Qtz) run 2402 weight = 0.05779 g, J value = 0.00652 ± 0.5%. Total gas age 469 ± 2.7 Ma (weight %K = 8.5, $*^{40}Ar = 1769 \times 10^{-7}$ cm^3 g^{-1}).
[4] Biotite from Chimoio Granite Gneiss (Cvl Gr/P), run 2403 weight = 0.07478 g, J value = 0.00651 ± 0.5%. Total gas age 553 ± 2.5 Ma (weight %K = 8.0, $*^{40}Ar = 2002 \times 10^{-7}$ cm^3 g^{-1}).
[5] Biotite from the Vumba Granite Gneiss (V Gr2/P), run 2404 weight = 0.07297 g, J value = 0.00651 ± 0.5%. Total gas age 1067 ± 4 Ma (weight %K = 7.5, $*^{40}Ar = 4250 \times 10^{-7}$ cm^3 g^{-1}).
[6] Biotite from the Nhansipfe Granite Gneiss (N Fg/P), run 2405 weight = 0.09418 g, J value = 0.00650 ± 0.5%. Total gas age 465 ± 2.1 Ma (weight %K = 6.0, $*^{40}Ar = 1242 \times 10^{-7}$ cm^3 g^{-1}).

small degree Kirwanveggen, Antarctica) preserve Rapakivi textures and charnockitic mineralogy (Jacobs *et al.* 1993). The Nhansipfe Megacrystic Granite Gneiss has been shown to be characterized by A-type chemistry (Manhica 1998). The emplacement of these plutons in Natal has been interpreted to have occurred in a sinistral transpressional setting along the eastern and southern margins of the Kalahari Craton (Jacobs *et al.* 1993). Intriguingly, the ages of these plutons appears to decrease from the north, with the Nhansipfe Megacrystic Granite Gneiss having an age of >1100 Ma, to the south, through Antarctica into Natal. In Antarctica, U–Pb SHRIMP ages are 1093 ± 6 Ma (Grantham & Armstrong unpubl. data, Sverdrupfjella); 1088 ± 10 Ma (Harris 1999, N. Kirwanveggen); 1073–1088 Ma (Arndt *et al.* 1991, Heimefrontfjella). In Natal, U–Pb SHRIMP ages vary between 1030 and 1070 Ma (B. Eglington, pers comm.). This decrease in age could record the diachronous transpressional indentation of the Kalahari Craton with a subduction related island-arc terrane proposed by Jacobs *et al.* (1993).

The next phase of geological evolution is interpreted to have involved extension, with related dyke emplacement, when Rodinia fragmented to form the Mozambique Ocean. The subsequent closure of the Mozambique Ocean is correlatable with the Pan African assembly of East and West Gondwana, which overprinted this part of the Archaean–Early Proterozoic Kalahari Craton resulting in the resetting of the $^{40}Ar/^{39}Ar$ system in mica from both the Kalahari and Mozambique Belt lithologies. The c. 1100 Ma Mozambique Belt lithologies of the Mozambique Belt show thermal reactivation/overprinting followed by cooling through c. 350°C at c. 553 Ma and c. 468 Ma. The younger c. 468 Ma ages appear to be confined to the N–S oriented zone of strong shear in structural domain 2. This zone of shear at the craton margin, could conceivably be argued as representing the suture formed when East and West Gondwana amalgamated to form Gondwana. However, this is viewed as being unlikely for the following reasons:

(1) No lithologies typical of sutures have been recognized in this vicinity, i.e. ultramafic ophiolites as in the Alpine Ivrea Zone; and

(2) it would be expected that extensive new crust, younger than <1100 Ma Mozambique Belt, should have been generated and found in the vicinity of the amalgamation zone between East and West Gondwana.

Consequently it is concluded that the crustal boundary, developed during the c. 1100 Ma belt accretion of the Mozambique Belt onto the Kalahari Craton, has experienced repeated deformation and reactivation at $c.550$ Ma and $c.470$ Ma. A suture defining the boundary between East and West Gondwana has not been recognized in the study area and is probably located further to the east as proposed by Jacobs *et al.* (1999).

Comparison with western Dronning Maud Land, Antarctica

Reconstructions of Gondwana (Fig. 1) suggest that the Mozambique Belt extends southwards into western Dronning Maud Land in Antarctica. Geochronological comparisons of the data presented here, with that from western Dronning Maud Land, show broad similarities with slight differences. The Vumba Granitic Gneisses are similar in age to the granitoids exposed at Annandagstoppane, which has yielded whole-rock Rb–Sr ages of $c.2800$ Ma (Barton *et al.* 1987). No equivalent of the Messica Granite Gneiss has been recognized in Antarctica. The crystallization age and chemistry of the Chimoio Granodiorite Gneiss is similar to the $c.1140$ Ma ages, and compositions of the Grey Gneiss Complex in Sverdrupfjella (Moyes & Barton 1990; Grantham 1992; Grantham & Armstrong unpubl. data) and the Kvervelnatten Granodiorite Gneiss in Kirwanveggen (Jackson & Armstrong 1997). The chemistry and field appearance of the Nhansipfe Megacrystic Granite Gneiss is similar to megacrystic gneisses at Salknappen, Sverdrupfjella and the Kirwanveggen Megacrystic Orthogneiss in Kirwanveggen (Grantham *et al.* 1995), although those units yield marginally younger crystallization ages of $c.1090$ Ma (Grantham unpubl. data; Harris *et al.* 1995; Jackson & Armstrong 1997; Harris 1999). No granitoids similar to the Tchinadzandze Granodiorite Gneiss are recognized in Antarctica. In contrast, the syn-tectonic sheeted Dalmatian Granite in Sverdrupfjella has yielded a Rb/Sr combined whole-rock/mineral age of $c.470$ Ma (Grantham *et al.* 1991), similar to the cooling ages related to the shearing at the craton margin recognized in this study. Mafic dykes in western Sverdrupfjella, which have been deformed by D_3 (Grantham 1992) and metamorphosed having typical medium grade assemblages of hornblende + plagioclase, have yielded metamorphic rim SHRIMP U/Pb zircon ages of $c.480$ Ma, and relict cores possibly of $c.900-$ 1000 Ma (Grantham & Armstrong unpubl. data). Consequently, if the mafic dykes in Mozambique are of the same generation, then the data from Antarctica would suggest a period of crustal extension possibly $c.900-1000$ Ma. This was followed later during the Pan Africa, by medium to high grade metamorphism with accompanying N–S oriented shearing. This Pan-African-age thermal pulse would then be responsible for the granitoid magmatism reported by Grantham *et al.* (1991) and Moyes *et al.* (1993*b*), as well as the $c.450-650$ Ma resetting of Rb/Sr and Sm/Nd radiogenic isotope systems in garnet, biotite, amphibole and feldspars reported in Moyes *et al.* (1993*a*) and Moyes & Groenewald (1996), and of Ar/Ar systems in amphibole and reported by Jacobs *et al.* (1999), Jacobs *et al.* (1995) and Harris (1999).

The research presented here constituted a significant portion of the MSc Thesis by A. D. S. T. Manhica at the University of Pretoria which was supervised by G. H. Grantham and co-supervised by Professor C. P. Snyman. The contribution by Professor C. P. Snyman is gratefully acknowledged. A. S. T. D. Manhica would like to thank Anglo American Corporation for their generous financial and logistical support. Additional financial support from the Mozambican Geological Survey to A. S. T. D. Manhica and a N.R.F. (F.R.D.) research grant to G. H. Grantham are gratefully acknowledged and contributed dominantly to analytical expenses and, to a lesser extent, subsistence costs. Bruce Eglington is thanked for his assistance with geochronological calculations as well as constructive comments on an earlier draft, which materially improved the manuscript. Joachim Jacobs and M. Krabbendam are thanked for critical and constructive reviews. This paper is a contribution to IGCP 368 on 'The Proterozoic of East Gondwana'.

References

ALEXANDER, E. C. JR., MICKELSON, G. M. & LANPHERE, M. A. 1978. *MMHb-1: A new $^{40}Ar-^{39}Ar$ dating standard.* U.S.G.S. Open File Report, **78-701**.

ARNDT, N. T., TODT, W., CHAUVEL, C., TAPFER, M. & WEBER, K. 1991. U–Pb zircon age and Nd isotopic composition of granitoids, charnockites and supracrustal rocks from Heimefrontfjella, Antarctica. *Geologische Rundschau*, **80**, 759–777.

BARTON, J. M., KLEMD, R., ALLSOPP, H. C., AURET, S. H. & COPPERTHWAITE, Y. E. 1987. The geology and geochronology of the Annandagstoppane granite, western Dronning Maud land, Antarctica. *Contributions to Mineralogy and Petrology*, **97**, 488–496.

BLANCKENBURG, F. V., VILLA, I. M., BAUR, H., MORTEANI, G. & STEIGER, R. H. 1989. Time calibration of a P–T path from the Tauern window Eastern Alps: the problem of closure temperatures. *Contributions to Mineralogy and Petrology*, **101**, 1–11.

COMPSTON, W., WILLIAMS, I. S., KIRSCHVINK J. L., ZHANG, Z. & MA, G. 1992. Zircon U–Pb ages for the Early Cambrian time-scale. *Journal of the Geological Society, London*, **149**, 171–184.

CUMMING, G. L. & RICHARDS, J. R. 1975. Ore lead isotope ratios in a continuously changing Earth. *Earth and Planetary Science Letters*, **28**, 155–171.

DALRYMPLE, G. B. & LANPHERE, M. A. 1971. $^{40}Ar/^{39}Ar$ technique of K–Ar dating: a comparison with the conventional technique. *Earth and Planetary Science Letters*, **12**, 300–315.

GRANTHAM, G. H. 1992. *Geological Evolution of western, H.U. Sverdrupfjella, Dronning Maud Land, Antarctica*. Unpublished PhD Thesis, University of Natal (Pietermaritzburg).

GRANTHAM, G. H., GROENEWALD, P. B. & HUNTER, D. R. 1988. Geology of the northern, H.U. Sverdrupfjella, western Dronning Maud land and implications for Gondwana reconstructions. *South African Journal of Antarctic Research*, **18**, 2–10.

GRANTHAM, G. H., MOYES, A. B. & HUNTER, D. R. 1991. The age, petrogenesis and emplacement of the Dalmatian Granite, H.U. Sverdrupfjella, Dronning Maud Land, Antarctica. *Antarctic Science*, **3**, 197–204.

GRANTHAM, G. H., STOREY, B. C., THOMAS, R. J. & JACOBS, J. 1997. The pre-breakup position of Haag Nunataks within Gondwana: possible correlatives in Natal and Dronning Maud Land. *In*: RICCI, C. A. (ed.) *The Antarctic Region: Geological Evolution and processes*. Proceedings of the VII International Symposium on Antarctic Earth Sciences, Siena, 13–20.

GRANTHAM, G. H., JACKSON, C., MOYES, A. B., GROENEWALD, P. B., HARRIS, P. D., FERRAR, G. & KRYNAUW, J. R. 1995. The tectonothermal evolution of the Kirwanveggan-H.U. Severdrupfjella areas, Dronning Maud Land, Antarctica. *Precambrian Research*, **75**, 209–230

GROENEWALD, P. B., GRANTHAM, G. H. & WATKEYS, M. K. 1991. Geological evidence for a Proterozoic to Mesozoic link between southeastern Africa and Dronning Maud land, Antarctica. *Journal of the Geological Society, London*, **148**, 1115–1123.

HAMES, W. E. & BOWRING, S. A. 1994. An empirical evaluation of the argon diffusion geometry in muscovite. *Earth and Planetary Science Letters*, **124**, 161–167.

HARMER, R. E. & EGLINGTON, B. M. 1990. A review of the statistical principles of geochronometry: towards a more consistent approach for reporting geochronological data. *South African Journal of Geology*, **93**, 845–856.

HARRIS, P. D. 1999. *The geological evolution of Neumayerskarvet in the northern Kirwanveggen, Western Dronning Maud Land, Antarctica*. Unpublished PhD Thesis, Rand Afrikaans University, Johannesburg.

HARRIS, P. D., MOYES, A. B., FANNING, C. M. & ARMSTRONG, R. A. 1995. Zircon ion microprobe results from the Maudheim High Grade Gneiss Terrane, Western Dronning Maud Land, Antarctica. *In*: *Geocongress '95 Abstracts*. R.A.U., Johannesburg, 240–243.

JACKSON, C. & ARMSTRONG, R. A. 1997. The tectonic evolution of the central Kirwanveggen, Dronning Maud Land, Antarctica: Temporal resolution of deformation episodes using SHRIMP U–Pb zircon geochronology. *Tectonics Division of the Geological Society of South Africa 13th Anniversary Conference*, Abstracts, 21–22.

JACOBS, J., THOMAS, R. J. & WEBER, K. 1993. Accretion and indentation tectonics at the southern edge of the Kaapvaal craton during the Kibaran (Grenville) orogeny. *Geology*, **21**, 203–206.

JACOBS, J., AHRENDT, H., KREUTZER, H. & WEBER, K. 1995. K–Ar, $^{40}Ar/^{39}Ar$ and apatite fission-track evidence for Neoproterozoic and Mesozoic basement rejuvenation events in the Heimefrontfjella and Mannefallknausane (East Antarctica). *Precambrian Research*, **75**, 251–262.

JACOBS, J., HANSEN, B. T., HENJES-KUNST, F., THOMAS, R. J., WEBER, K., BAUER, W., ARMSTRONG, R. A. & CORNELL, D. H. 1999. New Age constraints on the Proterozoic/Lower Paleozoic Evolution of Heimefrontfjella, East Antarctica, and its bearing on Rodinia/Gondwana Correlations. *Terra Antarctica*, **6**(4), 377–389.

JELSMA, H. A., VINYU, M. L., VALBRACHT, P. J., DAVIES, G. R., WIJBRANS, J. R. & VERDURMEN, ED. A. T. 1996. Constraints on Archaean crustal evolution of the Zimbabwe craton: a U–Pb zircon, Sm–Nd and Pb–Pb whole-rock isotope study. *Contributions to Mineralogy and Petrology*, **124**, 55–70.

LUDWIG, K. R. 1990. *A plotting and regression program for radiogenic-isotope data, for IBM-PC compatible computers*. U.S.G.S. Open-File Report **88-557**.

LUDWIG, K. R. 1999. *Isoplot/Ex version 2.00: A Geochronological Toolkit for Microsoft Excel*. Berkeley Geochronology Centre, Special Publication, **1a**, 46.

MANHICA, A. D. S. T. 1998. *The geology of the Mozambique Belt and the Zimbabwe Craton around Manica, western Mozambique*. Unpublished MSc Thesis, University of Pretoria.

MANUEL, I. R. V. 1992. *Geologie, Petrgraphie, Geochemie und Lagerstatten der Manica – Greenstone – Belt (Mozambique)*. PhD Thesis, Rheinisch – Westfalishen Technishen Hochschule Aachen.

MOYES, A. B. & BARTON, J. M. 1990. A review of isotopic data from western Dronning Maud Land, Antarctica. *Zentralblatt fur Geologie und Palaeontologie*, **1**(1/2), 19–31.

MOYES, A. B. & GROENEWALD, P. B. 1996. Isotopic constraints on Pan African Metamorphism in Dronning Maud Land, Antarctica. *Chemical Geology*, **129**, 247–256.

MOYES, A. B., BARTON, J. M. & GROENEWALD, P. B. 1993*a*. Late Proterozoic to early Palaeozoic tectonism in Dronning Maud Land, Antarctica: supercontinental fragmentation and amalgamation. *Journal of the Geological Society, London*, **150**, 833–842.

MOYES, A. B., GROENEWALD, P. B. & BROWN, R. W. 1993*b*. Isotopic constraints on the age and origin of the Brattskarvet intrusive suite, Dronning

Maud Land, Antarctica. *Chemical Geology*, **106**, 453–466.

OBERLHOLZER, W. F. 1964. *A geologia da Mancha de Manica*. Direccao dos servicos de Geologia e Minas, Maputo, Mozambique, 2–11.

OBRETENOV, N. 1977. *Regiao de Manica. Relatorio sobre os resultados do estudo geologico e dos trabalhos da prospeccao e pesquisa executados em 1976*. Direccao Nacional de Geologia, Maputo, Mozambique, 8–15.

PACES, J. B. & MILLER, J. D. 1989. Precise U–Pb ages of Duluth Complex and related mafic intrusions, Northeastern Minnesota: Geochronological insights to physical, petrogenic, paleomagnetic and tectonomagmatic processes associated with the 1.1 Ga continent rift system. *Journal of Geophysical Research*, **98B**, 13 997–14 013.

PHAUP, A. E. 1937. *The geology of the Umtali Gold Belt.* Geological Survey Bulletin Southern Rhodesia **45**.

PINNA, P., JOURDE, G., CALVEZ, J. Y., MROZ, J. P. & MARQUES, J. M. 1993. The Mozambique Belt in northern Mozambique: Neoproterozoic (1100–850 Ma) crustal growth and tectogenesis, and superimposed Pan-African (800–550 Ma) tectonism. *Precambrian Research*, **62**, 1–59.

REX, D. C. & GUISE, P. G. 1986. *Age of the Tinto felsite, Lanarkshire: a possible ^{39}Ar–^{40}Ar monitor.* Bulletin of Liaison and Information **6**.

RODDICK, J. C. 1983. High precision intercalibration of ^{40}Ar–^{39}Ar standards. *Geochimica et Cosmochimica Acta*, **47**, 887–898

STEIGER, R. H. & JAGER, E. 1977. Subcommission on geochronology: Convention on the use of decay constants in geo- and cosmochronlogy. *Earth and Planetary Science Letters*, **36**, 359–362.

TURNER, G., HUNEKE, J. C., PODOSEK, F. A. & WASSERBURG, G. J. 1971. ^{40}Ar–^{39}Ar ages and cosmic ray exposure ages of Apollo 14 samples. *Earth and Planetary Science Letters*, **12**, 19–35.

VAIL, J. R. 1965. Estrutura e geochronologia da parte oriental da Africa Central, com referencia a Mocambique. *Boletim dos Servicos de Geologia e Minas, Mozambique*, **33**, 15–31.

WAREHAM, C. D., PANKHURST, R. J., THOMAS, R. J., STOREY, B. C., GRANTHAM, G. H., JACOBS, J. & EGLINGTON, B. M. 1998. Pb, Nd, and Sr isotope mapping of Grenville-age Crustal Provinces in Rodinia. *Journal of Geology*, **106**, 647–659.

WILLIAMS, I. S. & CLAESSON, S. 1987. Isotopic evidence for the Precambrian provenance and Caledonian metamorphism of high grade paragneisses from the Seve Nappes, Scandinavian Caledonides. *Contributions to Mineralogy & Petrology*, **97**, 205–217.

Polymetamorphism of mafic granulites in the North China Craton: textural and thermobarometric evidence and tectonic implications

GUOCHUN ZHAO,[1,2] PETER A. CAWOOD,[1]
SIMON A. WILDE[1] & MIN SUN[2]

[1] *Tectonic Special Research Centre, School of Applied Geology, Curtin University of Technology, GPO Box U1987, Perth, W.A. 6001, Australia*
[2] *Department of Earth Sciences, The University of Hong Kong, Hong Kong*
(e-mail: gzhao@hkncc.hku.uk)

Abstract: The basement of the North China Craton can be divided into the Archaean Eastern and Western Blocks, separated by major Palaeoproterozoic terrane boundaries that roughly correspond with the limits of a 100–300 km wide zone, named the Trans-North China Orogen. Some mafic granulites from the orogen and adjoining areas in the Eastern and Western Blocks preserve textural evidence for two granulite facies events involving contrasting P–T paths. The first event is characterized by three distinct mineral assemblages, M_{1a} to M_{1c}. M_{1a} is represented by fine-grained orthopyroxene + clinopyroxene + plagioclase ± quartz, which is surrounded by the M_{1b} garnet + quartz symplectite, which itself is mantled by the M_{1c} plagioclase + biotite symplectite. These assemblages and their P–T estimates define an anticlockwise P–T path, with peak metamorphism of 7.0–8.0 kbar and 800–850°C (M_{1a}) followed by isobaric cooling to 700–750°C (M_{1b}) and pressure-decreasing cooling to 630–700°C (M_{1c}). The second event also includes three mineral assemblages, M_{2a} to M_{2c}. M_{2a} represents growths of garnet porphyroblasts and matrix orthopyroxene + plagioclase + clinopyroxene + quartz; M_{2b} consists of orthopyroxene + plagioclase ± clinopyroxene symplectites or coronas; and M_{2c} is represented by plagioclase + hornblende symplectites. These assemblages and their P–T estimates define a clockwise P–T path, with peak metamorphism of 9.2–9.8 kbar and 820–850°C (M_{2a}), followed by near-isothermal decompression (M_{2b}) of 7.0–7.6 kbar and 760–810°C and cooling (M_{2c}) to 690–760°C. The isobaric cooling, anticlockwise, P–T path of the first granulite facies event is similar to the P–T paths inferred for the *c.* 2.5 Ga metamorphosed mafic granulites from the Eastern and Western Blocks, whereas the near-isothermal decompression, clockwise, P–T path of the second granulite facies event is similar to the P–T paths inferred for the *c.* 1.8 Ga metamorphosed khondalite series in the Western Block and some mafic granulites in the Trans-North China Orogen. These relations suggest that the polymetamorphic granulites were derived from the reworking of the 2.5 Ga metamorphosed granulites during the 1.8 Ga collision between the Eastern and Western Blocks that resulted in the final amalgamation of the North China Craton.

Mafic granulites in the North China Craton record textural and compositional evidence for two distinct and locally overprinting granulite facies metamorphic events (Liu *et al.* 1993; Liu 1994; Wu & Zhong 1998; Zhao *et al.* 1999*a, b, c*). One occurred at *c.* 2.5 Ga, with anticlockwise P–T paths involving near-isobaric cooling, represented by the development of garnet + quartz coronas replacing orthopyroxene + clinopyroxene + plagioclase, whereas the other occurred at *c.* 1.8 Ga, with clockwise P–T paths involving near-isothermal decompression, represented by orthopyroxene/clinopyroxene + plagioclase symplectites replacing garnet porphyroblasts (Liu 1994; Wu & Zhong 1998; Zhao *et al.* 1999*c*). In the past decade, numerous investigations on

the tectonothermal evolution of the two granulite facies events have been carried out throughout the North China Craton and large amounts of data on the mineral assemblages, textural relations, P–T paths and timing of these events have been collected (Cui *et al.* 1991; Jin *et al.* 1991; Lu 1991; Zhai *et al.* 1992; Liu *et al.* 1993; Lu & Jin 1993; He *et al.* 1994; Liu & Liang 1997; Li *et al.* 1998; Mao *et al.* 1999; Zhao *et al.* 1999*c*; Liu *et al.* 2000). However, few studies have been undertaken on reworking of these two metamorphic events. Unravelling reworking of different metamorphic events is crucial in resolving the framework of large-scale orogenic belts, because they provide information of changing P–T conditions and P–T–t paths during the

From: MILLER, J. A., HOLDSWORTH, R. E., BUICK, I. S. & HAND, M. (eds) *Continental Reactivation and Reworking.* Geological Society, London, Special Publications, **184**, 323–342. 1-86239-080-0/01/$15.00 © The Geological Society of London 2001.

polycyclic orogenesis of mountain belts and consequently contribute to our understanding of tectonothermal processes and history of continental reactivation and recyclicity (Goscombe 1992; Liu 1994; Cesari 1999). This paper presents textural and thermobarometric data indicating that some mafic granulites in the North China Craton represent the products of the reworked Late Archaean basement rocks during the *c*. 1.8 Ga granulite facies event. These data in conjunction with structural and geochronological considerations place rigorous constraints on the Late Archaean to Palaeoproterozoic evolution of the North China Craton.

Regional setting

The North China Craton is a general term used to refer to the Chinese part of the Sino–Korea Platform, covering most of north China, the southern part of northeast China, Inner Mongolia, the Bohai Bay and the northern part of the Yellow Sea, with an area of approximately 1 500 000 km^2 (Fig. 1). The craton is bounded by faults and younger orogenic belts. Structurally, the North China Craton is dominated by

N–NE to NE-trending fault systems. An integrated examination of lithological, geochronological, structural and thermobarometric data suggests that two of these fault systems define major terrane boundaries, which roughly correspond with the limits of a 100 to 300 km wide zone called the Trans-North China Orogen that separates the craton into two distinct Archaean to Palaeoproterozoic blocks, called the Eastern and Western Blocks (Fig. 2; Zhao *et al.* 1998, 1999*a, b, c, d*). The blocks and the orogen can be subdivided into a number of tectonic domains, boundaries of which are either faults or covered by younger rock units.

Both the Eastern and Western Blocks consist predominantly of Late Archaean tonalitic–trondhjemitic–granodioritic (TTG) gneisses, with minor amounts of mafic igneous rocks and sedimentary rocks. Late Archaean basement rocks in both blocks were intruded by syn-tectonic granites and were metamorphosed from greenschist to granulite facies at *c*. 2.5 Ga (Pidgeon 1980; Jahn & Zhang 1984; Li *et al.* 1987; Shen *et al.* 1987; Cui *et al.* 1991; Kröner *et al.* 1998). Distinctive garnet + quartz symplectite and corona textures occur in the mafic granulites of both these blocks (Fig. 3), and are described more fully below. These textures indicate anticlockwise

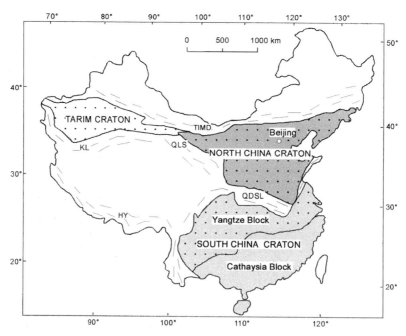

Fig. 1. Tectonic map of China showing the major Precambrian blocks and the Late Neoproterozoic and Palaeozoic fold belts. HY, Himalaya fold belt; KL, Kunlun fold belt; QDSL, Qinling–Dabie–Su–Lu fold belt; QLS, Qilianshan fold belt; TIMD, Tianshan–Inner Mongolia–Daxinganling fold belt.

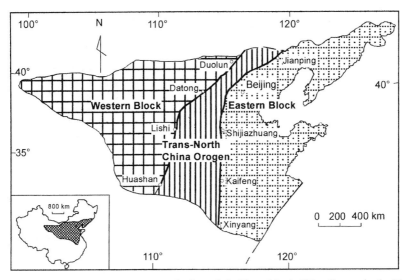

Fig. 2. Distribution of the Eastern and Western Blocks and the Trans-North China Orogen in the North China Craton.

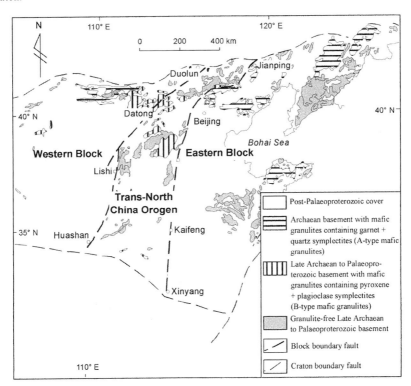

Fig. 3. Spatial distribution of A- and B-type mafic granulites in the North China Craton.

metamorphic P–T paths following the *c.* 2.5 Ga metamorphic event, and are interpreted to reflect underplating and intrusion of mantle-derived mafic magmas (Zhao *et al.* 1998; 1999*c, d*).

Local occurrences of Early Archaean (3.85 to 3.50 Ga) and Mid-Archaean (3.50 to 3.00 Ga) basement rocks in the Eastern Block may represent microcontinental fragments (Jahn *et al.* 1987;

Liu *et al.* 1992; Song *et al.* 1996). These Early to Middle Archaean granitic gneisses, mafic amphibolites, greenschists, and sedimentary supracrustal rocks are found as enclaves, boudins and sheets within the Late Archaean TTG gneisses and syn-tectonic granites. The original extent and tectonic history of these basement rocks are unclear owing to extensive reworking during the 2.5 Ga tectonothermal event, but is thought to include a tectonothermal event at *c.* 3.0 Ga (Bai & Dai 1998). The Archaean basement rocks of the Eastern Block are unconformably overlain by a Palaeoproterozoic lithotectonic assemblage, represented by the Liaohe Group, which has been interpreted to be an intracontinental rift sequence (Bai & Dai 1998). The Archaean rocks of the Western Block are overlain by, and interleaved with, Palaeoproterozoic khondalite sequences that consist of graphite-bearing, sillimanite–garnet gneisses, garnet quartzites, calc-silicate rocks and marbles, with minor amounts of felsic gneisses and mafic granulites (Lu & Jin 1993). These rocks record a metamorphic event at *c.* 1.8 Ga that involved a near-isothermal decompressional clockwise P–T path (Jin *et al.* 1991; Liu *et al.* 1993; Lu & Jin 1993; Zhao *et al.* 1999*d*).

Separating the two blocks is the Trans-North China Orogen, which extends as a roughly N–S-trending belt across the North China Craton (Fig. 2). The rocks in this orogen consist of reworked Archaean basement derived from the Eastern and Western Blocks, as well as Late Archaean to Palaeoproterozoic igneous and sedimentary rocks that include voluminous thick-layered marbles and calc-silicate rocks. The geological and geophysical data show that the major boundary faults along the east and west sides of the orogen are deep-seated, possibly reaching into the lower crust or upper mantle (Ren 1980).

The Trans-North China Orogen was metamorphosed at greenschist to granulite facies conditions and intruded by syn-tectonic S-type granites at *c.* 1.8 Ga. The pyroxene/hornblende + plagioclase symplectite and corona textures that occur in the mafic granulites of the Trans-North China Orogen and some immediately adjacent areas (Fig. 3) are interpreted to indicate nearly isothermal decompression resulting from the collision of the Eastern and Western Blocks at *c.* 1.8 Ga, probably in a subduction zone related to magmatic arcs (Bai & Dai 1998; Zhao *et al.* 1999*c*, 2000*a*).

Mafic granulite is one of the major lithologies in the North China Craton. Based on the features of symplectites and coronas, three textural types of mafic granulites, called A-, B- and AB-type, can be recognized from the craton (Liu 1994; Zhao *et al.* 1999*c*). The A-type mafic granulites crop out in the Eastern and Western Blocks (Fig. 3) and exhibit garnet + quartz symplectites enclosing orthopyroxene + clinopyroxene + plagioclase (Fig. 4a), whereas the B-type mafic granulites are mainly exposed in the Trans-North China Orogen and its immediately adjacent areas (Fig. 3) and display orthopyroxene + plagioclase ± clinopyroxene symplectites or coronas surrounding embayed garnet (Fig. 4b). The AB-type mafic granulites show a spatial distribution

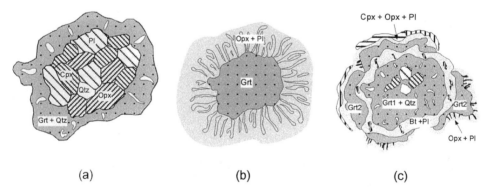

(a)   (b)   (c)

Fig. 4. Schematic diagram showing the representative textures of A-, B- and AB-type mafic granulites in the North China Craton. (**a**) A-type mafic granulite: garnet + quartz symplectic coronas surrounding the peak assemblage orthopyroxene + clinopyroxene + plagioclase. (**b**) B-type mafic granulite: orthopyroxene + plagioclase symplectites surrounding garnet. (**c**) AB-type mafic granulite: garnet + quartz intergrowth is surrounded by biotite + plagioclase symplectite, which itself is surrounded by inclusion-free garnet rims that are mantled by pyroxene + plagioclase symplectites or coronas.

similar to that of the B-type mafic granulites, but contain both the garnet + quartz symplectite, which characterizes the A-type mafic granulites, and the pyroxene/hornblende + plagioclase symplectite, which characterizes the B-type mafic granulite (Fig. 4c). Zhao *et al.* (1999*c*) have discussed the textural relations and P–T paths of the A- and B-type mafic granulites and their possible tectonic environments. This study focuses on the AB-type mafic granulite that is important for understanding tectonothermal processes and history of the polycyclic basement of the North China Craton.

Mineral assemblages and textures of the AB-type mafic granulites

The AB-type mafic granulites are composed mainly of orthopyroxene, clinopyroxene, garnet, plagioclase, quartz, hornblende and biotite, with minor magnetite, ilmenite and apatite.

A marked textural feature of the AB-type mafic granulites is their 'multiple-ring symplectites' texture that contains various mineral symplectic intergrowths, including garnet + quartz, plagioclase + biotite, pyroxene + plagioclase and/or hornblende + plagioclase symplectites. As the inner part of the multiple-ring symplectite texture is dominated by red garnet and the outer part is dominated by white symplectic plagioclase, this texture was also called 'red–white-ring' texture by some Chinese geological workers (Liu 1994).

The multiple-ring symplectite texture generally contains a core of garnet + quartz intergrowth that is surrounded by biotite + plagioclase symplectite, which itself is surrounded by inclusion-free garnet rims that are mantled by pyroxene/hornblende + plagioclase symplectites or coronas (Fig. 5a, b). In some rocks, the garnet occurring with quartz in the core of the multiple-ring symplectite texture contains fine-grained plagioclase + clinopyroxene + orthopyroxene ± quartz inclusions (Liu 1994). Thus, the garnet + quartz

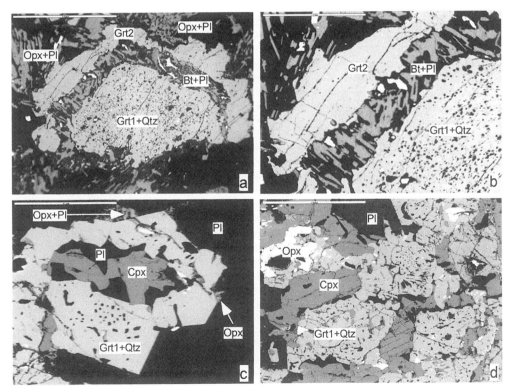

Fig. 5. Back-scattered electron images of textural features in the AB-type mafic granulites. (**a**) Overview of the multiple-ring symplectite texture (scale bar, 2.5 mm). (**b**) Enlargement of the left-upper part of (**a**), showing a garnet + quartz intergrowth surrounded by biotite + plagioclase symplectite which is mantled by inclusion-free garnet rims (scale bar, 1.0 mm). (**c**) Garnet + quartz symplectite around clinopyroxene + plagioclase (scale bar, 1.0 mm). (**d**) Garnet + quartz symplectite around clinopyroxene + orthopyroxene + plagioclase (scale bar, 1.0 mm). Mineral symbols are after Kretz (1983).

intergrowth in the core of the multiple-ring symplectite texture is texturally similar to those garnet + quartz symplectites which occur outside the multiple-ring symplectite texture (Fig. 5c); the former may have resulted from the further growth of the latter until the enclosed plagioclase + clinopyroxene + orthopyroxene ± quartz aggregates were completely consumed.

Based on microstructures and reaction relations between mineral phases present in the multiple-ring symplectite textures and matrixes or porphyroblasts, six distinct mineral assemblages, assigned to M_{1a-c} and M_{2a-c}, can be recognized from the AB-type mafic granulites (Table 1). The M_{1a} assemblage is the fine-grained orthopyroxene + clinopyroxene + plagioclase ± quartz aggregate which is surrounded by the M_{1b} assemblage of garnet + quartz symplectites (Fig. 5c, d). M_{1a} represents a typical two-pyroxene granulite facies assemblage, whereas M_{1b} is commonly interpreted as an isobaric cooling assemblage from the peak granulite facies metamorphism (Harley 1989, 1992). In most rocks, M_{1b} in the multiple-ring symplectite texture is surrounded by the M_{1c} assemblage of biotite + plagioclase symplectites of which biotites occur as skeleton-like crystals intergrown with plagioclase crystals (Fig. 5b). The M_{1c} symplectites are commonly mantled by inclusion-free garnet rims which are surrounded by orthopyroxene + plagioclase or hornblende + plagioclase symplectites (Fig. 5a).

The inclusion-free garnet rim in the multiple-ring symplectite textures is texturally similar to the porphyroblast-type garnet which is also inclusion-free and is surrounded by orthopyroxene + plagioclase ± clinopyroxene or hornblende + plagioclase symplectites or coronas, suggesting that they formed at the same metamorphic stage. Where the garnet porphyroblasts are not surrounded by clino- and orthopyroxene + plagioclase or hornblende + plagioclase symplectites, they generally show relatively straight contact boundaries with the matrix-type clinopyroxene, orthopyroxene, plagioclase and quartz. This indicates an equilibrium between these minerals; they are assigned to the M_{2a} assemblage. Following M_{2a} is the development of orthopyroxene + plagioclase ± clinopyroxene symplectites (M_{2b}) and hornblende + plagioclase symplectites (M_{2c}) surrounding garnet grains in M_{2a}. The M_{2b} and M_{2c} symplectites are texturally similar to orthopyroxene + plagioclase ± clinopyroxene and hornblende + plagioclase symplectites in the B-type mafic granulites from the Trans-North China Orogen, respectively (Zhao *et al.* 1999c).

Polymetamorphism

Of six distinct mineral assemblages recognized from the AB-type mafic granulites, the former three assemblages (M_{1a-c}) show an evolutionary sequence essentially similar to that of the A-type mafic granulites in the Eastern and Western Blocks (cf. Zhao *et al.* 1998, 1999d), whereas the latter three assemblages (M_{2a-c}) show an evolutionary sequence the same as that of the B-type mafic granulites in the Trans-North China Craton (cf. Zhao *et al.* 1999c, 2000a, b). These similarities led Liu (1994) to propose that the AB-type mafic granulites (wherein called mafic granulites with 'red–white-ring' texture) in the North China Craton represent the products of

Table 1. *Metamorphic stage, mineral assemblages, reaction textures and possible reactions in the AB-type mafic granulites*

Stages	Mineral assemblages	Reaction textures	Possible reactions
M_{1a}	Cpx + Opx + Pl ± Qtz	Enclosed by Grt + Qtz symplectites	
M_{1b}	Grt + Qtz	Symplectites surrounding the M_{1a} minerals	Cpx + Opx + Pl → Grt + Qtz
M_{1c}	Bt + Pl	Symplectites surrounding the M_{1b} minerals	Grt + Qtz + H_2O + K_2O → Bt + Pl
M_{1d}	Grt + Cpx + OPx + Pl + Qtz	Porphyroblast- and matrix-type minerals	
M_{1e}	Opx + Pl	Symplectites on the M_{2a} garnet	Grt + Qtz → Opx + Pl
	Cpx + Opx + Pl	Coronas around the M_{2a} garnet	Grt + Qtz → Opx + Cpx + Pl
M_{2c}	Hbl + Pl	Symplectites on the M_4 garnet	Grt + Cpx + Qtz + O_2 → Pl + Hbl + Mag
	Hbl	Retrograde rims replacing Cpx/Opx	Cpx + Opx + Pl + H_2O → Hbl + Qtz

reworking of the Archaean A-type mafic granulites during the 1.8 Ga tectonothermal event.

Petrographic data support the argument that the AB-type mafic granulites in the North China Craton underwent two distinct granulite facies events. As discussed above, of three mineral assemblages formed during the first granulite facies event, the M_{1a} and M_{1b} assemblages were essentially anhydrous, whereas the M_{1c} assemblage was hydrous (Table 1). This suggests that the breakdown of the M_{1b} minerals to form the M_{1c} assemblage was accompanied by an increase in a_{H2O}. The presence of symplectic biotite in the M_{1c} assemblage also suggests an increase in a_{K2O}. Generally, increases in aH_2O and other components (e.g. K_2O, Na_2O, etc.) in granulite facies metamorphic fluids occurred in the later cooling and retrogressive stage (Newton 1986; Hansen et al. 1987; Lamb & Valley 1988; Goscombe 1992). These retrogressive H_2O- and K_2O-bearing phases are considered to be locally expelled from post-peak partial melt (Goscombe 1992). This implies that the M_{1c} assemblage formed when the granulite facies terrain cooled sufficiently after M_{1a} and M_{1b} to have been thermally equilibrated before the second metamorphic cycle. This is supported directly by the geothermometric data (see below). During the second granulite facies metamorphic cycle, the M_{2a} and M_{2b} assemblages are essentially anhydrous, whereas the last assemblage (M_{2c}) was hydrous. This also implies an increase in a_{H2O} in metamorphic fluids during the later stage of the second metamorphic cycle.

Isotopic dating of mafic granulites in the North China Craton has been concentrated mostly in the A- and B-type mafic granulites, ages from the AB-type mafic granulites are very sparse. Recently, Hu et al. (1999) applied the laser-microprobe $^{40}Ar/^{39}Ar$ technique to the dating of high-pressure mafic granulites from the Huaian domain in the Trans-North China Orogen. The high-pressure mafic granulites that were dated are considered to be AB-type mafic granulites because as described by Hu et al. (1999), they contain garnets that enclose large amounts of pyroxene + plagioclase + quartz inclusions and are surrounded by orthopyroxene + plagioclase symplectites or coronas. Four microprobe analyses of garnets from one sample yielded a $^{40}Ar/^{39}Ar$ isochron age of 2510 ± 50 Ma, with initial $^{40}Ar/^{39}Ar$ values of 363, slightly higher than the present-day $^{40}Ar/^{39}Ar$ values of 295.5 (cf. Hanes 1991), and three microprobe analyses of symplectic plagioclases yielded a $^{40}Ar/^{39}Ar$ isochron age of 1968 ± 39, with initial $^{40}Ar/^{39}Ar$ values of 764, much higher than the present-day $^{40}Ar/^{39}Ar$ values. The for-

mer age was interpreted as the time of the first granulite facies event at c. 2.5 Ga, whereas the latter age can only be interpreted as the apparent age of the second granulite facies at c. 1.8 Ga, since the high initial $^{40}Ar/^{39}Ar$ values (764) suggest that the symplectic plagioclases in these mafic granulites have trapped excess argon during crystallization and subsequent cooling due to the presence of an external high partial-pressure of argon (Hu et al. 1999). It should be noted that Hu et al. (1999) interpreted the 2510 ± 50 Ma $^{40}Ar/^{39}Ar$ age as the peak metamorphic time of the high-pressure granulite facies event. Data from this study and a previous study cannot accommodate this interpretation. It is believed that the 2.5 Ga event resulted in the formation of metamorphic assemblages and textures in the A-type (IBC) mafic granulites in the Eastern and Western Blocks, and the 1.8 Ga event resulted in the development of metamorphic assemblages and textures in the B-type mafic granulites in the Trans-North China Orogen, whereas the metamorphic assemblages and textures in the AB-type mafic granulites resulted from the reworking of the A-type mafic granulites during the 1.8 Ga granulite facies event.

Mineral compositions

Selected minerals were analysed with a Link EDS system connected to a Jeol 6400 electron microprobe at the University of Western Australia, using Link's software for ZAF correction and data processing. Analyses were performed with a 15 kV accelerating voltage, c. 5 nA beam current and counting time of 30–40 s. Natural and synthetic minerals were used as standards. Representative analyses of garnet, plagioclase, orthopyroxene, clinopyroxene, biotite and hornblende are presented in Tables 2–6.

Garnet

Table 2 lists representative garnet analyses of: (1) inner rim of garnet occurring with quartz in symplectic coronas in contact with orthopyroxene + clinopyroxene + plagioclase inclusions; (2) outer rim of garnet occurring with quartz in symplectic coronas surrounded by plagioclase + biotite symplectite; (3) core of porphyroblast-type garnet surrounded by matrix plagioclase and/or quartz; (4) rim of porphyroblast-type garnet surrounded by matrix plagioclase and/or quartz; (5) rim of garnet surrounded by orthopyroxene + plagioclase + clinopyroxene coronas; (6) rim of garnet surrounded by orthopyroxene + plagioclase symplectites; and (7) rim of

G. ZHAO *ET AL.*

Table 2. *Microprobe analyses of garnet*

Sample	13201	13501	13514	13519	13520	13522	13201	13205	13207	13244
Texture*				Type 1				Type 2		
SiO_2	38.66	37.23	37.19	37.83	38.02	37.55	38.63	38.30	37.93	38.26
Al_2O_3	21.08	20.96	20.33	20.38	20.84	20.76	20.94	20.97	21.04	20.91
FeO	27.78	28.73	28.21	28.48	28.67	28.43	29.31	27.66	26.62	27.42
MnO	3.12	2.21	3.16	3.46	2.92	2.64	1.78	3.01	3.21	3.25
MgO	2.68	2.83	2.35	2.84	2.80	2.98	2.34	2.57	2.42	2.75
CaO	7.36	8.37	8.80	7.76	7.63	8.01	7.87	7.88	9.46	7.76
Totals	100.68	100.33	100.04	100.75	100.88	100.37	100.87	100.39	100.68	100.35
Si	3.044	2.948	2.962	2.990	3.000	2.972	3.034	3.027	2.988	3.027
Al	1.957	1.957	1.909	1.899	1.938	1.937	1.939	1.954	1.954	1.950
Fe^{3+}	0.000	0.147	0.154	0.117	0.061	0.121	0.000	0.005	0.062	0.000
Fe^{2+}	1.830	1.756	1.725	1.766	1.830	1.761	1.925	1.823	1.692	1.814
Mn	0.208	0.148	0.213	0.232	0.195	0.177	0.118	0.201	0.214	0.218
Mg	0.315	0.334	0.279	0.335	0.329	0.351	0.274	0.303	0.284	0.324
Ca	0.621	0.711	0.752	0.657	0.646	0.679	0.662	0.667	0.799	0.658
Total	7.975	8.001	7.994	7.996	7.999	7.998	7.952	7.980	7.993	7.991
Alm	0.615	0.595	0.581	0.591	0.610	0.593	0.646	0.609	0.566	0.602
Sps	0.070	0.050	0.072	0.078	0.065	0.060	0.040	0.067	0.072	0.072
Prp	0.106	0.113	0.094	0.112	0.110	0.118	0.092	0.101	0.095	0.107
Grs	0.209	0.166	0.175	0.161	0.185	0.168	0.222	0.220	0.236	0.218
Adr	0.000	0.074	0.077	0.059	0.031	0.061	0.000	0.003	0.031	0.000

Sample	133-1	13334	7533	7535	7564	7580	7533	7535	7564	7580
Texture*	Type 2			Type 3				Type 4		
SiO_2	37.33	38.00	38.13	38.76	38.09	38.92	38.03	39.51	38.61	38.49
Al_2O_3	20.83	20.65	20.82	21.59	21.18	21.47	20.79	20.10	20.80	21.12
FeO	29.82	27.42	26.53	24.88	25.78	24.70	28.53	27.45	28.05	28.46
MnO	3.37	2.98	1.41	0.45	0.62	0.60	2.49	1.82	1.92	2.30
MgO	2.92	3.06	3.30	5.11	4.75	5.20	3.03	3.20	3.39	3.25
CaO	6.96	7.39	10.03	9.72	9.65	9.18	6.76	8.34	8.38	7.84
Totals	101.23	99.50	100.22	100.51	100.07	100.07	99.63	100.42	101.15	101.46
Si	3.011	2.995	2.996	2.997	2.968	3.017	3.031	3.097	2.982	3.003
Al	1.929	1.919	1.929	1.968	1.945	1.962	1.953	1.857	1.985	1.943
Fe^{3+}	0.040	0.079	0.076	0.039	0.128	0.000	0.000	0.000	0.039	0.055
Fe^{2+}	1.860	1.818	1.667	1.570	1.552	1.601	1.902	1.799	1.773	1.802
Mn	0.224	0.199	0.094	0.029	0.041	0.039	0.168	0.121	0.126	0.152
Mg	0.342	0.359	0.386	0.589	0.552	0.601	0.360	0.374	0.390	0.378
Ca	0.586	0.625	0.845	0.805	0.806	0.763	0.578	0.700	0.694	0.655
Total	7.992	7.994	7.993	7.997	7.992	7.983	7.992	7.948	7.989	7.988
Alm	0.618	0.606	0.557	0.525	0.526	0.533	0.632	0.601	0.594	0.603
Sps	0.074	0.066	0.031	0.010	0.014	0.013	0.056	0.040	0.042	0.051
Prp	0.114	0.120	0.129	0.197	0.187	0.200	0.120	0.125	0.131	0.127
Grs	0.175	0.169	0.244	0.249	0.208	0.254	0.192	0.234	0.213	0.192
Adr	0.020	0.040	0.038	0.020	0.064	0.000	0.000	0.000	0.020	0.028

garnet surrounded by hornblende + plagioclase symplectites or coronas. All garnets are dominantly almandine (52–65%), with grossular (15–25%), pyrope (9–20%), and minor spessartine (1–8%) and andradite (0–7%) components.

Microprobe analyses reveal three types of variations: (1) between coronal and porphyroblast-type garnets; (2) core to rim variations within individual grains; and (3) between garnet rims surrounded by different symplectites or coronas.

Table 2. (continued)

Sample	7531	7606	7609	7567	7570	7571	7545	7606	7610	7611
Texture*	Type 5		Type 6				Type 7			
SiO_2	38.03	38.74	38.17	37.88	38.09	37.67	38.58	38.79	38.57	39.91
Al_2O_3	21.18	21.08	20.48	21.04	21.23	21.38	21.08	21.43	21.21	22.74
FeO	29.44	28.75	26.26	28.88	29.41	29.39	28.33	28.27	29.14	24.96
MnO	1.89	2.20	2.20	2.44	1.97	2.19	2.12	1.99	2.56	2.38
MgO	3.69	3.30	2.70	3.12	3.80	3.69	3.23	3.30	3.34	2.58
CaO	5.97	7.49	7.46	6.88	5.79	5.63	7.73	7.64	7.02	8.30
Totals	100.20	101.56	97.27	100.24	100.29	99.95	101.07	101.42	101.84	100.87
Si	3.008	3.015	3.080	2.999	3.007	2.986	3.023	3.022	3.005	3.075
Al	1.975	1.934	1.948	1.964	1.976	1.998	1.947	1.968	1.948	2.066
Fe^{3+}	0.007	0.020	0.000	0.024	0.011	0.021	0.000	0.000	0.037	0.000
Fe^{2+}	1.941	1.851	1.772	1.888	1.930	1.928	1.856	1.842	1.862	1.608
Mn	0.127	0.145	0.150	0.164	0.132	0.147	0.141	0.131	0.169	0.155
Mg	0.435	0.383	0.325	0.368	0.447	0.436	0.377	0.383	0.388	0.296
Ca	0.507	0.625	0.645	0.584	0.490	0.479	0.650	0.638	0.587	0.685
Total	8.000	7.973	7.920	7.991	7.993	7.995	7.994	7.984	7.996	7.885
Alm	0.645	0.616	0.613	0.628	0.644	0.645	0.614	0.615	0.619	0.586
Sps	0.042	0.048	0.052	0.055	0.044	0.049	0.047	0.044	0.056	0.056
Prp	0.145	0.127	0.112	0.123	0.149	0.146	0.125	0.128	0.129	0.108
Grs	0.165	0.198	0.223	0.182	0.158	0.150	0.215	0.213	0.177	0.250
Adr	0.004	0.010	0.000	0.012	0.006	0.011	0.000	0.000	0.019	0.000

*Texture: Type 1, inner rim of coronitic garnet in contact with orthopyroxene + clinopyroxene + plagioclase inclusions; Type 2, outer rim of coronitic garnet surrounded by plagioclase + biotite symplectites; Type 3, core of garnet surrounded by plagioclase and/or quartz; Type 4, rim of garnet surrounded by plagioclase and/or quartz; Type 5, rim of garnet surrounded by clinopyroxene + orthopyroxene + plagioclase coronas; Type 6, rim of garnet surrounded by orthopyroxene + plagioclase symplectites; Type 7, rim of garnet surrounded by hornblende + plagioclase symplectites. Alm = $Fe^{2+}/(Fe^{2+} + Mn + Mg + Ca)$; Sps = $Mn/(Fe^{2+} + Mn + Mg + Ca)$; Prp = $Mg/(Fe^{2+} + Mn + Mg + Ca)$; Adr = $Fe^{3+}/2$; Grs = $(Ca-3*Adr)/(Fe^{2+} + Mn + Mg + Ca)$. Fe^{3+} calculations following Droop (1987). Mineral symbols after Kretz (1983).

A pronounced compositional difference is present between the coronal and porphyroblast-type garnets. The cores of the porphyroblast-type garnet (type 3 in Table 2) have relatively higher grossular and pyrope and lower almandine and spessertine contents than the coronal garnets (types 1 and 2 in Table 2). Most cores of the porphyroblast-type garnet are >24% in grossular and 12% in pyrope and <55% in almandine and 3% in spessertine, whereas the coronal garnet are <23% in grossular and 12% in pyrope and >60% in almandine and 6% in spessertine (Table 2). Higher grossular and pyrope components in the porphyroblast-type garnet suggest that they may have formed under relatively higher P–T conditions than the coronal garnets.

Variations between the inner and outer rim compositions for garnet occurring with quartz in symplectic coronas are not pronounced, except that the outer rims contain higher grossular component than the inner rims (Table 2). There are clear core to rim variations for the porphyr-

oblast-type garnet. The cores have higher grossular and pyrope components, whereas the rims are higher in almandine and spessertine (Table 2). Most cores of the porphyroblast-type garnet have >24% grossular and 18% pyrope and <55% almandine and 3% spessertine components, but the rims have >60% almandine and 4% spessertine and <23% grossular and 13% pyrope (Table 2).

There are no pronounced variations between the garnet rims in contact with clinopyroxene + orthopyroxene + plagioclase coronas and those in contact with orthopyroxene + plagioclase symplectites (Table 2), but there is a slight compositional difference between the garnet rims in contact with orthopyroxene + plagioclase symplectites and those in contact with hornblende + plagioclase symplectites. The former is higher in almandine and lower in grossular than the latter (Table 2). These compositional variations imply that the two textural garnets are likely to have re-equilibrated under different conditions.

Plagioclase

Table 3 lists representative analyses of: (1) core and rim plagioclase occurring with orthopyroxene and clinopyroxene as inclusions enclosed by garnet + quartz symplectic coronas; (2) plagioclase occurring with biotite in symplectites surrounding garnet + quartz symplectic coronas; (3) core of matrix-type plagioclase; (4) rim of matrix-type plagioclase; (5) plagioclase occurring with clinopyroxene and orthopyroxene in coronas; (6) plagioclase occurring with orthopyroxene in symplectites; and (7) plagioclase occurring with hornblende in symplectites. The principal compositional features of the different textural plagioclases include:

(1) Plagioclase occurring with orthopyroxene and clinopyroxene as inclusions surrounded by the garnet + quartz symplectic coronas shows no pronounced core to rim compositional variations; both the core and the rim compositions are andesine, ranging from An_{30} to An_{45}.
(2) Plagioclase occurring with biotite as symplectites surrounding the garnet + quartz symplectic coronas is also andesine in composition, ranging from An_{40} to An_{50}, more calcic than the inclusion-type plagioclase, but less calcic than other symplectic plagioclases (Table 3).
(3) Matrix-type plagioclase shows a clear core to rim variation, with compositions ranging from An_{39-42} in the core to An_{47-52} in the rim (Table 3). The higher anorthite contents in matrix-type plagioclase are always present in the rims in contact with or close to garnet grains.
(4) Symplectic plagioclase is labradorite in composition, more calcic than matrix-type plagioclase. Compositional variations between plagioclase from the orthopyroxene + plagioclase symplectite and from the hornblende + plagioclase symplectite are not pronounced (Table 2).
(5) Plagioclase occurring with clinopyroxene and orthopyroxene as coronas is labradorite to bytownite in composition, slightly more calcic than symplectic plagioclases occurring with orthopyroxene and hornblende (Table 3).

Orthopyroxene

Representative analyses of orthopyroxene are listed in Table 4. These are: (1) core and rim compositions of orthopyroxene occurring with plagioclase and clinopyroxene as inclusions en-

closed by garnet + quartz symplectic coronas; (2) core and rim compositions of the matrix-type orthopyroxene; (3) compositions of orthopyroxene occurring with clinopyroxene and plagioclase in coronas; and (4) compositions of orthopyroxene occurring with plagioclase in symplectites. A clear compositional difference is present between core and rim compositions of the orthopyroxene occurring with plagioclase and clinopyroxene as inclusions. Cores are higher in MgO and lower in FeO, and thus higher in X_{Mg} [$= Mg/(Mg + Fe)$] than rims (Table 4). Matrix-type orthopyroxene also shows a similar compositional variation, with higher MgO and lower FeO and thus higher X_{Mg} in its core (Table 4). The symplectic and coronal orthopyroxenes show essentially similar chemical compositions; both have X_{Mg} values slightly higher than that of the matrix-type orthopyroxene (Table 4).

Clinopyroxene

Table 5 shows representative clinopyroxene compositions, including (1) core and rim compositions of clinopyroxene occurring with plagioclase and orthopyroxene as inclusions, enclosed by garnet + quartz symplectic coronas; (2) core and rim compositions of the matrix-type clinopyroxene; and (3) compositions of clinopyroxene occurring with orthopyroxene and plagioclase in coronas. The inclusion-type clinopyroxene shows no pronounced core to rim variations, except that most inclusion-type clinopyroxene grains have slightly higher X_{Mg} values in their rim. The matrix-type clinopyroxene shows a marked core to rim variation, especially with respect to MgO and FeO. The core compositions are higher in MgO and lower in FeO and thus, have relatively higher X_{Mg} values than the rim compositions (Table 5). The coronal clinopyroxene is unzoned and has compositions similar to those of the matrix-type clinopyroxene.

Biotite and hornblende

Representative analyses of biotite occurring with plagioclase in symplectite, surrounding the garnet + quartz symplectic coronas, and hornblende occurring with plagioclase in symplectites, surrounding the porphyroblast-type garnet, are listed in Table 6. The symplectic biotite has a narrow range of X_{Mg} values, from 0.43 to 0.46, and high TiO_2 contents (3.12–5.33 wt%). Symplectic hornblende compositions range from paragasitic hornblende to magnesiohastingsite,

Table 3. *Microprobe analyses of plagioclase*

Sample	13201	13501	13514	13519	13520	13522	13201	13501	13501	13519	13520	13522
Texture*	Type 1						Type 2					
SiO_2	60.33	61.02	57.83	60.26	59.30	60.04	58.11	60.09	57.61	57.45	59.20	59.19
Al_2O_3	24.77	24.56	26.22	24.59	25.44	25.71	26.86	26.05	26.51	25.92	25.47	26.18
FeO	0.13	0.01	0.74	0.14	0.52	0.14	0.43	0.34	0.60	0.21	0.00	0.07
CaO	6.70	6.56	8.61	6.45	7.77	7.42	7.81	6.83	7.97	7.55	8.00	7.89
Na_2O	7.40	7.75	6.15	7.65	7.00	6.84	7.12	6.85	6.79	7.63	6.78	6.95
K_2O	0.35	0.51	0.36	0.49	0.39	0.51	0.45	0.38	0.32	0.35	0.40	0.27
Totals	99.68	100.41	99.91	99.58	100.42	100.66	100.78	100.54	99.80	99.11	99.85	100.55
Si	2.692	2.706	2.595	2.692	2.642	2.659	2.585	2.658	2.587	2.600	2.648	2.629
Al	1.303	1.284	1.387	1.295	1.336	1.342	1.409	1.359	1.404	1.383	1.343	1.371
Ca	0.320	0.312	0.414	0.309	0.371	0.352	0.372	0.324	0.384	0.366	0.383	0.375
Na	0.640	0.667	0.535	0.663	0.605	0.587	0.614	0.588	0.591	0.669	0.588	0.599
K	0.020	0.029	0.021	0.028	0.022	0.029	0.026	0.021	0.018	0.020	0.023	0.015
Total	4.975	4.998	4.952	4.987	4.976	4.969	5.006	4.950	4.984	5.038	4.985	4.989
An	0.327	0.310	0.427	0.309	0.372	0.364	0.368	0.347	0.387	0.347	0.385	0.379

Sample	13201	13205	13207	13244	133-1	13334	7533	7535	7564	7580	7533	7535
Texture*	Type 3						Type 4				Type 5	
SiO_2	57.89	56.46	57.01	55.76	56.97	56.18	57.44	58.46	58.74	59.32	56.82	55.39
Al_2O_3	26.64	26.76	26.13	27.04	27.23	27.97	26.61	26.23	26.27	25.64	27.51	27.95
FeO	0.21	0.32	0.28	0.62	0.61	0.30	0.18	0.00	0.00	0.06	0.30	0.58
CaO	8.92	9.54	8.68	9.87	9.55	10.32	8.80	8.25	8.44	7.98	9.44	10.40
Na_2O	6.30	5.72	6.10	5.83	5.76	5.61	6.32	6.85	6.71	6.62	5.74	5.65
K_2O	0.21	0.20	0.28	0.10	0.32	0.23	0.34	0.33	0.31	0.28	0.22	0.29
Totals	100.17	99.00	98.48	99.22	100.44	100.61	99.79	100.12	100.47	99.90	100.03	100.26
Si	2.585	2.556	2.591	2.528	2.548	2.511	2.578	2.609	2.615	2.645	2.546	2.488
Al	1.402	1.428	1.400	1.445	1.436	1.474	1.408	1.380	1.379	1.348	1.453	1.480
Ca	0.427	0.463	0.423	0.479	0.458	0.494	0.423	0.395	0.403	0.381	0.453	0.501
Na	0.546	0.502	0.538	0.512	0.500	0.486	0.550	0.593	0.579	0.572	0.499	0.492
K	0.012	0.012	0.016	0.006	0.018	0.013	0.019	0.019	0.018	0.016	0.013	0.017
Total	4.972	4.961	4.968	4.970	4.960	4.978	4.978	4.996	4.994	4.962	4.964	4.978
An	0.434	0.474	0.433	0.480	0.469	0.497	0.426	0.392	0.403	0.393	0.469	0.496

Sample	7564	7580	7531	7606	7609	7567	7570	7571	7545	7606	7610	7611
Texture*	Type 5		Type 6			Type 7			Type 8			
SiO_2	56.53	55.78	48.19	52.39	53.02	54.39	53.40	55.91	55.60	55.59	55.65	56.56
Al_2O_3	27.49	27.78	32.49	30.12	29.21	27.43	27.74	27.73	27.71	27.96	28.05	27.52
FeO	0.12	0.40	0.13	0.14	0.18	1.84	1.26	0.36	0.18	0.33	0.04	0.10
CaO	9.99	10.47	16.55	12.91	12.53	11.12	12.17	10.74	10.72	10.57	10.47	10.15
Na_2O	5.80	5.29	2.24	4.06	4.17	4.44	4.15	5.09	5.15	5.40	5.37	5.52
K_2O	0.30	0.13	0.06	0.14	0.16	0.31	0.47	0.23	0.17	0.19	0.22	0.22
Totals	100.23	99.85	99.66	99.76	99.27	98.33	99.19	100.06	99.53	100.04	99.80	100.07
Si	2.534	2.509	2.214	2.378	2.415	2.468	2.430	2.511	2.510	2.501	2.503	2.536
Al	1.453	1.473	1.760	1.612	1.568	1.414	1.488	1.468	1.475	1.483	1.488	1.455
Ca	0.480	0.505	0.815	0.628	0.611	0.541	0.593	0.517	0.518	0.510	0.505	0.488
Na	0.504	0.461	0.200	0.357	0.368	0.391	0.366	0.443	0.451	0.471	0.468	0.480
K	0.017	0.007	0.004	0.008	0.009	0.018	0.027	0.013	0.010	0.011	0.013	0.013
Total	4.988	4.955	4.993	4.983	4.971	4.832	4.904	4.952	4.964	4.975	4.977	4.972
An	0.480	0.519	0.800	0.632	0.618	0.569	0.601	0.531	0.529	0.514	0.512	0.497

* Texture: Type 1, core of inclusion-type plagioclase surrounded by coronitic garnet; Type 2, rim of inclusion-type plagioclase in contact with coronitic garnet; Type 3, plagioclase from biotite + plagioclase symplectites surrounding coronitic garnet; Type 4, core of matrix-type plagioclase; Type 5, rim of matrix-type plagioclase; Type 6, plagioclase from clinopyroxene + orthopyroxene + plagioclase coronas; Type 7, plagioclase from orthopyroxene + plagioclase symplectites; Type 8, plagioclase from hornblende + plagioclase symplectites. An = Ca/(Ca + Na + K).

Table 4. *Microprobe analyses of orthorpyroxene*

Sample	13201	13501	13514	13519	13520	13522	13201	13501	13514	13519	13520	13522	7533
Texture			Type 1							Type 2			
SiO_2	51.03	50.55	49.94	51.62	50.98	51.07	49.46	50.61	51.08	50.63	51.93	50.54	51.17
TiO_2	0.00	0.05	0.00	0.00	0.09	0.03	0.14	0.00	0.00	0.00	0.00	0.00	0.01
Al_2O_3	0.64	0.70	0.50	0.59	0.76	1.01	0.50	0.61	0.62	0.52	0.41	0.75	0.72
FeO	32.36	32.34	32.78	33.71	33.88	33.68	33.93	33.40	33.59	34.28	33.11	33.49	32.33
MnO	0.32	1.30	0.23	1.06	0.74	0.68	1.11	1.36	1.27	1.45	1.22	1.31	0.54
MgO	15.03	12.79	14.27	14.28	14.28	14.43	12.22	13.20	13.84	13.49	13.49	13.53	15.28
CaO	0.67	0.66	0.60	0.13	0.69	0.17	1.20	0.69	0.67	0.45	0.73	0.76	0.65
Totals	100.05	98.39	98.32	101.39	101.42	101.07	98.56	99.87	101.07	100.82	100.89	100.38	100.70
Si	1.993	2.017	1.993	2.001	1.981	1.985	1.989	2.000	1.993	1.989	2.021	1.988	1.985
Al	0.029	0.033	0.024	0.027	0.035	0.046	0.024	0.028	0.029	0.024	0.019	0.035	0.033
Fe	1.057	1.079	1.094	1.093	1.101	1.095	1.140	1.104	1.096	1.126	1.077	1.102	1.049
Mn	0.011	0.044	0.008	0.035	0.024	0.022	0.038	0.046	0.042	0.048	0.040	0.044	0.018
Mg	0.875	0.760	0.849	0.825	0.827	0.836	0.732	0.778	0.805	0.790	0.782	0.793	0.884
Ca	0.028	0.028	0.026	0.005	0.029	0.007	0.052	0.029	0.028	0.019	0.030	0.032	0.027
Total	3.993	3.961	3.994	3.986	3.997	3.991	3.975	3.985	3.993	3.996	3.969	3.994	3.996
X_{Mg}	0.453	0.413	0.437	0.430	0.429	0.433	0.391	0.413	0.423	0.412	0.421	0.418	0.457

Sample	7535	7564	7580	7533	7535	7564	7580	7531	7606	7609	7567	7570	7571
Texture		Type 3				Type 4			Type 5			Type 6	
SiO_2	51.53	51.30	51.20	51.71	51.44	51.10	51.39	51.05	50.76	51.55	50.86	51.12	51.54
TiO_2	0.00	0.00	0.13	0.00	0.00	0.04	0.13	0.06	0.06	0.00	0.08	0.06	0.00
Al_2O_3	0.54	0.67	0.73	0.56	0.53	0.47	0.48	0.60	0.75	0.59	0.89	1.15	1.08
FeO	30.73	31.42	31.63	32.87	31.60	32.49	32.70	31.99	31.18	32.31	31.49	31.09	31.36
MnO	0.60	0.78	0.94	0.81	0.60	0.58	0.88	0.82	0.75	0.94	0.62	0.72	0.84
MgO	16.22	15.69	15.10	15.02	15.11	15.16	15.14	15.25	15.41	15.24	15.30	15.62	15.41
CaO	0.61	0.63	0.64	0.64	0.60	0.58	0.68	0.63	0.49	0.56	0.49	0.43	0.69
Totals	100.23	100.49	100.37	101.61	99.88	100.42	101.40	100.40	99.40	101.19	99.73	100.19	100.88
Si	1.994	1.989	1.991	1.993	2.003	1.991	1.985	1.988	1.988	1.992	1.985	1.982	1.985
Al	0.025	0.031	0.033	0.025	0.024	0.022	0.022	0.028	0.035	0.027	0.041	0.053	0.049
Fe	0.994	1.019	1.028	1.060	1.029	1.059	1.056	1.042	1.021	1.044	1.028	1.008	1.010
Mn	0.020	0.026	0.031	0.026	0.020	0.019	0.029	0.027	0.025	0.031	0.020	0.024	0.027
Mg	0.935	0.906	0.875	0.863	0.877	0.880	0.871	0.885	0.899	0.878	0.890	0.903	0.885
Ca	0.025	0.026	0.027	0.026	0.025	0.024	0.028	0.026	0.021	0.023	0.020	0.018	0.028
Total	3.993	3.997	3.985	3.993	3.978	3.995	3.991	3.996	3.989	3.995	3.984	3.988	3.984
X_{Mg}	0.485	0.471	0.460	0.449	0.460	0.454	0.452	0.459	0.468	0.457	0.464	0.473	0.467

* Texture: Type 1, core of inclusion-type orthopyroxene surrounded by coronitic garnet; Type 2, rim of inclusion-type orthopyroxene in contact with coronitic garnet; Type 3, core of matrix-type orthopyroxene; Type 4, rim of matrix-type orthopyroxene; Type 5, orthopyroxene from clinopyroxene + orthopyroxene + plagioclase coronas; Type 6, orthopyroxene from orthopyroxene + plagioclase symplectites. $X_{Mg} = Mg/(Mg + Fe^{2+})$.

with X_{Mg} values of 0.58–0.63. Most symplectic hornblende grains are unzoned towards the boundaries with embayed garnet.

Estimates of P–T conditions

The P–T conditions of the AB-type mafic granulite cannot be estimated from petrogenetic grids because of the absence of P–T sensitive univariant equilibria in these rocks. Estimates of pressures and temperatures must therefore rely on application of geothermometry and geobarometry.

The M_{1a} mineral assemblage represents the growth of metamorphic clino- and orthopyroxene + plagioclase ± quartz in the absence of garnet. The two-pyroxene thermometers of Fonarev & Graphchikov (1991) and Wood &

Table 5. *Microprobe analyses of clinopyroxene*

Sample	13201	13501	13514	13519	13520	13522	13201	13501	13514	13519	13520	13522
Texture*			Type 1						Type 2			
SiO_2	52.24	51.03	51.64	52.96	52.08	52.76	51.22	51.16	53.06	51.02	52.09	51.35
Al_2O_3	1.33	1.32	1.53	1.30	1.30	1.29	1.10	1.23	0.85	1.13	1.33	1.01
FeO	13.78	14.74	13.74	13.49	14.98	14.74	14.64	14.65	12.61	13.30	12.94	14.31
MnO	0.51	0.74	0.23	0.24	0.60	0.16	0.57	0.48	0.41	0.71	0.46	0.52
MgO	11.02	10.92	10.27	10.90	10.62	10.61	10.46	10.60	11.83	11.30	11.59	11.11
CaO	21.96	21.90	21.74	22.06	21.04	21.08	21.59	21.91	22.42	21.13	21.71	21.08
Na_2O	0.13	0.19	0.42	0.39	0.38	0.16	0.00	0.29	0.18	0.12	0.29	0.12
Totals	100.97	100.84	99.57	101.34	101.00	100.80	99.58	100.42	101.36	98.71	100.41	99.50
Si	1.972	1.930	1.973	1.986	1.968	1.993	1.968	1.943	1.983	1.965	1.962	1.967
Al	0.059	0.059	0.069	0.057	0.058	0.057	0.050	0.055	0.037	0.051	0.059	0.046
Fe^{3+}	0.007	0.086	0.020	0.001	0.036	0.000	0.020	0.069	0.007	0.023	0.029	0.023
Fe^{2+}	0.428	0.380	0.419	0.423	0.437	0.466	0.451	0.396	0.387	0.405	0.379	0.435
Mn	0.016	0.024	0.007	0.008	0.019	0.005	0.019	0.015	0.013	0.023	0.015	0.017
Mg	0.620	0.615	0.585	0.609	0.598	0.597	0.599	0.600	0.659	0.649	0.651	0.634
Ca	0.888	0.887	0.890	0.886	0.852	0.853	0.889	0.892	0.898	0.872	0.876	0.865
Na	0.010	0.014	0.031	0.028	0.028	0.012	0.000	0.021	0.013	0.009	0.021	0.009
Total	4.000	3.995	3.994	3.998	3.996	3.992	3.996	3.991	3.997	3.997	4.010	3.996
X_{Mg}	0.592	0.618	0.583	0.590	0.578	0.562	0.570	0.602	0.630	0.616	0.632	0.593

Sample	7533	7535	7564	7580	7533	7535	7564	7580	7531	7606	7609
Texture*		Type 3				Type 4				Type 5	
SiO_2	52.79	52.22	52.44	52.88	52.77	52.59	52.33	52.18	52.77	52.87	52.26
Al_2O_3	0.97	0.82	1.20	0.69	0.67	1.16	0.79	0.75	0.97	1.07	1.14
FeO	11.37	11.22	11.17	11.17	12.71	12.91	13.12	13.17	11.58	13.40	14.21
MnO	0.13	0.34	0.30	0.13	0.47	0.19	0.17	0.24	0.25	0.26	0.25
MgO	12.25	12.00	12.45	12.41	11.54	11.93	11.53	11.57	12.24	11.60	11.86
CaO	22.52	22.56	22.39	22.57	22.13	22.06	21.78	21.65	22.33	21.53	21.05
Na_2O	0.03	0.09	0.09	0.00	0.00	0.15	0.06	0.04	0.00	0.43	0.04
Totals	100.06	99.25	100.04	99.85	100.29	100.99	99.78	99.60	100.14	101.16	100.81
Si	1.986	1.984	1.971	1.989	1.995	1.968	1.984	1.987	1.985	1.978	1.969
Al	0.043	0.037	0.053	0.031	0.030	0.051	0.035	0.034	0.043	0.047	0.051
Fe^{3+}	0.000	0.000	0.006	0.000	0.000	0.011	0.000	0.000	0.000	0.029	0.005
Fe^{2+}	0.358	0.357	0.345	0.351	0.402	0.393	0.416	0.420	0.364	0.390	0.442
Mn	0.004	0.011	0.010	0.004	0.015	0.006	0.005	0.008	0.008	0.008	0.008
Mg	0.687	0.680	0.697	0.696	0.650	0.665	0.652	0.657	0.686	0.647	0.666
Ca	0.908	0.919	0.902	0.910	0.896	0.885	0.885	0.884	0.901	0.863	0.850
Na	0.002	0.007	0.007	0.000	0.000	0.011	0.004	0.003	0.000	0.031	0.003
Total	3.988	3.995	3.991	3.981	3.988	3.990	3.981	3.993	3.987	3.993	3.994
X_{Mg}	0.657	0.656	0.669	0.665	0.618	0.629	0.610	0.610	0.653	0.624	0.601

* Texture: Type 1, core of inclusion-type clinopyroxene surrounded by coronitic garnet; Type 2, rim of inclusion-type clinopyroxene in contact with coronitic garnet; Type 3, core of matrix-type clinopyroxene; Type 4, rim of matrix-type clinopyroxene; Type 5, clinopyroxene from clinopyroxene + orthopyroxene + plagioclase coronas. $X_{Mg} = Mg/(Mg + Fe^{2+})$. Fe^{3+} calculations following Droop (1987).

Banno (1973) and the clinopyroxene–plagioclase–quartz barometer of McCarthy & Patiño Douce (1998) were applied to estimate the temperatures and pressures of the M_{1a} mineral assemblage. Based on the core compositions of the M_{1a} clino- and orthopyroxene and plagioclase from six samples, the peak P–T conditions of M_{1a} were estimated with these thermobarometers at 800–850°C and 7.0–8.0 kbar (Table 7), with an average uncertainty of ±50°C and

Table 6. *Microprobe analyses of symplectic biotite and amphibole*

Sample	13201	13205	13207	13244	133-1	13334	7545	7606	7610	7611
Mineral			Biotite*					Hornblende†		
SiO_2	34.47	35.48	35.59	34.24	35.45	35.20	44.01	41.57	43.43	42.74
TiO_2	4.04	3.12	3.54	3.15	5.33	4.83	1.18	1.20	1.47	1.41
Al_2O_3	13.85	15.06	14.98	14.90	13.13	14.31	11.14	13.23	12.56	11.62
Cr_2O_3	0.09	0.03	0.00	0.07	0.03	0.06	0.09	0.06	0.26	0.16
FeO	21.37	21.30	21.70	21.57	21.09	21.28	17.38	17.78	17.87	18.16
MnO	0.10	0.19	0.00	0.00	0.00	0.02	0.09	0.11	0.13	0.16
MgO	9.17	10.09	10.44	10.25	8.56	9.67	10.62	9.72	10.3	10.61
CaO	0.00	0.07	0.14	0.23	0.00	0.08	11.60	11.80	11.80	11.77
Na_2O	0.00	0.00	0.00	0.30	0.00	0.05	1.45	1.50	1.17	1.14
K_2O	10.47	10.20	10.76	9.88	10.37	10.67	1.04	0.95	0.87	0.94
Totals	93.56	95.54	97.15	94.59	93.96	96.17	98.60	97.92	99.86	98.71
Si	2.749	2.750	2.724	2.694	2.802	2.723	6.475	6.173	6.289	6.268
Ti	0.242	0.182	0.204	0.186	0.317	0.281	0.131	0.134	0.160	0.156
Al	1.302	1.376	1.352	1.382	1.223	1.305	1.932	2.316	2.144	2.009
Cr	0.006	0.002	0.000	0.004	0.002	0.004	0.010	0.007	0.030	0.019
Fe^{3+}	0.000	0.000	0.000	0.000	0.000	0.000	0.526	0.688	0.726	0.893
Fe^{2+}	1.425	1.381	1.389	1.419	1.394	1.377	1.612	1.52	1.438	1.334
Mn	0.007	0.012	0.000	0.000	0.000	0.001	0.011	0.014	0.016	0.020
Mg	1.090	1.166	1.191	1.202	1.008	1.115	2.329	2.151	2.223	2.319
Ca	0.000	0.006	0.011	0.019	0.000	0.007	1.829	1.878	1.831	1.849
Na	0.000	0.000	0.000	0.046	0.000	0.007	0.414	0.432	0.329	0.324
K	1.066	1.010	1.052	0.993	1.047	1.054	0.195	0.180	0.161	0.176
Total	7.887	7.885	7.923	7.945	7.793	7.867	15.464	15.493	15.347	15.367
X_{Mg}	0.433	0.458	0.462	0.459	0.420	0.447	0.591	0.586	0.607	0.635

* Symplectic biotite associated with symplectic plagioclase surrounding coronitic garnets.
† Symplectic hornblende associated with symplectic plagioclase surrounding garnets. $X_{Mg} = Mg/(Mg + Fe^{2+})$.
Fe^{3+} calculations following Droop (1987).

±1.0 kbar. These P–T estimates are consistent with medium-pressure granulite facies metamorphic conditions.

The P–T conditions for the M_{1b} assemblage are estimated using the THERMOCALC program, based on rim compositions of the re-equilibrated M_{1a} clino- and orthopyroxene and plagioclase and the coronal garnet, assuming that these compositions have been in equili-

Table 7. *P–T estimates for M_{1a}*

Sample	T (°C)*	T (°C)†	P (kbar)‡
13201	845	828	7.99
13501	855	850	7.25
13514	834	808	7.96
13519	841	817	7.34
13520	847	820	7.23
13522	846	852	7.04

* Fonarev & Graphchikov (1991).
† Banno & Wood (1973).
‡ McCarthy & Patiño Douce (1998).

brium. Mineral activities were calculated for pyroxenes following Holland & Powell (1998), using an ideal two-site mixing model; for garnet following Berman (1990), using the ternary mixing model; for plagioclase following Holland & Powell (1992), using Darken's quadratic formulism. Quartz was assumed to be pure. The selected end-members are listed in Table 8 (a). The average P–T conditions are estimated at 7.0–8.0 kbar and 700–750°C (Table 9). The results overlap at relatively small uncertainty, and all of their σ_{fit} values are within statistical limits (Powell & Holland 1994).

The temperature of M_{1c} can be estimated using the updated biotite–garnet thermometer of Patio Douce *et al.* (1993), based on compositions of symplectic biotite and the outer rim compositions of coronal garnet that is in contact with the biotite + plagioclase symplectite. This thermometer yielded a wide temperature range from 630°C to 700°C. The pressure of M_{1c} cannot be quantitatively estimated because of the lack of suitable geobarometers. Considering the features

Table 8. *Independent equilibria used in THERMO-CALC calculations*

(a) Independent equilibria used for P–T estimates of M_{1b}, M_{2a} and M_{2b}

1) q + cats = an
2) 3an + 3en = 3q + 2py + gr
3) 3q + gr + 3mgts = 3an + py
4) 3an + 3di = 3q + py + 2gr
5) 3an + 3fs = 3q + gr + 2alm
6) 3an + 3hed = 3q + 2gr + alm

(b) Independent equilibria used for P–T estimates of M_{2c}

1) 5ts + 11q + 3gr = 3tr + 13an + 2H₂O
2) 3ts + 12q + 2py + 4gr = 3tr + 12an
3) 6fact + 21an = 6H₂O + 27q + 11gr + 10alm
4) 40ts + 11gl + 24gr = 13tr + 22parg + 82an + 16H₂O
5) 2parg + 8q = tr + ts + 2ab

Abbreviations: ab, albite; alm, almandine; an, anorthite; cats, Ca-tschermak's di, diopside; en, enstatite; fact, Fe-actinolite; fs, ferrosilite; gr, grossular; gl, glaucophane; hed, hedenbergite; mgts, Mg-tschermak's; py, pyrope; para, pargasite; q quartz; tr, tremolite; ts, tschermakite.

of mineral assemblages, pressures for the biotite + plagioclase symplectite should not be higher than pressures for the garnet + quartz symplectic coronas.

THERMOCALC was used to calculate the P–T conditions of M_{2a} based on the core and rim compositions of porphyroblast-type garnet and matrix-type clinopyroxene, orthopyroxene and plagioclase. The same end-members and their solid solution models used for the M_{1b} P–T calculation were selected (Table 8a). The average P–T conditions are 9.2–9.8 kbar and 820–850°C for the core compositions, and 7.7–8.7 kbar and 800–830°C for the rim compositions (Table 9). The P–T estimates from the core compositions are considered to represent the peak P–T conditions of M_{2a}, whereas the estimates from the rim compositions are considered to reflect the resetting P–T conditions during M_{2b} and M_{2c}.

To calculate the pressure and temperature of the orthopyroxene/clinopyroxene + plagioclase symplectites (M_{2b}), mineral analyses were taken from the rims of garnet porphyroblasts and associated symplectic plagioclase, orthopyroxene and clinopyroxene grains that were devoid of chemical zoning. The selected end-members are the same as those used for the P–T estimates of the M_{1b} and M_{2a} assemblages (Table 8a). The results of the average P–T calculations estimated for M_{2b} are 7.0–7.6 kbar and 760–810°C (Table 9), with small uncertainties and low σ_{fit} values.

Table 9. *P–T estimates for M_{1b} and M_{2a-c} by THERMOCALC*

Sample	T (°C)	P (kbar)	s.d. (T)	s.d. (P)	correl.	fit
(a) P–T estimates for M_{1b}						
13201	748	7.6	93	1.3	0.719	0.57
13501	715	7.6	111	1.6	0.821	0.63
13514	744	7.7	82	1.2	0.656	1.02
13519	735	7.4	88	1.3	0.702	0.53
13520	742	7.9	81	1.2	0.668	0.86
7580	727	7.1	94	1.4	0.737	0.49
(b) P–T estimates for M_{2a}						
Core						
7533	853	9.8	99	1.3	0.762	0.77
7535	832	9.6	109	1.5	0.786	0.63
7564	844	9.5	101	1.3	0.719	0.73
7580	820	9.2	95	1.2	0.696	0.68
Rim						
7533	830	8.7	101	1.2	0.741	0.49
7535	777	7.7	98	1.2	0.740	0.25
7564	800	7.9	104	1.3	0.775	0.36
7580	807	7.7	98	1.2	0.734	0.39
(c) P–T estimates for M_{2b}						
7531	790	7.4	95	1.1	0.685	0.89
7606	810	7.6	96	1.1	0.700	0.33
7609	796	7.0	104	1.2	0.733	0.50
7567	762	7.3	236	2.7	0.940	0.41
7570	781	7.2	187	2.2	0.895	0.58
7571	771	7.4	194	2.3	0.908	0.54
(d) P–T estimates for M_{2c}						
7545	738	5.2	75	1.1	0.462	0.63
7606	764	5.0	80	1.2	0.418	0.97
7610	729	5.5	96	1.6	0.469	1.36
7611	690	4.2	92	1.2	0.339	1.14

Note: s.t. (T), standard deviations on temperature; s.t. (P), standard deviations on pressure; correl., correlation between the uncertainties on P and T; fit, a goodness-of-fit parameter describing the averaging of n equilibria (Powell & Holland 1988); NR, the number of independent equilibria calculated by THERMOCALC.

Mineral analyses for the P–T estimates of the hornblende + plagioclase symplectite (M_{2c}) were also taken from the rims of garnet porphyroblasts and associated chemically-unzoned hornblende and plagioclase symplectites. Hornblende activities were calculated following Holland & Blundy (1994), using a non-ideal mixing model. The end-members are listed in Table 8b. THERMOCALC calculation on four samples yielded consistent results that fall in the range 4.2–5.5 kbar and 690–760°C (Table 9). These results also overlap with small uncertainties, and σ_{fit} values are within statistical limits.

P–T paths

The combination of mineral assemblages, reaction textures and geothermobarometric data defines a polymetamorphic evolution of the AB-type mafic granulites to have involved two contrasting P–T paths, as shown in Figure 6. Figure 6 also shows a portion of P–T pseudosection of the NCFMAS grid of Holland & Powell (1998), which shows a sequence of mineral assemblages similar to those observed in the AB-type mafic granulites of this study, except for the absence of symplectic hornblende + plagioclase and biotite + plagioclase assemblages because this P–T pseudosection only involves anhydrous relations. Holland & Powell (1998) suggested that under wet conditions, various hydrous assemblages would become stable in lower temperature domains of this P–T pseudosection.

The early prograde metamorphic history of the first metamorphic cycle was not preserved because of subsequent reworking during peak and post-peak metamorphic events. The oldest metamorphic assemblage (M_{1a}) consists of orthopyroxene + clinopyroxene + plagioclase that is consistent with Opx + Cpx + Pl field in Figure 6. The M_{1a} P–T conditions of 7.0–8.0 kbar and 800–850°C further confine the M_{1a} assemblage to a relatively low pressure and high temperature granulite field (Fig. 6). P–T conditions of 7.0–8.0 kbar and 700–750°C calculated for the

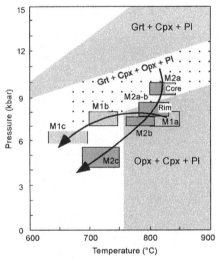

Fig. 6. Contrasting P–T paths inferred for the first (M_{1a}, M_{1b}, M_{1c}) and second (M_{2a}, M_{2b}, M_{2c}) metamorphic events recognized in the AB-type mafic granulites in the North China Craton. The P–T fields for Opx + Cpx + Pl, Grt + Opx + Cpx + Pl and Grt + Cpx + Pl are after Holland & Powell (1998).

garnet + quartz symplectic corona (M_{1b}) suggest an isobaric cooling process following the M_{1a} metamorphism (Fig. 6). The calculated P–T conditions define the M_{1b} assemblage partly to the Grt + Opx + Cpx + Pl field in Figure 6 suggesting the existence of local equilibrium between the M_{1b} garnet + quartz symplectic corona and the M_{1a} mineral rims. The M_{1c} assemblage characterized by the presence of hydrous phase (symplectic biotite) should be located in the lower temperature domains in Figure 6 indicating further cooling. This has been supported by the M_{1c} temperature estimates of 600–700°C. Although the pressure conditions of M_{1c} cannot be quantitatively estimated because of the lack of suitable barometers, the cooling was most probably accompanied with decompression because the breakdown of garnet into biotite commonly occurs under pressure-decreasing conditions (Goscombe 1992). The whole pressure–temperature evolution of the first metamorphic cycle is characterized most probably by an anticlockwise path, since the isobaric cooling occurred at pressures of 7–8 kbar.

The second metamorphic cycle commenced with the growths of garnet porphyroblasts and matrix clino- and orthopyroxene, plagioclase and quartz grains (M_{2a}). This assemblage corresponds to the Grt + Opx + Cpx + Pl field in Figure 6. P–T conditions of M_{2a} were estimated at 9.2–9.8 kbar and 820–850°C for the core compositions and 7.7–8.7 kbar and 780–830°C for the rim compositions (Table 9). This core to rim change indicates a nearly isothermal decompressional process following peak metamorphism (Fig. 6). The M_{2b} assemblage corresponds to the Opx + Cpx + Pl field in Figure 6. P–T conditions of this assemblage was estimated at 7.0–7.6 kbar and 760–810°C, which indicate further nearly isothermal decompression from the rim P–T conditions of the M_{2a} assemblage. The P–T conditions of 4.5–6.0 kbar and 680–790°C estimated with THERMOCALC for hornblende + plagioclase symplectite (M_{2c}) indicate cooling accompanied with further decompression (Fig. 6). Therefore, the whole pressure–temperature evolution of the second metamorphic cycle is characterized by a clockwise P–T path involving near-isothermal decompression and cooling accompanied with decompression (Fig. 6).

Tectonic implications

The tectonic evolution of the North China Craton has not been well constrained. Traditionally, it has been considered to be composed of a relatively uniform Archaean to Palaeoproterozoic crystalline basement, partially overlain by a younger cover sequence, and its tectonic history was explained using a pre-plate tectonic geosynclinal-style model (Huang 1977). Terrane accretion and collisional tectonic models have only recently been applied (Li *et al.* 1990; Sun *et al.* 1992; Zhai *et al.* 1992, 1995, 1998; Wang *et al.* 1996, 2000; Bai & Dai 1998; Cawood *et al.* 1998; Wu & Zhong 1998; Zhao *et al.* 1999c,d, 2000a,b), including recognition of the Trans-North China Orogen (Zhao *et al.* 1998, 1999d). P–T–t determinations reveal that most Archaean basement rocks from the Eastern and Western Blocks are characterized by anticlockwise P–T paths involving near-isobaric cooling, reflecting underplating and intrusion of mantle-derived mafic magmas (Zhao *et al.* 1998, 1999d), whereas the Archaean to Palaeoproterozoic basement rocks from the Trans-North China Orogen are characterized by clockwise P–T paths involving near-isothermal decompression, reflecting continental collisional environments (Wu & Zhong 1998; Zhao *et al.* 1998, 1999c, 2000a,b). On the basis of P–T data and in conjunction with lithological, structural, geochemical and geochronological data, Zhao *et al.* (1999c) proposed that in the late Archaean, the Eastern and Western Blocks in the North China Craton represented two separate continental blocks that developed through the interaction of large volumes of mantle-derived magmas with the lithosphere, whereas the Trans-North China Orogen formed by the collision of these two blocks in the late Palaeoproterozoic at *c.* 1.8 Ga. This study provides a further support for the above tectonic model.

Mineral assemblages, reaction textures, mineral compositions, geothermobarometry and geochronological data suggest that the AB-type mafic granulites encountered two tectonically distinct granulite facies metamorphic events that involved contrasting P–T paths. The first metamorphic event occurred at *c.* 2.5 Ga (Hu *et al.* 1999) and is characterized by an isobaric cooling, anticlockwise, P–T path, which is essentially similar to those P–T paths established for the A-type granulites from the Eastern and Western Blocks, whereas the second granulite facies event occurred at *c.* 1.8 Ga and involved a near-isothermal decompressional, clockwise, P–T path, which is remarkably similar to those P–T paths defined for the B-type granulites from the Trans-North China Orogen (Zhao *et al.* 1998, 1999c, 1999d). These similarities suggest that the AB-type mafic granulites were the products of reworking of the Archaean A-type mafic granulites during the 1.8 Ga collisional event.

Based on the data presented here and previous studies, the following tectonic scenario for the late Archaean to Palaeoproterozoic evolution of the North China Craton is proposed. In the late Archaean, the North China Craton consisted of two separate continental blocks, the Eastern and Western Blocks. At *c.* 2.5 Ga, the intrusion and underplating of large volumes of mantle-magmas occurred and resulted in widespread granulite facies metamorphism (M_{1a} to M_{1c}) involving an isobaric cooling P–T path, forming the A-type mafic granulites in the Eastern and Western Blocks. In the early Palaeoproterozoic, the Eastern Block was bordered along its western edge by an active-type continental margin on which continental arcs and intra-arc basins developed, whereas the Western Block had a passive-type continental margin on which stable continental margin sediments were deposited, forming the protoliths of the khondalitic rocks. Separating the two blocks was an ocean basin which was subducted beneath the western margin of the Eastern Block. At around 1.9–1.8 Ga, the old ocean between the two blocks completely disappeared and led to the collision of the Eastern and Western Blocks and the formation of the Trans-North China Orogen. The collision caused crustal folding, thrusting and thickening, and resulted in peak high- and medium-pressure granulite facies metamorphism (M_{2a}), forming the B-type mafic granulites within the orogen. The collision also resulted in the reworking of the Archaean A-type mafic granulites to form the AB-type mafic granulites in the Trans-North China Orogen and its neighbouring areas. Following the peak metamorphism, the thickened crust with the B- and AB-type mafic granulites underwent exhumation, resulting in decompression (M_{2b}). Finally, cooling accompanied with further decompression (M_{2c}) took place when the crust was exhumed to shallow levels. These tectonic processes led to the final assembly of the North China Craton at *c.* 1.8 Ga.

We would like to express our thanks to J. A. Miller and I. S. Buick for organizing the stimulating meeting, *Orogenesis in the Outback*, at which a preliminary version of this paper was presented, and to the journal reviewers J. Vry, R. White and J. A. Miller for their critical but constructive comments that led to substantial improvements in this paper. This research was supported in part by an ARC Large Grant (No. A39532446) to S. A. Wilde and P. A. Cawood and a RGC Grant (HKU 7100/97P) to Min Sun. This is Tectonics Special Research Centre Publication No 112.

References

BAI, J. & DAI, F. Y. 1998. Archaean crust of China. *In*: MA, X. Y. & BAI, J. (eds) *Precambrian Crust Evolution of China*. Springer–Geological Publishing House, Beijing, 15–86.

BERMAN, R. G. 1990. Mixing properties of Ca–Mg–Fe–Mn garnets. *American Mineralogist*, **75**, 328–344.

CAWOOD, P., WILDE, S. A., WANG, K. Y. & NEMCHIN, A. 1998. Integrated geochronology and field constraints on subdivision of the Precambrian in China: Data from the Wutaishan. *Chinese Science Bulletin*, **43**, 17.

CESARI, B. 1999. Multi-stage pseudomorphic replacement of garnet during polymetamorphism: 1. Microstructures and their interpretation. *Journal of Metamorphic Geology*, **17**, 723–734.

CUI, W. Y., WANG, C. Q. & WANG, S. G. 1991. Geochemistry and metamorphic P–T–t path of the Jianping Complex in the western Liaoning Province. *Acta Petrologica Sinica*, **7**, 13–26 (in Chinese).

FONAREV, V. I. & GRAPHCHIKOV, A. A. 1991. Two-pyroxene thermometry: a critical evaluation. *In*: PERCHUK, L. L. (ed.) *Progress in Metamorphic And Magmatic Petrology*. Cambridge University Press, Cambridge, UK, 65–92.

GOSCOMBE, B. 1992. High-grade reworking of central Australian granulites: Metamorphic evolution of the Arunta Complex. *Journal of Petrology*, **33**, 917–962.

HANES, J. A. 1991. K–Ar and ^{40}Ar–^{39}Ar geochronology: methods and applications. *In*: HEAMAN, L. & LUDDEN, J. N. (eds) *Short Course Handbook on Applications of Radiogenic Isotope Systems to Problems in Geology*. Mineralogical Association of Canada, Toronto, 27–58.

HANSEN, E. C., JANARDHAN, A. S., NEWTON, R. C., PRAME, W. K. B. N. & RAVINDRA KUMAR, G. R. 1987. Ar-rested charnockite formation in southern India and Sri Lanka. *Contributions to Mineralogy and Petrology*, **96**, 225–244.

HARLEY, S. L. 1989. The origins of granulites: a metamorphic perspective. *Geological Magazine*, **126**, 215–247.

HARLEY, S. L. 1992. Proterozoic granulite terranes. *In*: CONDIE, K. C. (ed.) *Proterozoic Crustal Evolution*. Elsevier, Amsterdam, New York, 301–360.

HE, G. P., YIE, H. W. & XIA, S. L. 1994. The evolution of metamorphism in a granulite facies terrane in Miyun, Beijing, China. *Acta Petrologica Sinca*, **10**, 14–24.

HOLLAND, T. J. B. & BLUNDY, J. 1994. Non-ideal interactions in calcic amphiboles and their bearing on amphiboleplagioclase thermometry. *Contributions to Mineralogy and Petrology*, **116**, 433–447.

HOLLAND, T. J. B. & POWELL, R. 1992. Plagioclase feldspar activity-composition relations based on Darken's Quadratic Formalism and Landau theory. *American Mineralogist*, **77**, 53–61.

HOLLAND, T. J. B. & POWELL, R. 1998. An internally consistent thermodynamic data set for phases of petrological interest. *Journal of Metamorphic Geology*, **16**, 309–343.

HU, S. L., GUO, J. H., DAI, T. M. & PU, Z. P. 1999. Continuous laser-probe ^{40}Ar–^{39}Ar age dating on garnet and plagioclase: constraints on metamorphism of high-pressure granulites from Sang-gan area, North China Craton. *Acta Petrologica Sinica*, **15**, 518–523.

HUANG, J. Q. 1977. The basic outline of China tectonics. *Acta Geologica Sinica*, **52**, 117–135 (in Chinese).

JAHN, B. M. & ZHANG, Z. Q. 1984. Archaean granulite gneisses from eastern Hebei Province, China: rare earth geochemistry and tectonic implications. *Contributions to Mineralogy and Petrology*, **85**, 224–243.

JAHN, B. M., AUVRAY, B., CORNICHET, J., BAI, Y. D., SHEN, Q. H. & LIU, D. Y. 1987. 3.5 Ga old amphibolites from eastern Hebei Province, China: field occurrence, petrography, Sm–Nd isochron age and REE geochemistry. *Precambrian Research*, **34**, 311–346.

JIN, W., LI, S. X. & LIU, X. S. 1991. The Metamorphic dynamics of Early Precambrian high- grade metamorphic rocks series in Daqing-Ulashan area, Inner Monglia. *Acta Petrologica Sinica*, **7**, 27–35 (in Chinese with English abstract).

KREZE, R. 1983. Symbols for rockforming minerals. *American Mineralogist*, **68**, 277–279.

KRÖNER, A., CUI, W. Y., WANG, W. Y., WANG, C. Q. & NEMCHIN, A. A. 1998. Single zircon ages from high-grade rocks of the Jianping Complex, Liaoning Province, NE China. *Journal of Asian Earth Science*, **16**, 519–532.

LAMB, W. M. & VALLEY, J. W. 1988. Granulite facies amphibole and biotite equilibria, and calculated peak-metamorphic water activities. *Contributions to Mineralogy and Petrology*, **100**, 349–360.

LI, S. X., LIU, X. S. & ZHANG, L. Q. 1987. Granite greenstone belt in Sheerteng area, Inner Mongolia, China. *Journal of Changchun University of Science & Technology*, **17**, 81–102 (in Chinese).

LI, J. L., WANG, K. Y., WANG, C. Q., LIU, X. H. & ZHAO, Z. Y. 1990. Early Proterozoic collision orogenic belt in Wutaishan area, China. *Scientia Geologica Sinica*, **25**, 1–11.

LI, J. H., ZHAI, M. G., LI, Y. G. & ZHAN, Y. G. 1998. Discovery of Late Archaean high-pressure granulites in Luanping-Chengde area, Northern Hebei Province: tectonic implications. *Acta Petrologica Sinica*, **14**, 34–41 (in Chinese with English abstract).

LIU, X. S. 1994. Characteristics of reworked basement rocks and their implications for evolution of the Daqingshan Orogenic belt. *Acta Petrologica Sinica*, **10**, 413–426.

LIU, S. W. & LIANG, H. H. 1997. Metamorphism of Al-rich gneisses from the Fuping Complex, Taihang Mountain, China. *Acta Petrologica Sinica*, **13**, 303–312.

LIU, X. S., JIN, W., LI, S. X. & XU, X. C. 1993. Two types of Precambrian high-grade metamorphism, Inner Mongolia, China. *Journal of Metamorphic Geology*, **11**, 499–510.

LIU, S. W., LIANG, H. H., ZHAO, G. C., HUA, Y. G. & JIAN, A. H. 2000. Geochronology and geological events of Early Precambrian Complex in Taihuangshan area, China. *Science in China*, **30**, 18–24.

LIU, D. Y., NUTMAN, A. P., COMPSTON, W., WU, J. S. & SHEN, Q. H. 1992. Remnants of 3800 crust in the Chinese Part of the Sino-Korean craton. *Geology*, **20**, 339–342.

LU, L. Z. 1991. Metamorphic P–T–t path of the Archaean granulite-facies terrains in Jining area, Inner Mongolia and its tectonic implications. *Acta Petrologica Sinica*, **8**, 1–12.

LU, L. Z. & JIN, S. Q. 1993. P–T–t paths and tectonic history of an early Precambrian granulite facies terrane, Jining district, southeastern Inner Mongolia, China. *Journal of Metamorphic Geology*, **11**, 483–498.

MAO, D. B., ZHONG, C. T., CHEN, Z. H., LIN, Y. X., LI, H. M. & HU, X. D. 1999. The isotopic ages and their geological implications of high-pressure basic granulites in the northern Chengde area, Hebei Province, China. *Acta Petrologica Sinca*, **15**, 524–531.

MCCARTHY, T. C. & PATIÑO DOUCE, A. E. 1998. Empirical calibration of the silica–Ca–tschermak anorthite (ScAn) geobarometry. *Journal of Metamorphic Geology*, **16**, 675–686.

NEWTON, R. C. 1986. Fluids of granulite facies metamorphism. *In*: WALTHER, J. V. & WOOD, B. J. (eds) *Fluid–Rock Interactions During Metamorphism, Advances in Physical Geochemistry, 5*. Springer, New York, 36–59.

PATIÑO DOUCE, A. E., JOHNSTON, A. D. & RICE, J. M. 1993. Octahedral excess mixing properties in biotite: a working model with applications to geothermometry and geobarometry. *American Mineralogist*, **78**, 113–131.

PIDGEON, R. T. 1980. Isotopic ages of the zircons from the Archaean granulite facies rocks, Eastern Hebei, China. *Geology Review*, **26**, 198–207.

POWELL, R. & HOLLAND, T. J. B. 1994. Optimal geothermometry and geobaromemetry. *American Mineralogist*, **79**, 120–133.

REN, J. S. 1980. *Tectonics and evolution of China*. Science Press, Beijing.

SHEN, Q. H., LIU, D. Y., WANG, P. & GAO, J. F. 1987. U–Pb and Rb–Sr isotopic ages of m metamorphic rock series from the Jining Group, southern Inner Mongolia. *Journal of Chinese Institute of Geology*, **16**, 165–178 (in Chinese).

SONG, B., NUTMAN, A. P., LIU, D. Y. & WU, J. S. 1996. 3800 to 2500 Ma crustal evolution in Anshan area of Liaoning Province, Northeastern China. *Precambrian Research*, **78**, 79–94.

SUN, M., ARMSTRONG, R. L. & LAMBERT, R. St. J. 1992. Petrochemistry and Sr, Pb and Nd isotopic geochemistry of Early Precambrian rocks, Wutaishan and Taihangshan areas, China. *Precambrian Research*, **56**, 1–31.

WANG, K. Y., HAO, J., WILDE, S. A. & CAWOOD, P. A. 2000. Reconsideration of some key problems of Late Archaean–Early Proterozoic basement in the Wutaishan–Tiahangshan area: Constraints from SHRIMP U–Pb zircon data. *Scientia Geologica Sinica*, **35**, 175–184.

WANG, K. Y., LI, J. L., HAO, J., LI, J. H. & ZHOU, S. P. 1996. The Wutaishan mountain belt within the Shanxi Province, Northern China: a record of late Archaean collision tectonics. *Precambrian Research*, **78**, 95–103.

WOOD, B. J. & BANNO, S. 1973. Garnet–orthopyroxene and orthopyroxene–clinopyroxene relationships in simple and complex systems. *Contributions to Mineralogy and Petrology*, **42**, 109–142.

WU, C. H. & ZHONG, C. T. 1998. The Palaeoproterozoic SW–NE collision model for the central North China Craton. *Progress in Precambrian Research*, **21**, 28–50 (in Chinese).

ZHAI, M. G., GUO, J. H. & YAN, Y. H. 1992. Discovery and preliminary study of the Archaean high-pressure granulites in the North China. *Science in China*, **12B**, 1325–1330.

ZHAI, M. G., GUO, J. H., LI, Y. G. & YAN, Y. H. 1995. Discovery of Archaean retrograded eclogites in the North China Craton and their tectonic implications. *China Science Bulletin*, **40**, 1590–1594.

ZHAI, M. G., GUO, J. H., LI, Y. G., YAN, Y. H. & LIU, W. J. 1998. Eclogitehigh-pressure granulite belt in the northern edge of the Archaean North China craton. *Earth Science*, **9**, 6–13.

ZHAO, G. C., CAWOOD, P. A. & WILDE, S. A. 1999a. Polymetamorphism of Archaean mafic granulites from the Trans-North China Orogen: textural evidence and tectonic implications. *In*: MILLER, J. A. & BUICK, I. S. (eds) *Orogenesis in the Outback*. Geological Society of Australia, Abstracts, **54**, 113–114.

ZHAO, G. C., CAWOOD, P. A. & LU, L. Z. 1999b. Petrology and P–T history of the Wutai amphibolites: implications for tectonic evolution of the Wutai Complex, China. *Precambrian Research*, **93**, 181–199.

ZHAO, G. C., WILDE, S. A., CAWOOD, P. A. & LU, L. Z. 1998. Thermal evolution of Archaean basement rocks from the eastern part of the North China craton and its bearing on tectonic setting. *International Geology Review*, **40**, 706–721.

ZHAO, G. C., WILDE, S. A., CAWOOD, P. A. & LU, L. Z. 1999c. Thermal evolution of two textural types of mafic granulites in the North China craton: evidence for both mantle plume and collisional tectonics. *Geological Magazine*, **136**, 223–240.

ZHAO, G. C., WILDE, S. A., CAWOOD, P. A. & LU, L. Z. 1999d. Tectonothermal history of the basement rocks in the western zone of the North China Craton and its tectonic implications. *Tectonophysics*, **310**, 223–240.

ZHAO, G. C., CAWOOD, P. A., WILDE, S. A. & LU, L. Z. 2000a. Metamorphism: implications for Palaeoproterozoic tectonic evolution. *Precambrian Research*, **103**, 55–88.

ZHAO, G. C., WILDE, S. A., CAWOOD, P. A. & LU, L. Z. 2000b. Petrology and P–T path of the Fuping mafic granulites: Implications for tectonic evolution of the central zone of the North China craton. *Journal of Metamorphic Geology*, **18**, 375–391.

Pervasive Pan-African reactivation of the Grenvillian crust and large igneous intrusions in central Dronning Maud Land, East Antarctica

H.-J. PAECH

Bundesanstalt für Geowissenschaften und Rohstoffe (BGR), Postfach 510153, D-30631 Hannover, Germany (e-mail: hans.paech@t-online.de)

Abstract: The geological history of central Dronning Maud Land (cDML) is described on the basis of age data, recent observations, and published data. The ages of both the protoliths of the metamorphic rocks of cDML and their primary metamorphism are Grenvillian (1200 to 1000 Ma). Following a long hiatus for which geochronological data are lacking, the Grenvillian crust of cDML was intruded by abundant Pan-African partly charnockitic granitoids and anorthosites of the Dronning Maud Land igneous province. These intrusions occurred predominantly in two Pan-African igneous episodes. The early igneous episode (about 600 Ma), in which the Grubergebirge anorthosite intruded, was followed by Pan-African high-grade metamorphism, intense ductile deformation and then medium-grade retrogression during later Pan-African events (570 to 520 Ma). The Grenvillian structures and metamorphic signature were pervasively overprinted or completely eradicated. The Pan-African metamorphism and tectonism were completed prior to the late Pan-African igneous episode (500 Ma), in which predominantly post-kinematic granitoids, as well as a small anorthosite, intruded. Thus, the present tectonic structure of the crust in cDML was formed during Pan-African events by overprinting of Grenvillian basement and intrusion of granitoids/syenites and anorthosites.

Central Dronning Maud Land (cDML) is located in Antarctica between 8° and 16°E on the Indian segment of the Southern Ocean. Formerly, Dronning Maud Land (DML) was thought to be composed of Grenvillian (1200 to 1000 Ma, Moores 1991) and older crust (Harley & Hensen 1990) lacking penetrative younger reworking, despite the overwhelmingly young ages, i.e. Pan-African (600 to 500 Ma) mostly K–Ar whole rock data (Krylov *et al.* 1961; Ravich *et al.* 1962; Ravich & Solov'ev 1966), that were thought to be not very reliable. Later, based on the new sensitive high-resolution ion microprobe (SHRIMP) data set of Jacobs *et al.* (1998), more details of the pervasive Pan-African reworking of the Grenvillian crust in cDML became clear, and only on the basis of this data could the Pan-African reworking of the Grenvillian crust of cDML be recognized. Thus, the crust of cDML has much in common with both the adjacent areas in western and eastern DML with the Gondwana fragments around the actual Indian Ocean (Fig. 1). They were formerly in direct contact within the Gondwana supercontinent during its Pan-African amalgamation, which was completed in Early Palaeozoic times. In western DML, an Archaean craton (Paech 1985) is flanked by a Grenvillian mobile belt (Arndt *et al.* 1991) which shows an increasing amount of Pan-African reworking towards the east (Jacobs *et al.* 1996). In eastern DML, the timing of the crust-forming events is not well constrained. Either the Grenvillian (Shiraishi & Kagami 1992) or the Pan-African ages are over-emphasized (Shiraishi *et al.* 1994). Nevertheless, the data allow the assumption of a huge mobile belt characterized by Pan-African reworking of older crust to be followed across the entire DML. It has been called the East Antarctic mobile belt (e.g. Jacobs *et al.* 1998).

Crust reworked during the Pan-African, similar to that in the Antarctic mentioned above, is also known from other Gondwana fragments now bordering the Indian Ocean. In southeastern Africa, the Mozambique Belt, first described by Holmes (1951), exhibits similar features: a Grenvillian crust (Sacchi *et al.* 1984) reworked during Pan-African events. The aim of this case study is to review the geological history of cDML as part of Gondwana, particularly from the perspective of Pan-African processes, that reactivated older crust and formed new igneous crust.

Geographical setting of central Dronning Maud Land

Geographically, cDML can be subdivided into ice shelves in the north, an ice slope in the middle,

From: MILLER, J. A., HOLDSWORTH, R. E., BUICK, I. S. & HAND, M. (eds) *Continental Reactivation and Reworking.* Geological Society, London, Special Publications, **184**, 343–355. 1-86239-080-0/01/$15.00 © The Geological Society of London 2001.

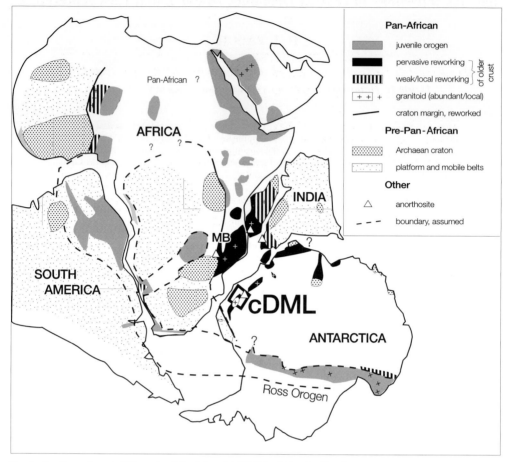

Fig. 1. Position of central Dronning Maud Land (cDML) within the Gondwana supercontinent (MB, Mozambique Belt).

where the glacier surface ascends from almost zero to almost 1000 m above sea level, and a mountain chain in the south. The mountain chain is characterized by more or less N–S-trending ridges with elevations to nearly 3000 m above sea level, making up the mountain groups Orvinfjella (Orvin Mts.), the Wohlthatmassiv with the mountains of the Humboldtgebirge and Hoelfjella (Fig. 2). The mountainous ridges are intersected by glaciers draining the Wegener-Inlandeis.

The mountain ridges and nunataks in cDML were discovered during the air-borne German Schwabenland Expedition in 1939 and were mapped topographically on the basis of oblique aerial photographs (Ritscher 1942) taken during the expedition. Several German geographical names were introduced at this time. Later, more detailed maps were made by Norwegian (in the 1950s and 1960s) and Soviet (mostly in the 1960s) surveys. These were conducted independently, without

Fig. 2. Geological overview map of central Dronning Maud Land (cDML) (mostly based on results of the 1995/1996 GeoMaud expedition). (Names of intrusive bodies are used in this paper according to precedence; secondary names are also given, [geological names according Ravich & Solov'ev 1966 in parentheses]:
(1) Grubergebirge anorthosite; (2) granitoid and anorthosite on the Eckhörner peak, [massiv Insel]; (3) granitoid of the eastern Petermannkette, [massiv Zavaritskogo and massiv Curie]; (4) biotite granite in the Skaly Gubkina rocks or Oddenskjera; (5) granitoid at Gora Titova or Gjeruoldsenhögda, [massiv Titova]; (6) granitoid occurrence in the Khrebet Shcherbakova or Småskeidrista; (7) granitoid of Söre Petermannfjella and adjacent regions; (8) granitoid of Gora Lodochnikova, [part of the massiv Lodochnikova]; (9) orthogneiss Sandhö.

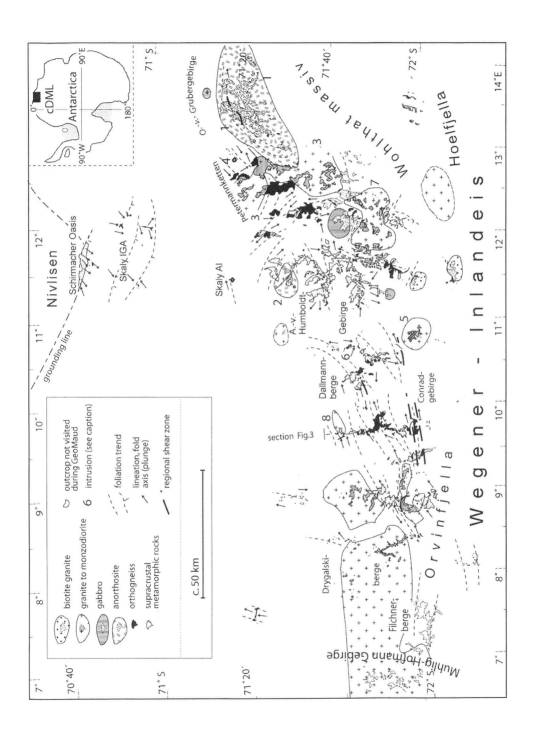

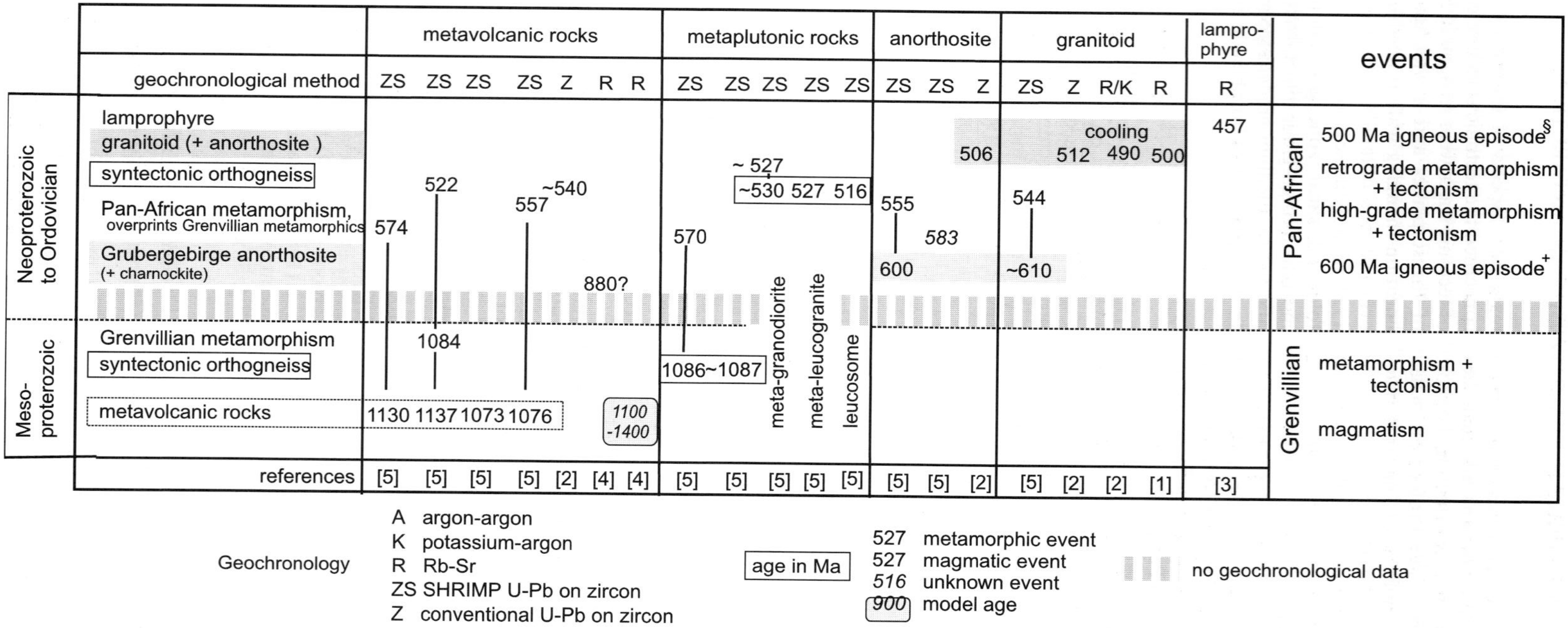

Fig. 3. Geochronological data on rocks from central Dronning Maud Land based on Ohta *et al.* (1990) [1], Mikhalsky *et al.* (1997) [2], Dayal & Hussain (1997) [3], Rameshwar Rao *et al.* (1997) [4], Jacobs *et al.* (1998) [5], in chronological order. +, early Pan-African igneous episode; §, late Pan-African igneous episode.

containing igneous zircons or igneous zircon cores) and sedimentary origin (now metasediments). The metavolcanic rocks are bimodal, i.e. an alternation of acid (felsic) and basic (mafic) layers up to ten metres thick. Their extrusion age is constrained by SHRIMP data on zircons and zircon cores (Jacobs *et al.* 1998) at about 1100 Ma (ranging from 1137 ± 21 to 1073 ± 9 Ma, Fig. 3), i.e. the late Mesoproterozoic age is constrained. The metasedimentary rocks are represented by garnet-rich schists, aluminous schists (sillimanite-bearing), calc-silicate rocks and marbles. In the metasedimentary sequences, mafic intercalations are also present. Besides the metasupracrustal sequences, the metamorphic basement of cDML contains metaplutonic bodies, mostly of felsic composition. They include garnet-bearing leucogneiss, and granitic and tonalitic gneiss. Probably, they intruded mostly in late Mesoproterozoic times (Jacobs *et al.* 1998, see Fig. 3), exemplified by those in the southern Conradgebirge (1086 ± 20 Ma) and Dallmannberge (1087 ± 28 Ma). An exception is the Pan-African orthogneiss in the Conradgebirge (see below).

In terms of structure, there is no obvious evidence for Grenvillian structures in cDML. However, it is likely that transversely oriented foliation planes in mafic intercalations and boudins described by Sengupta (1993) and several intrafolial folds (shown schematically in Fig. 4) can be assigned to the Grenvillian deformation. The Grenvillian structures may be represented by lineations and intrafolial fold axes (stereogram of the northern part of the Conradgebirge; Fig. 4). However, it must be carefully checked that they are not Pan-African stretching lineations or sheath folds.

In summary, the only evidence of for the Grenvilllian age of metamorphic assemblages is constrained by SHRIMP zircon data (Jacobs *et al.* 1998).

Pan-African events

New observations and age data (Jacobs *et al.* 1998) provide evidence of Pan-African reactivation (tectonism and metamorphism) of the Grenvillian crust in cDML. This evidence for Pan-African reactivation of Grenvillian metamorphic rocks is corroborated by metamorphic zircon ages ranging between 574 and 540 Ma, obtained for Mesoproterozoic metaplutonic (granitic to quartzdiorite) and metavolcanic rocks (Jacobs *et al.* 1998). It seems that the Grenvillian metamorphic mineral assemblages were completely recrystallized during the Pan-African events.

Only resistant minerals such as zircons survived and preserved the Grenvillian signature. Moreover, abundant Pan-African igneous intrusions belonging to the Dronning Maud Land igneous province (Paech 1997) pervasively modified the pre-existing crust.

Pan-African magmatism. The Pan-African igneous bodies cover about 30% of the outcrop area of cDML (Fig. 2). The DML igneous province contains two anorthosite intrusions (Grubergebirge and Eckhörner anorthosites in the Humboldtgebirge, intrusions 1 and 2 in Fig. 2) and numerous granitoids (Ravich & Solov'ev 1966) represented by granite, monzonite, quartz monzonite and syenite and its charnockitic equivalents. They are often coarse grained, even containing kalifeldspar megacrysts up to 5 cm across. In part, they derive from primary charnockitic (high-temperature) intrusions that have suffered autometamorphism (Ravich & Solov'ev 1966; Markl & Piazolo 1998). In a few cases, they have preserved their primary high-temperature signature (charnockitic rock types) characterized by orthopyroxene, which mostly was subsequently transformed to amphibole. Often, the postkinematic character of the granitoids is striking; however, a few indications of compressive deformation are present in some igneous bodies (see below). The ages of the granitoids vary between 500 and 600 Ma (Ohta *et al.* 1990; Mikhalsky *et al.* 1997; Jacobs *et al.* 1998) (Fig. 3). However, this data suggests two igneous episodes, an early Pan-African igneous episode around 600 Ma ago and a late Pan-African igneous episode around 500 Ma ago. Both episodes are represented by anorthosite and granitoid associations, including their charnockitic equivalents. Evidently, the Grubergebirge anorthosite and a few granitoids belong to the early episode (600 ± 12 Ma, Jacobs *et al.* 1998) and the Eckhörner anorthosite (506 ± 2.4 Ma, Mikhalsky *et al.* 1997) plus most other granitoids belong to the late Pan-African post-kinematic igneous episode.

Coevally with the granitoid magmatism, a patchy charnockitization of the metamorphic rocks (coarsening of feldspar and crystallizsation of orthopyroxene) occurred. The charnockization is younger than the main metamorphism but is older than hydrothermal alteration or 'bleaching' of the country rocks adjacent to felsic and intermediate dykes (Markl & Piazolo 1998), and along minor thrusts (Fig. 4). This 'bleaching' reflects the waning igneous activity of the late Pan-African igneous episode in cDML during Ordovician times. Lamprophyres also intruded during this final phase ($<500 \pm$ Ma, Dayal & Hussain 1997) (Fig. 3).

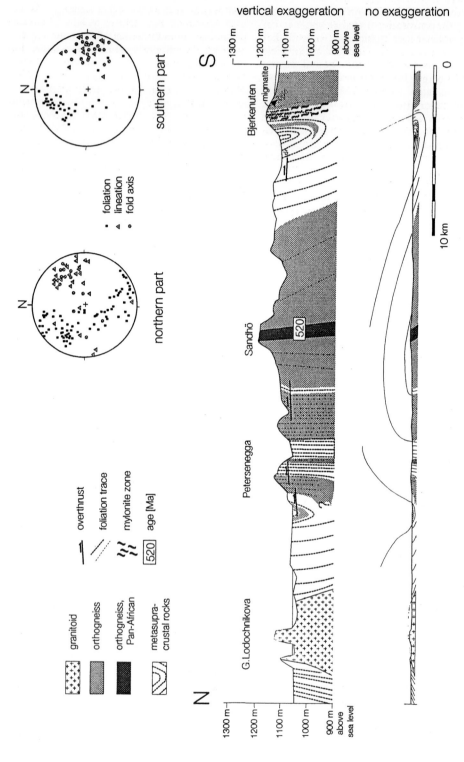

Fig. 4. Geological section along the Conradgebirge in the Orvinfjella (for location see Fig. 2).

Pan-African tectonism. Evidence for Pan-African reactivation of the Grenvillian crust is supported by the conformity of the foliation trend in the overprinted metamorphic basement with that of the adjacent Pan-African intrusive rocks of the 600 Ma igneous episode. Thus, the outer parts of the Grubergebirge anorthosite, which have an intrusion age of about 600 Ma, suffered ductile deformation (indicated by flattened and stretched rafts of metamorphic rocks oriented along the foliation), and were intensely foliated and transformed by pervasive recrystallization (Ravich & Kamenev 1972) under high-grade metamorphic conditions (Kämpf *et al.* 1995) to meta-anorthosite about 555 Ma ago (Jacobs *et al.* 1998) (Fig. 3). Subsequent weak retrograde metamorphism transformed the pyroxenes marginally to amphiboles (Ravich & Kamenev 1972). The foliation trend within the meta-anorthosite shows remarkable conformity with that in the adjacent metamorphic country rocks (Fig. 2), indicating that the foliation in the originally Grenvillian metamorphic country rock is an overprint structure of Pan-African age.

In terms of structure, the metamorphic basement of cDML is characterized by polyphase tectonism consisting of at least three folding episodes, which are probably largely Pan-African in age (see below). In the Orvinfjella, the dominant tectonic trend is ENE–WSW (Fig. 2), whereas in the Wohlthatmassiv and Hoelfjella, the tectonic trend varies considerably. Here in the Wohlthatmassiv, the variable trend is associated with the predominance of calcareous and pelitic sequences characterized by high tectonic mobility and, in contrast, the rigid behaviour of the 600 Ma Grubergebirge anorthosite and coeval granitoid intrusions.

Granitoids assumed to be post-kinematic also locally show effects of Pan-African compressive deformation. Locally, the granitoids contain a high proportion of megacrysts (up to 5 cm across), e.g. at Gora Titova or Gjeroldhögda (intrusion 5 in Fig. 2). These megacrysts were locally, and only weakly, transformed to augens typical of orthogneiss. And a small granitoid body in the Khrebet Shcherbakova (intrusion 6 in Fig. 2, too small to be drawn on the map) contain megacrysts that have locally been transformed during ductile deformation to real augens, leading to the formation of orthogneiss.

In the middle part of the Conradgebirge (Conrad Mts.) in the Sandhö area (9 in Fig. 2), a stratiform Pan-African metaplutonic sheet about 1 km thick occurs in a large northward-vergent antiform with a wavelength of about 14 km, which belongs to a series of megafolds (Fig. 4). These megafolds are younger than the Pan-African metamorphism (about 525 Ma) and older than the 500 Ma post-kinematic granitoids, i.e. their formation is Cambrian in age. The metaplutonic sheet consists primarily of a Pan-African meta-granodiorite (530 ± 8 Ma), a younger, thin meta-leucogranite (527 ± 6 Ma), and a late-tectonic leucosome (516 ± 5 Ma) (Jacobs *et al.* 1998). This Pan-African sheet-like metaplutonic intrusion is syntectonic. It was – at least the metamorphic parts –involved in the formation of the antiform (Fig. 4). Additionally, this Pan-African granitoid was foliated and metamorphosed, as reflected by the clinopyroxene–garnet–plagioclase ± hornblende–quartz assemblage, at high-grade (relatively high pressure upper amphibolite to granulite facies, Jacobs *et al.* 1998). The metamorphism was polyphase in at least three stages. At the beginning it attained high-grade and culminated in low-grade retrograde metamorphism. Retrogression is also confirmed by different generation of fluid inclusions within the quartz (Jacobs *et al.* 1998). The foliation is pervasive and its formation was followed by the above-mentioned formation of synforms and antiforms (Fig. 4). These relationships show that intrusion, high-grade metamorphism and deformation of the metaplutonic rock suite occurred in a relatively short time interval around 530–515 Ma. Only the orthogneiss was affected by Pan-African deformation and metamorphism.

Furthermore, the foliation trend in relation to the margins of rigid intrusive bodies provides some indication of the relative ages, i.e. in the case of early Pan-African intrusives, the Pan-African deformation age is constrained. On Söre Petermannfjella and adjacent areas to the west (intrusion 7 in Fig. 2), the foliation trend conspicuously wraps around the western and southern sides of a granitoid body. This is interpreted to indicate that the Pan-African intrusive body is older than the final episode of the Pan-African deformation of the country rocks.

However, most of the so-called post-kinematic granitoids are undeformed. They belong to the late Pan-African igneous episode (500 Ma). The ages of the Eckhörner anorthosite and associated granitoid are constrained by a conventional zircon age of about 506 ± 2.4 Ma, which is slightly younger than the associated granitoid of 512 ± 2.3 Ma (Mikhalsky *et al.* 1997). Thus, the granitoids of the late Pan-African igneous episode are post-kinematic with respect to the Pan-African deformation. Evidently, the Eckhörner anorthosite (intrusion 2 in Fig. 2) is also completely undeformed.

Pan-African metamorphism. The metamorphic rocks in cDML are characterized by coexistence of granulite-facies and amphibolite-facies features (Rameshwar Rao *et al.* 1997, Markl & Piazolo 1998, Piazolo & Markl 1999). The metamorphic grade was determined from the garnet–pyroxene–plagioclase–quartz assemblage in mafic granulite as 837 ± 26 C at 7.1 ± 0.2 kbar (granulite facies), and from the garnet–pyroxene–amphibole–biotite–quartz (amphibolite facies) as 652 ± 33 C at 5.9 ± 0.3 kbar (Rameshwar Rao *et al.* 1997). This geothermometry is supported by mineral inclusion studies, which revealed CO_2 inclusions of high density in high-grade minerals from the mafic granulite and hydrous fluid admixture in CO_2 inclusions in medium-grade minerals from an associated leucogranite.

Similar metamorphic conditions with slightly lower values have been published by Markl & Piazolo (1998) and Piazolo & Markl (1999), who demonstrated high-grade metamorphic conditions of 830 ± 20 C at 6.8 ± 0.5 kbar at the beginning of the metamorphic event and $650°C$ at 4.5 ± 0.7 kbar at the end, which was assumed to be coeval with the intrusion of the post-kinematic granitoids.

Discussion

The geological history of cDML is today better constrained in general terms (Fig. 3, event column, right). It is now clear that the main crustal forming events in cDML belong to the Grenvillian and Pan-African orogenies. Previous assumptions of a simple Grenvillian circum-Antarctic mobile belt (covering also cDML) were highly speculative (e.g. Moores 1991) and were needed for the SWEAT hypothesis (linkage of North America-Antarctica). These speculations can be disproved on the basis of a new geochronological data summarized here. The geological history of cDML is a long process, starting during the Grenvillian Orogeny and ending during the Pan-African events, which transformed pervasively the Grenvillian crust. Only some relics of Grenvillian structures and metamorphic assemblages are preserved and geochronological data is restricted to resistent minerals such as zircon. Thus, many detailed problems of the structural and metamorphic history of the area remain unsolved.

The time interval around 1100 Ma for the extrusion of the Mesoproterozoic volcanic rocks and the intrusion of the stratiform acid rocks is so short that they may belong to the same igneous province. Similarly, the intrusion of the stratiform Pan-African acidic rocks (now orthogneiss) and intrusion of the granitoids of the DML igneous province, which are both inferred to be post-kinematic, coincide in age. Thus, the magmas may

have a common source. Geochemical analyses currently being undertaken on the granitoids of cDML should help to resolve this problem. The present crust of cDML is the result of pervasive Pan-African reactivation of a Grenvillian crust and resembles the Mozambique Belt in southeast Africa (Holmes 1951; Shackleton 1996; Jacobs *et al.* 1998).

From the geotectonic point of view, the polyphase formation of the crust of cDML may be the result of a continent–continent collision or a continental reworking and reactivation of pre-existing older crust comparable with orogenies under in intracratonic conditions. A comparison between the formation history of the cDML crust with that of southeastern Africa indicates many common features (e.g. Pan-African reactivation of Grenville-age crust and formation of post-kinematic granitoids and consanguineous anorthosite during Pan-African events; Fig. 1). Hence, the continuation of the Mozambique Belt (Holmes 1951) as part of the East African Orogen (Stern 1994) from Africa to DML (here part of the East Antarctic mobile belt; Dirks & Wilson 1995; Jacobs *et al.* 1998) is now generally accepted (Groenewald *et al.* 1991; Grunow *et al.* 1996; Jacobs *et al.* 1998; Fitzsimons 2000). However, the suture zone running at the western boundary of the East African Orogen from Arabia to South Africa as the contact zone between East and West Gondwana (Shackleton 1996) has features characteristic of continent–continent collision, with relics of the Mozambique Ocean occurring only as ophiolites in Arabia and adjacent areas. Further to the south, in the Mozambique Belt (Fig. 1), this suture zone separates crustal material highly reworked during Pan-African events and affected crust that is undeformed or scarcely reactivated crust. In Antarctica, this suture zone is only preserved in western DML (Jacobs *et al.* 1996). In other Antarctic regions it's existence and position are more or less speculative.

Therefore, alternative explanations of the crustal setting of cDML as intracratonic reactivation of older crust have to be addressed. In Australia, crust characterized by subsequent pervasive continental reworking and reactivation is described. Hand *et al.* (1999) reported an early Ordovician amphibolite-facies overprint of a Mesoproterozoic high-grade metamorphic basement in the Harts Range which culminated in the Devonian-Carboniferous Alice Springs Orogeny (400–300 Ma) under upper amphibolite-facies conditions ($>600°C$ at 6 kbar) associated with a pervasive foliation (D2). Subsequently it experienced high-temperature decompression. Additionally, the age of the Alice Springs Orogeny overprint is evidenced by the fact that the Lower Palaeozoic succession of the northern flank of the Amadeus Basin covering the metamorphic basement was affected by this deformation. At the southern margin of the Amadeus Basin, Scrimgeour & Close (1999) described a similar structural pattern in the Mann Range. A Mesoproterozoic metamorphic basement was overprinted during the Petermann Range Orogeny, which in age resembles the Pan-African events in Antarctica. The overprint started under high-pressure (12–13 kbar) followed by peak metamorphism at c. 700°C at 9–10 kbar. The existence of a relatively high geothermic regime is evidenced by the small amount of migmatization along shear zones seen in Th- and K-rich granites, which are considered to generate the radiogenic heat. Thus, in central Australia an intracratonic overprint of older crust is widespread and not associated with a short time span. However, the complexity and differing tectonic character and the scarcity of igneous activity are evident.

The dominance of high-temperature igneous activity in cDML during the Pan-African events has no real equivalent in central Australia. Here, the processes affecting the older crust not only differ in character from those in cDML but are also extremely complex. It is most likely that underplating occurred coevally with the Pan-African metamorphic overprint in cDML; this is not incompatible with the presence of large igneous bodies.

Conclusions

- The volcanic and probably also sedimentary protoliths of the metamorphic crust of central Dronning Maud Land (cDML) were formed at the end of the Mesoproterozoic (1100 Ma) (Fig. 3). Remnants of older crust are unknown.
- These protoliths were subjected to high-grade metamorphism and intruded by syntectonic stratiform granitoids during the Grenvillian orogeny (1085 Ma).
- The Pan-African events are characterized by pervasive metamorphic overprinting of the Grenvillian metamorphic basement and large intrusions.
- Large intrusions of primary mostly high-temperature (charnockitic) granitoids and associated anorthosite and gabbro are characteristic of the Pan-African events in cDML and belong to the Dronning Maud Land igneous province. These intrusions caused considerable modification of the Grenvillian

metamorphic crust. In isotopic signature no differences exist between Grenvillian crust and Pan-African igneous bodies (Jacobs et al. 1998). Thus, the Pan-African magma represents partial melt from the Grenvillian crust.
- The formation of the DML igneous province began with an early Pan-African igneous episode (about 600 Ma), followed by a late Pan-African igneous episode (about 500 Ma) and subsequent cooling. Most likely, the syntectonic Pan-African magmatism belongs to this igneous province as well.
- The formation of the DML igneous province was accompanied or interrupted by metamorphic and polyphase tectonic processes. After the early Pan-African igneous episode (after the intrusion of the Grubergebirge anorthosite), high-grade metamorphism and ductile tectonism took place (570 Ma), decreasing in intensity in the course of Pan-African, and ending prior to the late Pan-African post-kinematic igneous episode.
- Judging from the intensity of the Pan-African high-grade metamorphism and ductile deformation observed in metaplutonic rocks and the correlation of the Pan-African foliation trend in the Grubergebirge anorthosite with that in the adjacent country rocks, it can be concluded that the present tectonic structure of cDML was formed during the Pan-African. The pre-existing Grenvillian structures and metamorphic signature were almost completely overprinted or completely eradicated, mostly under high-grade conditions.
- There is no record that supracrustal rocks were deposited during the Pan-African.
- The tectonic setting of the crust in cDML during the Pan-African events is discussed controversially: continent–continent collision or intracratonic reworking? The interpretation of an intracratonic reworking and reactivation is favoured.

References

ARNDT, N. T., TODT, W., CHAUVEL, C., TAPFER, M. & WEBER, K. 1991. U–Pb age and Nd isotopic composition of granitoids, charnockites and supracrustal rocks from Heimefrontfjella, Antarctica. Geologische Rundschau, 80(3), 759–777.

DAYAL, A. M. & HUSSAIN, S. M. 1997. Rb–Sr ages of lamprophyre dykes from Schirmacher Oasis, Queen Maud Land, East Antarctica. Journal of Geological Society of India, 50(4), 457–460.

DIRKS, P. H. G. M. & WILSON, C. J. L. 1995. Crustal evolution of the East Antarctic mobile belt in Prydz Bay: continental collision at 500 Ma? Precambrian Research, 75, 189–207.

FITZSIMONS, I. C. W. 2000. A review of tectonic events in the East Antarctic Shield and their implication for Gondwana and earlier supercontinents. *Journal of African Earth Sciences*, **31**(1), 3–23.

GROENEWALD, P. B., GRANTHAM, G. H. & WATKEYS, M. K. 1991. Geological evidence for a Proterozoic to Mesozoic link between southeastern Africa and Dronning Maud Land, Antarctica. *Journal of the Geological Society, London*, **148**, 1115–1123.

GRUNOW, A., HANSON, R. & WILSON, T. 1996. Were aspects of Pan-African deformation linked to Iapetus opening. *Geology*, **24**(12), 1063–1066.

HAND, M., MAWBY, J., KINNY, P. & FODEN, J. 1999. U–Pb ages from the Harts Range, central Australia: evidence for early Ordovician extension and constraints on Carboniferous metamorphism. *Journal of the Geological Society, London*, **156**, 715–130.

HARLEY, S. L. & HENSEN, B. J. 1990. Archaean and Proterozoic high.grade terranes of East Antarctica (40–80°E): a case study of diversity in granulite facies metamorphism. *In*: ASHWORTH, J. R. & BROWN, M. (eds) *High-temperature metamorphism and crustal anatexis*. Unwin Hyman, London, 320–370.

HOLMES, A. 1951. The sequence of Pre-Cambrian orogenic belts in south and central Africa. *Report 18th International Geological Congress, London*, **14**, 254–269.

JACOBS, J., BAUER, W., SPAETH, G., THOMAS, R. J. & WEBER, K. (1996). Lithology and structure of the Grenville-aged (~1.1 Ga) basement of Heimefrontfjella (East Antarctica). *Geologische Rundschau*, **85**, 800–821.

JACOBS, J., FANNING, C. M., HENJES-KUNST, F., OLESCH, M. & PAECH, H.-J. 1998. Continuation of the Mozambique Belt into East Antarctica: Grenville-age metamorphism and polyphase Pan-African high-grade events in central Dronning Maud Land. *The Journal of Geology, Chicago*, **106**, 385–406.

JOSHI, A., PANT, N. C. & PARIMOO, M. L. 1995. Petrology, geochemistry and evolution of the charnockite suite of the Petermann Ranges, East Antarctica. *In*: YOSHIDA, M. & SANTOSH, M. (eds) *India and Antarctica during the Precambrian*. Geological Society of India, Bangalore, Memoir, **34**, 241–258.

KÄMPF, H., STACKEBRANDT, W., HAHNE, K., PAECH, H.-J. & LEPIN, V. S. 1995. Wohlthat Massif. *In*: BORMANN, P. & FRITZSCHE, D. (eds) *Schirmacher Oasis, Queen Maud Land, Antarctica, and its surrounding*. Petermanns Geographische Mitteilungen, Ergaenzungsheft, **287**, 133–170.

KRYLOV, A. YA., VORONOV, P. S. & SILIN, YU. I. 1961. Absolyutnyjy vozrast kristallicheskogo fundamenta vostochno–antarkticheskojy platformy (Absolute age of the crystalline basement of the East-Antarctic platform). *Doklady Akademii Nauk SSSR*, **143**, 18–21. (in Russian)

MARKL, G. & PIAZOLO, S. 1998. Halogen-bearing minerals in syenites and high-grade marbles of Dronning Maud Land, Antarctica: monitors of fluid compositional changes during late-magmatic fluid

rock interaction processes. *Contribution Mineralogy and Petrology*, **132**, 246–268.

MIKHALSKY, E. V., BELIATSKY, B. V., SAVVA, E. V., WETZEL, H.-U., FEDEROV, L. V., WEISER, TH. & HAHNE, K. 1997. Reconnaissance geochronologic data on polymetamorphic and igneous rocks of the Humboldt Mountains, central Queen Maud Land, East Antarctica. *In*: RICCI, C. A. (ed.) *The Antarctic Region: Geological evolution and processes*. Proceedings of the VII International Symposium on Antarctic Earth Sciences, Siena 1995, 45–54.

MOORES, E. M. 1991. Southwest, US–East Antarctic (SWEAT) connection: A hypothesis. *Geology*, **19**(5), 425–428.

OHTA, Y., TORUDBAKKEN, B. O. & SHIRAISHI, K. 1990. Geology of Gjelsvikfjella and western Mühlig-Hofmannfjella, Dronning Maud Land, East Antarctica. *Polar Research*, **8**, 99–126.

PAECH, H.-J. 1985. Comparison of the geologic development of southern Africa and Antarctica. *Zeitschrift fuer geologische Wissenschaften*, **13**(3), S.399–415.

PAECH, H.-J. 1997. Central Dronning Maud Land: Its history from amalgamation to fragmentation of Gondwana. *Terra Antartica*, **4**, 1 41–49.

PAECH, H.-J. & STACKEBRANDT, W. 1995. Geology. *In*: BORMANN, P. & FRITZSCHE, D. (eds) *Schirmacher Oasis, Queen Maud Land, East Antarctica and adjacent areas*. Petermanns Geographische Mitteilungen, Ergaenzungsheft, Justus Perthes, Gotha, **287**, 59–169.

PIAZOLO, S. & MARKL, G. 1999. Humite- and scapolite-bearing assemblages in marbles and calcsilicates of Dronning Maud Land, Antarctica: new data for Gondwana reconstructions. *Journal of Metamorphic Geology*, **17**, 91–107.

RAMESHWAR RAO, D., SHARMA, R. & GURURAJA, N. C. N. S. 1997. Mafic granulites of the Schirmacher region, East Antarctica: fluid inclusion and geothermobarometric studies focusing on the Proterozoic evolution of the crust. *Transactions Royal Society Edinburgh: Earth Science*, **88**, 1–17.

RAVICH, M. G. 1982. The lower Precambrian of Antarctica. *In*: CRADDOCK, C. (ed.) *Antarctic Geoscience*. Madison, 421–427.

RAVICH, M. G. & KAMENEV, E. N. 1972. *Kristallicheskij fundament arkticheskojy platformy*. Gidrometeoizdat, Leningrad. (in Russian). (Translation: Crystalline basement of the Antarctic platform, Jerusalem 1975, 574).

RAVICH, M. G. & SOLOV'EV, D. S. 1966. *Geologiya i petrologiya tsentral'noy chasti gor Zemli Korolevy Mod (Vostochnaya Antarktida) (Geology and petrology of the mountains of central Queen Maud Land, eastern Antarctica)*. Nedra, Leningrad (in Russian).

RAVICH, M. G., KRYLOV, A. YA., SOLOV'EV, D. S. & SILIN, YU. I. 1962. Absolyutnyj vozrast porod tsentral'noj chasti Zemli Korolevy Mod (Vostochnaya Antarktida). *Doklady Akademii Nauk*, **147**, 130–133 (in Russian). (Translation: Age of the rocks in the central part of the mountains of Queen Maud Land, East Antarctica. *Doklady Akademii Nauk*, **147**(6), 1433–1436.)

RITSCHER, A. 1942. *Wissenschaftliche und fliegerische Ergebnisse der deutschen Antarktischen Expedition 1938/1939*. Köhler & Amelang, Leipzig.

SACCHI, R., MARQUES, J., COSTA, M. & CASATI, C. 1984. Kibaran events in the southernmost Mozambique Belt. *Precambrian Research*, **25**, 141–159.

SCRIMGEOUR I. & CLOSE, D. 1999. Regional high pressure metamorphism during intracratonic deformation: the Petermann Orogeny, central Australia. *Journal of Metamorphic Geology*, **17**, 557–572.

SENGUPTA, S. 1993. Tectonothermal history recorded in mafic dykes and enclaves of gneissic basement in the Schirmacher Hills, East Antarctica. *Precambrian Research*, **63**, 273–291.

SHACKLETON,, R. M. 1996. The final collision zone between East and West Gondwana: where is it? *Journal African Earth Science*, **23**, 271–287.

SHIRAISHI, K. & KAGAMI, H. 1992. Sm–Nd and Rb–Sr ages of metamorphic rocks from the Sor Rondane Mountains, East Antarctica. *In*: YOSHIDA, Y., KAMINUMA, K. & SHIRAISHI, K. (eds) *Recent progress in Antarctic Earth Science*, TERRAPUB, Tokyo, 29–35.

SHIRAISHI, K., ELLIS, D. J., HIROI, Y., FANNING, C. M., MOTOYOSHI, Y. & NAKAI, Y. 1994. Cambrian orogenic belt in East Antarctica and Sri Lanka: implications for Gondwana assembly. *Journal Geology*, **102**, 47–65.

STERN, R. J. 1994. Arc assembly and continental collision in the Neoproterozoic East African Orogen: Implication for the consolidation of Gondwanaland. *Annual Revues Earth Planetary Sciences*, **22**, 319–351.

WETZEL, U. 1995. Lithostratigraphy and tectonics of the nunataks region. *In*: BORMANN, P., FRITZSCHE, D. (eds.) *The Schirmacher Oasis, Queen Maud Land, East Antarctica and its surroundings. Petermanns Geographische Mitteilungen Ergaenzungsheft*, **287**, 115–126.

Fluid–rock interaction in the Reynolds Range, central Australia: superimposed, episodic, and channelled fluid flow systems

IAN CARTWRIGHT[1], IAN S. BUICK[2] & JULIE K. VRY[3]

[1] *Victorian Institute of Earth and Planetary Sciences, Department of Earth Sciences, Monash University, Clayton, Victoria 3800, Australia*
(e-mail: Ian.Cartwright@sci.monash.edu.au)
[2] *Victorian Institute of Earth and Planetary Sciences, School of Earth Sciences, La Trobe University, Bundoora, Victoria 3083, Australia*
[3] *School of Earth Sciences, PO Box 600, Victoria University, Wellington, New Zealand*

Abstract: The Reynolds Range, central Australia, is a polymetamorphic Proterozoic terrain within the Arunta Inlier. The terrain comprises a diverse sequence of metasedimentary rocks (including pelites, psammites, quartzites, marls and marbles) intruded by with two generations of granites. The Reynolds Range preserves evidence of undergoing several metamorphic events, including: phases of contact metamorphism at 1.82 Ga and 1.78 Ga; regional metamorphism at 1.6 Ga that varied in grade from greenschist facies ($c. 400°C$) to granuite facies (750–800°C) at 400–500 GPa; and metamorphism at up to amphibolite facies (550–600°C at 500–600 MPa) in the Alice Springs Orogeny at $c. 334$ Ma, the affects of which are recorded mainly within shear zones. Fluid flow occurred during both contact metamorphic events, cooling from the peak of regional metamorphism at 1.57–1.58 Ga, and additionally during Alice Springs shearing. By contrast, there is little fluid–rock interaction that can be attributed to the prograde stages of regional metamorphism, implying that fluids generated by metamorphic devolatilization at that time escaped relatively rapidly and did not interact with the rocks as a whole.

Fluid flow changed the mineralogy and stable isotope ratios of rocks, and locally caused extensive metasomatism. During cooling from the peak of regional metamorphism, the fluids were derived from crystallization of partial melts and reflect internal fluid recycling. However, at least some fluid flow during contact metamorphism and shearing involved external fluids. In both contact metamorphic episodes, igneous and locally surface-derived fluids interacted with the country rocks adjacent to the granite plutons. During Alice Springs shearing, large volumes of surface-derived fluids infiltrated the middle crust. Much of the fluid flow was channelled on scales ranging from hecto- to millimetres as a result of variations in intrinsic permeability caused by deformation or reaction enhancement.

Constraining crustal fluid–rock interaction is important as the infiltration of large volumes of fluid may: cause significant metasomatism; form mineral deposits; transfer heat; promote partial melting; and affect crustal rheology, thus controlling deformation. Studying fluid–rock interaction has been one of the major goals of metamorphic petrologists in the past two decades. Stable isotopes (especially O and H) are a major tool in these studies as they directly trace the fluids whereas other elements (e.g. C isotopes, major elements, radiogenic isotopes, halogens, or the rare earth elements) trace the behaviour of solutes. Recently, advection–dispersion fluid flow models have been applied to crustal fluid flow systems in order to quantify fluid–rock interaction (e.g. Bickle & McKenzie 1987; Baumgartner & Rumble 1988; Bickle & Baker 1990a, b; Baumgartner & Ferry 1991; Dipple & Ferry 1992a, b; Bowman et al. 1994; Cartwright & Weaver 1997). However, due to the large amount of data required, individual episodes of fluid flow are often considered in isolation, and there have been relatively few studies of the history of fluid–rock interaction in large, complex, metamorphic terrains (e.g. Valley et al. 1990; Cartwright & Valley 1991; Oliver et al. 1994). This is in contrast to structural or metamorphic studies that tend to emphasize evolution with time.

Fluids and metamorphism

The role of fluids in metamorphism has been the subject of considerable debate. Proposals

From: MILLER, J. A., HOLDSWORTH, R. E., BUICK, I. S. & HAND, M. (eds) *Continental Reactivation and Reworking.* Geological Society, London, Special Publications, **184**, 357–379. 1-86239-080-0/01/$15.00 © The Geological Society of London 2001.

that granulites result from the pervasive infil-tration of CO_2-rich fluids (e.g. Newton 1986) are now largely discounted due to the very reduced nature of most granulite-facies assem-blages (Lamb & Valley 1984) and the preserva-tion of stable isotope and other geochemical heterogeneities in granulite-facies terrains (e.g. Valley *et al.* 1990; Vry *et al.* 1990; Cartwright & Valley 1991, 1992). However, in a number of terrains (e.g. Mount Lofty Ranges, Australia; New England, USA; Trois Seigneurs, France-Spain; Lizzies Basin, USA; Naxos, Greece) the progress of mineral reactions, the systematic low-ering of $\delta^{18}O$ values with metamorphic grade, and/or the occurrence of migmatites formed by fluid-present melting suggest that large vol-umes of water-rich fluids infiltrated many rock types during regional metamorphism (Wickham & Taylor 1985; Symmes & Ferry 1991; Léger & Ferry 1993; Stern *et al.* 1992; Cartwright *et al.* 1995; Cartwright & Buick 1998). By con-trast, regional metamorphism in other terrains (e.g. Adirondack Mountains, USA; Omeo Zone, Australia) probably occurred with small vol-umes of fluid (Valley *et al.* 1990; Cartwright & Harper 1998; Cartwright & Buick 1998). It is unclear why there is this divergence of behaviour, and it is thus critically important to document fluid flow during metamorphism.

Unmetamorphosed sedimentary and crystal-line rocks may contain connate and fracture-hosted water (e.g. Pearson *et al.* 1991); however, much of this fluid will be lost prior to meta-morphism during diagenesis and compaction. Fluids present during prograde metamorphism will be those generated by devolatilization to-gether with those from other sources (e.g. igneous bodies, the Earth's surface, overlying sedimen-tary basins, or the mantle). Pelitic, semipelitic, and intermediate to granitic igneous rocks dominate most of the crust. Thompson (1997), Yardley (1997), and Cartwright & Oliver (2000), amongst others, have presented general models of fluid production from such rocks during regional metamorphism. Crustal rocks under-go dehydration between *c.* 200°C and the onset of melting (*c.* 650°C). Initial reactions that breakdown minerals with high water contents (clays with 15–20 wt% H_2O or chlorite with 10–12 wt% H_2O) may liberate more fluid per unit volume of rock than later reactions that break down minerals with lower water contents (biotite and muscovite with 3–4 wt% H_2O, or staurolite and cordierite with *c.* 2 wt% H_2O). At temperatures in excess of 650–700°C many crustal rocks will melt. The melts represent a temporary sink for the fluids until they crystal-lize during cooling at 600–650°C when much of

that fluid is exsolved (Corbett & Phillips 1981; Cartwright 1988).

This paper summarizes the history of fluid-rock interaction in the Reynolds Range, cen-tral Australia. Discussion is restricted to the Reynolds Range itself as adjacent areas (e.g. Mount Stafford and the Anmatajira Ranges) have somewhat different tectonometamorphic histories. The episodes of fluid–rock interaction that have been discussed individually in other publications are placed into the structural, meta-morphic, and geochronological framework of the terrain, and the fluid flow systems that oper-ated at the various stages are identified. The Reynolds Range is an ideal terrain for the study of fluid flow as it contains a wide variety of rock types that may be traced from low to high regional metamorphic grade. Thus, many often contentious questions, such as whether fluid–rock interaction in uniformly high-grade metamorphic terrains occurred during or prior to regional metamorphism, may be addressed. The Reynolds Range has a metamorphic his-tory that is similar to many low- to medium-pressure metamorphic terrains, and the ideas discussed will be generally applicable to other regions. This summary represents work over a seven year period that generated some 2900 stable isotope analyses, 2000 mineral analyses, 80 radiogenic isotope analyses, and 80 whole-rock major and trace element analyses (as well as the associated field, petrographic, and struc-tural observations). A compilation of the stable isotope data and a locality map is available as a supplement to this paper archived at http://www.monash.edu.au/~icart/reynolds/.

Local geology

The Reynolds Range (Fig. 1) is a multiply deformed and metamorphosed Proterozoic ter-rain within the Arunta Inlier of central Australia (Stewart *et al.* 1984; Dirks 1990; Black *et al.* 1991). It occupies an area of 100×20 km, and consists of two stratigraphic associations. The Lander Rock Beds, together with early granites, form the base-ment to metasediments of the Reynolds Range Group. The Reynolds Range Group includes: a basal quartzite; a unit of dominantly calc-pelites (the Lower Calcsilicate Unit); a unit of pelitic and psammitic rocks (the Pelite Unit); and the Upper Calcsilicate Unit, a discontinuous lens within the Pelite Unit dominated by mar-ble. A later suite of granites intruded both the Reynolds Range Group and Lander Rock Beds.

The metamorphic and structural history of the Reynolds Range has been described by Dirks

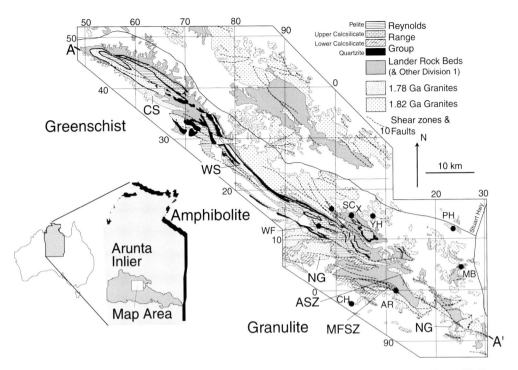

Fig. 1. Geological map of the Reynolds and Anmatajira Ranges simplified from the 1:100,000 Reynolds Range Map Sheet (Northern Territory Geological Survey, Darwin). Localities discussed in text: AR, Anna Reservoir; CH, Conical Hill; MB, Mount Boothby; PH, Peaked Hill; SC, Sandy Creek; WF, Woodford River; X, 'X-Locality'; YH, Yaningidjara Hills. CS, Conniston Schist; NG, Napperby Gneiss; WS, Warrimbi Schist. ASZ, Aileron Shear Zone; MFSZ, Mount Freeling Shear Zone. A–A' shows the line of section for Figure 3. A more detailed locality map is available as a supplement to this paper.

& Wilson (1990), Dirks *et al.* (1991), Clarke *et al.* (1990), Clarke & Powell (1991), Buick *et al.* (1994, 1997, 1998, 1999), Vry & Cartwright (1994), Collins & Williams (1995), Collins & Shaw (1995), and Cartwright *et al.* (1999), although interpretations have changed markedly as the database of radiometric ages has increased. In this study the geochronological framework of Vry *et al.* (1996), Williams *et al.* (1996), Vry & Cartwright (1998) and Buick *et al.* (1998, 1999) is used (Table 1). SHRIMP ages from detrital zircons from the Lander Rock Beds indicate that it was deposited after 1840 Ma (Vry *et al.* 1996). The early granites were emplaced at 1.82–1.80 Ga (Collins & Williams 1995; Vry *et al.* 1996). Inclusion trails of quartz and muscovite in contact metamorphic andalusite porphyroblasts around these early granites indicate some deformation and low-grade metamorphism of the Lander Rock beds occurred prior to 1.82–1.80 Ga (Vry & Cartwright 1998). This may have been contemporaneous with high-grade metamorphism at Mount Stafford, some

20 km N of the northwestern Reynolds Range. Deposition of the Reynolds Range Group sediments followed this contact metamorphism. Contact metamorphism of both the Lander Rock Beds and Reynolds Range Group rocks around the younger granites occurred at 1.78 Ga. The main regional tectonometamorphic event to affect the Reynolds Range caused the formation of tight to isoclinal northwest-trending upright folds and a penetrative steep northwest-trending fabric. This event occurred at 400–500 MPa, and grades vary from greenschist (*c.* 400°C) in the northwest to granulite (750–800°C) in the southeast at approximately the same structural level (Dirks *et al.* 1991; Vry & Cartwright 1994; Buick *et al.* 1998). SHRIMP U–Pb zircon ages suggest that the peak of regional metamorphism was at *c.* 1.6 Ga (Vry *et al.* 1996; Williams *et al.* 1996), contemporaneous with the Chewings Orogeny that affected rocks within the southern Arunta Inlier (Collins & Shaw 1995). The Reynolds Range is cut by a system of major northeast-dipping thrusts and associated smaller

Table 1. *Summary of fluid flow episodes in the Reynolds Range*

Age	XCO_2	Fluid sources	Conditions	Mineralogical changes	Stable isotope resetting	Channelling
Early contact metamorphism						
1.8–1.82 Ga (SHRIMP U–Pb zircon & monazite ages)	Low ?	Metamorphic, magmatic, locally surface (combined internal/external)	<500°C, 250 MPa	• Relict contact metamorphic assemblages (esp andalusite) • Mg-rich pods (now kornerupine + sapphirine) in SE of terrain	Low $\delta^{18}O$ values near granites and in Mg-rich pods	Not obvious, except for pods
Later contact metamorphism						
1.78 Ga (SHRIMP U–Pb zircon & monazite ages)	0.1–0.3	Metamorphic, magmatic, locally surface (combined internal/external)	<500°C, 250 MPa	• Relict contact metamorphic assemblages (esp andalusite) • Grandite-bearing layers in Lower Calcsilicate rocks • Scapolite in Upper Calcsilicate marbles	Low $\delta^{18}O$ & $\delta^{13}C$ values at granite-metasediment contacts. Resetting of $\delta^{18}O$ values along whole range	Not obvious
Prograde regional metamorphism						
1.6 Ga (SHRIMP U–Pb zircon & monazite ages)	Increase with grade	No widespread fluid flow	400°C (NW), 740±40°C (SE), 400–500 MPa	• Formation of highest-grade mineral assemblages • Partial melting in SE	None recorded	
Early stages of cooling from regional metamorphism						
1.59–1.55 Ga (?1.47 Ga) (SHRIMP U–Pb zircon ages, Pb–Pb step leach garnet & epidote ages)	≤0.2–0.3	Segregated partial melts of metapelites crystallizing during cooling (mainly internal)	575–650°C, 300–400 MPa	• Clinohumite & forsterite in dolomite-marbles • Wollastonite in calcite-marbles • Clinozoisite in marls • Anthophyllite and • Cummingtonite in metapelites • Epidotisation of calcsilicates • Garnet-epidote veins	Resetting of $\delta^{18}O$ & $\delta^{13}C$ values around pegmatites and veins	Distinct channelling by reaction-enhanced permeability and fractures (?)
Alice Springs orogeny						
c. 335 Ma (Ar–Ar & Rb–Sr ages of sheared rocks)	Low ?	Mainly surface, possibly with sources within the terrain (mainly external)	up to 600–650°C, 500–600 MPa	• Growth of staurolite & kyanite in sheared pelites • Muscovite-quartz-chlorite assemblages in highly-sheared granites	Very low $\delta^{18}O$ values in some highly-sheared granites	Strong channelling in shear zones

shear zones that form part of an extensive net-work of structures within the Arunta Inlier. Most major structures in the Reynolds Range have a reverse sense of movement with south-west transport directions (Collins & Teyssier 1989), and several of the kilometre-scale struc-tures (e.g. the Aileron and Mount Freeling Shear Zones: Fig. 1) extend into the deep crust (Lambeck et al. 1988). The metamorphic grade of the Reynolds Range shear zones is locally as high as amphibolite facies, with kyanite, staur-olite, and sillimanite present in sheared pelites in the southeast of the terrain (Dirks & Wilson 1990; Dirks et al. 1991). Rb–Sr and ^{40}Ar–^{39}Ar ages of amphibolite- and greenschist-facies shear zones in the southeastern Reynolds Range are c. 334 Ma (Cartwright et al. 1999), implying that they are Alice Springs age (cf. Collins & Shaw 1995).

Petrology

This section briefly describes the petrology of the Reynolds Range rocks as it is interpreted to have been at the peak of 1.6 Ga regional meta-morphism. Subsequent mineralogical changes associated with fluid flow are discussed later.

Lander Rock Beds

In the northwest of the terrain, Lander Rock Bed psammites comprise mainly quartz with beards of white mica and locally-abundant detrital zir-con and tourmaline (Vry & Cartwright 1998). Pelitic rocks are fine-grained slates with less quartz, and biotite that may be partially detri-tal. Biotite and andalusite porphyroblasts occur within contact aureoles of granites. Southeast-wards, biotite, cordierite and sillimanite are pro-gressively developed in the pelitic rocks. The highest-grade Lander Rock Bed pelites contain quartz + biotite + cordierite ± K-feldspar ± silli-manite ± garnet and have locally undergone mig-matization. Rocks from within 2–3 km of the granites commonly contain andalusite and cor-dierite porphyroblasts, or their pseudomorphs. These porphyroblasts were probably formed dur-ing 1.78 Ga contact metamorphism.

Reynolds Range Group: Lower Calcsilicate Unit

The Lower Calcsilicate Unit, which is up to 200 m thick, spans the amphibolite- to granulite-facies transition in the Reynolds Range and is everywhere directly underlain by the Napperby Gneiss, one of the 1.78 Ga granites (Fig. 1). The Lower Calcsilicate Unit comprises calcite-poor calcsilicate rock with minor intercalations of marble and impure calcareous quartzite. The calcsilicate rock comprises a finely-banded clin-opyroxene + feldspar-rich rock interlayered on a centi- to decimetre scale with massive grandite-rich layers (Buick et al. 1994; Cartwright et al. 1996). The clinopyroxene + feldspar-rich rocks contain diopside + anorthitic plagioclase + alkali feldspar + quartz ± grandite garnet ± biotite ± hornblende ± epidote ± scapolite. Milli- to centi-metre-thick layers of clinopyroxene and plagio-clase define the regional metamorphic fabric. There is little change in mineral assemblage or proportions with increasing regional metamor-phic grade, except that hornblende is most com-mon at amphibolite-facies. The grandite-rich layers typically comprise 10 to 20% of the rock and their outcrop volume and mineralogy is also independent of regional metamorphic grade. They are laterally continuous on scales from 1 to 100 cm, vary in thickness between c. 0.5 and 15 cm, and are concordant with layering. They are characterized by the assemblage 70–80% grandite garnet, with minor diopside + anorthite ± calcite ± quartz ± alkali feldspar ± late epi-dote. Rare 2 to 5 cm-thick marble layers that locally occur at granulite-facies contain calcite + anorthite + diopside + alkali feldspar ± scapo-lite ± wollastonite ± quartz ± grandite garnet ± epidote. At their margins, centimetre-thick gran-dite-rich layers with similar assemblages to those that occur elsewhere in this unit are common.

Reynolds Range Group: Pelite Unit

Pelitic rocks in the northeast of the terrain are slates that contain: quartz + muscovite + chlorite + rutile + haematite (Dirks et al. 1991). Southeast of the biotite-in isograd, the pelites contain: quartz + muscovite + biotite + chlorite ± magnetite ± andalusite. Upper amphibolite-facies pelitic gneisses are best preserved at Anna Reservoir (Fig. 1) and contain quartz + biotite + sillimanite + cordierite + alkali feldspar ± pla-gioclase ± magnetite ± rutile ± ilmenite ± gar-net. At granulite-facies grades, the pelites are cordierite-rich migmatites that locally contain garnet, or rarely orthopyroxene, which probably formed via incongruent dehydration melting reactions that consumed biotite and sillimanite (Dirks et al. 1991). Pelites from within a kilome-tre or so of the 1.78 Ga granites (notably the Conniston and Warrimbi Schists and the Nap-perby Gneiss; Fig. 1) commonly contain relict porphyroblasts of andalusite and/or cordierite that are wrapped by the regional metamorphic fabric and which probably formed during contact metamorphism (Dirks *et al.* 1991; Buick & Cartwright 1996).

Reynolds Range Group: Upper Calcsilicate Unit

The Upper Calcsilicate Unit generally comprises fine-grained, interlayered calcite- and dolomite-rich marbles that are locally interlayered with milli- to centimetre-thick, fine-grained, calcite-poor metamorphosed marls, and metre- to tens of metre-thick layers of siliceous psammites and rare pelites. At greenschist facies, both dolomite- and calcite-rich marbles contain calcite + dolomite + quartz + phlogopite ± muscovite ± chlorite ± rutile (Buick & Cartwright 1994). As regional metamorphic grade increases, anorthite, K-feldspar, and titanite occur. Relict millimetre-sized scapolite porphyroblasts (now composed of the same minerals as in the marble matrix) are interpreted to have formed by contact metamorphism at higher temperatures (540–575°C) than the regional metamorphism in that area (400–450°C; Buick & Cartwright 1994). The granulite-facies calcite-rich marbles comprise 60–90% calcite, diopside, phlogopite, alkali feldspar, anorthite, quartz, titanite and, locally, dolomite (Cartwright & Buick 1995; Buick & Cartwright 1996). The dolomitic marbles at that grade comprise dolomite, calcite, forsterite, spinel and phlogopite, and may additionally contain anorthite, magnetite, alkali feldspar, and diopside. The metamorphosed marls have similar assemblages to the calcite-rich marbles, but much lower total carbonate contents (typically 10–15%; Cartwright & Buick 1994, 1995; Buick & Cartwright 1996). Granulite facies Upper Calcsilicate unit psammites are cordierite-rich with quartz, biotite, plagioclase, cordierite, and ilmenite. Temperatures at 1.6 Ga do not appear to have been high enough for these rocks to melt. Minor pelitic layers contain similar assemblages to pelites in the Pelite Unit, and may have partially melted at granulite grades (Dirks *et al.* 1991).

Granites

Both younger (1.78 Ga) and older (1.8–1.82 Ga) granites were originally megacrystic S-type plutons. Many of the bodies preserve relict igneous minerals (quartz, K-Feldspar, plagioclase, biotite garnet) and, while the regional metamorphic fabric is locally developed, it is much weaker than

in the surrounding rocks. The 1.78 Ga Conniston and Warrimbi Schists (Fig. 1), however, commonly comprise quartz augen that are wrapped by a strong regional metamorphic fabric defined by muscovite and biotite. Within these two bodies zones of relict granite (up to several metres in diameter) are locally preserved. These granites have been altered, most probably by fluid flow during contact metamorphism at 1.78 Ga (Buick & Cartwright 1996).

The tectonometamorphic history described above provides a framework in which to discuss fluid flow in the Reynolds Range. The episodes of fluid flow may be divided both spatially and temporally.

Fluid flow during 1.82 Ga contact metamorphism

Quartzites, psammites, and pelites of the Lander Rock Beds underwent low-grade contact metamorphism at 250 MPa and <500°C around the older granites at 1.8–1.82 Ga (Vry & Cartwright 1998). Oxygen isotope ratios of the Lander Rock Beds range from 13.4 ± 0.8‰ to as low as 6.7‰ near the contacts of the larger plutons, and to 10.3 ± 1.1‰ around the smaller plutons (Fig. 2). The higher $\delta^{18}O$ values are close to those expected for pelitic or psammitic rocks (e.g. Hoefs 1997), and these are interpreted as being little

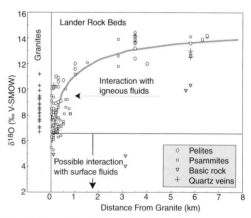

Fig. 2. Plot of Lander Rock Beds $\delta^{18}O$ values v. distance from contacts of 1.8–1.82 Ga granite plutons (data from Vry & Cartwright 1998). The higher $\delta^{18}O$ values are close to those expected for metamorphosed pelitic or psammitic rocks, while the occurrence of rocks with low $\delta^{18}O$ values adjacent to granites suggests that they were affected by fluid flow during contact metamorphism.

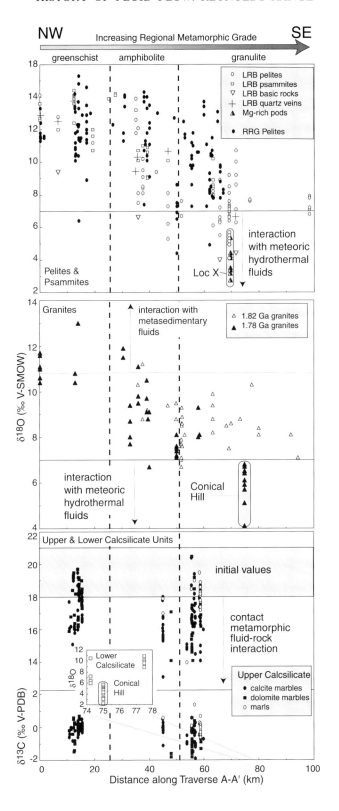

altered. The occurrence of rocks with low $\delta^{18}O$
values adjacent to granites suggests that they were
affected by contact metamorphic fluid flow.
Biotites in all the major rock types in the aureoles
have δ^2H values between -52 and $-69‰$, probably reflecting interaction with a cooling igneous and metamorphic fluid near the plutons.
Sapphirine-bearing and other Mg- and Al-rich
rock types that occur as rare layers and pods
within the Lander Rock Beds at locality X in
Figure 1 (Vry & Cartwright 1994) have low $\delta^{18}O$
values ($4.0 \pm 0.7‰$: Fig. 3). The precursors to
these rocks were probably low-temperature dia-
genetic-hydrothermal deposits of Mg-rich chlorite. The fluids associated with this early contact
metamorphism were probably mainly metamorphic and igneous; however, formation of the
Mg-and Al-rich rock types with very low $\delta^{18}O$
values, and the local occurrence of low $\delta^{18}O$
Lander Rock Bed pelites throughout the terrain
(Fig. 3), indicates at least local involvement of
surface water. The incursion of surface water
may have occurred relatively late in the contact
metamorphism as the hydrothermal system expanded and cooled (Vry & Cartwright 1998).

Fluid flow during 1.78 Ga contact metamorphism

Infiltration of igneous and surface-derived fluids into the Lower Calcsilicate Unit

The grandite-rich layers in the Lower Calcsilicate Unit described above are deformed by, and
thus predate, folds formed during regional metamorphism at 1.6 Ga. These layers were most
probably formed by fluid flow at 1.78 Ga associated with the emplacement of the Napperby
Gneiss and were closed chemical systems during
the subsequent regional metamorphism (Buick
et al. 1994). The mineral assemblages in the
grandite-rich layers imply that they were formed
by the infiltration of oxidized (log $fO_2 > -16$
to -14), water-rich ($XCO_2 < 0.1$ to 0.3; Fig. 4)
fluids. Calcite from the grandite-rich layers has
$\delta^{13}C$ and $\delta^{18}O$ values of -4.2 to $-0.8‰$ and 10.5

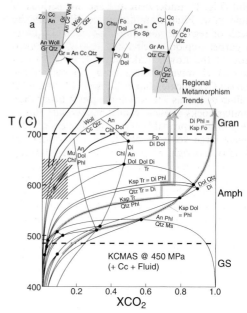

Fig. 4. Summary T-XCO$_2$ diagram in the K$_2$O–CaO–
MgO–Al$_2$O$_3$–SiO$_2$–H$_2$O–CO$_2$ system (KCMAS) at
450 MPa (after Buick & Cartwright 1996, figs 2–4) for
calcite-bearing rocks. Contact metamorphic fluid flow
in the Lower Calcsilicate Unit (**a**) and retrogression of
the dolomite marbles (**b**) and calcite marbles (**c**) of the
Upper Calcsilicate Unit occurred at low XCO$_2$
conditions. By contrast, XCO$_2$ values in
unretrogressed Upper Calcsilicate Unit marbles
increase with increasing regional metamorphic grade
from greenschist (GS) through amphibolite (Amph) to
granulite (Gran) facies and a there is predominance of
internally-buffered assemblages. For clarity not all
reactions are labelled. Abbreviations: An, anorthite;
Chl, chlorite; Chu, clinohumite; Di, diopside;
Dol, dolomite; Fo, forsterite; Ksp, K-feldspar;
Ms, muscovite; Phl, phlogopite; Qtz, quartz; Sp, spinel;
Tr, tremolite; Woll, wollastonite; Zo, zoisite.

to $14.0‰$, respectively and the $\delta^{18}O$ of the silicate fraction is 6.1 to $10.8‰$ (Fig. 3). These data
suggest that the infiltrating fluid was from the
underlying Napperby Gneiss.

The Lower Calcsilicate Unit metasediments
and the underlying Napperby Gneiss at Conical

Fig. 3. (*previous page*) $\delta^{18}O$ values of major Reynolds Range rock types that preserve peak metamorphic
mineralogies projected onto the section A–A' (Fig. 1). Most metasedimentary units show little variation with
metamorphic grade but do show distinct lowering of $\delta^{18}O$ values adjacent to granites that is due to contact
metamorphic fluid flow with igneous and locally low-$\delta^{18}O$ surface-derived fluids. Some granites have high $\delta^{18}O$
values that probably reflect interaction with fluids derived from adjacent metasediments during contact
metamorphism. The lack of systematic change of $\delta^{18}O$ values with regional metamorphic grade suggests that
these rocks were infiltrated not by significant volumes of fluid at that time. Metamorphic grade divisions are
approximate as the amphibolite–granulite transition is not exactly orthogonal to A–A'; in particular, Conical Hill
is probably uppermost amphibolite facies. Data from Buick & Cartwright (1996), Buick *et al.* (1994, 1997, 1998),
Cartwright & Buick (1995, 1999, 2000), Cartwright *et al.* (1996), Vry & Cartwright (1994, 1998).

Hill (Fig. 1), underwent at least two periods of fluid flow. Fluid flow during cooling from the peak of regional metamorphism formed quartz + garnet + epidote veins and caused epidotization of the Lower Calcsilicate Unit (discussed below). However, the Napperby Gneiss and the Lower Calcsilicate Unit at Conical Hill have $\delta^{18}O$ values as low as 2‰ which are much lower than similar rocks elsewhere in the Reynolds Range (Fig. 3) that probably reflect the ingress of relatively large volumes of heated surface-derived fluid (Cartwright *et al.* 1996). For the following reasons it is unlikely that the later epidotization event produced the widespread low $\delta^{18}O$ values at Conical Hill: (1) the local source of fluid for this event was probably the Napperby Gneiss that itself has lower than expected $\delta^{18}O$ values; (2) minerals in the Lower Calcsilicate Unit that predate epidotization (plagioclase and clinopyroxene) also have low $\delta^{18}O$ values; and (3) there is no correlation between the degree of epidotization and $\delta^{18}O$ values (Cartwright *et al.* 1996). Thus the fluid–rock interaction that produced the low $\delta^{18}O$ values occurred prior to epidotization, probably during the 1.78 Ga contact metamorphism.

Contact metamorphic fluid flow in the Pelite Unit

The $\delta^{18}O$ values of the pelites from the Pelite Unit vary between 4.4 and 15.3‰, and there is little variation in the average or range of $\delta^{18}O$ values with metamorphic grade (Fig. 3). The higher $\delta^{18}O$ values are likely to be only slightly different from their sedimentary precursors, while the lower $\delta^{18}O$ values are commonly, but not always, from rocks that are adjacent to 1.78 Ga granites and which contain relict contact metamorphic porphyroblasts. Many greenschist-facies pelites close to the Connsiton Schist have $\delta^{18}O$ values of 10–12‰, which are similar to those of that granite (10.4–13.0‰). However, greenschist-facies pelites with $\delta^{18}O$ values as low as 6.4‰ occur well away from exposures of the Conniston Schist. Mid amphibolite-facies pelites from close to the Warrimbi Schist have $\delta^{18}O$ values of 7.4–14.0‰, which are similar to the $\delta^{18}O$ values of that granite (6.7–13.1‰). Upper amphibolite-facies pelites within 200 m of the Napperby Gneiss have $\delta^{18}O$ values of 4.4 to 8.6‰ that are similar to those of the outer margin of the Napperby Gneiss (5.1 to 8.1‰). The pelites with lower $\delta^{18}O$ values (especially those adjacent to the granites) are interpreted as interacting with fluids during contact metamorphism at 1.78 Ga. The lowest $\delta^{18}O$ values are lower than those

typical of the granites in this terrain, implying that, as in the Lower Calcsilicate Unit, fluid flow locally involved surface-derived fluids. Fluid flow at this stage probably also altered the Conniston and Warrimbi Schists and may account for the wide range of $\delta^{18}O$ values developed in those rock types.

Fluid flow during regional metamorphism

There is little evidence of significant pervasive fluid flow during the 1.6 Ga regional metamorphism. The ranges of stable isotope ratios in all rock units are largely independent of metamorphic grade (Fig. 3). Stable isotope ratios that had been altered by fluid flow during contact metamorphism (discussed above) are preserved, precluding significant later fluid flow at those localities. The observed trend of increasing XCO_2 values in the Upper Calcsilicate Unit marbles from <0.05 at greenschist facies to 0.7–1.0 at granulite facies (Fig. 4) is also most consistent with internal buffering of fluids during metamorphism (Tracy & Frost 1991). Further, centimetre- to decimetre-size granitic segregations that also contain cordierite, garnet, or spinel within granulite-facies pelites and psammites of the Lander Rock Beds and Reynolds Range Group probably formed via fluid-absent melting reactions (Dirks *et al.* 1991; Buick *et al.* 1998; Cartwright & Buick 1998). Even at the amphibolite to granulite-facies transition, there are no signs of a 'minimum melting' zone where significant fluid-present anatexis may have occurred. Clearly, many of the rocks would have undergone devolatilization during regional metamorphism (especially those that escaped significant contact metamorphism); however, those fluids do not appear to have interacted with the rocks as a whole. Additionally, there are no indications of the infiltration of external fluids at this stage.

Fluid flow during early stages of cooling from 1.6 Ga regional metamorphism

The southeastern Reynolds Range contains a series of retrogressed rocks that are interpreted as being infiltrated by fluids during the early stages of cooling from the peak of 1.6 Ga regional metamorphism with fluids exsolved from crystallizing partial melts. Constraints on the relative timing of fluid flow, include: (1) the macroscopic fluid flow zones and the pegmatites and veins that represent the fluid sources and pathways cut, and hence postdate, the regional metamorphic foliation that is defined by peak metamorphic minerals; (2) the minerals formed during fluid flow are generally not aligned in the regional metamorphic fabric; and (3) the mineral assemblages associated with fluid flow formed at lower temperatures than those of the peak of regional metamorphism, commonly at 600–650°C, which is close to the temperatures at which fluids are exsolved from crystallizing initially water-undersaturated granitic melts (Cartwright 1988; Cartwright & Buick 1995; Buick *et al.* 1997). Radiometric dating (Williams *et al.* 1996; Buick *et al.* 1999) confirm the field and petrographic interpretations.

Fluid flow in the Upper Calcsilicate Unit

Retrogressed equivalents of the granulite-facies calcite and dolomitic marbles occur in the southeastern Reynolds Range. At the Woodford River locality (Fig. 1), retrogressed dolomitic rocks are forsterite or clinohumite marbles, and a variety of high-variance calcsilicate rocks. The clinohumite marbles are less common and locally occur close to pegmatites. The forsterite marbles contain: calcite $\pm$ dolomite $+$ forsterite $+$ spinel $\pm$ phlogopite $\pm$ tremolite $\pm$ chlorite $\pm$ ilmenite $\pm$ titanite. The most iron-rich assemblages are spinel rich and carbonate poor, otherwise the minerals are Mg-rich. The XCO_2 of the retrogressed mineral assemblages are <0.2 (Buick *et al.* 1997), which contrasts with the very high (0.9–1.0) XCO_2 values in unretrogressed granulite-facies dolomitic marbles. Compared with their unretrogressed equivalents, the retrogressed dolomitic marbles are also much coarser grained and have lower total carbonate contents (*c.* 50 wt%). They also contain no anorthite or diopside. The forsterite marbles have carbonate (calcite $\pm$ dolomite) $\delta^{18}O$ and $\delta^{13}C$ values of 10.5 to 23.2‰ and −3.9 to 2.5‰, respectively while clinohumite marbles have a similar range of carbonate $\delta^{18}O$ and $\delta^{13}C$ values (9.8 to 17.5‰ and −4.0 to −0.6‰, respectively; Buick *et al.* 1997). Average $\delta^{18}O$ and $\delta^{13}C$ values of the retrogressed marbles are *c.* 3 and *c.* 2‰, respectively, lower than those of unretrogressed equivalents (Fig. 5). The retrogressed marbles also have lower carbonate contents than their unretrogressed equivalents and $\delta^{13}C$ values broadly decrease with decreasing carbonate contents (Fig. 5).

The forsterite and, less commonly, clinohumite marbles contain layers of coarse-grained (several centimetres diameter) diopside, pargasite, or tremolite that are up to several metres across and tens of metres long. The clinopyroxene layers are locally discordant to compositional layering in the host rocks and are associated with diopside vein networks, suggesting that they formed from fracture-controlled fluid flow. These layers also contain relic forsterite or spinel and minor calcite, suggesting that they developed at the expense of the forsterite marble. The formation of these silica-rich high-variance layers from silica-poor dolomitic marbles (typically <10 wt% SiO_2) probably involved the introduction of aqueous silica by the infiltrating fluid. These metasomatic clinopyroxene layers have a relatively small range of calcite $\delta^{18}O$ and $\delta^{13}C$ values (12.0 to 13.2‰ and −3.1 to −1.9‰, respectively) that are similar to the most reset stable isotope compositions in the forsterite and clinohumite marbles. The oxygen isotope ratios of clinopyroxene from the metasomatic layers vary between 7.4 and 12.5‰, with most values below 11‰.

The altered equivalents of the calcite-rich marbles are wollastonite-bearing rocks with up to 70% wollastonite and variable quantities of quartz (in rocks with lower wollastonite contents) calcite, grossular garnet, anorthite, alkali feldspar, diopside, and titanite (Cartwright & Buick 1994, 1995, 2000). Wollastonite appears to postdate the regional metamorphic fabric. By contrast with the unretrogressed Upper Calcsilicate calcite-rich marbles that define high XCO_2 values, the mineralogy of these rocks suggest that they were in equilibrium with water-rich fluids (XCO_2 *c.* 0.1) at *c.* 650°C (Fig. 4). These rocks have calcite $\delta^{18}O$ and $\delta^{13}C$ values that are generally lower than those of unretrogressed equivalents (Fig. 5). Metamorphosed marls within the wollastonite-bearing rocks also have lower calcite $\delta^{18}O$ and $\delta^{13}C$ values and lower carbonate contents than unretrogressed equivalents (Fig. 5). The retrogressed marls with the lowest $\delta^{18}O$ values are invariably clinozoisite-bearing. The wollastonite-rich rocks contain far more wollastonite than can be accounted for by reaction of the small volume (*c.* 10%) of quartz that occurs in their unretrogressed counterparts (Cartwright & Buick 1994, 1995). The additional wollastonite probably formed through the introduction of aqueous silica during fluid infiltration (Cartwright & Buick 1994, 1995). Millimetre- to centimetre- wide grandite-dominated zones commonly occur within the wollastonite marbles (especially at the contacts with the marl layers). The grandite garnet overgrows the wollastonite, grossular garnet and diopside in the marbles, and is interpreted as small skarn zones where addition of Fe occurred relatively late in the fluid flow episode. Small (<0.1 mm wide) late veins of calcite, grandite garnet, and locally epidote locally cut the wollastonite marbles. The marl layers within the

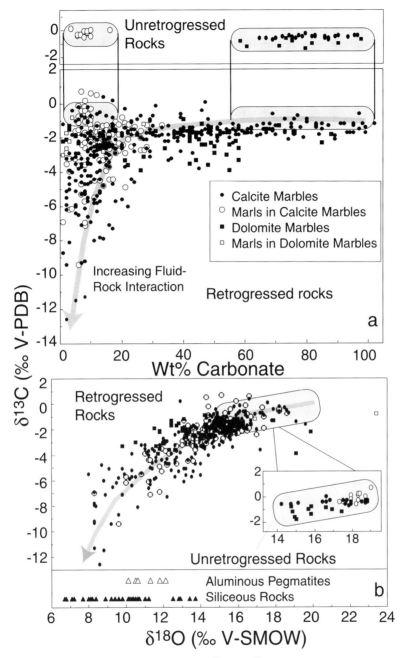

Fig. 5. Summary of $\delta^{13}C$ v. wt% carbonate (**a**) and the $\delta^{13}C$ and $\delta^{18}O$ values (**b**) of calcite and dolomite marbles and associated aluminous pegmatites and siliceous rocks from the Upper Calcsilicate Unit at Woodford River (Fig. 1). The main figures show data from the retrogressed rocks only. The insets show the range of values in unretrogressed rocks; these are shown as shaded boxes on the main figures. The reduction in carbonate content and the alteration of stable isotope values in the retrogressed rocks is due to fluids emanating from the pegmatites. Data from Cartwright & Buick (1995, 2000) and Buick *et al.* (1997).

wollastonite marbles comprise quartz, calcite, diopside, anorthite, and K-feldspar, with minor titanite. There is little difference in mineral assemblage between the marls in the calcite marbles and those in the wollastonite marbles, and both contain calcite and quartz with no wollastonite. However, the marls in the wollastonite marbles are considerably coarser grained (up to 0.5 mm) than those in the calcite marbles (<0.1 mm) and late epidote or clinozoisite is locally present.

Psammites in the Upper Calcsilicate Unit also show evidence of undergoing fluid–rock interaction. These rocks are interlayered with, and pass along strike into, quartz-poor anthophyllite or cummingtonite-bearing gneisses (Fig. 6). In these gneisses, the amphiboles are typically randomly oriented and grow in rosettes; however, a relic regional metamorphic foliation is locally preserved as inclusion trails of quartz and ilmenite in cordierite (Buick *et al.* 1998). Discontinuous biotite quartz-bearing lenses are completely surrounded by quartz-poor, ortho-amphibole- and cordierite-rich domains, suggesting that the orthoamphibole has grown at the expense of the biotite and quartz, as has been invoked in other terrains (Arnold & Sandiford 1990). The psammitic rocks have $\delta^{18}O$ values between 6.7 and 12.8‰ (Fig. 5), with little difference on an outcrop scale between unretrogressed

psammites and the orthoamphibole- or cummingtonite-bearing gneisses. The lower $\delta^{18}O$ values generally come from samples with extensive, late orthoamphibole or cummingtonite and that lack matrix quartz, and hence may reflect bulk compositional differences.

The zones of retrogression contain decimetre- to several metre-thick, coarse-grained pegmatitic rocks (Fig. 6) of two main varieties: (1) feldspar-poor, quartz-rich (60–80%) veins that locally contain cordierite, biotite, and sillimanite; and (2) two feldspar pegmatites that may also contain tourmaline, biotite, and sillimanite. Locally, the quartz veins can be traced into pegmatites over several metres. The pegmatites are commonly semi-concordant to gross lithological layering and the regional metamorphic fabric, whereas the quartz veins are typically discordant (Williams *et al.* 1996). Several types of skarns comprising unoriented, centimetre-sized crystals commonly separate the pegmatites and veins from the metasediments, including sillimanite, biotite and cordierite-rich layers at pegmatite–psammite contacts and clinopyroxene or amphibole-rich layers at pegmatite–marble contacts. Clinohumite in the dolomitic marbles is also more abundant within several metres of pegmatites. The pegmatites probably represent fluid sources while the veins represent fluid conduits within the retrograde zones. SHRIMP ages of zircon and

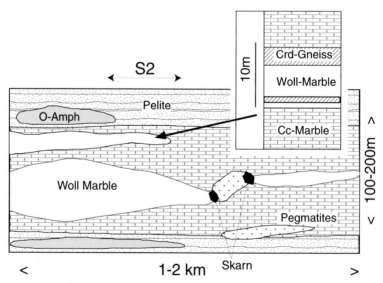

Fig. 6. Generalized map of part of the Woodford River locality summarizing distribution of zones of retrogression (wollastonite marble and orthoamphibole metapelites). Fluids derived from the late pegmatites were channelled along strike through both pelites and marbles producing elongate zones of retrogression that are parallel to large-scale lithological layering. Modified from Cartwright & Buick (1994, fig. 1). Cc, calcite; Crd, cordierite; O-amph, orthoamphibole; Woll, wollastonite.

monazite from pegmatites and veins in the south-east Reynolds Range are 1.58–1.57 Ga (Williams *et al.* 1996) confirming the age of fluid flow as being soon after the peak of regional metamorphism. The aluminous quartz veins and pegmatites have $\delta^{18}O$ values of 10.1–12.1‰ (Fig. 5), and their constituent minerals have concordant, high-temperature, oxygen isotope fractionations (Cartwright & Buick 1995). These $\delta^{18}O$ values are within the range recorded from granulite-facies Reynolds Range Group pelites (10.0 ± 1.8‰; Buick & Cartwright 1996), but are generally higher than those of adjacent 1.78 Ga granites (6.1 ± 2.5‰; Buick & Cartwright 1996).

Fluid flow in the Lower Calcsilicate Unit

Post peak regional metamorphic fluid flow is recorded within the Lower Calcsilicate Unit near the contact with the Napperby Gneiss at Conical Hill (Cartwright *et al.* 1995; Fig. 1). Fluids exsolved from partial melts within the Napperby Gneiss (Hand & Dirks 1992) during cooling from the peak of regional metamorphism formed quartz + garnet + epidote veins within the Lower Calcsilicate Unit and also caused extensive epidotization that dies out over a few metres away from the contact with the Napperby Gneiss (Fig. 7). Pb–Pb step-leaching ages of 1566–1577 Ma from garnets in the veins (Buick *et al.* 1999) confirm the timing of fluid flow. Epidote from the veins yields Pb–Pb ages of 1469 ± 26 Ma, suggesting that vein formation may have

occurred in several phases (Buick *et al.* 1999). Fluid inclusions and the displaced position of mineral equilibria indicate that fluid flow occurred at 575–625°C and the fluids were water rich ($XCO_2 \leq 0.1$: Fig. 4) and saline (Cartwright *et al.* 1996).

At Conical Hill, the Napperby Gneiss has relatively low $\delta^{18}O$ values of 5.7–7.5‰ and pegmatite layers within the Napperby Gneiss have $\delta^{18}O$ values of 4.1–7.6‰ (Fig. 8). The Lower Calcsilicate rocks also have $\delta^{18}O$ values of 2.1–5.7‰ that are much lower than recorded elsewhere in this unit. There is no correlation of $\delta^{18}O$ values with the degree of epidotization and, as discussed earlier, the low $\delta^{18}O$ values probably reflect the affects of contact metamorphic fluid flow at 1.78 Ga. Minerals in the Napperby Gneiss and pegmatites have $\delta^{18}O$ values that are close to those expected at 500–600°C (Cartwright *et al.* 1996). Quartz from quartz + garnet + epidote veins that cut the Lower Calcsilicate has very similar $\delta^{18}O$ values (6.2–6.9‰) as quartz in the Lower Calcsilicate rocks (6.5–6.7‰). Likewise, the $\delta^{18}O$ values of epidote from the veins (3.1–3.7‰) and the Lower Calcsilicate rocks (3.3–3.6‰) are closely similar. Epidote and quartz in the veins and in the epidotized Lower Calcsilicate rocks have $\delta^{18}O$ values suggesting equilibration 560–625°C (Cartwright *et al.* 1996). These temperatures are close to

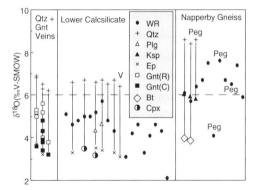

Fig. 8. Summary of $\delta^{18}O$ values of the Lower Calcsilicate Unit calcpelites and underlying Napperby Gneiss at Conical Hill (Fig. 1). The overall low $\delta^{18}O$ values suggest infiltration by heated surface-derived fluids, probably during contact metamorphism at 1.78 Ga. Epidotization during cooling from regional metamorphism has not destroyed the overall isotopic systematics at this locality due to the fluids being internally derived from pegmatite crystallization. Mineral $\delta^{18}O$ values are concordant, suggesting that late low-temperature fluid flow following epidotization has not affected these rocks. Data from Cartwright *et al.* (1996).

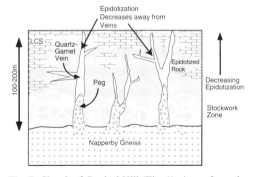

Fig. 7. Sketch of Conical Hill (Fig. 1). Away from the Napperby Gneiss–Lower Calcsilicate (LCS) contact, pegmatite veins grade into garnet + quartz veins that cut the regional metamorphic fabric. Epidotization also decreases away from the contact and occurs preferentially along some layers. The cross-cutting relationships suggest that epidotization occurred after the regional metamorphism and was caused by fluid emanating from crystallizing pegmatite segregations in the Napperby Gneiss. After Cartwright *et al.* (1996, fig. 2).

370   I. CARTWRIGHT *ET AL.*

those implied by the mineral assemblages in these rocks. By contrast, garnet and quartz in the veins yield anomalously-high isotopic temperatures (720–880°C: Cartwright *et al.* 1996) and the garnets are zoned in $\delta^{18}O$ values (Fig. 8). These data may reflect isotopic disequilibrium within the veins. If garnet grew early during vein formation in an environment where fluid compositions were changing, it may not have continued to exchange with the fluid. By contrast quartz and epidote that grew later, or which continued to recrystallize in the presence of the fluid, may have maintained isotopic equilibrium.

δ^2H values of epidote from the quartz + garnet veins within the Lower Calcsilicate and biotite from the Napperby Gneiss range from −145 to −130‰ (Cartwright *et al.* 1996). At the estimated temperatures of fluid flow (560–625°C) the fractionation of $^1H/^2H$ between biotite and epidote is <10‰ (Suzuoki & Epstein 1976); hence, the similarity in δ^2H values is consistent with the veins being derived from fluids produced from crystallizing melts within the Napperby Gneiss. However, as for the oxygen isotope ratios, the δ^2H values are much lower than is typical for normal igneous rocks (−55 to −85‰) and are in the range often recorded from igneous rocks that have interacted with surface-derived fluids (Taylor & Sheppard 1986). The low δ^2H values thus also probably reflect fluid flow during contact metamorphism at *c.* 1.78 Ga.

Fluid flow during the Alice Springs Orogeny

The Reynolds Range is cut by major NE-dipping thrusts and shear zones (Fig. 1) that Ar–Ar and Rb–Sr data suggest are Alice Springs age (Cartwright *et al.* 1999). The grade of the assemblages in the shear zones are locally as high as amphibolite facies, with kyanite and staurolite or garnet and sillimanite developed in sheared metapelites (Dirks *et al.* 1991).

Surface-derived fluid flow in granite-hosted shear zones. Shear zones from the Mount Airy Orthogneiss at Sandy Creek, the Yaningidjara Orthogneiss at Yaningidjara Hills, the Boothby Orthogneiss at Peaked Hill, and the Napperby Gneiss at Anna Reservoir (Fig. 1) cut the regional metamorphic fabric and contain a strong schistosity oriented *c.* 270–300°/70–90° NE and a down-dip lineation. The two Sandy Creek shear zones are 30–40 m wide and exhibit decimetre- to metre-scale alternation of more- and less-highly sheared zones. The other shear zones are <10 m wide and have fabrics of more uniform intensity. With increasing intensity of shear,

the unsheared granitic orthogneiss that contains 1–5 cm megacrysts of microcline in a coarse-grained (0.5–1 cm) plagioclase + quartz + biotite + microcline matrix is transformed into a micaceous quartz + muscovite ± biotite ± chlorite schist with very little feldspar. The mineralogical changes require a change in bulk rock composition, implying that the shear zones hosted fluid flow. Chemical changes that accompanied shearing were determined for the Sandy Creek and Peaked Hill shear zones from the relative concentrations of elements in the unaltered and metasomatized rocks assuming that Al, Ti, Zr, and Nb were immobile (Gresens 1967). Most sheared rocks gained SiO_2 (0.6–11.5 g/100 g) and K_2O (0.3–2.3 g/100 g) but lost Na_2O (0.2–3.8 g/100 g) and CaO (0.2–0.8 g/100 g) relative to their unsheared counterparts (Fig. 9). The two Sandy Creek shear zones showing larger average elemental changes than the Peaked Hill shear zone. Changes in MgO were variable, and changes in FeO were relatively minor.

From the mineral chemistry of the sheared rocks, oxygen isotope fractionations, and estimates of likely geothermal gradients, shearing occurred at 420–535°C and 400–650 MPa (Cartwright *et al.* 1999). The sheared granitic rocks in the Sandy Creek and Anna Reservoir shear zones have $\delta^{18}O$ values as low as 0 that are much lower than those of their unsheared counterparts (7.0–9.3‰: Figs 10 and 11). The calculated

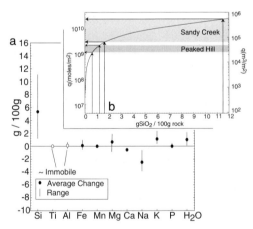

Fig. 9. (a) Summary of geochemical changes in Alice Springs shear zones cutting granites at Sandy Creek (Fig. 1). Data from Cartwright & Buick (1999). (b) Relationship between silica increases and time-integrated fluid flux, assuming that Si is deposited due to decreasing temperature and pressure along the flowpath. The recorded increase in silica within the Sandy Creek and Peaked Hill shear zones implies time-integrated fluid fluxes of 5×10^4–5×10^5 m³/m².

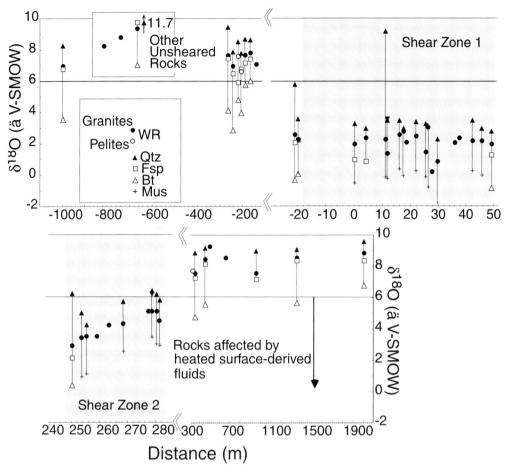

Fig. 10. $\delta^{18}O$ values of granites, metapelites, and quartz veins from the Sandy Creek shear zones (Fig. 1). Unsheared granites have $\delta^{18}O$ values that are similar to those typical of S-type granites from this region and $\delta^{18}O$ values of screens of metapelites away from the shear zones are similar to those of the granites, probably due to fluid–rock interaction soon after granite emplacement. Within, and immediately adjacent to the shear zones, the granites and metapelites have anomalously low $\delta^{18}O$ values due to interaction with heated surface-derived fluids. Data from Cartwright & Buick (1999).

fluid $\delta^{18}O$ values (−1 to −2‰) lie within the range of surface-derived fluids (either meteoric water or evolved basinal brines: e.g. Sheppard 1986). By contrast with the Anna Reservoir and Sandy Creek shear zones, both sheared and unsheared granitic rocks from Yaningidjara Hills and Peaked Hill have similar $\delta^{18}O$ values (6.2–8.9‰), which are within the range of typical granite $\delta^{18}O$ values in this area (Fig. 11).

The increase in Si and K coupled with the decrease in Na and Ca during metasomatism are trends predicted to occur during down-temperature fluid flow (Dipple & Ferry 1992a: Cartwright & Buick 1999). Thus, fluid flow in these shear zones was probably toward the surface driven by buoyancy. From the changes in silica con-

centration, and assuming a temperature gradient of 25°C/km (0.025°C/m), time-integrated fluid fluxes for the Peaked Hill and Sandy Creek shear zones of 5×10^4–5×10^5 m³/m² were estimated using the approach of Dipple & Ferry (1992a), as summarized in Figure 9. These are similar to time-integrated fluid fluxes estimated for shear zones in other terrains (e.g. Selverstone et al. 1991; Dipple & Ferry 1992a).

Some of the shear zones that cut the granites record, therefore, the influx of surface-derived water that paradoxically was flowing toward the surface. There are several tectonic models that can resolve this dilema, including: (1) thrusting of high-grade gneisses onto unmetamorphosed sediments promoting dewatering (e.g. Lobato et al.

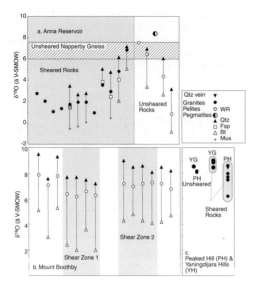

Fig. 11. $\delta^{18}O$ values of granites, Lander Rock Bed pelites, and quartz veins from the Anna Reservoir, Mount Boothby, Peaked Hill, and Yaningdijara Hills shear zones (Fig. 1). At Anna Reservoir, the sheared rocks have anomalously low $\delta^{18}O$ values, suggesting interaction with heated surface-derived fluids. However, at the other localities, there is little difference between the $\delta^{18}O$ values of the sheared and the unsheared rocks, possibly due to lower fluid volumes or fluids from different sources. Data from Cartwright & Buick (1999).

1983); (2) convection of surface-derived fluids around igneous bodies; and (3) topographically driven fluid flow similar to that documented in recent mountain belts (e.g. the Himalayas: Chamberlain *et al.* 1995). The first two models are unlikely given the geometry of the Alice Springs orogen and the lack of contemporaneous igneous activity (Collins & Shaw 1995), and topographically-driven fluid flow generally occurs above the brittle-ductile transition zone. Alternatively, rupturing on large low-angle faults or shear zones (e.g. the Aileron Shear Zone) may have led to a transient reduction in fluid pressures in the fault zone allowing fluid to be drawn downward into the crust. Following such seismic events, the fluid may be expelled upwards along the fault itself or along intersecting shear zones such as the Sandy Creek or Anna Reservoir shear zones. McCaig (1988) proposed a similar model to account for the transport of surface-derived fluid into the ductile crust in the Pyrenees and such a scenario would explain both the infiltration of surface-derived water and the local down-temperature direction of fluid flow in the shear zones.

Shear zones in pelites

Shear zones that cut pelites at Mount Boothby (Fig. 1) range from a few centimetres to a few metres wide and are NW-striking with steep, generally NE-dipping fabrics and down dip lineations. The sheared pelites are coarse-grained rocks with predominantly unaligned, centimetre-long kyanite and staurolite porphyroblasts in a biotite + quartz + muscovite matrix. By comparison with petrogenetic grids, Dirks *et al.* (1991) estimated that the shear zones in the Mount Boothby pelites formed at amphibolite facies conditions (600–650°C and 500–600 MPa). Most rocks in the shear zones are moderately sheared, although zones of more intense shearing exist. The presence of quartz segregations with up to decimeter-long kyanite crystals that contain abundant fluid inclusions attests to the presence of fluids during shearing. However, the sheared pelites do not appear to have undergone the intense metasomatism that was experienced by many of the greenschist facies sheared granites, implying that lower volumes of fluid flowed through these shear zones or that fluids were more highly channelled, possibly within fractures, and did not interact with the bulk of the rock.

Sheared and unsheared pelites at Mount Boothby show little variation in $\delta^{18}O$ values. The unsheared pelites have $\delta^{18}O$ values of 6.8–7.9‰ that are lower than expected for pelitic rocks, but similar to other high grade pelites in both the Reynolds Range Group and Lander Rock Beds that are interpreted as resulting from fluid flow during contact metamorphism (Vry & Cartwright 1994, 1998; Buick & Cartwright 1996). The sheared pelites have $\delta^{18}O$ values of 6.2–7.3‰ that are similar to their unsheared counterparts. Hence, as for the sheared granites at Yaningdijara Hills and Peaked Hill, shearing at Mount Boothby did not produce significant shifts in $\delta^{18}O$ values, probably again reflecting lower fluid volumes within these shear zones.

Discussion

Rocks in the Reynolds Range underwent fluid–rock interaction in a number of discrete events both prior to and after the 1.6 Ga regional metamorphism (Table 1, Fig. 12). The events that have been discerned are localized and at least three (contact metamorphic fluid flow at 1.82 and 1.78 Ga and Alice Springs age fluid flow in the shear zones) locally involved fluids from the surface.

Variation in time integrated fluid fluxes

One of the major quests of the last few years has been to calculate the volumes of fluid involved in metamorphic fluid flow systems. However, the advection–dispersion models that may theoretically be used to calculate time-integrated fluid fluxes from isotopic, mineralogical, or major element changes (e.g. Bickle & McKenzie 1987; Bickle & Baker 1990a, b; Dipple & Ferry 1992a, b; Bowman et $al.$ 1994) are often difficult to apply as they require the fluid flow system to be well defined. In addition, they make a number of assumptions (e.g. equilibrium being maintained, one-dimensional fluid flow, unchanging temperature gradients, presence of a fluid-filled porosity) that may not be the case in reality (Cartwright & Buick 1996; Cartwright 1997). Nevertheless, some order-of-magnitude estimates may be made. Using methods described by Dipple & Ferry (1992a), the silica addition within greenschist-facies shear zones yields time-integrated fluid fluxes of 5×10^4–5×10^5 m^3/m^2 (Cartwright & Buick 1999). High-temperature retrogression in the marbles that altered the mineralogy and the stable isotopes over 10–200 m required time-integrated fluid fluxes of 20–400 m^3/m^2 (Buick & Cartwright 1998). Silica metasomatism in these rocks, which occurs on smaller (1–20 m) lengthscales, required time-integrated fluid fluxes of c. 100–1000 m^3/m^2. These volumes of fluids are within the ranges that are typically recorded in metamorphic terrains (Thompson 1997; Yardley 1997).

Internal v. external fluid sources

Understanding whether fluids are generated from within, or external to, a terrain is important in order to document the overall fluid flow systems. In some cases (e.g. rocks with low δ^{18}O values), the oxygen isotopes clearly indicate an external source. However, in some cases oxygen isotope values do not unambiguously constrain fluid origins as they cannot distinguish between similar sources within or external to a terrain. Additionally, where only moderate fluid volumes are involved, stable isotopes may be altered by fluid–rock interaction. Additional constraints on fluid origins may be made by considering what internal fluid sources may have been present and the spatial distribution of fluid–rock interaction.

High-temperature retrogression of the Upper Calcsilicate Unit is interpreted to have been caused by fluids from crystallizing pegmatites that were derived from the underlying pelites during regional metamorphism (Cartwright & Buick 1994, 1995; Buick et $al.$ 1997). The outcrop-scale pattern of retrogression (Fig. 6), and the stable isotope data (Fig. 5) are consistent with this model. The temperatures of retrogression (650–700°C) also support this idea as granitic melts crystallize and yield water-rich fluids at those temperatures (Cartwright 1988). In the southeastern Reynolds Range, the total area of the Upper Calcsilicate Unit is c. 2×10^7 m^2 of which c. 10% (or 2×10^6 m^2) is retrogressed. The underlying Pelite Unit is c. 500 m thick and thus has a volume of c. 10^{10} m^3. The Pelite Unit rocks underwent 15–25% partial melting during granulite facies metamorphism to form melts with 3–5 wt% water (Buick et $al.$ 1997). These melts thus contained c. 1–4×10^8 m^3 water which, if it were channelled through 2×10^6 m^2 of rock would, produce fluid flow with an average time-integrated fluid flux of 50–200 m^3/m^2. While this calculation is relatively crude, it does illustrate that sufficient fluid could have been produced from internal sources to account for the volumes of fluids required to effect the alteration of mineralogy and stable isotopes. Further support for this model is that the Upper Calcsilicate Unit at low grades in the northwestern Reynolds Range is not underlain by migmatitic rocks and did not undergo retrogression during cooling from the peak of regional metamorphism. High-temperature retrogression in the Lower Calcsilicate Unit also shows a close spatial correlation with granitic veins in the underlying Napperby Gneiss, consistent with fluid produced internally by melt crystallization.

The fluids that were present during the two episodes of contact metamorphism were derived from several sources. Igneous fluids were present as were fluids produced by devolatilization of the country rocks, and there are indications of localized infiltration of surface-derived fluids. However, the igneous activity does not appear to have driven large-scale meteoric-hydrothermal fluid circulation cells, as are present around some high-level granitic plutons (e.g. Criss & Taylor 1986).

Fluid flow pathways

Fluid flow is rarely truly pervasive but rather occurs preferably along specific lithological layers (often due to reaction-enhanced permeability) or along structures such as faults or shear zones that have high permeabilities (e.g. Graham et $al.$ 1983; Rumble & Spear 1983; Oliver et $al.$ 1993; Cartwright 1994; Davis & Ferry 1994; Cartwright & Buick 1995; Skelton et $al.$ 1995;

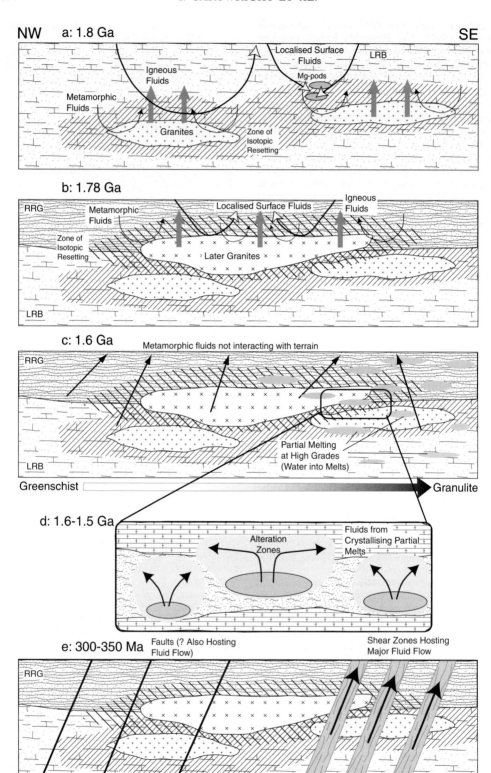

Cartwright et al. 1995, 1996, 1997; Cartwright & Weaver 1997). The Reynolds Range provides examples of fluid flow channelled both along structures and along lithological layering.

At the macroscopic scale, fluid flow during high-temperature retrogression after the peak of regional metamorphism was largely focussed along strike. The patterns of mineralogical and isotopic alteration in retrogressed Upper Calcsilicate Unit marbles and pelites from the southeast Reynolds Range show that fluid flow was dominantly layer parallel (Fig. 6), implying that time-integrated fluid fluxes, and hence intrinsic permeabilities, varied significantly on a centimetre to metre scale between different layers (Cartwright & Buick 1994, 1995, 2000; Buick et al. 1997). However, the microscopic patterns of fluid flow in the retrogressed calcite-rich marbles are more complicated. Cartwright & Buick (1994, 1995, 2000) showed that the distribution of wollastonite, small andraditic skarn zones, and the magnitude of stable isotope resetting was highly-variable on a centimetre scale (Fig. 13), implying that fluid–rock interaction was also highly variable at that scale. Thus, the pattern of fluid flow in these rocks probably reflect processes at two different scales. The macroscopic strike-parallel fluid flow may reflect large-scale differences in intrinsic permeability between layers or, as the fluids originated from pegmatites that only occur in a few places, the distribution of fluid sources. The small-scale channelling reflects feedback effects within the rocks. Infiltration of a water-rich fluid into marbles drives mineralogical reactions that have a negative ΔV. This results in fluid flow localizing due to the distribution of reactants and causes fluid flow channels to persist once initiated (Cartwright 1997). Metasomatic reactions (such as the ones that create 'excess' wollastonite in marbles) have a positive ΔV that can reduce the permeability and cause migration of the fluid flow zones.

Fluid flow adjacent to quartz veins in the Lower Calcsilicate Unit at Conical Hill also was channelled along specific layers. Boudins of early (1.78 Ga) grandite-bearing skarn largely escaped the effects of later fluid flow presumably due to their low permeability. However, over much of the outcrop, fluid flow in the dominant clinopyroxene + feldspar-bearing calcsilicates was nearly pervasive. This implies that

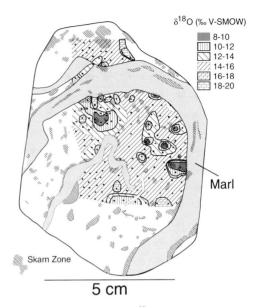

δ^{18}O (‰ V-SMOW)

▨	8-10
▥	10-12
▨	12-14
▨	14-16
▨	16-18
▤	18-20

Marl

Skarn Zone

5 cm

Fig. 13. Patterns of calcite δ^{18}O values in a single retrogressed calcite marble sample from Woodford river locality. (Fig. 1). The lowest δ^{18}O values are often centred on skarn zones that represent regions of Fe metasomatism. Aside from some concentration of skarn zones at the marble–marl contact that probably reflects fluid channelling along that boundary, neither the pattern of δ^{18}O variation nor the distribution of skarn zones appear to be closely related to the layering nor to the fold. Rather, the 'bulls eye' patterns of low δ^{18}O values indicates that fluid flow was one-dimensional at a high angle to the face of the sample. From Cartwright & Buick (2000, fig 3).

these rocks did not have appreciable differences in intrinsic permeability at the time of fluid infiltration. The epidote-forming reactions in these rocks are metasomatic, and few, if any, solely mineralogical reactions appear to have affected these rocks. Thus, reaction-enhanced permeabilities are unlikely to have developed. On a large-scale the intensity of alteration decreases away from the ultimate fluid source, the crystallizing pegmatites within the underlying Napperby Gneiss.

Fluid flow during the Alice Springs event was largely channelled through shear zones, with few effects observed away from those structures.

Fig. 12. Summary of the fluid flow history of the Reynolds Range showing location and sources of fluids.
(**a**) Contact metamorphic fluid flow in the Lander Rock Beds around the older granites at 1.8 Ga. (**b**) Contact metamorphic fluid flow around the younger granites at 1.78 Ga following the deposition of the Reynolds Range Group. (**c**) 1.6 Ga regional metamorphism with fluids generated by devolatilization escaping. Partial melts at high metamorphic grades act as stores for fluid. (**d**) Fluids derived from partial melts during cooling causing localized retrogression at 1.6–1.5 Ga. (**e**) Highly-channelled fluid flow during the Alice Springs Orogeny (300–350 Ma).

Shear zones represent sites of enhanced permeability due to a combination of active deformation and mineralogical reaction. Thus, it is not
surprising that fluid flow during this event was
very focussed. Highly-focussed fluid flow is also
evident in the quartz veins that are formed during
the early stages of cooling from the regional
metamorphic peak in the Upper and Lower
Calcsilicate Units.

The fluid flow events for which we have the
poorest spatial constraints are those during the
1.82 and 1.78 Ga contact metamorphic events.
This is not entirely due to the affects of later
metamorphism and deformation as the contact
aureoles are easily distinguished in the field and in
many cases have escaped significant later deformation, probably due to strain partitioning (Vry
& Cartwright 1998). Rather, aside from features
such as the Al-rich pods, fluid flow during this
event, which is marked by zones of oxygen isotope depletion, appears to have occurred over
large areas centred on the granites. It does not
appear to have been constrained by lithological
layering nor obvious macroscopic structures.
This pattern of fluid flow is typical of that documented around high-level granite plutons in
other terrains (e.g. Criss & Taylor 1986). The
temperatures of contact metamorphism were not
excessively high (andalusite is the characteristic
mineral), and the rocks were at shallow levels
and may have been brittle at that time. If fluid
flow through the country rocks was via micro-
or mesoscopic fractures that resulted from the
emplacement of the granites, widespread fluid–
rock interaction around the granite is expected.
Additionally, the pelites are relatively uniform
and there may not have been significant variation in reaction enhanced permeability between
different layers.

Overall, the spatial distribution of fluid–rock
interaction in the Reynolds Range reflects several
factors. The first-order control is the distribution
of the fluid sources (best illustrated by the
distribution of the zones of contact metamorphic
fluid flow and high-temperature retrogression).
Mesoscopic fluid channelling was often important, especially where fluid flow drove mineralogical reactions (e.g. the Upper Calcsilicate Unit
marbles) or was dominantly through discrete
structures (e.g. the Alice Springs shear zones).
Small-scale fluid channelling is evident where
heterogenous reactions occurred that created and
maintained zones of high-permeabilities (e.g. the
Upper Calcsilicate Unit marbles). Non-structu-
rally confined fluid flow was channelled where
mineralogical reactions were important, but more
pervasive where metasomatism occurred or where
fluids flowed through microfractures.

Contrasting fluid flow during regional and contact metamorphism

The Reynolds Range shows a distinct contrast in
the pattern of fluid–rock interaction during regional and contact metamorphism. As discussed
above, there was widespread fluid flow during
contact metamorphic fluid flow. By contrast,
there is little evidence of widespread fluid–rock
interaction during the prograde stages of regional
metamorphism even though, aside from some
rocks previously affected by contact metamorphism, devolatilization at that time would have
produced fluids. There is no evidence of either
across-strike fluid flow (e.g. between marbles
and surrounding pelites) nor large-scale strike-
parallel metamorphic fluid flow, as has been
documented in terrains such as the Mount Lofty
Ranges, Australia (Cartwright *et al.* 1995) or
Arcadia, USA (Stern *et al.* 1992). These observations are similar to ones made in other terrains
(e.g. the Adirondacks, USA: Valley *et al.* 1990;
Omeo Zone, Australia: Cartwright & Harper
1998), and imply a fundamental difference in the
fluid flow regimes operating during contact and
regional metamorphism. The lack of large-scale
strike-parallel fluid flow is probably the result
of several factors, including: (1) earlier contact
metamorphism dehydrating parts of the terrain,
which would inhibit development of large-scale
fluid flow pathways; (2) fluid generation may
not have been synchronous at different metamorphic grades; and (3) an along-strike fluid
pressure gradient may not have existed. The lack
of fluid–rock interaction between layers probably
reflects non-synchronous fluid production. Due
to differences in bulk composition, individual
units may generate fluid at different temperatures
(and, hence, different times) to their neighbours.
Thus, permeable layers may be confined within
less-permeable layers, inhibiting across-strike
fluid–rock interaction.

Overall, the Reynolds Range shows superposition of a number of fluid flow systems that
are tied into specific tectonometamorphic episodes within the terrain. The fluid flow history
documented here is probably typical of many
low-pressure regional metamorphic terrains, but
may have not been as obvious without the particular variation in regional metamorphic grade
that allows discrimination of earlier fluid flow
episodes at lower grades.

We would like to thank the many colleagues, referees,
and editors who gave us advice, showed us localities,
and commented on manuscripts over the several years
that the studies summarized here took to complete.
The comments of A. Matthews and C. Harris helped

improve this manuscript. In addition, we thank the many technical staff who helped in sample preparation and data generation. The support of grants and fellowships from the Australian Research Council and Monash and La Trobe Universities is also gratefully acknowledged.

References

ARNOLD, J. & SANDIFORD, M. 1990. Petrogenesis of cordierite-orthoamphibole assemblages from the Springton region, South Australia. *Contributions to Mineralogy and Petrology*, **105**, 100–109.

BAUMGARTNER, L. P. & FERRY, J. M. 1991. A model for coupled fluid-flow and mixed-volatile mineral reactions with applications to regional metamorphism. *Contributions to Mineralogy and Petrology*, **106**, 273–285.

BAUMGARTNER, L. P. & RUMBLE, D. III. 1988. Transport of stable isotopes: 1: development of a kinetic continuum theory for stable isotope transport. *Contributions to Mineralogy and Petrology*, **98**, 417–430.

BICKLE, M. J. & BAKER, J. 1990a. Advective-diffusive transport of isotopic fronts: an example from Naxos, Greece. *Earth and Planetary Science Letters*, **97**, 78–93.

BICKLE, M. J. & BAKER, J. 1990b. Migration of reaction and isotopic fronts in infiltration zones: assessments of fluid flux in metamorphic terrains. *Earth and Planetary Science Letters*, **97**, 1–13.

BICKLE, M. J. & MCKENZIE, D. M. 1987. The transport of heat and matter by fluids during metamorphism. *Contributions to Mineralogy and Petrology*, **95**, 384–392.

BLACK, L. P., SHAW, R. D. & STEWART, A. J. 1991. Rb–Sr geochronology of Proterozoic events in the Arunta Inlier, central Australia. *Bureau of Mineral Resources Journal of Australian Geology and Geophysics*, **8**, 129–138.

BOWMAN, J. R., WILLETT, S. D. & COOK, S. J. 1994. One-dimensional models of oxygen isotopic transport and exchange during fluid flow: Constraints from spatial patterns of oxygen isotopic compositions on fluid fluxes and flow parameters. *American Journal of Science*, **295**, 1–55.

BUICK, I. S. & CARTWRIGHT, I. 1994. The significance of early scapolite in greenschist facies marbles from the Reynolds Range Group, central Australia. *Journal of the Geological Society, London*, **151**, 803–812.

BUICK, I. S. & CARTWRIGHT, I. 1996. Fluid–rock interaction during low pressure polymetamorphism of the Reynolds Range Group, central Australia. *Journal of Petrology*, **37**, 1097–1124.

BUICK, I. S., CARTWRIGHT, I. & HARLEY, S. L. 1998. The retrograde P–T–t path for low pressure granulites from the Reynolds Range, central Australia: petrological constraints and implications for LP/HT metamorphism. *Journal of Metamorphic Geology*, **16**, 511–529.

BUICK, I. S., CARTWRIGHT, I. & WILLIAMS, I. S. 1997. High-temperature retrogression of granulite-facies marbles from the Reynolds Range. *Journal of Petrology*, **38**, 877–910.

BUICK, I. S., FREI, R. & CARTWRIGHT, I. 1999. The timing of high-temperature retrogression in the Reynolds Range, central Australia: constraints from single mineral Pb–Pb dating. *Contributions to Mineralogy and Petrology*, **135**, 244–254.

BUICK, I. S., CARTWRIGHT, I., HAND, M. & POWELL, R. 1994. Evidence for pre-regional metamorphic fluid infiltration of the Lower Calcsilicate Unit, Reynolds Range Group (central Australia). *Journal of Metamorphic Geology*, **12**, 789–810.

CARTWRIGHT, I. 1988. Melt crystallization, pegmatite intrusion and the Inverian retrogression of the Scourian complex of NW Scotland. *Journal of Metamorphic Geology*, **6**, 77–93.

CARTWRIGHT, I. 1994. The two-dimensional pattern of metamorphic fluid flow at Mary Kathleen, Australia: fluid focusing, transverse dispersion, and implications for modeling fluid flow. *American Mineralogist*, **79**, 526–535.

CARTWRIGHT, I. 1997. Permeability and fluid flow in metamorphic rocks: Implications for advection-dispersion models. *Contributions to Mineralogy and Petrology*, **129** 198–208.

CARTWRIGHT, I. & BUICK, I. S. 1994. Channelled fluid flow in wollastonite calc-silicates, Reynolds Range, central Australia. *Journal of the Geological Society, London*, **151**, 583–586.

CARTWRIGHT, I. & BUICK, I. S. 1995. Formation of wollastonite-bearing marbles during late regional metamorphic channelled fluid flow in the Upper Calcsilicate Unit of the Reynolds Range Group, Central Australia. *Journal of Metamorphic Geology*, **13**, 397–418.

CARTWRIGHT, I. & BUICK, I. S. 1996. Determining the direction of contact metamorphic fluid flow: an assessment of mineralogical and stable isotope criteria. *Journal of Metamorphic Geology*, **14**, 289–306.

CARTWRIGHT, I. & BUICK, I. S. 1998. Oxygen isotope resetting, partial melting, and fluid flow in metamorphic terrains. *Terra Nova*, **10**, 81–85.

CARTWRIGHT, I. & BUICK, I. S. 1999. Fluid flow in Alice Springs age shear zones, Reynolds Range, central Australia. *Journal of Metamorphic Geology*, **17**, 397–414.

CARTWRIGHT, I. & BUICK, I. S. 2000. Generation and destruction of millimetre-scale metamorphic permeability in marbles: an example from the Reynolds Range, central Australia. *Contributions to Mineralogy & Petrology*, **140**, 163–179.

CARTWRIGHT, I. & HARPER, S. 1998. Oxygen isotope geochemistry of the Omeo Metamorphic Complex, Victoria: implications for metamorphic fluid flow mineralisation and anatexis. *Australian Journal of Earth Sciences*, **45**, 963–970.

CARTWRIGHT, I. & OLIVER, N. H. S. 2000. Metamorphic fluids and their relationship to the formation of metamorphosed and metamorphogenic ore deposits. *In*: SPRY, P. G., MARSHALL, B. & VOKES, F. M. (eds). *Metamorphosed and Metamorphogenic Ore Deposits*. Reviews in Economic Geology, **11**, 81–96. Society of Economic Geologists, Boulder.

CARTWRIGHT, I. & VALLEY, J. W. 1991. Steep oxygen-isotope gradients at marble-metagranite contacts in the northwest Adirondack Mountains, New York, USA: products of fluid-hosted diffusion. *Earth and Planetary Science Letters*, **107**, 148–163.

CARTWRIGHT, I. & VALLEY, J. W. 1992. Oxygen-isotope geochemistry of the Scourian complex, NW Scotland. *Journal of the Geological Society, London*, **149**, 115–126.

CARTWRIGHT, I. & WEAVER, T. R. 1997. Patterns of metamorphic fluid flow and isotopic resetting in layered and fractured rocks. *Journal of Metamorphic Geology*, **15**, 497–512.

CARTWRIGHT, I., BUICK, I. S. & MAAS, R. 1997. Fluid flow in marbles at Jervois, central Australia: Oxygen isotope disequilibrium and zoning produced by decoupling of mineralogical and isotopic resetting. *Contributions to Mineralogy and Petrology*, **128**, 335–351

CARTWRIGHT, I., BUICK, I. S. & VRY, J. K. 1996. Polyphase metamorphic fluid flow in the Lower Calc-silicate Unit, Reynolds Range, central Australia. *Precambrian Research*, **77**, 211–229.

CARTWRIGHT, I., VRY, J. K. & SANDIFORD, M. 1995. Changes in stable isotope ratios of metapelites and marbles during regional metamorphism, Mount Lofty Ranges, South Australia: implications for crustal fluid flow. *Contributions to Mineralogy and Petrology*, **120**, 292–310.

CARTWRIGHT, I., BUICK, I. S., FOSTER, D. & LAMBERT, D. D. 1999. Alice Springs age shear zones from the Reynolds Range, central Australia: implications for regional tectonics. *Australian Journal of Earth Sciences*, **46**, 355–363.

CHAMBERLAIN, C. P., ZEITLER, P. K., BARNETT, D. E., WINSLOW, D., POULSON, S. R., LEAHY, T. & HAMMER, J. E. 1995. Active hydrothermal systems during the recent uplift of Nanga Parbat, Pakistan Himalaya. *Journal of Geophysical Research, B*, **100**, 439–495.

CLARKE, G. L. & POWELL, R. 1991. Proterozoic granulite facies metamorphism in the southeastern Reynolds Range, central Australia; geological context, P–T path and overprinting relationships. *Journal of Metamorphic Geology*, **9**, 267–281.

CLARKE, G. L., COLLINS, W. J. & VERNON, R. H. 1990. Successive overprinting granulite facies metamorphic events in the Anmatjira Range, central Australia. *Journal of Metamorphic Geology*, **8**, 65–88.

COLLINS, W. J. & SHAW, R. D. 1995. Geochronological constraints on orogenic events in the Arunta Inlier: a review. *Precambrian Research*, **71**, 315–346.

COLLINS, W. J. & TEYSSIER, C. 1989. Crustal-scale ductile fault systems in the Arunta Inlier, central Australia. *Tectonophysics*, **158**, 49–66.

COLLINS, W. J. & WILLIAMS, I. S. 1995. SHRIMP ion probe dating of short-lived Proterozoic tectonic cycles in the northern Arunta Inlier, central Australia. *Precambrian Research*, **71**, 69–89.

CORBETT, G. J. & PHILLIPS, G. N. 1981. Regional retrograde metamorphism of a high grade terrain; the Willyama Complex, Broken Hill, Australia. *Lithos*, **14**, 59–73.

CRISS, R. E. & TAYLOR, H. P. 1986. Meteoric-hydrothermal systems. *In*: VALLEY, J. W., TAYLOR, H. P. & O'NEIL, J.R (eds) *Stable Isotopes in High-Temperature Geological Processes*. Mineralogical Society of America, Reviews in Mineralogy, **16**, 373–424.

DAVIS, S. R. & FERRY, J. M. 1994. Fluid infiltration during contact metamorphism of interbedded marble and calc-silicate hornfels, Twin Lakes area, central Sierra Nevada, California. *Journal of Metamorphic Geology*, **11**, 71–88.

DIPPLE, G. M. & FERRY, J. M. 1992*a*. Metasomatism and fluid flow in ductile fault zones. *Contributions to Mineralogy and Petrology*, **112**, 149–164.

DIPPLE, G. M. & FERRY, J. M. 1992*b*. Fluid flow and stable isotopic alteration in rocks at elevated temperatures with applications to metamorphism. *Geochimica et Cosmochimica Acta*, **56**, 3539–3550.

DIRKS, P. H. G. M. 1990. Intertidal and subtidal sedimentation during a mid-Proterozoic marine transgression, Reynolds Range Group, Arunta Block, central Australia. *Australian Journal of Earth Sciences*, **37**, 409–422.

DIRKS, P. H. G. M., HAND, M. & POWELL, R. 1991. The P–T deformation path for a mid-Proterozoic, low pressure terrane: the Reynolds Range, central Australia. *Journal of Metamorphic Geology*, **9**, 641–661.

DIRKS, P. H. G. M. & WILSON, C. J. L. 1990. The geological evolution of the Reynolds Range, central Australia: evidence for three distinct structural-metamorphic cycles. *Journal of Structural Geology*, **12**, 651–665.

GRAHAM, C. M., GREIG, K. M., SHEPPARD, S. M. F. & TURI, B. 1983. Genesis and mobility of the $H_2O–CO_2$ fluid phase during regional greenschist and epidote amphibolite facies metamorphism: A petrological and stable isotope study in the Scottish Dalradian. *Journal of the Geological Society, London*, **140**, 577–599.

GRESENS, R. L. 1967. Composition-volume relationships of metasomatism. *Chemical Geology*, **2**, 47–65.

HAND, M. & DIRKS, P. H. G. M. 1992. The influence of deformation on the formation of axial-planar leucosomes and the segregation of small melt bodies within the migmatitic Napperby Gneiss, central Australia. *Journal of Structural Geology*, **14**, 591–604.

HOEFS, J. 1997. *Stable isotope geochemistry*. Springer, Berlin.

LAMB, W. & VALLEY, J. W. 1984. Metamorphism of reduced granulites in low-CO_2 vapour-free environment. *Nature*, **312**, 56–58.

LAMBECK, K., BURGESS, G. & SHAW, R. D. 1988. Teleseismic travel-time anomalies and deep crustal structure in central Australia. *Geophysical Journal of the Royal Astronomical Society*, **94**, 105–124.

LÉGER, A. & FERRY, J. M. 1993. Fluid infiltration and regional metamorphism of the Waits River Formation, North-east Vermont, USA. *Journal of Metamorphic Geology*, **11**, 3–29.

LOBATO, L. M., FORMAN, J. M. A., FAZIKAWA, K., FYFE, W. S. & KERRICH, R. 1983. Uranium in overthrust Archaean basement, Bahia, Brazil. *Canadian Mineralogist*, **21**, 647–654.

MCCAIG, A. M. 1988. Deep fluid circulation in fault zones. *Geology*, **16**, 867–870.

NEWTON, R. C. 1986. Fluids of granulite facies metamorphism. *In*: WALTHER, J. V. & WOOD, B. J. (eds) *Fluid–rock interactions during metamorphism*. Advances in physical geochemistry, **5**, 36–59, Springer, New York.

OLIVER, N. H. S., CARTWRIGHT, I., WALL, V. J. & GOLDING, S. D. 1993. The stable isotope signature of large-scale fracture-dominated metamorphic fluid pathways, Mary Kathleen, Australia. *Journal of Metamorphic Geology*, **11**, 705–720.

OLIVER, N. H. S., RAWLING, T. J., CARTWRIGHT, I. & PEARSON, P. J. 1994. High-temperature fluid–rock interaction and scapolitization in an extension-related hydrothermal system, Mary Kathleen, Australia. *Journal of Petrology*, **35**, 1455–1491.

PEARSON, F. J., BALDERER, W., LOOSLI, H. H., LEHMANN, B. E., MATTER, A., PETERS, T., SCHMASSMANN, H. & GAUTSCHI, A. 1991. Applied isotope hydrogeology; a case study in northern Switzerland. *Studies in Environmental Science*, **43**, 439.

RUMBLE, D. & SPEAR, F. S. 1983. Oxygen-isotope equilibration and permeability enhancement during regional metamorphism. *Journal of the Geological Society, London*, **140**, 619–628.

SELVERSTONE, J., MORTEANI, G. & STAUDE, J. M. 1991. Fluid channelling during ductile shearing: transformation of granodiorite into aluminous schist in the Tauern Window, eastern Alps. *Journal of Metamorphic Geology*, **9**, 419–431.

SHEPPARD, S. M. F. 1986. Characterization of isotopic variations in natural waters. *In*: VALLEY, J. W., TAYLOR, H. P. & O'NEIL, J. R. (eds) *Stable Isotopes in High-Temperature Geological Processes*. Mineralogical Society of America, Reviews in Mineralogy, **16**, 165–184.

SKELTON, A. D. L., GRAHAM, C. L. & BICKLE, M. J. 1995. Lithological and structural controls on regional 3-D fluid flow patterns during greenschist facies metamorphism of the Dalradian of the SW Scottish Highlands. *Journal of Petrology*, **38**, 563–586.

STERN, L. A., CHAMBERLAIN, C. P., BARNETT, D. E. & FERRY, J. M. 1992. Stable isotope evidence for regional-scale fluid migration in a Barrovian metamorphic terrain, Vermont, USA. *Contributions Mineralogy and Petrology*, **112**, 475–489.

STEWART, A. J., SHAW, R. D. & BLACK, L. P. 1984. The Arunta Inlier: a complex ensialic mobile belt in central Australia. Part 1: stratigraphy, correlations and origin. *Australian Journal of Earth Sciences*, **31**, 445–455.

SYMMES, G. H. & FERRY, J. M. 1991. Evidence from mineral assemblages for infiltration of pelitic schists by aqueous fluids during metamorphism. *Contributions to Mineralogy and Petrology*, **108**, 419–438.

SUZUOKI, T. & EPSTEIN, S. 1976. Hydrogen isotope fractionation between OH-bearing minerals and water: *Geochimica et Cosmochimica Acta*, **35**, 1229–1240.

TAYLOR, H. P. & SHEPPARD, S. M. F. 1986. Igneous rocks: I. Processes of isotopic fractionation and isotope systematics. *In*: VALLEY, J. W., TAYLOR, H. P. & O'NEIL, J. R. (eds) *Stable Isotopes in High-Temperature Geological Processes*. Mineralogical Society of America, Reviews in Mineralogy, **16**, 227–272.

THOMPSON, A. B. 1997. Flow and Focusing of Metamorphic Fluids. *In*: JAMVEIT, B. & YARDLEY, B. W. D. (eds). *Fluid Flow and Transport in Rocks*. Chapman-Hall, London, 297–314.

TRACY, R. J. & FROST, B. R. 1991. Phase equilibria and thermobarometry of calcareous, ultramafic and mafic rocks, and iron formations. *In*: KERRICK, D. M. (ed.) *Contact Metamorphism*. Mineralogical Society of America, Reviews in Mineralogy, **26**, 207–290.

VALLEY, J. W., BOHLEN, S. R., ESSENE, E. J. & LAMB, W. 1990. Metamorphism in the Adirondacks: II The role of fluids. *Journal of Petrology*, **31**, 555–596.

VRY, J. K. & CARTWRIGHT, I. 1994, Sapphirine-kornerupine rocks from the Reynolds Range, central Australia: constraints on the uplift history of a Proterozoic low pressure terrain. *Contributions to Mineralogy and Petrology*, **115**, 78–91.

VRY, J. K. & CARTWRIGHT, I. 1998. Stable isotopic evidence for early fluid infiltration in a multiply-metamorphosed terrane: the Reynolds Range, Arunta Block, central Australia. *Journal of Metamorphic Geology*, **16**, 749–766.

VRY, J. K., BROWN, P. E. & VALLEY, J. W. 1990. Cordierite volatile content and the role of CO_2 in high-grade metamorphism. *American Mineralogist*, **75**, 71–88.

VRY, J., COMPSTON, W. & CARTWRIGHT, I. 1996. SHRIMP II dating of zircons and monazites: reassessing the timing of high-grade metamorphism and fluid flow in the Reynolds Range, northern Arunta Block, Australia. *Journal of Metamorphic Geology*, **14**, 335–350.

WICKHAM, S. M. & TAYLOR, H. P. 1985. Stable isotopic evidence for large-scale seawater infiltration in a regional metamorphic terrane; the Trois Seigneurs Massif, Pyrenees, France. *Contributions to Mineralogy and Petrology*, **91**, 122–137.

WILLIAMS, I. S., BUICK, I. S. & CARTWRIGHT, I. 1996. An extended episode of early Mesoproterozoic metamorphic fluid flow in the Reynolds Range, central Australia. *Journal of Metamorphic Geology*, **14**, 29–47.

YARDLEY, B. W. D. 1997. Evolution of fluids through the metamorphic cycle. *In*: JAMVEIT, B. & YARDLEY, B. W. D. (eds) *Fluid Flow and Transport in Rocks*. Chapman-Hall, London, 99–122.

The effects of early Cambrian metamorphism in western Dronning Maud Land, East Antarctica: a carbon and oxygen isotope study of fluid–rock interaction in the Sverdrupfjella Group

WARREN P. JOHNSTONE & CHRIS HARRIS

Department of Geological Sciences, University of Cape Town, Rondebosch 7700, Republic of South Africa (email: charris@geology.uct.ac.za)

Abstract: The Sverdrupfjella Group in western Dronning Maud Land is a 1200 to 900 Ma orogenic belt that experienced a thermal overprint in the early Cambrian. Evidence for distinct episodes of fluid–rock interaction is found in calc-silicate rocks, veins and retrograde mineral assemblages. A sequence of altered but undeformed basalts exhibit extreme ^{18}O depletion ($\delta^{18}O$ as low as -1.8‰) apparently due to interaction with meteoric water during regional (possibly early Cambrian) metamorphism. In the central Kirwanveggen, metasomatic calc-silicates, and retrograde mineral assemblages are associated with late high-strain zones of probable Cambrian age. The former have $\delta^{13}C$ values which overlap those of massive metacarbonate units 150 km to the NE, and imply regional scale movement of fluids. The latter record ^{18}O depletion relative to unretrogressed equivalent rocks and suggest interaction with externally derived fluids.

The understanding of crustal evolution in many parts of Gondwana is complicated by the difficulty in distinguishing the effects of c. 500 Ma overprinting of c. 1000 Ma orogenic belts. The Mesoproterozoic high-grade Circum East Antarctic Mobile Belt (Yoshida 1992) has been variably affected by an early Palaeozoic (550–500 Ma) metamorphic event (Groenewald 1991; Grantham *et al.* 1995; Groenewald *et al.* 1995; Krynauw 1996). Geochronological studies in central and eastern Dronning Maud Land (Fig. 1) show that this early Cambrian high-grade metamorphism was associated with magmatism and deformation, and reached granulite facies conditions in places (e.g. Shiraishi *et al.* 1992; Dirks *et al.* 1993; Fitzsimons 1998). The Maud Belt of western Dronning Maud Land is the westernmost part of the Circum East Antarctic Mobile Belt, where the effects of the Cambrian metamorphism are least well developed (Fig. 1). Although early Cambrian Rb–Sr and K–Ar cooling ages are recorded in the Maud Belt (e.g. Moyes *et al.* 1993; Krynauw 1996), these appear to be associated with a greenschist–amphibolite facies thermal overprint accompanying the emplacement of granitoid bodies. Cambrian deformation appears to be re-stricted largely to mylonite zones of limited development, which contain hydrous mineral assemblages and thus record the passage of syn- and possibly post-deformation fluids. In terms of understanding the overall effects of Cambrian metamorphism, it is important to determine to what extent the thermal overprint recorded in the Cambrian was accompanied by pervasive fluid–rock interaction associated with the developments of the mylonite zones. A study of the fluid–rock interaction enables an assessment to be made of the extent of crustal 'reworking' in the absence of significant structural features.

In the Maud Belt, as in the case of many other metamorphic terranes, there is abundant field evidence for fluid–rock interaction, in the form of veins, hydrous retrograde mineral assemblages, and metasomatic rocks. However, when the terrane has undergone a complex geological history involving multiple episodes of deformation and metamorphism, it can be difficult to ascertain when fluid–rock interaction occurred and whether it resulted from a single or multiple periods of metamorphism. This is exacerbated by the problems of dating fluid–rock interaction and it may not be easy to determine the relationship between different metamorphic and/or deformation episodes, and different episodes of fluid–rock interaction.

In this study oxygen and carbon isotopes have been used to investigate the nature and extent of these episodes of fluid–rock interaction recorded in the Sverdrupfjella Group. Stable isotopes alone cannot provide any age constraints, but in combination with previously published accounts of the crustal evolution of the area (Grantham

From: MILLER, J. A., HOLDSWORTH, R. E., BUICK, I. S. & HAND, M. (eds) *Continental Reactivation and Reworking.* Geological Society, London, Special Publications, **184**, 381–394. 1-86239-080-0/01/$15.00 © The Geological Society of London 2001.

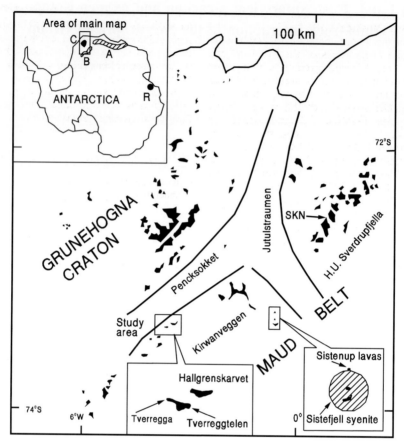

Fig. 1. Regional Geological Map of western Dronning Maud Land, East Antarctica. The Grunehogna Craton is separated from the Maud Belt by the Jutulstraumen–Pencksokket glaciers. The metacarbonate samples (WJS1–6) from the Fuglefjellet Formation were collected at Skarsnuten (SKN) in the H.U. Sverdrupfjella. The inset geological map of Antarctica shows the position of parts of the East Antarctic Mobile Belt which were affected by high-grade Cambrian metamorphism (A); those parts of the mobile belt which were not affected by Cambrian high-grade metamorphism (B); the Grunehogna Craton (C); and the location of the Rauer Group (R). The shape of the Mesozoic Sistefjell syenite is interpreted from aeromagnetic data (Corner 1994). Note that H.U. Sverdrupfjella refers to the mountain range to the west of the Jutulstraumen Glacier; the Sverdrupfjella Group comprises rocks from both the H.U. Sverdrupfjella and the Kirwanveggen montain range to the east of the Jutulstraumen.

et al. 1995; Groenewald *et al.* 1995), can be used to construct the extent of fluid–rock interaction during the complex metamorphic history of western Dronning Maud Land. Stable isotope studies have been successfully applied to similar aged 'Pan-African' reworking of older high-grade metamorphic rocks of Madagascar by Pili *et al.* (1997).

Regional geological setting

Western Dronning Maud Land (Fig. 1) is characterized by two areas with distinctly different geological histories. To the west of the Jutulstraumen–Penksokket glaciers (Fig. 1), the major rock types are relatively undeformed sedimentary sequences and magmatic rocks of uncertain age (but >1000 Ma) that overlie Archaean granite (Groenewald *et al.* 1995). This western area has been termed the Grunehogna Craton and is interpreted as a fragment of the Kalahari Craton (Wolmarans & Kent 1982; Krynauw 1996). To the east of the Jutulstraumen–Penksokket glaciers, high-grade metamorphic rocks make up the Sverdrupfjella Group (Roots 1969; Hjelle 1974). This terrane has been termed the Maud Belt (Groenewald *et al.* 1995) and it forms part of the Mesoproterozoic Circum East Antarctic Mobile Belt (Yoshida 1992).

Most of this paper concerns the amphibolite to granulite facies rocks of the Sverdrupfjella Group in the Kirwanveggen mountain range in the Maud Belt. The metamorphic history of these rocks are summarized by Grantham et al. (1995) and Groenewald et al. (1995). The earliest metamorphism involved high-grade conditions (12–15 kbar, 750°C) during prograde metamorphism between 1200 Ma and 900 Ma followed by decompression and thermal relaxation (8 kbar, 850°C). These conditions are attributed by Groenewald et al. (1995) to continental collision (the 'Maud Orogeny') involving deep burial with an elevated geotherm of a marginal basin. The Cambrian metamorphism reached amphibolite facies conditions (5–6 kbar, 600°C). This later orogenic episode has been correlated with 'Pan African' orogenesis (Grantham et al. 1995; Krynauw 1996) and resulted in tectonic inversion of the metamorphic profile by thrusting, followed by rapid uplift and exhumation (Groenewald et al. 1995). Apart from Mesozoic intrusions, the only undeformed rocks in the area are the late Precambrian Sistenup lavas (Ravich & Solov'ev 1969). These lavas are pervasively altered to amphibolite facies assemblages that, because of the lack of associated deformation, may be Cambrian in age.

Petrography

Sistenup lavas

The Sistenup lavas (Fig. 1) are considered to be correlatives of the Precambrian Richtersflya Supergroup, which is exposed on the Grunehogna Craton (Ravich & Solov'ev 1969). The lavas are dominantly basaltic with minor evolved types (SiO_2 up to 66 wt.%), and it is important to note that they are undeformed except that flows dip at about 30°S. They are highly altered with hornblende as the dominant secondary mineral. Some samples contain abundant large (up to 3 cm long) laths of plagioclase that in thin section appear comparatively unaltered. The plagioclase exhibits well-developed igneous zoning and is set in a groundmass of amphibole. The abundance of secondary hornblende rather than chlorite and epidote, and the comparative stability of plagioclase, is evidence of alteration at elevated temperatures. The temperature at which hornblende rather than chlorite forms in mafic systems is strongly dependent on oxygen fugacity, PH_2O and Ptotal (e.g. Apted & Liou

1983), but is probably at least 450°C. Experimental work by Apted & Liou (1983) and Moody et al. (1983) showed that the relict calcic plagioclase persists up to 600°C at QFM buffer conditions of 7 kbar (PH_2O = Ptotal). Although the Sistenup lavas are exposed only at the contact with a Mesozoic syenite intrusion, it is suggested that the amphibole resulted from much earlier fluid–rock interaction (see below).

Calc silicates and metacarbonates

Regionally extensive units (>300 m thick and 5 km strike length) consisting of marbles, and quartzofeldspathic and calc-silicate gneisses are exposed in the H.U. Sverdrupfjella mountain range (Fig. 1) where they collectively make up the Fuglefjellet Formation (Hjella 1974; Grantham et al. 1995; Groenewald et al. 1995). According to Groenewald et al. (1995), these carbonates represent a shallow-marine sequence. Although they are best preserved at Fuglefjellet, substantial thicknesses are also exposed at Skarsnuten further south (Fig. 1), where a 200 m high cliff reveals a package of alternating bands of sparry calcite (up to 3 m thick) and quartzofeldspathic gneiss punctuated by strings of boudinaged calc-silicate gneiss.

In contrast, in the central Kirwanveggen (Fig. 1) calc-silicate mineral assemblages are rare. Scattered calc-silicate boudins in quartz–feldspar–biotite gneisses (Fig. 2a) conform to the early regional fabric and are hosted in an extensive horizon that forms part of the Tverregga Banded Gneiss described by Grantham et al. (1995). Individual boudins range from 5 to 20 cm in size, are ellipsoidal in shape with the long axis parallel to the mineral stretching lineation within the plane of the foliation (Fig. 2a). They show a mineral zonation from cores which are relatively rich in mafic minerals and comprise quartz–K-feldspar–plagioclase–diopside–hornblende–epidote with minor calcite and sphene to more felsic rims of plagioclase–quartz–K-feldspar with minor calcite. A second type of calc-silicate rock occurs in the form of a metasomatic alteration body at Tverreggtelen. This body extends for at least 150 m along strike and is hosted within a late shear zone (Fig. 2b). The metasomatic paragenesis comprises calcite–diopside with variable amounts of hornblende–actinolite ± chlorite ± serpentinite ± tremolite ± talc ± epidote. The body consists of a hornblende and diopside-rich core surrounded by an outer zone consisting of calcite–quartz–diopside–tremolite–talc–sphene, with a rim of quartz–calcite–hornblende–talc–serpentinite–actinolite and diopside.

Calc-silicate types   ## Vein types

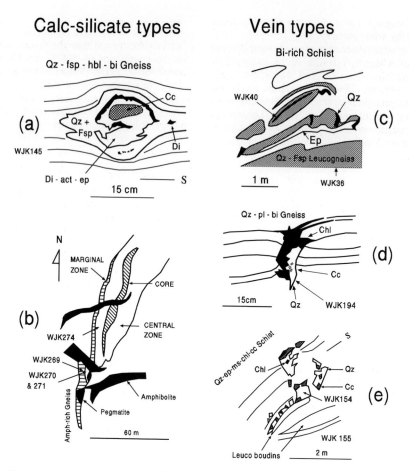

Fig. 2. Calc-silicate rocks and vein types sampled in the central Kirwanveggen. (**a**) Field sketch of a single zoned calc-silicate boudin. *S* denotes the foliation of the host gneiss, the orientation of the line is parallel to the direction of the mineral stretching lineation within the plane of the foliation. (**b**) Sketch map of the metasomatic calc-silicate body at Tverreggtelen. The core comprises hbl–di; the central zone comprises cc–qz–di–trem–tc–sph; and the marginal zone comprises qz–cc–hbl–tc–serp–act–di. Mineral abbreviations: act, actinolite; bi, biotite; cc, calcite; di, diopside; ep, epidote; fsp, feldspar; hbl, hornblende; serp, serpentinite; sph, sphene; tc, talc; trem, tremolite; qz, quartz. (**c–e**) Vein types in the central Kirwanveggen (**c**) synfolial, (**d**) boudin-eye and (**e**) discordant. *S* denotes the foliation of the host gneiss.

Veins

Veins are ubiquitous throughout the central Kirwanveggen and attest to the presence of fluids during the evolution of this part of the Maud Belt. However, the scale of fluid-flow during vein formation remains a subject of considerable debate (e.g. Gray *et al.* 1991; Henry *et al.* 1996) with possible end-member behaviour ranging from percolation of large volumes of fluid on the scale of an orogenic belt, to closed-system behaviour on the scale of a geological formation, where fluid occupies pore spaces and is essentially static. In the central Kirwanveggen, the majority of

veins consist of quartz, a minority of which also contain calcite. Some of the veins show a more complex mineralogy, e.g. quartz–chlorite–calcite ± feldspar ± biotite and quartz–hornblende–feldspar ± epidote ± calcite. The veins can be classified as: synfolial, boudin-eye and discordant types according to their form and structural disposition in the field (Fig. 2c, d and e). Synfolial veins regularly indicate a sense of shear while boudin-eye veins fill dilational sites within the boudinaged fabric. These two vein types are likely to have been formed by pressure solution processes during fabric development. The discordant veins show a variety of geometrical forms, these include irregular vug-filling and fracture-filling varieties as well as tension gash arrays. The process by which discordant veins form is distinct from that which is responsible for the formation of synfolial veins because they reflect more brittle conditions. The synfolial and boudin-eye vein types precipitated from a fluid phase present during regional fabric development whereas the discordant veins are probably the result of later fluid–rock interaction at a shallower level in the crust.

Hydrous retrograde mineral assemblages

Greenschist facies overprinting of upper amphibolite and granulite facies mineral assemblages is widespread in the central Kirwanveggen (e.g. Krynauw 1996). Some of the best examples of this overprinting occur in banded gneisses at Halgrenskarvet and Tverregga (Fig. 1). These rocks display a variable degree of hydrous retrogression characterized by retrograde assemblages of biotite, chlorite and epidote with lesser amounts of sericite, titanite and calcite. The retrograde mineral assemblages are associated with high strain zones within the banded gneiss and often exhibit mylonitic textures. These high strain zones define the moderately SE-dipping regional planar fabric of the central Kirwan-veggen. Lineations are generally down-dip, and field criteria indicate a top-to-the-north-west sense of shear. Grantham et al. (1995) described retrogressed diopside–epidote–plagioclase calc silicate gneiss, chlorite phyllite, actinolite–epidote–chlorite–calcite schist and tremolite–serpentine schist components from what they call the Tverregga Banded Gneiss. Similar assemblages are observed around the margins of an intrusive leucogranite at nearby Tverreggtelen, and within a thick sequence of banded gneisses at Hallgrenskarvet (Fig. 1). These zones consist of three distinctive lithological components (Fig. 2b), namely quartz–chlorite–epidote–biotite schist, quartz–K-feldspar–plagioclase–sericite–chlorite schist, and quartz–epidote–muscovite–chlorite–calcite–titanite schist. Quartzo–feldspathic gneisses in these zones have epidote and biotite as major phases whereas magnetite, titanite, apatite and allanite are common accessory phases in most of the samples from this area. On the basis of the similarity of these features with those that cut 500 Ma mafic dykes in the Heimefrontfella to the southwest, Grantham et al. (1995, pp. 217–218) concluded that the high-strain zones and the associated hydrous retrogression are Cambrian in age.

Analytical techniques

The majority of the analytical data reported in this paper were produced at the University of Cape Town (UCT). Some of the silicate $\delta^{18}O$ values were obtained at Monash University. Mineral separates were obtained by a combination of hand-picking and magnetic separation. Oxygen isotope ratios determined in both laboratories were obtained by conventional methods using ClF_3 as the oxidizing reagent (Borthwick & Harmon 1982). Data are reported in δ notation where $\delta^{18}O = (R_{sample}/R_{standard} - 1) * 1000$ and $R = {}^{18}O/{}^{16}O$. At UCT, a quartz standard (MQ) calibrated against NBS-28 was analysed in duplicate with each run of eight samples and used to convert the raw data to the SMOW scale using the value of 9.64‰ for NBS-28 recommended by Coplen et al. (1983). The average observed difference between MQ duplicates during the course of this work was 0.13‰. Further details of the methods used for the extraction of oxygen from silicates at UCT are given by Vennemann & Smith (1990) and Harris & Erlank (1992). Standards run at Monash gave $\delta^{18}O$ values within 0.2‰ of their accepted values and the long-term average $\delta^{18}O$ value of NBS-28 is 9.58 ± 0.12‰.

Carbon and oxygen isotope ratios of calcite were obtained at UCT according to the method of McCrea (1950). A fractionation factor of 1.01025 was used to correct the $\delta^{18}O$ value of the acid liberated CO_2 to that of calcite. The $\delta^{18}O$ and $\delta^{13}C$ values are reported with respect to the SMOW and PDB reference standards, respectively, to which they were calibrated using an internal carbonate standard calibrated using NBS-19 ($\delta^{18}O$, 28.64‰; $\delta^{13}C$, 1.95‰).

Results

Sistenup lavas

The Sistenup lavas show a range in whole-rock $\delta^{18}O$ values from -1.8 to $+4.4$‰ (Table 1). The very low values occur in both aphyric and plagioclase–phyric varieties of the basalt. In two highly porphyritic samples where plagioclase–whole-rock pairs were analysed, the plagioclase and whole-rock $\delta^{18}O$ values were of -0.1 and -0.6‰, and 4.7 and 4.4‰ respectively. The small difference between plagioclase and whole-rock $\delta^{18}O$ values suggests approximate oxygen isotope equilibrium between phenocrysts and groundmass. There is no systematic variation of whole-rock $\delta^{18}O$ value with distance from the intrusion and/or stratigraphic height at Sistenup.

Table 1. *Whole–rock oxygen isotope data for retrogressed rocks*

Sample	Rock type	$\delta^{18}O$	Sample	Rock type	$\delta^{18}O$
Tverregga			Hallgrenskarvet		
WJK155	qz–ep–chl–ms schist	5.4	WJK20	qz–fsp–ms–bi gneiss	3.5
WJK156	qz–fsp–ser–chl schist	12.0	WJK21	qz–fsp–ms–bi gneiss	4.0
WJK159	qz–ep–bi gneiss	8.7	WJK22	qz–fsp–ms–bi gneiss	5.3
WJK160	qz–ksp–bi–ms gneiss	9.3	WJK23	qz–fsp–ms–bi gneiss	4.1
WJK180	qz–ksp–ep leucosome	10.4	WJK24	qz–fsp–ms–bi gneiss	6.3
WJK193	qz–pl–bi–ep gneiss	4.9	WJK36	qz–pl–ksp gneiss	7.7
WJK196	qz–pl–bi gneiss	7.8	WJK37	qz–bi–ep schist	6.8
WJK199	qz–fsp pegmatite	8.8	WJK50	qz–fsp–ms–bi gneiss	6.3
WJK200	qz–pl–bi gneiss	7.9	WJK68	qz–fsp–bi gneiss	10.5
WJK201	qz–ksp–bi–ep agmatite	7.5	WJK74	qz–fsp–bi schist	4.2
WJK202	qz–fsp–bi gneiss	10.2	WJK80	qz–ksp–pl–ms gneiss	9.3
WJK205	qz–ksp–bi gneiss	7.8	WJK82	qz–fsp–hbl–bi schist	6.3
WJK207	qz–ksp–bi–ep agmatite	8.2	WJK85	qz–fsp–bi schist	4.6
WJK208	qz–fsp–bi gneiss	8.2	WJK95	qz–pl–bi schist	6.1
WJK209	qz ultramylonite	9.7	WJK97	qz–pl–bi schist	8.0
WJK210	qz–fsp–bi gneiss	9.1	WJK20	qz–fsp–ms–bi gneiss	3.5
WJK211	qz–fsp–bi gneiss	7.7	WJK21	qz–fsp–ms–bi gneiss	4.0
WJK212	qz–fsp–bi gneiss	8.9	WJK22	qz–fsp–ms–bi gneiss	5.3
WJK213	qz–fsp–bi gneiss	9.0	WJK23	qz–fsp–ms–bi gneiss	4.1
WJK214	qz–fsp–bi gneiss	9.0	WJK24	qz–fsp–ms–bi gneiss	6.3
WJK215	qz–fsp–bi schist	8.1	WJK36	qz–pl–ksp gneiss	7.7
WJK220	qz–pl–bi gneiss	7.9	WJK37	qz–bi–ep schist	6.8
			WJK50	qz–fsp–ms–bi gneiss	6.3
Tverregatelen			WJK68	qz–fsp–bi gneiss	10.5
WJK255	qz–bi–fsp gneiss	5.2	WJK74	qz–fsp–bi schist	4.2
WJK260	qz–bi–hbl–ep schist	4.7	WJK80	qz–ksp–pl–ms gneiss	9.3
WJK262	qz–fsp–ms gneiss	6.7	WJK82	qz–fsp–hbl–bi schist	6.3
WJK278	qz–ksp–ms–bi–pl–fl gneiss	8.1	WJK85	qz–fsp–bi schist	4.6
WJK282	qz–ksp–ms–bi–pl–fl gneiss	5.6	WJK95	qz–pl–bi schist	6.1
WJK283	qz mylonite	4.5	WJK97	qz–pl–bi schist	8.0
WJK285	qz–ksp–ms–bi–pl–fl gneiss	7.0	WJK99	qz–pl–bi–ep schist	2.9
WJK292	qz–ksp–ms–bi–pl–fl gneiss	10.0	WJK103	bi–qz–hbl–chlt schist	3.6
			WJK113	qz–bi–chl schist	1.7
Sistenup			WJK121	qz–fsp–bi pegmatite	9.6
WJK328	Tuff?	2.7	WJK123	qz–pl–bi gneiss	9.5
WJK329	Felsic lava	1.8			
WJK330	Felsic lava	3.5			
WJK336	Basalt	1.0			
WJK337	Plag–phyric basalt	−0.6 (plag −0.1)			
SIS5	Basalt	−1.8			
SIS8	Plag–phyric basalt	4.4 (plag 4.7)			
SIS48	Plag–phyric basalt	1.5			
SIS49	Basalt	0.7			
SIS51	Felsic lava	0.6			

Notes: qz, quartz; chl, chlorite; ksp, alkali feldspar; bi, biotite; pl, plagioclase; hbl, hornblende; ep, epidote. Samples WJK199–220 collected from vicinity of shearzone containing ultramylonite (WJK209).

Calc-silicate rocks

Calcite oxygen and carbon isotope data for the calc-silicate rocks are summarized in Table 2. Whole-rock samples of marble from the Fugle-fjellet Formation at Skarsnuten range in $\delta^{18}O$ and $\delta^{13}C$ value from 13.3 to 26.8‰ and −2.4 to +2.8‰ respectively. The highest $\delta^{18}O$ and $\delta^{13}C$ values are from a homogeneous marble horizon

(3 m thick; >90% sparry calcite). The samples with much lower $\delta^{18}O$ and $\delta^{13}C$ values were collected adjacent to intrusive undeformed pegmatite bodies of unknown age (WJS1 and 2), or from a calc-silicate boudin in the same area (WJS4).

The calcite from the calc-silicate boudins in the Kirwanveggen shows lower $\delta^{18}O$ (9.0 to 12.8‰) and much lower $\delta^{13}C$ (−7.4 to −11.1‰) values than the Skarsnuten marbles (Fig. 3).

Table 2. *Oxygen and carbon isotope data for carbonate rocks*

Sample	Location	$\delta^{18}O$ cc	$\delta^{13}C$ cc	$\delta^{18}O$ qz	wt.% cc
Boudins					
WJK81	Hall	10.0	−11.1		0.4
WJK83	Hall	9.6	−9.1		0.6
WJK145 B rim	Tva	10.4	−7.6		3.6
WJK145 C core	Tva	12.8	−7.4		51
WJK148	Tva	9.0	−8.6		4.3
Metasomatic					
CJK 96	Tvn	13.9	1.9		24
CJK 97	Tvn	11.4	2.9		41
CJK 98	Tvn	10.7	1.0		60
WJK269	Tvn	11.0	1.5		58
WJK270	Tvn	10.6	1.2		45
WJK271	Tvn	11.4	1.7		54
WJK 273 A vein	Tvn	12.6	−1.7		98
WJK 273 B boudin	Tvn	10.2	0.6		2.2
WJK 274 A	Tvn	13.4	0.9		20
WJK 274 B	Tvn	14.4	0.6		11
WJK 274 C	Tvn	13.5	0.9		32
WJK 274 D	Tvn	13.9	1.3		19
WJK 274 E	Tvn	14.9	2.6		83
WJK 274 F	Tvn	14.5	1.3		15
Fuglefjellet Formation					
WJS 1 contact with pegmatite	Skn	15.8	0.9		36
WJS 2 contact with pegmatite	Skn	13.3	−2.4	11.7	6.1
WJS 4 boudin	Skn	14.6	0.1	13.6	12
WJS 6 thick unit; cc-rich	Skn	26.8	2.8		81

Notes: cc, calcite; qz, quartz. Hall, Hallgrenskarvet; Tva, Tverrega; Tvn, Tverreggtelen; Skn, Skarsnuten.

A zoned boudin (WJK 145; Fig. 2a) consists of a calcite-rich core with $\delta^{13}C$ and $\delta^{18}O$ values of −7.4 and 12.8‰, respectively, and a silicate-rich outer zone with $\delta^{13}C$ and $\delta^{18}O$ values of −7.6 and 10.4‰, respectively. Similar values are seen in the other calc-silicate boudins.

The calcite from the late metasomatic body (Fig. 2b) shows a range in $\delta^{13}C$ and $\delta^{18}O$ values of −1.7 to 2.6‰ and 10.2‰ to 14.9‰ respectively (Fig. 3). Samples from the central zone have slightly higher $\delta^{18}O$ values than samples from the marginal zone. Sample WJK 274 consists of five generations of calc-silicate material (A, host calcite–amphibole–talc; B, coarse-grained amphibole–talc layer; C, calcite–amphibole–biotite layer; D, calcite boudin; E, a coarse calcite pod; and F, a single amphibole porphyroblast). The range in $\delta^{18}O$ value of all these components is 13.4 to 14.9‰ and the range in $\delta^{13}C$ values is from 0.6 to 2.6‰. Sample WJK273 from within the metasomatic body was taken from a boudin which contains a calcite-bearing vein. This sample has a similar $\delta^{18}O$ value (10.2‰) to the early calc-silicate boudins, but a much higher $\delta^{13}C$ value (0.6‰) which is similar to those of the metasomatic body.

Carbonate $\delta^{13}C$ and $\delta^{18}O$ values are plotted against wt.% calcite in Figure 3a, b. Apart from the core of one sample, the boudins contain <5 wt.% calcite (Table 2). The metacarbonate rocks from the Fuglefjellet Formation at Skarsnuten show reasonable positive correlations between both $\delta^{13}C$ and $\delta^{18}O$ values and wt.% calcite. The metasomatic carbonate rocks show no correlation between either $\delta^{13}C$ and $\delta^{18}O$ value and wt.% calcite.

Veins

The $\delta^{18}O$ values of quartz from the three different vein types (Fig. 2c–e) are presented in Table 3. There is a substantial range in vein quartz $\delta^{18}O$ values from 3.7 to 11.0‰, which is similar in magnitude to the range seen in the host rocks (1.7 to +10.5‰). The host rock $\delta^{18}O$ values show a positive correlation with those of the vein quartz (Fig. 4), with the majority of the veins having higher $\delta^{18}O$ values than the host rocks (average $\Delta_{\text{vein–host}} = 1.8‰$). A minority of veins, however have lower quartz $\delta^{18}O$ values than their immediate host rocks. Synfolial and

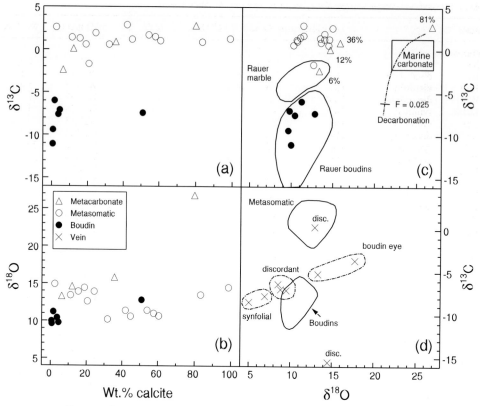

Fig. 3. Plot of $\delta^{13}C$ and $\delta^{18}O$ v. wt.% calcite (**a, b**) in the whole-rock for calc-silicate rocks from the central Kirwanveggen and the H.U. Sverdrupfjella. Plot of $\delta^{13}C$ v. $\delta^{18}O$ for calc-silicate (**c**) and calcite-bearing veins (**d**). The dashed curve in (**c**) is the path followed by an impure carbonate rock undergoing Rayleigh decarbonation starting from an arbitrary composition within the field for typical limestones assuming a α_{CO_2-rock} of 1.0022 for C and a α_{CO_2-rock} of 1.012 for O. F (carbon), fraction of carbon remaining (Valley 1986). The fields for Rauer Group marbles and boudins (Buick *et al.* 1994) are shown as is a rectangle representing the typical values expected for marine carbonate (e.g. Viezer & Hoefs 1976). The types of veins are indicated on (**d**) where disc., discordant.

boudin-eye veins show a similar range in $\delta^{18}O$ values whereas the discordant veins (and their host rocks) tend to have higher $\delta^{18}O$ values (from 7.4 to 11.0‰).

In calcite-bearing quartz veins, the range in $\delta^{18}O$ values of the quartz is similar to that observed in all quartz veins (Fig. 4). The host-rock and vein quartz $\delta^{18}O$ values show a similar correlation to that seen in the quartz veins (Fig. 4). The calcite $\delta^{18}O$ values are much more variable than those of the quartz and with $\Delta_{quartz-calcite}$ ranging from 0.5 to −11.0‰. Apart from two samples, values of $\Delta_{quartz-calcite}$ are negative, whereas equilibrium fractionation between quartz and calcite should always be positive (e.g. Chiba *et al.* 1989). The correlation between vein calcite $\delta^{18}O$ and host rock $\delta^{13}C$ and $\delta^{18}O$ value is poor (Fig. 4). Synfolial calcite-bearing

veins exhibit lower calcite $\delta^{18}O$ values than the boudin-eye veins whereas the discordant veins have intermediate calcite $\delta^{18}O$ values. Vein $\delta^{13}C$ v. $\delta^{18}O$ values are plotted on Figure 3d. Apart from two discordant veins, the samples show a good positive correlation but it is possible that this is an artefact of the small number of samples analysed.

Retrogressed rocks

Data from samples of banded gneiss, schist and other samples which show features of enhanced hydrous retrogression are presented in Table 1 and Figure 5. The range of $\delta^{18}O$ values for samples from Tverregga and Tverreggtelen is similar (4.9 to 12.0‰ and 4.5 to 10‰). Hallgrenskarvet

Table 3. *Oxygen and carbon isotope data for quartz and quartz–calcite veins*

Sample	Type	Location	$\delta^{18}O$ qz	$\delta^{18}O$ cc	$\delta^{13}C$ cc	Host rock	$\delta^{18}O$ wr
Quartz veins							
WJK4	synfolial	Hall	9.6			WJK3	8.5
WJK8	discordant	Hall	9.6			WJK1	7.8
WJK19	discordant	Hall	8.8			WJK15	8.3
WJK38	discordant	Hall	7.4			WJK37	6.8
WJK40	synfolial	Hall	7.5			WJK36	7.7
WJK56	synfolial	Hall	8.2			WJK55	6.8
WJK57	discordant	Hall	8.5			WJK55	6.8
WJK71	synfolial	Hall	9.2			WJK68	10.5
WJK75	synfolial	Hall	4.3			WJK74	4.2
WJK76	boudin eye	Hall	5.3			WJK74	4.2
WJK86	synfolial	Hall	6.7			WJK85	4.6
WJK87	boudin eye	Hall	5.8			WJK85	4.6
WJK96	synfolial	Hall	10.8			WJK97	8.0
WJK100	boudin eye	Hall	5.8			WJK99	2.9
WJK108	synfolial	Hall	6.6			WJK103	3.6
WJK111	boudin eye	Hall	8.5			WJK95	6.1
WJK122	boudin eye	Hall	10.4			WJK123	9.5
WJK164	synfolial	Tva	10.3			WJK159	8.7
WJK165	boudin eye	Tva	10.3			WJK159	8.7
WJK167	discordant	Tva	10.1			WJK160	9.3
WJK194	boudin eye	Tva	7.7			WJK193	4.9
WJK226	discordant	Tvn	9.9			WJK224	7.8
WJK244	synfolial	Tvn	8.1			WJK243	6.0
WJK257	synfolial	Tvn	7.9			WJK255	5.2
WJK261	synfolial	Tvn	6.9			WJK260	4.7
WJK263	synfolial	Tvn	7.8			WJK262	6.7
WJK279	synfolial	Tvn	6.7			WJK278	8.1
WJK283	synfolial	Tvn	4.5			WJK282	5.6
WJK286	discordant	Tvn	7.9			WJK285	7.0
WJK293	discordant	Tvn	9.2			WJK292	10.0
Calcite-bearing quartz veins							
WJK46	synfolial	Hall	5.2	5.0	−8.5	WJK44	2.5
WJK51	boudin eye	Hall	6.7	17.7	−3.6	WJK50	6.3
WJK79	discordant	Hall	11.0	14.5	−15.5	WJK80	9.3
WJK109	discordant	Hall	9.0	8.5	−6.4	WJK95	6.1
WJK110	discordant	Hall	8.8	9.5	−7.1	WJK95	6.1
WJK114	synfolial	Hall	3.7	6.9	−7.8	WJK113	1.7
WJK154	discordant	Tva	8.1	8.8	−7.0	WJK155	5.4
WJK166	boudin eye	Tva	10.8	13.3	−5.2	WJK159	8.7
WJK182	discordant	Tva	9.3	12.9	0.3	WJK180	10.4

Notes: cc, calcite; qz, quartz; wr, whole-rock. Hall, Hallgrenskarvet; Tva, Tverregga; Tvn, Tverreggtelen.

retrogressed rocks have generally lower $\delta^{18}O$ values with approximately one third of the samples having values <5.0‰. The lowest $\delta^{18}O$ value observed is 1.7‰ in a quartz–biotite–chlorite schist. Although the retrogressed rocks have generally lower $\delta^{18}O$ values than similar rock types not associated with the shear zones, there is no systematic change in $\delta^{18}O$ with distance across the shear zone. One of the mylonitic rocks from a high strain zone at Tverregga (WJK209) has slightly a higher $\delta^{18}O$ value (9.7‰) than the surrounding rocks, a similar mylonite at Tverregatelen has a much lower $\delta^{18}O$ value (4.5‰).

Discussion

The relationship between the materials analysed (Tables 1–3) and the metamorphic evolution of western Dronning Maud Land can be summarized as follows. Because the boudins are evidently associated with deformation, they must have formed during the Maud Orogeny (1200–900 Ma). Retrogression and rehydration during late fabric development in the central Kirwanveggen which appears to be associated with shear zones and the formation of mylonite are considered to be Cambrian in age, as is the

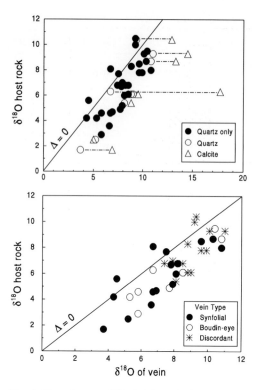

Fig. 4. Plot of quartz $\delta^{18}O$ (vein) v. whole-rock $\delta^{18}O$ (host) for veins from the central Kirwanveggen (lower diagram). The upper diagram compares the composition of quartz veins to quartz + calcite veins. Tie-lines join the compositions of coexisting quartz and calcite. The diagonal line indicates a value of $\Delta_{vein-host} = 0$ in both diagrams.

metasomatic calc-silicate body at Tverregtelen, which is associated with a shear zone. The lack of deformation of the Sistenup lavas means that the alteration must post-date the Maud Orogeny and for reasons discussed below, the alteration is most likely to be Cambrian in age. The veins possibly represent a wide range of ages. The boudin eye and synfolial veins are related to fabric development (see above) and must have formed during the Maud Orogeny, the discordant veins could have formed during either of the orogenic events.

Fluid–rock interaction during the Maud Orogeny

The calc silicate boudins in the Sverdrupfjella Group rocks have lower $\delta^{18}O$ and $\delta^{13}C$ values compared to the more massive metacarbonate samples from Skarsnuten (Fig. 3). Because the thick marble sequence is close to pure calcite

(81%), decarbonation must have been limited and would not have had a significant effect on the carbon and oxygen isotope composition of this sample (WJS6, Table 2). The $\delta^{13}C$ value of WJS6 is slightly higher than normal for marine limestones, and it is possible that the original carbonate formed during a 'positive $\delta^{13}C$ excursion' (e.g. Viezer & Hoefs 1976). The other samples of Fuglefjellet marble plot on a curved path on Figure 3, with progressively lower $\delta^{18}O$ and $\delta^{13}C$ as the calcite content decreases. The Kirwanveggen boudins which, apart from the calcite-rich core of one sample, have <5 wt.% calcite lie on the extension of this curve. This type of curved path is typical for impure carbonate rocks which have undergone extensive decarbonation during metamorphism (e.g. Valley 1986). The theoretical decarbonation curve in Figure 3 represents Rayleigh decarbonation starting from the composition of the massive marble and assumes that the fluid was pure CO_2. If a mixed CO_2–H_2O fluid was present, H_2CO_3 would be more likely to be the dominant carbon-bearing species which has a smaller carbon isotope fractionation with calcite than CO_2 (e.g. Cartwright & Buick 1999), and the effects would be less pronounced.

Whereas the low $\delta^{13}C$ values and low calcite contents (<5 wt.%; Fig. 3a) of the boudins can be explained by decarbonation, it is clear that closed system decarbonation via loss of CO_2 cannot produce $\delta^{18}O$ values as low as those observed. The oxygen and carbon isotope patterns of these Kirwanveggen boudins are similar (Fig. 3) to those observed for zoned wollastonite–garnet boudins from the Rauer Group in East Antarctica (Buick et al. 1994). In the Rauer Group, fluid-flow is interpreted to be prior, and possibly unrelated to, peak metamorpism which was probably Cambrian in age (e.g. Sims & Wilson 1997). In the case of the Kirwanveggen, the calc-silicate bodies sampled were of the order of a few decimetres in size. Thus infiltration of fluid, which equilibrated isotopically at high-temperatures, with the surrounding silicate rocks (bulk rock $\delta^{18}O$ values typically between 7 and 10‰) could account for the low $\delta^{18}O$ values of the boudin calcite. The small volume of carbonate material relative to the surrounding silicates means that the fluid/rock ratio need not have been large, provided that the fluids were rock-buffered on a scale greater than the size of the boudins.

The synfolial and boudin eye quartz vein show a relatively good positive correlation between vein and host-rock $\delta^{18}O$ values, with an average difference between vein and host $\delta^{18}O$ value ($\Delta_{vein-host}$) of 1.8‰. The relationship between quartz vein and host rock $\delta^{18}O$ values can be used

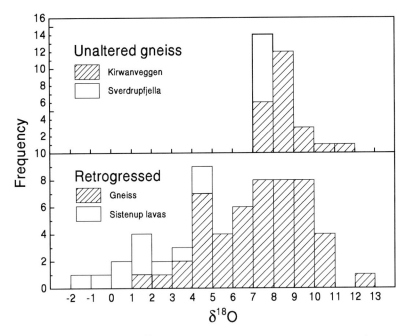

Fig. 5. Histogram showing the range of $\delta^{18}O$ values of retrogressed gneiss and schist samples, unretrogressed gneisses from the Kirwanveggen and H.U. Sverdrupfjella and the Sistenup lavas. Data from Table 3 (retrogressed rocks) and from Harris and Johnstone (unpublished data).

to distinguish between low fluid/rock ratio (rock-buffered) systems and those with greater relative volumes of fluid (fluid-buffered) (e.g. Gray *et al.* 1991; Henry *et al.* 1996). The synfolial and boudin eye vein/host relationship is generally consistent with rock-buffered conditions but samples where vein < host $\delta^{18}O$ value are more consistent with fluid-buffered conditions (Gray *et al.* 1991, p. 19 695). Consideration of only the synfolial quartz veins and their host rocks suggests that any lowering of $\delta^{18}O$ value, such as might have been experienced by the host rocks with $\delta^{18}O$ values, $<c.$ 6‰ occurred during the early metamorphic history with synfolial quartz veins forming from fluids in oxygen isotope equilibrium with the host rocks.

The discordant veins are the youngest vein type analysed but are otherwise of uncertain age. They show a more restricted range of generally higher $\delta^{18}O$ values than the synfolial and boudin-eye veins and the data array on Figure 4 also suggests that they formed from fluids whose $\delta^{18}O$ values were generally rock-buffered. The relatively restricted range of $\delta^{18}O$ values is most likely a mineralogical effect in that only phyllosilicate-poor rocks (e.g. gneisses) were brittle enough to allow the formation of dicordant veins. Such rocks would have inherently higher

$\delta^{18}O$ values because of the relatively large fractionation for oxygen between phyllosilicates and quartz/feldspar.

For veins containing both quartz and calcite, the quartz $\delta^{18}O$ value also correlates positively with the whole-rock $\delta^{18}O$ value. The co-existing calcite shows no correlation between its $\delta^{18}O$ value and that of the host and, as pointed out above, calcite $\delta^{18}O$ values are too high to be in equilibrium with the quartz (this applies to all three vein types; Fig. 4). Henry *et al.* (1996) reported similar disequilibrium between quartz and calcite in veins which they explained as being caused by later recrystallization of quartz during subsequent metamorphism. In the Kirwanveggen, the calcite has generally higher $\delta^{18}O$ values values than the quartz which suggests later equilibration (or precipitation) of calcite from lower temperature, possibly unrelated fluids.

^{18}O-depletion in the Sistenup lavas

The very low oxygen isotope ratios in the Sistenup lavas almost certainly resulted from the alteration episode that produced the groundmass amphibole, and are unlikely to be the original magmatic $\delta^{18}O$ values. Most unaltered basalts have $\delta^{18}O$ values >5.7‰ (Taylor & Sheppard 1987). A decrease in $\delta^{18}O$ value during metamorphism is normally interpreted (e.g. Valley 1986; Cartwright & Buick 1999) as indicating fluid–rock interaction at high fluid/rock ratios at high temperatures (>400°C). There are two possible causes of high temperature fluid–rock interaction at Sistenup. The lavas are exposed adjacent to a syenite intrusion which has been dated at 173 ± 2 Ma (Rb–Sr internal mineral isochron; Harris 1995). Fluid–rock interaction in the basalts could have occurred either as a result of hydrothermal activity associated with the intrusion of the syenite, or during regional metamorphism at some time prior to intrusion of the syenite. The latter explanation is favoured because:

(a)   quartz and alkali feldspar within the syenite are in oxygen isotope equilibrium (Harris 1995), which is not consistent with circulation of an external fluid;
(b)   although the whole-rock $\delta^{18}O$ values of the lavas are variable, they appear to be in internal oxygen isotope equilibrium ($\delta^{18}O$ plag $\sim \delta^{18}O$ whole-rock), which is not typical of hydrothermal aureoles (e.g. Forester & Taylor 1977); and
(c)   the minerals within the syenite, even right at the contact, have $\delta^{18}O$ values consistent with equilibrium at magmatic temperatures, whereas lavas a few metres from the contact are highly depleted in ^{18}O.

The oxygen isotope fractionation factor between basalt and the altering fluid can be assumed to be approximately that of plagioclase (An_{50}) and water (that is 1.2‰ at 450°C and 0.3‰ at 600°C). If oxygen isotope equilibrium between the lavas and fluid were attained with the lava $\delta^{18}O$ values around zero, then the alteration fluid would have to have had a $\delta^{18}O$ value of -1.2‰, which is consistent with meteoric rather than magmatic water. It is suggested that the substantial fluid–rock interaction recorded at Sistenup is the result of a regional metamorphic/hydrothermal event post-dating the Maud Orogeny and pre-dating the intrusion of the Sistefjell syenite. This interpretation is supported by preliminary Ar–Ar dating of hornblende in the basalt (D. Phillips, pers. comm., 2000). Given the lack of deformation, the alteration is likely to be Cambrian in age.

Cambrian fluid–rock interaction in the Sverdrupfjella Group

Calcite from the metasomatic calc silicate body has $\delta^{18}O$ values between 9 and 15‰ and is con-sistent with the oxygen isotope composition being buffered by the surrounding silicate assemblages. The $\delta^{13}C$ values are similar to those of the meta-carbonates of the Fuglefjellet Formation which may have been the source of carbon in the fluid responsible for the metasomatism. These rocks are the only volumetrically important carbonate rocks exposed in the area and are situated some 150 km NE of the central Kirwanveggen. If the Fuglefjellet rocks are the source of carbon this implies regional-scale movement of fluids.

Petrographical evidence for hydrous retrogression in rocks associated with high strain zones indicates possible infiltration of fluids via these late structural features which are considered to be early Cambrian in age (Grantham et al. 1995; Johnstone et al. 1998; Jackson 1999). The $\delta^{18}O$ values of the retrogressed rocks are compared with samples of unaltered gneiss in Figure 5. The retrogressed rocks show a negative skewed distribution, with the main peak at the same values as the unaltered gneisses, but with a significant number of rocks having much lower $\delta^{18}O$ values. In part this might be due to the mineralogy of the materials analysed. The gneisses samples are generally leucocratic and are dominated by quartz and feldspar, minerals that concentrate ^{18}O, whereas some of the retrogressed samples are rich in biotite, and/or amphibole, and/or chlorite which would lead to lower $\delta^{18}O$ values even in assemblages in equilibrium with the gneisses. Although in some rocks mineralogical effects do appear to be the cause of 'low' $\delta^{18}O$, a significant proportion of samples with $\delta^{18}O <$ 6‰ are quartz–feldspar dominated gneisses (Table 1). Felsic gneiss samples with the low $\delta^{18}O$ values indicate exchange with external fluids which themselves have a low $\delta^{18}O$ value (meteoric water), as discussed above for the Sistenup lavas. None of the retrogressed samples, however, has as low $\delta^{18}O$ values as the Sistenup lavas (with the lowest $\delta^{18}O$ values) and the lack of systematic shifts in $\delta^{18}O$ value across the high-strain zones indicates relatively limited influx of such fluids.

Conclusions

The nature of the Cambrian overprinting in western Dronning Maud Land is poorly understood. Regional resetting of the Rb–Sr, Sm–Nd and K–Ar systems in the area (Moyes et al. 1993) suggests more than the passive thermal effects associated with the intrusion of a limited volume of 460–500 Ma granitoids. The stable isotope data presented in this paper documented multiple episodes of fluid–rock interaction. The calc-silicate boudins in the Sverdrupfjella Group record fluid–rock interaction between 1200 and 900 Ma during the Maud Orogeny. Various generations of quartz and quartz–calcite veins appear to have formed from fluids whose isotope compositions were largely rock-buffered with low fluid/rock ratios. Extensive fluid–rock interaction with externally derived fluids are required by the very low $\delta^{18}O$ values of the Sistenup lavas. Their lack of deformation indicates that ^{18}O-depletion occurred during the Cambrian. The high-strain zones are the features in the Sverdrupfjella group most likely to be related to early Cambrian metamorphism. A significant number of samples from within and adjacent to high strain zones that show petrographical evidence of hydrous retrogression exhibit ^{18}O-depletion, which requires interaction with externally derived fluids. Although the ^{18}O-depletion in the high-strain zones was not to the same extent as seen at Sistenup, it is suggested that the two events are related. Also associated with a late shear zone is the metasomatic calc-silicate body at Tverreggtelen in the central Kirwanveggen. This body shows very different carbon isotope character to the calc-silicate boudins in nearby nunataks. The overlap in $\delta^{13}C$ values between this body and the Fuglefjellet 150 km to the NE suggests a possible source of carbon and movement of fluids on a regional scale. Granitic plutons of Cambrian age are present in the H.U. Sverdrupfella (Grantham *et al.* 1995; Groenewald *et al.* 1995) and are more abundant further east (e.g. Shiraishi *et al.* 1992). It is suggested that these intrusions provided the driving mechanism for infiltration of dominantly meteoric fluids along high-strain zones.

The authors would like to thank Warwick Board and Branko Corner for their contributions to this research. We greatly appreciate the contribution of Chris Jackson who read and commented on an earlier version of this paper and provided some of the samples used in this study. Most of the analyses were done by Fayrooza Rawoot at the University of Cape Town and we thank Ian Cartwright and Marlen Yanni of Monash University for additional analyses. Ian Buick, Ian Cartwright and Jodie Miller provided comments which helped to improve the manuscript substantially. This study was funded by the South African National Antarctic Programme.

References

APTED, M. J. & LIOU, J. G. 1983. Phase relations among greenschist, epidote–amphibolite, and amphibolite in a basaltic system. *American Journal of Science*, **283-A**, 328–354.

BORTHWICK, J. & HARMON, R. S. 1982. A note regarding ClF₃ as an alternative to BrF₅ for oxygen isotope analysis. *Geochemica et Cosmochemica Acta*, **46**, 1665–1668.

BUICK, I. S., HARLEY, S. L., CARTWRIGHT, I. & MATTEY, D. 1994. Stable isotopic signatures of superposed fluid events in granulite-facies marbles of the Rauer Group, East Antarctica. *Journal of Metamorphic Geology*, **12**, 285–299.

CARTWRIGHT, I. & BUICK, I. S. 1999. The flow of surface-derived fluids through Alice Springs age middle-crustal ductile shear zones, Reynolds Range, central Australia. *Journal of Metamorphic Geology*, **17**, 397–414.

CHIBA, H., CHACKO, T., CLAYTON, R. N. & GOLDSMITH, J. R. 1989 Oxygen isotope fractionations involving diopside, forsterite, magnetite, and calcite: applications to geothermometry. *Geochimica et Cosmochimica Acta*, **53**, 2985–2995.

COPLEN, T. B., KENDALL, C. & HOPPLE, J. 1983. Comparison of stable isotope reference samples. *Nature*, **302**, 236–238.

CORNER, B. 1994. *Geophysics Report*. Final project report, South African National Antarctic Programme, Pretoria.

DIRKS, P. H. G. M., CARSON, C. J. & WILSON, C. J. L. 1993. The deformational history of the Larsemann Hills, Prydz Bay: the importance of the Pan-African (500 Ma) in East Antartica. *Antarctic Science*, **5**, 179–192.

FORSETER, R. W. & TAYLOR, H. P. 1977. $^{18}O/^{16}O$ D/H and $^{13}C/^{12}C$ studies of the tertiary igneous complex of Skye, Scotland. *American Journal of Science*, **277**, 136–177.

FITZSIMONS, I. C. W. 1998. Early Cambrian tectonism in East Antarctica: implications for Gondwana assembly and earlier supercontinents. Special abstracts issue Gondwana 10: Event stratigraphy of Gondwana. *Journal of African Earth Sciences*, **27**, 74–75.

GRANTHAM, G. H., JACKSON, C., MOYES, A. B., GROENEWALD, P. B., HARRIS, P. D., FERRAR, G. & KRYNAUW, J. R. 1995. The tectonothermal evolution of the Kirwanveggen–H.U. Sverdrupfjella areas, Dronning Maud Land, Antarctica. *Precambrian Research*, **75**, 209–229.

GRAY, D. R, GREGORY, R. T. & DURNEY, D. W. 1991. Rock-buffered fluid-rock interaction in deformed quartz-rich turbidite sequences, eastern Australia. *Journal of Geophysical Research*, **96** 19 681–19 704.

GROENEWALD, P. B. 1991. Metamorphic evolution of the, H.U. Sverdrupfjella in the context of Maudheim orogenic province, western Dronning Maud Land, Antarctica. *In: 6th International Symposium on Antarctic Earth Sciences*. NIPR, Tokyo.

GROENEWALD, P. B., MOYES, A. B., GRANTHAM, G. H. & KRYNAUW, J. R. 1995. East Antarctic crustal evolution: geological constraints and modeling in western Dronning Maud Land. *Precambrian Research*, **75**, 231 250.

HARRIS, C. 1995. *Petrogenesis of the Sistefjell syenite complex, Dronning Maud land, Antarctica: generation of low $\delta^{18}O$ magmas by contamination of rift zone magmas*. Abstract Volume, Centennial Geocongress, Johannesburg, 235–239.

HARRIS, C. & ERLANK, A. J. 1992. The production of large volume low-$\delta^{18}O$ rhyolites during the rifting

of Africa and Antarctica: The Lebombo mono-
cline, southern Africa. *Geochemica et Cosmochimica Acta*, **56**, 3561–3570.

HENRY, C., BURKHARD, M. & GOFFÉ, B. 1996. Evolution of synmetamorphic veins and their wallrocks
through a Western Alps transect: no evidence for
large-scale fluid-flow. Stable isotope, major and
trace-element systematics. *Chemical Geology* **127**,
81–109.

HJELLE, A. 1974. Some observations on the geology of
the, H.U. Sverdrupfjella, Dronning Maud Land,
Antarctica. *Arbor 1972*, 7–22. Norsk Polarinsti-
tutt, Oslo, Norway.

JACKSON, C. 1999. *Characterization of Mesoprotero-
zoic to Palaeozoic crustal evolution of Western
Dronning Maud Land*. Final project report, South
African National Antarctic Programme.

JOHNSTONE, W. P., HARRIS, C. & JACKSON, C. 1998.
Fluid-rock interaction related to deformation and
regional metamorphism of the circum East Antarctic Mobile Belt in western Dronning Maud
Land, Antarctica. Special abstracts issue Gondwana 10: Event stratigraphy of Gondwana. *Journal of African Earth Sciences*, **27**, 122–123.

KRYNAUW, J. R. 1996. A Review of the Geology of
East Antarctica, with special Reference to the
c. 1000 Ma and *c.* 500 Ma Events. *Terra Antarctica*, **3**(2), 77–89.

McCREA, J. M. 1950. On the isotopic chemistry of
carbonates and a paleotemperature scale. *Journal
of Chemical Physics*, **18**, 849–857.

MOODY, J. B., MEYER, D. & JENKINS, J. E. 1983.
Experimental characterization of the greenschist/
amphibolite boundary in mafic systems. *American
Journal of Science*, **283**, 48–92.

MOYES, A. B., BARTON, J. M. & GROENEWALD, P. B.
1993. Late Proterozoic to early Palaeozoic tectonism in Dronning Maud Land, Antarctica: its
bearing on supercontinental fragmentation and
amalgamation. *Journal of the Geological Society,
London*, **150**, 833–842.

PILI, E., SHEPPARD, S. M. F., LARDEAUX, J.-M., MAR-
TELAT, J.-M. & NICOLLET, C. 1997. Fluid flow vs.
scale of shear zones in the lower continental crust
and the granulite paradox. *Geology*, **25**, 15–18.

RAVICH, M. G. & SOLOV'EV, D. S. 1969. *Geology and
petrology of the mountains of central Queen Maud
Land*. Israel Programme for Scientific Transla-
tions, Jerusalem.

ROOTS, E. F. 1969. *Geology of western Dronning Maud
Land*. Explanation of Plate VI, Folio **12**, Antarctic Map Series, American Geophysical Union.

SIMS, J. P. & WILSON, C. J. L. 1997. Strain localisation
and texture development in a granulite-facies shear
zone – the Rauer Group, East Antarctica. *In:*
RICCI, C. A. (ed.) *The Antarctic Region: Geological
Evolution and Processes*. Terra Antarctica, Siena,
Italy, 131–138.

SHIRAISHI, K., HIROI, Y., ELLIS, D. J., FANNING, C. M.,
MOTOYOSHI, Y. & NAKAI, Y. 1992. The first report
of a Cambrian orogenic belt in East Antarctica – an
ion microprobe study of Lutzow-Holm complex.
In: YOSHIDA, Y., KAMINUMA, K. & SHIRAISHI, K.
(eds) *Recent progress in Antarctic Earth Science*.
Terrapub, Toyko, 29–36.

TAYLOR, H. P. & SHEPPARD, S. M. F. 1987. Igneous
rocks I. Processes of isotopic fractionation and
isotope systematics. *In:* VALLEY, J. W., TAYLOR,
H. P. & O'NEIL, J. R. (eds) *Stable Isotopes in High
Temperature Geological Processes*. Reviews in
Mineralogy, **16**, 227–271.

VALLEY, J. W. 1986. Stable isotope geochemistry of
metamorphic rocks. *In:* VALLEY, J. W., TAYLOR,
H. P. & O'NEIL, J. R. (eds) *Stable Isotopes in High
Temperature Geological Processes*. Reviews in
Mineralogy, **16**, 445–490.

VENNEMANN, T. W. & SMITH, H. S. 1990. The rate and
temperature of reaction of ClF_3 with silicate
minerals, and their relevance to oxygen isotope
analysis. *Chemical Geology*, **86**, 83–88.

VIEZER, J. & HOEFS, J. 1976. The nature of $^{18}O/^{16}O$
and $^{13}C/^{12}C$ secular trends in sedimentary carbonate rocks. *Geochimica et Cosmochimica Acta*, **40**,
1387–1395.

WOLMARANS, L. G. & KENT, K. E. 1982. Geological
investigations in western Dronning Maud Land,
Antarctica – a synthesis. *South African Journal of
Antarctic Research*, **2**, 93.

YOSHIDA, M. 1992. Late Proterozoic to Early Palaeozoic events in East Gondwanian crustal fragments. *In:* 29th International Geological Congress,
Abstracts. Kyoto, **2/3**, 265.

Index

Note: Page numbers in *italic* type refer to illustrations; those in **bold** type refer to tables.